Fungal Biology

Series editor

Vijai Kumar Gupta
National University of Ireland Galway
School of Natural Sciences
Galway, Ireland

Maria G. Tuohy
National University of Ireland Galway
School of Natural Sciences
Galway, Ireland

Fungal biology has an integral role to play in the development of the biotechnology and biomedical sectors. It has become a subject of increasing importance as new fungi and their associated biomolecules are identified. The interaction between fungi and their environment is central to many natural processes that occur in the biosphere. The hosts and habitats of these eukaryotic microorganisms are very diverse; fungi are present in every ecosystem on Earth. The fungal kingdom is equally diverse, consisting of seven different known phyla. Yet detailed knowledge is limited to relatively few species. The relationship between fungi and humans has been characterized by the juxtaposed viewpoints of fungi as infectious agents of much dread and their exploitation as highly versatile systems for a range of economically important biotechnological applications. Understanding the biology of different fungi in diverse ecosystems as well as their interactions with living and non-living is essential to underpin effective and innovative technological developments. This Series will provide a detailed compendium of methods and information used to investigate different aspects of mycology, including fungal biology and biochemistry, genetics, phylogenetics, genomics, proteomics, molecular enzymology, and biotechnological applications in a manner that reflects the many recent developments of relevance to researchers and scientists investigating the Kingdom Fungi. Rapid screening techniques based on screening specific regions in the DNA of fungi have been used in species comparison and identification, and are now being extended across fungal phyla. The majorities of fungi are multicellular eukaryotic systems and therefore may be excellent model systems by which to answer fundamental biological questions. A greater understanding of the cell biology of these versatile eukaryotes will underpin efforts to engineer certain fungal species to provide novel cell factories for production of pro

More information about this series at http://www.springer.com/series/11224

Pradeep Kumar • Vijai Kumar Gupta
Ajay Kumar Tiwari • Madhu Kamle
Editors

Current Trends in Plant Disease Diagnostics and Management Practices

Springer

Editors
Pradeep Kumar
Department of Biotechnology Engineering
Ben Gurion University of The Negev
Beersheva
Israel

Vijai Kumar Gupta
MGBG, Biochemistry
National University of Ireland Galway
Galway
Ireland

Ajay Kumar Tiwari
Uttar Pradesh Council of Sugarcane
Shahjahnapur, Uttar Pradesh
India

Madhu Kamle
Dryland Biotechnology
Ben Gurion University of The Negev
Beersheva
Israel

ISSN 2198-7777 ISSN 2198-7785 (electronic)
Fungal Biology
ISBN 978-3-319-27310-5 ISBN 978-3-319-27312-9 (eBook)
DOI 10.1007/978-3-319-27312-9

Library of Congress Control Number: 2016938525

Printed on acid-free paper

This Springer imprint is published by Springer Nature
The registered company is Springer International Publishing AG Switzerland

Preface

Plant diseases are of significant concern due to the intimate relationship between plant health and the welfare of people, animals, and the environment. The ability to provide adequate food and fiber has become increasingly strained, and continuous improvement for sustainable plant disease management is required to help meet these demands. Plant diseases incited by fungal pathogens are one of the major concerns in agriculture commodities worldwide, and some of plant diseases are symptomatic and are caused by biotic and abiotic factors or in combination. Symptoms are usually the results of a morphological change and alteration or damage to plant tissue and/or cells due to an interference of the plant's metabolism. Typically, the appearance of a biotic symptom will indicate the relatively late stage of an infection and colonization of a pathogen. Pathogens pose a threat to plants in natural communities, horticultural commodities, and cultivated crops. Risks of pathogen spread have increased with increased human mobility and the globalization of trade. The failure in accurate disease diagnosis and management may lead to huge losses in plant production and related commodities that cause nutritional food scarcity. Plant health testing is an essential management tool for the control of fungal diseases, pathogen detection, and the implementation of effective management strategies. Diagnosis of plant pathogens can be even more difficult with infected seeds, soils, and asymptomatic infected plant materials. It is thus, significant to have expert diagnosis strategies and integrated management practices advanced with molecular diagnostic techniques to retain plants free from pathogens.

This volume consists of 20 chapters and basically illustrates various fungal plant diseases and management strategies. The chapters in this volume include expert view on new approaches, update knowledge and worthy information of plant diseases which compiles "Molecular Diagnosis of Killer Pathogen of Potato: *Phytophthora infestans* and Its Management" (Chap. 1), "Biocontrol Mechanism of *Bacillus* for Fusarium Wilt Management in Cumin (*Cuminum cyminum* L.)" (Chap. 2), "Eco-friendly Management of Damping-Off Disease of Chilli Caused by *Pythium aphanidermatum* (Edson) Fitzp" (Chap. 3), "Biological Agents in Fusarium Wilt Diagnostics for Sustainable Pigeon Pea Production, Opportunities and Challenges" (Chap. 4), "Recent Diagnostics and Detection Tools: Implications for Plant

Pathogenic *Alternaria* and Their Disease Management" (Chap. 5), "Molecular Prospecting: An Advancement in Diagnosis and Control of *Rhizoctonia solani*" (Chap. 6), "*Fusarium moniliforme* Associated with Sugarcane Leaf Binding Disease in India and Its Possible Management" (Chap. 7), "*Macrophomina phaseolina*: The Most Destructive Soybean Fungal Pathogen of Global Concern" (Chap. 8), "*Colletotrichum gloeosporioides*: Pathogen of Anthracnose Disease in Mango *(Mangifera indica L.)*" (Chap. 9), "Current Scenario of Mango Malformation and Its Management Strategies: An Overview" (Chap. 10), "Paradigm Shift in Plant Disease Diagnostics: A Journey from Conventional Diagnostics to Nano-diagnostics" (Chap. 11), "Detection of Plant Pathogenic Fungi: Current Techniques and Future Trends" (Chap. 12), "Molecular Diagnostics and Application of DNA Markers in the Management of Major Diseases of Sugarcane" (Chap. 13), "Physiological and Molecular Signaling Involved in Disease Management Through *Trichoderma*: An Effective Biocontrol Paradigm" (Chap. 14), "Biotechnology in the Diagnosis and Management of Infectious Diseases" (Chap. 15), "Antimycotic Potential of Fungal Endophytes Associated with *Schima wallichii* by Synthesizing Bioactive Natural Products" (Chap. 16), "The Biological Control Possibilities of Seed-Borne Fungi" (Chap. 17), "Mycotoxin Menace in Stored Agricultural Commodities and Their Management by Plant Volatiles: An Overview" (Chap. 18), "Diagnostics of Seed-Borne Pathogens in Quarantine and Conservation of Plant Genetic Resources" (Chap. 19), "Aflatoxigenic Fungi in Food Grains: Detection, Its Impact on Handlers and Management Strategies" (Chap. 20).

In this book, expert researchers share their research knowledge and key literature which are vital towards the diagnosis of fungal plant diseases across the globe, addressing traditional plant pathology techniques, as well as advanced molecular diagnostic approach. We are extremely grateful to all the authors who have made the production of this volume edition possible. Authors are experts in their field, and we appreciate their willingness to contribute to this book which will be useful for all scientists and researchers around the globe gave compilation of this book *Current Trends in Plant Disease Diagnostics and Management Practices*.

We are grateful to the Series Editors Dr. V. K. Gupta and Dr. Maria Tuohy, Editor (Botany) Eric Stannard, and others concerned with Springer for their help in various ways. Many thanks to Springer Project Coordinator (Books) Rekha Udaiyar for her continuous efforts and immense support in the preparation of this volume to which we are highly indebted. We are grateful to all the contributors for their concern and concerted effects in making this volume. It is our intense hope that information presented in this book/volume will make a valuable contribution to the science of plant disease diagnosis and management. We believe and trust that it will stimulate further discussions in the pursuit of new knowledge. We also hope that it will be useful to all concerned.

Beersheva, Israel — Pradeep Kumar
Galway, Ireland — Vijai Kumar Gupta
Shahjahanpur, India — Ajay Kumar Tiwari
Beersheva, Israel — Madhu Kamle

Acknowledgments

Our sincere thanks are extended to all the contributing authors for gladly agreeing to share their research knowledge and key literature which are vital towards the plant disease diagnosis and management practices in the form of chapters and timely cooperation in preparation of the manuscripts. We are grateful to Springer Publishing, for their determined efforts to publish the book on schedule. We would like to express our sincere thanks to Springer, the publisher of our book, for giving us an opportunity to share our knowledge and skills with our research community. In particular, we would like to give our sincere thanks Eric Stannard, Editor, Botany, Springer, and Rekha Udiyar, Springer Project Coordinator (Books), for their editorial and production guidance, who worked with us from beginning to end in compilation of this book.

Pradeep Kumar and Madhu Kamle (Eds.) would like to thank Ben-Gurion University of the Negev, Israel, which made it possible to write this book. We are grateful to be part of this visionary and academic organization. We would like to especially acknowledge Professor Ariel Kushmaro (The Department of Biotechnology Engineering); Professor Arieh Zaritsky (Faculty of Natural Sciences); Professor Yaron Sitrit (French Associates Institute for Agriculture and Biotechnology of Drylands) Ben-Gurion University of the Negev, Israel; and Professor Eitan Ben-Dov, Achava Academic College, Israel, for their endeavor and motivations at all stage of the progress.

The Series Editors are gratefully acknowledged for their support, encouragement, and trust during the process of writing this book.

Beersheva, Israel — Pradeep Kumar
Galway, Ireland — Vijai Kumar Gupta
Shahjahanpur, India — Ajay Kumar Tiwari
Beersheva, Israel — Madhu Kamle

Contents

Molecular Diagnosis of Killer Pathogen of Potato: *Phytophthora infestans* and Its Management 1
Touseef Hussain and Bir Pal Singh

Biocontrol Mechanism of *Bacillus* for *Fusarium* Wilt Management in Cumin (*Cuminum cyminum* L.) 29
H.P. Gajera, Disha D. Savaliya, Darshana G. Hirapara, S.V. Patel, and B.A. Golakiya

Eco Friendly Management of Damping-off of Solanaceous Crops Caused by *Pythium* Species 49
A. Muthukumar, R. Udhayakumar, and R. Naveenkumar

Biological Agents in *Fusarium* Wilt (FW) Diagnostic for Sustainable Pigeon Pea Production, Opportunities and Challenges 91
Deepak Kumar Verma, Shikha Srivastava, Vipul Kumar, and Prem Prakash Srivastav

Recent Diagnostics and Detection Tools: Implications for Plant Pathogenic *Alternaria* and Their Disease Management 111
Dipak T. Nagrale, Lalan Sharma, Sunil Kumar, and Shailesh P. Gawande

Molecular Prospecting: Advancement in Diagnosis and Control of *Rhizoctonia solani* Diseases in Plants 165
Hemant J. Patil and Manoj K. Solanki

***Fusarium moniliforme* Associated with Sugarcane Leaf Binding Disease in India and Its Possible Management** 187
Ajay Kumar Tiwari, Kavita Kumari, Surendra Pratap Singh, Nishtha Mishra, Atul Singh, and B.L. Sharma

***Macrophomina phaseolina*: The Most Destructive Soybean Fungal Pathogen of Global Concern** 193
Vibha

Colletotrichum gloeosporioides: Pathogen of Anthracnose Disease in Mango (_Mangifera indica_ L.) 207
Madhu Kamle and Pradeep Kumar

Current Scenario of Mango Malformation and Its Management Strategies: An Overview 221
Pradeep Kumar and Ashok Kumar Misra

Paradigm Shift in Plant Disease Diagnostics: A Journey from Conventional Diagnostics to Nano-diagnostics 237
Prachi Sharma and Susheel Sharma

Advance Detection Techniques of Phytopathogenic Fungi: Current Trends and Future Perspectives 265
Pardeep Kumar, Jameel Akhtar, A. Kandan, Sandeep Kumar, Ritu Batra, and S.C. Dubey

Molecular Diagnostics and Application of DNA Markers in the Management of Major Diseases of Sugarcane 299
Sangeeta Srivastava

Physiological and Molecular Signalling Involved in Disease Management Through _Trichoderma_: An Effective Biocontrol Paradigm 317
Veena Pandey, Alok Shukla, and J. Kumar

Biotechnology in the Diagnosis and Management of Infectious Diseases 347
Amit Kumar, R.S. Sengar, Raj Singh, Anju Rani, Vineet Girdharwal, and Gyanika Shukla

In Vitro Antimycotic and Biosynthetic Potential of Fungal Endophytes Associated with _Schima Wallichii_ 367
Vineet Kumar Mishra, Ajit Kumar Passari, and Bhim Pratap Singh

The Biological Control Possibilities of Seed-Borne Fungi 383
Nuray Özer and Arzu Coşkuntuna

Mycotoxin Menace in Stored Agricultural Commodities and Their Management by Plant Volatiles: An Overview 405
Abhay K. Pandey, Satish K. Sain, Pooja Singh, Uma T. Palni, and N.N. Tripathi

Diagnostics of Seed-Borne Plant Pathogens for Safe Introduction and Healthy Conservation of Plant Genetic Resources 429
Jameel Akhtar, A. Kandan, Baleshwar Singh, Pardeep Kumar, Z. Khan, B.H. Gawade, Sundeep Kumar, and S.C. Dubey

Aflatoxigenic Fungi in Food Grains: Detection, Its Impact on Handlers and Management Strategies 441
Sana Ali

Index 459

Contributors

Jameel Akhtar Division of Plant Quarantine, ICAR-National Bureau of Plant Genetic Resources, New Delhi, India

Sana Ali Department of Microbiology, Janakpuri Superspecialty Hospital, New Delhi, India

Ritu Batra Molecular Biology, Biotechnology and Bioinformatics Department, CCS Haryana Agricultural University, Hisar, India

A. Coşkuntuna Department of Plant Protection, Faculty of Agriculture, Namık Kemal University, Tekirdağ, Turkey

S.C. Dubey Division of Plant Quarantine, ICAR-National Bureau of Plant Genetic Resources, New Delhi, India

H.P. Gajera Department of Biotechnology, College of Agriculture, Junagadh Agricultural University, Junagadh, India

B.H. Gawade Division of Plant Quarantine, ICAR-National Bureau of Plant Genetic Resources, New Delhi, India

Shailesh P. Gawande ICAR- Central Institute for Cotton Research, Panjari, Nagpur, Maharashtra, India

Vineet Girdharwal Department of Zoology, Swami Vivekanand Subharti University, Meerut, Uttar Pradesh, India

B.A. Golakiya Department of Biotechnology, College of Agriculture, Junagadh Agricultural University, Junagadh, India

Darshana G. Hirapara Department of Biotechnology, College of Agriculture, Junagadh Agricultural University, Junagadh, India

Touseef Hussain Department of Life Science, Uttarakhand Technical University, Dehradun, UK, India

Madhu Kamle Department of Dryland Agricultural and Biotechnology, Ben Gurion University of the Negev, Beersheva, Israel

A. Kandan Division of Plant Quarantine, ICAR-National Bureau of Plant Genetic Resources, New Delhi, India

Kavita Kumari Central Lab, UP Council of Sugarcane Research, Shahjahnapur, Uttar Pradesh, India

Z. Khan Division of Plant Quarantine, ICAR-National Bureau of Plant Genetic resources, New Delhi, India

Amit Kumar Department of Biotechnology, Swami Vivekanand Subharti University, Meerut, Uttar Pradesh, India

J. Kumar Department of Plant Pathology, College of Agriculture, G.B. Pant University of Agriculture and Technology, Pantnagar, Uttarakhand, India

Pardeep Kumar Division of Plant Quarantine, ICAR-National Bureau of Plant Genetic Resources, New Delhi, India

Pradeep Kumar Department of Biotechnology Engineering, Ben Gurion University of the Negev, Beersheva, Israel

Sandeep Kumar Germplasm Evaluation Division, ICAR-National Bureau of Plant Genetic Resources, New Delhi, India

Sundeep Kumar Division of Plant Quarantine, ICAR-National Bureau of Plant Genetic Resources, New Delhi, India

Sunil Kumar ICAR- National Bureau of Agriculturally Important Microorganisms, Kushmaur, Maunath Bhanjan, Uttar Pradesh, India

Vipul Kumar Department of Plant Pathology, C. S. Azad University of Agriculture and Technology, Kanpur, Uttar Pradesh, India

Ashok Kumar Misra Division of Crop Protection, Central Institute for Subtropical Horticulture, Lucknow, Uttar Pradesh, India

Nishtha Mishra Central Lab, UP Council of Sugarcane Research, Shahjahnapur, Uttar Pradesh, India

Vineet Kumar Mishra Molecular Microbiology and Systematics Laboratory, Department of Biotechnology, Mizoram University, Aizawl, Mizoram, India

A. Muthukumar Department of Plant Pathology, Faculty of Agriculture, Annamalai University, Chidambaram, Tamil Nadu, India

Dipak T. Nagrale ICAR- Central Institute for Cotton Research, Panjari, Nagpur, Maharashtra, India

R. Naveenkumar Department of Plant Pathology, Faculty of Agriculture, Annamalai University, Chidambaram, Tamil Nadu, India

Nuray Özer Department of Plant Protection, Faculty of Agriculture, Namık Kemal University, Tekirdağ, Turkey

Uma T. Palni Department of Botany, DSB Campus, Kumaun University, Nainital, Uttarakhand, India

Abhay K. Pandey National Institute of Plant Health Management, Ministry of Agriculture and Co-operation, Hyderabad, Telangana, India

Veena Pandey Department of Plant Physiology, College of Basic Sciences and Humanities, G.B. Pant University of Agriculture and Technology, Pantnagar, Uttarakhand, India

Ajit Kumar Passari Molecular Microbiology and Systematics Laboratory, Department of Biotechnology, Mizoram University, Aizawl, Mizoram, India

S.V. Patel Department of Biotechnology, College of Agriculture, Junagadh Agricultural University, Junagadh, India

Hemant J. Patil Institute of Soil, Water and Environmental Sciences, Volcani Center, Agricultural Research Organization, Bet Dagan, Israel

Anju Rani Department of Botany, Swami Vivekanand Subharti University, Meerut, Uttar Pradesh, India

Satish K. Sain ICAR - Central Institute for Cotton Research, Regional Station Sirsa, Haryana, India

Disha D. Savaliya Department of Biotechnology, College of Agriculture, Junagadh Agricultural University, Junagadh, India

R.S. Sengar Department of Agricultural Biotechnology, Sardar Vallabhbhai Patel University of Agriculture and Technology, Meerut, Uttar Pradesh, India

B.L. Sharma Central Lab, UP Council of Sugarcane Research, Shahjahnapur, Uttar Pradesh, India

Lalan Sharma ICAR- National Bureau of Agriculturally Important Microorganisms, Kushmaur, Maunath Bhanjan, Uttar Pradesh, India

Prachi Sharma Division of Plant Pathology, Sher-e-Kashmir University of Agricultural Sciences and Technology of Jammu (SKUAST-J), Jammu, Jammu and Kashmir, India

Susheel Sharma School of Biotechnology, Sher-e-Kashmir University of Agricultural Sciences and Technology of Jammu (SKUAST-J), Jammu, Jammu and Kashmir, India

Alok Shukla Department of Plant Physiology, College of Basic Sciences and Humanities, G.B. Pant University of Agriculture and Technology, Pantnagar, Uttarakhand, India

Gyanika Shukla Department of Biotechnology, Swami Vivekanand Subharti University, Meerut, Uttar Pradesh, India

Atul Singh Division of Plant Pathology, UP Council of Sugarcane Research, Shahjahnapur, Uttar Pradesh, India

Baleshwar Singh Division of Plant Quarantine, ICAR – National Bureau of Plant Genetic resources, New Delhi, India

Bhim Pratap Singh Molecular Microbiology and Systematics Laboratory, Department of Biotechnology, Mizoram University, Aizawl, Mizoram, India

Bir Pal Singh Central Potato Research Institute, Shimla, HP, India

Surendra Pratap Singh UP Council of Sugarcane Research, Shahjahnapur, Uttar Pradesh, India

Pooja Singh Bacteriology and Natural Pesticide Laboratory, Department of Botany, DDU Gorakhpur University, Gorakhpur, Uttar Pradesh, India

Raj Singh Department of Botany, Swami Vivekanand Subharti University, Meerut, Uttar Pradesh, India

Manoj K. Solanki Institute of Soil, Water and Environmental Sciences, Volcani Center, Agricultural Research Organization, Bet Dagan, Israel

Guangxi Crop Genetic Improvement and Biotechnology Lab, Guangxi Academy of Agricultural Sciences, Nanning, China

Prem Prakash Srivastav Agricultural and Food Engineering Department, Indian Institute of Technology (IIT), Kharagpur, West Bengal, India

Sangeeta Srivastava Division of Crop Improvement, ICAR-Indian Institute of Sugarcane Research, Lucknow, India

Shikha Srivastava Department of Botany, Deen Dyal Upadhyay Gorakhpur University, Gorakhpur, Uttar Pradesh, India

Ajay Kumar Tiwari Central Lab, UP Council of Sugarcane Research, Shahjahnapur, Uttar Pradesh, India

N.N. Tripathi Bacteriology and Natural Pesticide Laboratory, Department of Botany, DDU Gorakhpur University, Gorakhpur, Uttar Pradesh, India

R. Udhayakumar Department of Plant Pathology, Faculty of Agriculture, Annamalai University, Chidambaram, Tamil Nadu, India

Deepak Kumar Verma Agricultural and Food Engineering Department, Indian Institute of Technology (IIT), Kharagpur, West Bengal, India

Vibha Department of Plant Pathology, JNKVV, Jabalpur, MP, India

Molecular Diagnosis of Killer Pathogen of Potato: *Phytophthora infestans* and Its Management

Touseef Hussain and Bir Pal Singh

Potato (*Solanum tuberosum* L.) a unique tuber crop and a member of the Solanaceae family, which is often quoted as the "Future food" & "Apple of soil" finds its roots of origin in the Andes of South America, also the original home of Late blight (Widmark 2010), which was introduced in India, in early seventeenth century and since then, it has been adopted through the length and breadth of the country. After Rice, and wheat, Potato is third most important non-grain food crop in the world, from the point of human consumption and is central to global food security. Looking at its good gustative as well as high nutritional qualities and can be grown under various climates thats why Food and Agriculture Organization (FAO) has declared the year 2008 the international year of the potato. Indeed, potato can helpfull-fill the first United Nations Organization's (UNO) millennium development goal that aims at eradicating extreme poverty and hunger in the world. Potato production and consumption is accelerating in most of the developing countries including India primarily because of increasing industrialization. In fact, Potato production in developing countries surpassed that of developed countries. India is the second largest potato producer in the world after China. In India, Potatoes are grown under varied climatic conditions ranging from tropics, subtropics to temperate highlands. India also export (both seed and table potato), through which foreign exchange is also generated. Potatoes belong to the crops that grow even in unfavourable conditions and at high altitudes. Moreover, a few other crops produce a comparably high yield of nutrients per cultivated area – a quality that is particularly welcome in regions where land is scarce. This is why potatoes are highly important for many farming families in the world's mountain regions. Consequently, the spectrum of

T. Hussain (✉)
Department of Life Science, Uttarakhand Technical University, Dehradun 248001, UK, India
e-mail: Hussaintouseef@yahoo.co.in

B.P. Singh
Central Potato Research Institute, Shimla 171001, HP, India

P. Kumar et al. (eds.), *Current Trends in Plant Disease Diagnostics and Management Practices*, Fungal Biology, DOI 10.1007/978-3-319-27312-9_1

insect-pests and diseases is very large. All of them put together have the potential to limit potato production upto 85 % depending upon the weather/region. Under favourable conditions for potato, diseases development are frequently the same as the conditions needed for potato growth: temperature ranges from 10 to 25 °C, with high humidity, medium pH, etc. In most cases, potato pathogens develop specific survival forms, dissemination ways and host penetration methods. The genetic variability of the pathogens implies the use of adapted diagnostic and control methods.

Throughout the World, potato is becoming a more and more important foodstuff it is therefore essential to control diseases which cause direct yield losses and decrease of farmer's incomes due to downgrading the quality of affected tubers in seed industry. That's why, knowledge about the pathogens as well as factors which are influencing disease severity is needed to setup efficient control strategies.

Detection of plant pathogens has enabled to setup adapted control measures and avoid disease expansion and helps in yield losses, even if the infestation level is low. Infected seeds are the most important carrier of pathogens for trans-regional and long-distance dissemination. Diagnosis and management of seed borne diseases through quality control programmes helps in monitoring of seeds which are from harvest to purchase, for marketing and sowing in the field, to ensure high quality, pathogen free and which are genetically pure seed. Healthy potato tuber/seeds are the base of the potato production chain and the most critical factor for success.

P. infestans is the most devastating diseases of potatoes in the world. This epidemic had totally destroyed potato crops in the 1840s which led to mass starvation in Europe. For example, in the Great Irish Famine of 1845–1847, up to one million people died and a similar number of people emigrated to the rest of Europe and the USA. There were, of course, factors that contributed to the starvation, including the land-tenure system in Ireland at that time, and the almost total dependence of the poorer working population on potatoes as their source of food. Nevertheless, potato blight ranks as one of the most devastating diseases in human history. Even today it is one of the major pathogens that chemical companies target in their search for new fungicides.

However, the relatively recent migration of the A2 mating type of *P. infestans* from its presumed centre of origin, Central Mexico (Niederhauser 1991; Goodwin et al. 1994a, b), to different parts of the world during 1970s and 1980s has resulted in increased diseases severity and has refocused its attention on the disease emergence (Fry et al. 1993; Goodwin 1997). The occurrence of both mating types of *P. infestans* has resulted in the emergence of sexual oospores (Goodwin and Drenth 1997), that allow survival of the pathogen in the soil (Singh et al. 1994).

Losses upto 85 % have been reported if crop (susceptible cultivar) remains unprotected. Disease appears every year in epiphytotic forms in hills as well as in plains. The causal organism of late blight is *Phytophthora infestans* (Mont.) de bary. *Phytophthora* in greek word means: Plant destroyer, *Phytophthora* has branched sporangiophore that produced lemon shaped sporangia at their tips. At the places where sporangia are produced, spornagiophores form swellings that are characteristics of *P. infestans* (Hussain et al. 2015).

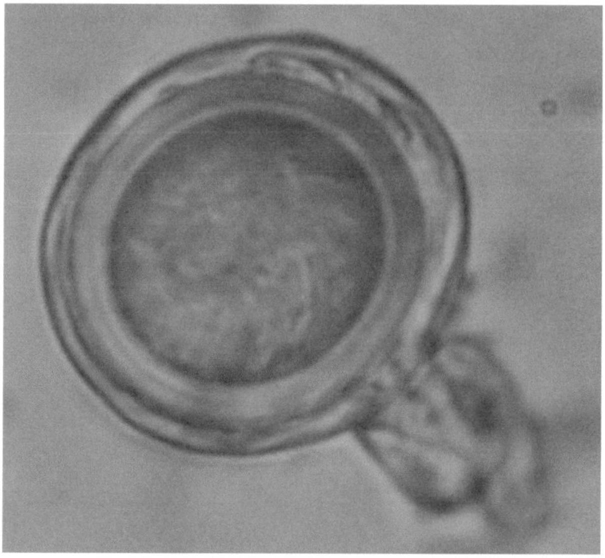

Fig. 1 *P. infestans* sexual Oospore (A2 mating type)

Morphology

The mycelium is hyaline, branched, co-enocytic and produces indeterminate sympodially-branched sporangiophores distinguishable from somatic hyphae. The thin walled, lemon–shaped, hylanie sporangium (21–38 × 12–23 m), having an apical papilla, is borne at the tip of the branched and, as it matures, the tip of the branch swells and continues to elongate, resulting in the sporangium being turned laterally. A sporulating hypha is characterized by periodic swelling that marks the points at which sporangia had been attached. Most strains are heterothallic and production of the sexual spore (oospore) requires two compatibility groups (mating types). Reproductive structures are anthredia and oogonia from opposite mating types. During the development of these structures, the antherdium is punctured by oogonium, which grows through it and matures into a round body above the antheridium-an arrangement termed amphygynous. The oospores has a thicked walled that renders it resistant to unfavourable conditions (Fig. 1). It germinates to form a zoosporangium.

Losses

The losses form the diseases in potatoes are of two different kinds: First losses caused by foliage infection which leads to premature death of the plant and consequently in tuber yield; and another caused by tuber infection and loss through rottage of infected tubers in the field and storage (Bhattacharyya et al. 1990; Robertson 1991). Losses are more in hilly regions where crop is grown under rain-fed condition in comparison to plains. Late blight attacks both the foliage and tubers of the potato crops. The diseases has potential of quickly developing into an epiphytotic form and often responsible for heavy loss in yield resulting even complete crop failure. Cox and Large (1960) have reviewed that blight losses reported in different countries. The loss yield due to defoliation may be negligible in countries like Australia while it may vary on average form 10–20 % in most countries. Currenlty causing losses estimated at US $6 billion around the world every year, www.huttonac.uk james hutton institute newsletter, 10 Feb 2016. In India, losses are much larger in unsprayed crop which go as high as 90 % (Bhattacharyya et al. 1990). In the plains (irrigated crop) losses mainly depends on the time of diseases appearance and its subsequent build-up. The tomato crop is also affected because of late blight diseases.

Symptoms

The diseases affect all the plant parts viz. leaves, stem and tubers. Symptoms appear at first as water soaked spot, usually at edges of lower leaves. The lesion starts as a pale green spot that turns brown in 1 or 2 days. These spots develop and may become nearly black. On the under surface of the leaf a white mildews growth develops around the lesion and is characteristics features of disease (Fig. 2a, b). On stem, blight brown elongated lesion are formed which encircle stems as well as Tubers develop reddish brown, shallow, to deep dry rot lesions in the potato plant. The affected tuber fleshes become caramelised with sugary texture. A tan brown, dry granular rot characteristically extend into the infected tuber upto 1.5 cm depth

Fig. 2 (**a**) Late blight infected Potato leaf. (**b**) Late blight infected potato crop in field

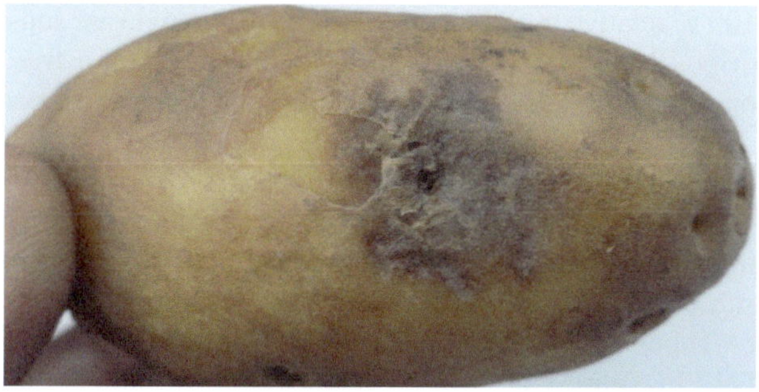

Fig. 3 Late blight infected potato tuber

(Fig. 3). In cold store, the pathogen does not advance inside the tubers but such tubers become susceptible for soft rot infection.

Appearance and build-up of late blight depend solely on weather conditions. There are specific requirements of temperature and humidity for initiation and further build-up of disease. Temperature requirements are different for fungus growth (16–20 °C), spore production (18–22 °C), spore germination (10–20 °C) and for infection and disease development (10–22 °C with 18 + 1 °C). Spore germination and infection requires 100 % humidity and spores get killed under low humidity (<75 %). Fungal spores are produced during the night and are sensitive to light.

Temperature Maximum: 18 ± 2 °C

Minimum: 10 ± 2 °C

Relative humidity: 80–95 %

Thick clouds/fog during day intermittent rain, heavy dew during night and light wind velocity

Spread sporangia are responsible for fresh infection. The infected crop under favourable weather is generally killed within 10–15 days.

Special advice if the crop is 65–70 days old and the diseases has damaged on fourth crop than cut the haulm and remove them from the field. Remove diseases infected tubers at the time of harvest before storage.

Epidemiology

Appearance of late blight of potato and its subsequent build-up and spread depends on many other factors also. They include sources of inoculum (Karolev 1978), host resistance per se (Fry 1977; Van der Plank 1984), Plant protection support provided through fungicides sprays, deployment of host resistance varieties in the field (Bhattacharyya et al. 1990) and weather conditions (Cook 1947). Field infection is the most successful under cool, moist conditions. However, infections takes place over a range of environment conditions, sporangial production is most rapid and prolific at

100 % relative humidity and at 20 °C. Sporangia are sensitive to desiccation and, after disposal by wind or splashing water, they require free water for germination. The optimal temperature for indirect germination via zoospores is 10 °C, whereas that for direct germination of sporangia via germ-tubes is 24 °C. In both types of germinations occur at by overlapping temperatures. Zoospores are quickly killed by drying. In the presence of free water they produces germ tube and penetration occurs at temperatures between 10 and 29 °C. Once penetration has occurred, in the plant infection and subsequent development of the diseases is most rapid at 21 °C.

Tubers, particularly those inadequately covered by soil, may be infected in the field by spore that have been washed form infected leaves into the soil by rain or irrigation (Fig. 3). Rapid tubers growth frequently causes soil to crack, exposing tubers infection. Tuber infection may also occur during wet harvest conditions via contact between tubers and sporangia.

Today one of the major challenges for plant pathologist is to develop detection techniques for plant pathogens, so that appropriate measures are taken for their management. Accurate detection and easy identification of plant pathogens is a pre-request of diseases management to sustain highly yield potential crops. Therefore continuous efforts are being made to develop simple, reliable, raid and safe methods for disease diagnosis. The use of diagnostics to investigate the extent of pathogen contamination in potato stocks, the factors affecting disease development during growth and storage and the relationship between inoculums load and diseases risk is considered.

Looking into the economic importance of potato, early pathogen detection in the field and storage facilities is a crucial step because sometime few contaminated tubers plants are enough to spread the infection and causing severely compromise production and reduction in yield (Trout et al. 1997; Judelson and Tooley 2000; Jayan et al. 2002; Hussain et al. 2005, 2013, 2013c, 2014). Besides being a destructive plant pathogen, *P. infestans* is also a model organism for oomycetes group (Haas et al. 2009). Therefore, producers as well as plant breeders and researchers, benefits form methodologies by allowing the accurate measurement of *P. infestans* growth (Dorrance and Inglis 1997).

The early, accurate reliable detection and identification of fungal and oomycete pathogens are required for quick plant quarantine decisions and effective plant diseases management. Traditionally, identification, and detection of fungal pathogens were mainly relied on culture based morphological approaches (Tsao and Guy 1977; Jeffers and Aldwinkle 1987). The identification of plant pathogenic fungi based on morphology is a laborious, time – consuming and mycological expertise – intensive task, therefore, a fats, rapid, sensitive and robust diagnostic tools are very important for the production of clean, safe planting, quarantine inspections and safe conservation and exchange of germplasm and tissue culture materials.

In the last three decades, molecular detection of plant pathogens has seen a very significant changes. The advents of Immunology based detection, i.e. the monoclonal antibodies and the Enzyme Linked Immunosorbent Assay (ELISA), which was an important turning point in virology and bacteriology. Then came the DNA based technology, like PCR, which totally changes the face of molecular diagnostics.

Furthermore, diagnostic PCR has been greatly improved by the introduction of second generation of PCR, which are known as Real Time PCR. The DNA Microarray technology, originally designed to study gene expression Single Nucleotide Polymorphism (SNP) is currently a new emerging pathogen diagnostic technology, which in theory, opens a new platform for unlimited multiplexing capability. It is viewed as a technology that fundamentally alter molecular diagnostics.

Sometimes molecular methods cant' provide all the results but have proved a valuable tools that consisted of teams of experienced plant pathologist, mycologist and ecologist Despite the technology advances reported in this chapter, there is considerable scope for improvement. There is an ongoing need for filed kits for rapid and accurate on–site detection of specific pathogens and advances in automation and miniaturisation.

Several diagnostic tests are being used to fulfil the diagnostic mandate. Such tests should be ideally be:

1. Sensitive – able to detect low numbers of pathogens
2. Specific – able to differentiate between pathogenic and harmless organisms,
3. Rapid – so that results can be obtained quickly,
4. Automatable – amenable to being used as a large-scale screening method,
5. Affordable – the price cannot significantly affect production costs of the crop produced.

Diagnosis Methods for Seed Borne Pathogens

The selection of a method depends upon the purpose of the test i.e. whether the seeds are to be tested for seed certification, seed treatment, quarantine etc. If the purpose is for quarantine, then a highly sensitive methods are preferred, which can detect even traces of inoculum.

Diagnostics methods based on nucleic acid began to develop a real momentum after the introduction of PCR in the mid-1980s. New approaches based on DNA are highly specific rapid, reliable were developed for accurate diagnosis of seed borne pathogens. These Diagnostic assays are based on PCR technique which have been developed for major seed borne diseases of potato.

Before any PCR protocol, three major points are very important:

1. Target region of DNA specific to the pathogen
2. DNA extraction from pathogen infected planting material, and
3. A good protocol for the diagnosis of the specific/targeted pathogen present in the sample

In each case, specific primers were designed within the internal transcribed spacer (ITS) regions of the rDNA repeat. ITS regions are most commonly used targets regions widely used in diagnostic assays of *Phytophthora* spp. because they are present in multiple copies, can be easily amplified, sequenced with universal primers and possess conserved as well as variable sequences (White et al. 1990; Bonants et al. 1997). Moreover, the availability of ITS sequence database easily

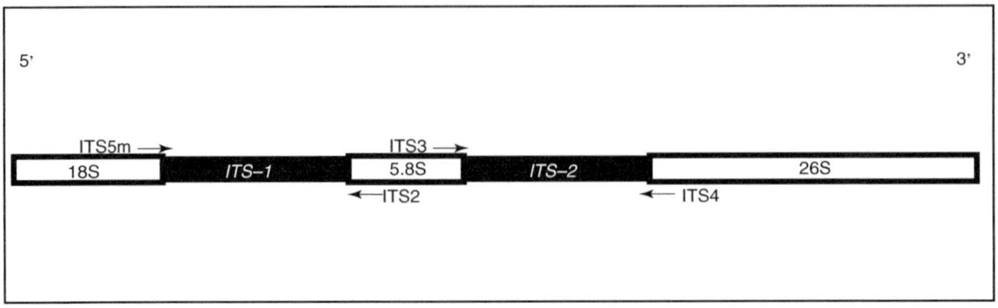

Fig. 4 The three coding and two internal transcribed spacer regions of the nuclear ribosomal DNA repeat unit. *Arrows* indicate approximate locations of the four primers used for PCR amplification

facilitates the alignment of sequences of a wide range of species for the selection of species-specific primers. Its means that, the ITS sequence database will help in designing of species specific primers for the specific detection of every phytophthora spp. to differentiate from other species.

Molecular Identification

One of the most widely used molecular approach to differentiating different species of plant pathogenic fungi is sequencing analysis of the internally, transcribed spacer region (ITS) of the ribosomal RNA gene cluster (Fig. 4). It has no protein product but is transcribed and may be significant in the folding and secondary structure of the rRNA molecule (Gottschling and Plötner 2004). Sequence variation is generally greatest at the 3′ and 5′ termini of the spacers and although sequence variation can define species boundaries, variation, within species is not consistent.

A phylogenetic analysis of 50 described *Phytophthora* taxa, based on ITS1 and ITS2 rDNA sequences, clustered the taxa into eight main lineages designated Clades 1–8 (Cooke et al. 2000). Several species specific markers were designed from ITS region by several workers (Kong et al. 2004; Chowdappa et al. 2003b; Hussain et al. 2010, 2013, 2013a), for detection and identification of *Phytophthora* sp. Mitochondrial genes are another set of genes used to identify and determine phylogenetic analysis of *Phytophthora*. TrnG-TrnY region (mitochondrial genome region between gene tranG (gcc) and gene trnY (gua)), Atp9-Nad9 (Mitochondrial genome region between gene Atp9 and gene Nad9), Cox2-Cox-1 region (mitochondrial genome region between cox2 and gene cox1) and TrnY-Rns region (mitochondrial genome between gene trnY(gua) and gene RNA) etc. were analysed by Schena and Cooke (2006). In Cox2-Cox1 region were more appropriate for identification, taxonomic and phylogenetic studies. That can amplified easily and aligned as the total length is quite similar in all *Phytophthoras*.

The intergeneric (IGS1& IGS2) regions of rDNA seem to which have great potential as alternative in comparison to ITS regions. Because, they are present in multicopy (up to 200 copies per haploid genome) but their large length (4000–5000 bp) which also provides a considerable scope for primers development.

Specific primers to detect *P. medicaginis* were developed by using IGS2 region because of difficulties in discriminating the closely related species on the basis of ITS regions which were unable to differentiate them (Liew et al. 1998). Earlier, a new set of Universal primers using IGS regions were developed and utilized for PCR amplification of approximately 450 bp in IGS regions very close to the 28S rDNA gene (Schena and Cooke 2006).

This technique of RAPD has been used as an auxiliary tool to determine the genetic analysis, classification or identification of soil-borne pathogens such as *Fusarium, Rhizoctonia*, or *Colletrichum* and for other ptahogens also (Angela et al. 1996; Kim et al. 1998). Dobrowolski (1998) reported that many of the product generated by RAPD-PCR are derived from repetitive DNA sequences and are frequently species-specific; in fact in recent years the SCAR markers were successfully used to produce species-specific probes and PCR primers in *Phytophthora* (Schubert et al. 1999; Hussain et al, in press) and in other fungal species (Hermosa et al. 2001; Mercado-Blanco et al. 2001; Taylor et al. 2001).

Nuclear-Encoded Genes

The nuclear-encoded ribosomal RNA genes (rDNA) provide attractive targets which are used to design specific primers as they are highly stable, which can be amplified and sequenced with universal primers also and occur in a multiple copies, and possess conserved as well as variable sequences (White et al. 1990). Another example of nuclear genes used for taxonomic studies of *Phytophthora* is ras-related protein (Ypt1) gene. They are highly polymorphic in nature and contain conserved coding regions flanking very variable introns. The Ypt1gene helps in the differentiation of all closely related species such as *P. pseudosyringae*, *P. nemorosa*, *P. psychrophila*, and *P. ilicistaht* which have almost an identical ITS regions. This region is free from any intraspecific variation that could cause problems for diagnostic programme when Compared to other available targets sequences So, Ypt1 gene has the enormous advantages to enable the design specific primers in short DNA segment but it is having the disadvantages of being single copy gene (Chen and Roxby 1996).

DNA Probes

DNA probes or also known as gene probe or genetic probe. These probes are Short, specific (complementary to desired gene) artificially produced segments of DNA which are combined with and detect the presence of specific segment of genes with a chromosomes. If a DNA probes of known composition and length is mingled with pieces of DNA(gene) from a known composition, the probe will matched to its exact counterpart in the chromosomal DNA pieces (genes), by forming a stable-double stranded hybrid. DNA probes were among the first molecular techniques applied in the detection, identification and phylogenetic analysis of fungal pathogens

(Manicon et al. 1987; Rollo et al. 1987). Now a days generation of DNA probes is based on sequences amplified by PCR (Williams et al. 2001a, b). If DNA of soil borne pathogens is extracted directly from soil samples, attention must be drawn to co-extraction of humic acids or other inhibitory substances. In this case, appropriate modifications of DNA isolation protocols (e.g. use of resin columns) have been developed (Cullens and Hirsch 1998). DNA probes have been widely applied in molecular fungal diagnostics especially before the development of the PCR techniques. Among the numerous examples about the use of DNA probes in plant pathogens is the detection of several species of the *P. parasitica* both in soil and in host tissues (Goodwin et al. 1989).

DNA array technology has been used for detection and identify multiple pathogens simultaneously to species and intra-species levels with highly detectors oligonucleotides (Bodrossy et al. 2004; Lievens and Thomma 2005). The microarray enables higher oligonucleotide density where thousands of detector oligonucleotides can be spotted on a single slide. Therefore, one microarray is capable of detecting hundreds of different pathogens, but each array can be used only once.

Nested PCR

The Nested PCR assay was used to detect *P. cactorum* infected strawberry plant because of it higher sensitivity and flexibility, by using two different sets of primers. This assays provide detection even in early stages of diseases development (Glen et al. 2007). Nested PCR increased the detection sensitivity of *P. melonis* 1000 fold, in comparison to a single round PCR, non-nested PCR (Wang et al. 2007). This PCR assay also provides facilitates for use of relatively non-specific PCR primers in the first round of PCR for amplification of multiple pathogen specific primers in the second round of PCR (Hussain et al. 2005). These two PCR main steps has been used for the detection of *P. fragariae* var. *Fragariae*, *V. dahliae*, etc. Affecting strawberry (Bhat and Browne 2007). One of the disadvantage of using nested PCR is that it is not worthy for quantifying target DNA concentrations. Along with sensitivity to DNA concentrations, therefore appropriate precautions (Coleman and Tsongalis 1997) should be taken while carried out this technique.

Multiplex PCR

To overcome the shortcoming of PCR and also to increase the diagnostic capacity of PCR, a variant termed multiplex PCR has been described. In multiplex PCR more than one target sequence can be amplified by including more than one pair of primers in single reaction tube (Hussain et al. 2014). Multiplex PCR has the potential to produce considerable savings of time and effort within the laboratory without compromising test utility. Since, its introduction, multiplex PCR has been successfully applied in many areas of nucleic acid diagnostics, including gene deletion analysis (Browine et al. 1997; Chamberlian et al. 1988), mutation and polymorphism analysis, quantitative analysis, and RNA detection.

Real Time PCR Assay

Real-Time PCR represents a recent advances in PCR methodology. Real Time testing is less time consuming that conventional PCR tests that requires post-PCR manipulations such as visualizing products on a gel. Several studies have demonstrated that the sensitivity of Real Time PCR may also exceed that of conventional PCR by up to 100 times (Mumford et al. 2004; Ratti et al. 2004). Addition, real time PCR can be used to quantify targets nucleic acids in a range of sample types. Real time reverse transcription PCR has been used previously to investigate the levels of defence associated gene transcripts within plants in the presence of non-pathogenic pathogens (Wen et al. 2005).

Several detection systems are now available and are either based on hybridization probes such as Taqman probes (Holland et al. 1991), Molecular Beacons (Tyagi and Kramer 1996), and Scorpions (Whitecombe et al. 1999) or on intercalation by fluorescent dyes, such as the ds-DNA-binding dye SYBR® Green I (Roche Diagnostics, Penzberg, Germany) (Morrison et al. 1998; Wittwer et al. 1997). While hybridization probes offer the advantage of target sequence specificity, a specific probe is required which required an both additional manipulation as well as extra cost for the probe.

The SYBR Green dye is the widely used intercalating dye for fungal disease diagnosis and detection is based on emitting of fluorescent light when intercalated with into double stranded DNA (dsDNA) (Hussain et al. 2013b). This dye is a simple and reliable low cost method for fungal pathogen progression, detection and quantification. Flurophores such as FAM, TE, TAMRA, HEX, JOE, ROX, CY5 and Texas Red and quenchers such as TAMRA, DABCYL and Methyl Red are the commonly used in fluorescent resonance energy transfer (FRET) probes. The ability to quantify pathogen populations using quantitative real-time PCR holds great potential for epidemiological studies, for determining seed-borne or plant-borne inoculum and for establishing and monitoring transmission thresholds as bases for improved disease management (Gitaitis and Walcott 2007).

Detecting Multiple Species

Many other techniques has been used and described for detection of more than one *Phytophthora* spp. within a sample. One such approach is PCR-ELISA in which digoxygenin-11-UTP (DIG) is incorporated into the amplicon during the amplification phase (Bailey et al. 2002). These amplicons are hybridized to oligonucleotides capture probe immobilied into microtiter plate wells after that amount of DIG retained in the well is determined through ELISA type reaction (Bailey et al. 2002). Capture probes for different species of *Phytophthora* can be immobilized in different wells, and hence multiple species can be detected simultaneously by this technique. Similarly, Bonants et al. 2004, found the sensitivity PCR-ELISA for detection of *P. fragariae* compared with that of Taqman PCR.

DNA Chips or DNA arrays is set of unabled oligonucleotides detectors specific for the taxa under study are immobilized using a solid support (usually a nylon membrane or glass slide). DNA is subsequently prepared from the samples (e.g., soil, irrigation water or plant tissues) and labelled during or after PCR amplification, the resulting complex labelled during mixture representing an array of different microorganisms is then hybridized to the membrane under stringent conditions.

Now a days with emergence of new machinery ideas has attracted an attention for identification of microbial species through MALDI-TOF mass spectroscopy (MALDI-TOF MS) (Leushner and Chiu 2000; Kim et al. 2005; Jackson et al. 2007). By using species specific primers annealed to template DNA are extended by a single nucleotide, in this technique, dissociated from the template and separated by mass. Since each of the four nucleotides has a different mass, four products are obtained for each primer, and since multiple primers can be annealed to the template the resolving power of the technique is very high. Its attractive features draw the attentions of Researchers because it is very high throughput, cost effective, and amenable to automation. Although the application of MALDI-TOF MS to diagnosis of plant pathogens is in its infancy, preliminary studies have revealed that it can different *Phytophthora* spp. can be differentiated using soil DNA sample (Siricord and O'Brien 2008).

DNA Array Hybridization

DNA array hybridization, also known as Reverse Dot Blot Hybridization (RDBH) or macroarray, is a technique based on hybridization of amplified and labelled genome regions of interest to immobilized oligonucleotides which is spotted on a solid support platform. It was originally used developed to detect mutations in human genes, and is still an important diagnostic tool in clinical research (Chehab and Wall 1992; Zhang et al. 1991). It is now considered a powerful and practical technique for the detection and identification of fungi and other microbes, such as bacteria, from a very dense complex samples collected form the environment with no the need of isolation in culture (Chen et al. 2009; Lévesque et al. 1998; Tambong et al. 2006; Uehara et al. 1999; Zhang et al. 2007, 2008). For this type of application, oligonucleotides, or microcodes (Summerbell et al. 2005), are designed from a taxonomically complete dataset of suitable genome region(s) (Chen et al. 2009; Tambong et al. 2006). The oligonucleotides can be selected manually, by analysing different multi-sequence alignments, or computer programs, such as SigOli and Array Designer (Premier Biosoft International, Palo Alto, CA). Synthesized oligonucleotides with 5′-amine modifications are then spotted onto a supporting platform, such as a nylon membrane or glass slide, either manually or robotically. Robotic spotting can significantly increase the maximum density of the array which can favour the detection of broader taxonomic groups (Chen et al. 2009). Amplicons of the target gene region(s) are amplified by PCR, labelled with digoxygenin (DIG) and subjected to the DNA hybridization procedure previously described (Fessehaie et al. 2003). A positive reaction between an amplicon and a perfectly matched (PM) oligonucleotide generates a chemiluminescent signal which can be detected by X-ray film or by using a digital camera in dark room. Captured images are then analysed on a computer program such as GenePix Pro (Molecular Devices, Sunnyvale, CA).

DNA arrays was developed for the diagnosis of plant pathogens which has been large scale of environmental samples, like as greenhouse crops (Le Floch et al. 2007; Lievens et al. 2003), potatoes (Fessehaie et al. 2003), ginseng (Punja et al. 2007), and fruits (Robideau et al. 2008; Sholberg et al. 2005, 2006). Macroarrays are also effective diagnostic tools for the detection of phytopathogenic bacteria (Fessehaie et al. 2003), fungi and fungus-like organisms (Chen et al. 2009; Tambong et al. 2006), nematodes (Uehara et al. 1999), and viruses (Agindotan and Perry 2007, 2008). In a recent study, DNA arrays were constructed from multiple loci of *Phytophthora* species, including *ITS*, *cox*1 and the intergenic region (*cox*2-1 spacer, *CS*) between cytochrome c oxidase 2 (*cox*2) and *cox*1 (Chen and Roxby 1996). In comparison to the *cox*1 region, the length variation and the presence of indelsin both sequence alignments of *ITS* and *CS* provided better opportunities to select highly specific oligonucleotides. The combination of all three arrays increased the discrimination potential, detection accuracy, and redundancy of the assay.

Multiplex Tandem PCR Assay

Multiplexed-tandem PCR (MT-PCR) is a technology platform developed for highly multiplexed gene expression profiling and the rapid identification of clinically important pathogens (Stanley and Szewczuk 2005). This assay consists of two rounds of amplification. In the first round a multiplex PCR is performed at 10–15 cycles which allow enrichment of the target DNA without creating competition between amplicons (Lau et al. 2009). Now this product is further diluted and used as template for the second amplification that consists of multiple individual quantitative PCR reactions with primers nested within those used in the multiplex PCR. Up to 72 different PCR reactions can be multiplexed and performed simultaneously. Fluorescence is measured by SYBR green technology at after final extension cycle, and melt-curve analysis data provides species-specific or gene-specific identification. The incorporation of two sets of species-specific primers for each target ensures correct amplification and detection, thus avoiding the expense of DNA probes. SYBR green detection also increases the multiplexing and quantitative capacity of real-time PCR systems, which are usually limited by the availability of fluorescent channels and the need to optimize each individual multiplex PCR.

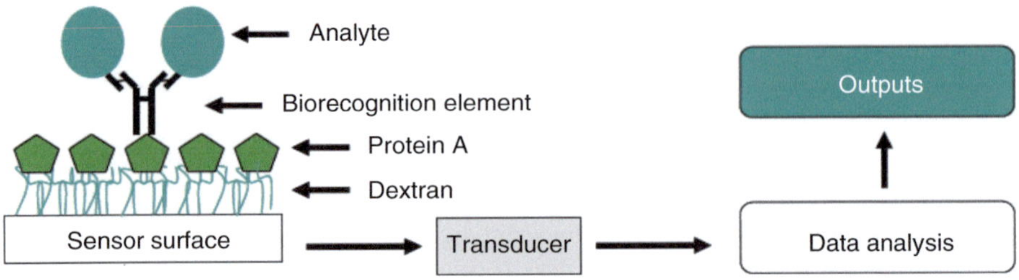

Fig. 5 Diagrammatic display of a biosensor. Here, a full-length antibody is captured on protein A immobilised on a carboxymethylated dextran-coated sensor surface and is used for the capture of an analyte. This interaction produces a specific physicochemical change, such as a change in mass, temperature or electrical potential. This is then converted (via a transducer) to a signal which the user can interpret

Biosensors

Biosensors contains an analytical devices which have combination of a biological recognition ligand with physical or chemical signaling devices called (transducers). This recorded biomolecular interactions are transformed into digital signals which are then interpreted by a computer-aided readout, thereby providing the user with are presentation of the interaction that occurs between the bound (ligand) and free (analyte) entities (Fig. 5). Different formats of sensor have been utilised for pathogen analysis using antibodies; common namely are electrochemical, mass-based, magnetic and optical. Their sensitivities of these assays are dependent on the properties of the transducer and the quality of the antibody. An overview of each sensor type and an explanation of how antibodies can be incorporated for pathogen detection follows.

Lateral Flow Devices

Since the first report of plant virus detection by enzyme-linked immunosorbent assay (ELISA) (Clark and Adams 1977), the incorporation of serological methods into routine diagnosis for plant pathogens has improved the sensitivity and reliability of diseases diagnosis. ELISA has generally superseded most other immunological approaches, such as gel diffusion and agglutination assays, to become the standard laboratory method. ELISA is particularly suited to large-scale testing of field samples, and antibodies have been developed for major plant pathogens. However, there remains a need for the detection of pathogens on site, in field conditions, using a test that can rapidly and reliably confirm the presence or absence of a particular pathogen in symptomatic tissue.

However, serological techniques to detect fungal pathogens of plants have not been widely reported until recently. Harrsion et al. (1990) developed a polyclonal antiserum, which reacted with crushed mycelial extracts of *Phytophthora* spp. But not cross react with other pathogens of potato. *P. infestans* was readily detected by

ELISA using both plate trapped antigen or F (ab') antibody fragment techniques. This indicates that, although polyclonal antibodies may be useful for detecting a range of *Phytophthora* spp., it is likely that monoclonal antibodies will be required to identify individual species. Commercial ELISA based kits are available and currently are used by some growers and state seed certification laboratories in USA. However, such kits are only genus specific and do not allow differentiation of *P. infestans* from other species such as those causing pink rot. The PCR allows a greater level of species specificity than available with ELISA based products and is comparable in terms of cost and therefore it is an ideal tool to identify *P. infestans*. However, for quantification, we have to rely on ELISA test only.

Variations include competitive ELISAs in which surface-bound antigen and antigen in solution competes for antibody binding. In this system, comparison of signal antigen signals form known antigen standards allow very accurate quantification. In double antibody sandwich ELISA (DAS-ELISA), surface-bound antibody is used to capture the antigen, followed by detection using a second enzyme-labelled antibody. Sandwich ELISAs are extremely specific as the antigen must react within second antibodies to be detected. Typically, antibodies are conjugated to alkanephosphatase or peroxidise and the signal observed as a colour development following substrate incubation. ELISAs are run in 96 well plates and scanned by automated devices. Several commercial DAS-ELSIA's exist for plant pathogen detection and these are mostly used for detection of pathogens in homogenised plant tissues.

Immunofluorescence Microscopy Assays

In direct immunofluorescence assays (IFA), the pathogen sample is fixed onto a microscope slide, and analysed with a drop of pathogen specific antibody labelled with a suitable flurochromeisothiosyanate, FITHC), unbound FITC-conjugate is rinsed off, and the slide is examined under an epifluorescent microscope. If antibody has bound, the sample will display a great fluorescence signal, not present in control samples. Indirect IFA's can also be performed, in which pathogen specific antibody is detected by FITC-labelled anti-species immunoglobin (Goldsby et al. 2003). IFA methods are laboratory-based and require an epifluorescent microscope equipped with appropriate excitation and emission filters. Also, relatively few samples can be examined per day, and the analysis of results is subjective and requires experienced personal.

Later Flow Immunoassays

Lateral Flow immunoassays (LFIA) is becoming increasingly popular as on –site diagnostic tools for plant pathogen detection. The assay consists of an immunoreactions and a chromatography step (often termed immune chromatography) and the power of the technique lies in the speed of analysis. The technology has been

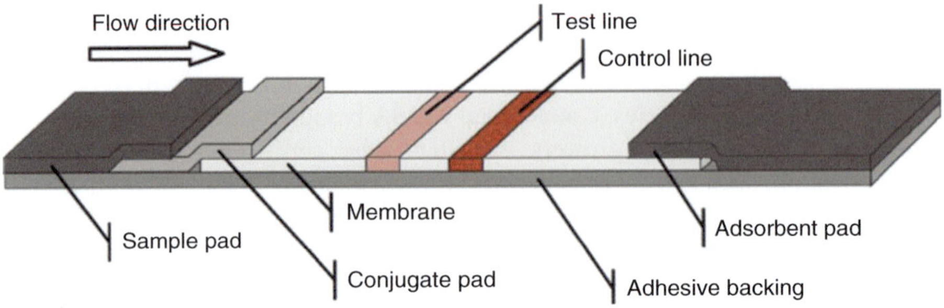

Fig. 6 Lateral flow immunoassay (Source: mpdi.com)

available for many years in clinical diagnostics. The most popular and well known is the home pregnancy test, which detects the glycoprotein hormone, human chorionic Gonadotrophin (Hcg) in the urine of pregnant women. The assays involve the unidirectional flow of particles (e.g. latex, silica, carbon) coated with analyte-specific antibody along a nitrocellulose membrane. A basic sandwich type LFIA exemplified by viral detection is seen in Fig. 6. Applied samples are transferred along the membrane by capillary flow, which allows for good sample separation as the reactant site is different from the application site. The antibody-coated particles are present in the application site, where the particles capture the antigen as they are applied to the strip. As the antigen-particle complex migrates in the membrane, the complex is selectively bound by an immobilized antigen antibody, thereby generating a solid line visible to the naked eye. Particles that did not bind antigen continue to migrate up the membrane, where they are bound by an anti-species antibody. For a positive result both the test line and the control line must appear (Danks and Barker 2000). In addition to the sandwich type assay described here, competitive and inhibition type assays exist. LFIA's are user friendly, relatively inexpensive, and suited for on-site testing by minimally trained personal. However, LFIA's have major limitations as only qualitative results are obtained (i.e. it gives "yes" or "No" answer).

Dipstick Assays

Different dipstick assays type exist. One of the widely used consists of a dipstick (typically nitrocellulose) coated with pathogen specific antibody. The dipstick is emerged into the sample and if present, the pathogen is specifically captured by the antibody. Following extensive washing of the dipstick, a second pathogen-specific enzyme conjugate antibody is added. After incubation in enzyme substrate, a colour development is indicative of pathogen contamination. Another format consists of a blank dipstick which is emerged into the sample. Pathogen will bind the dipstick and enzyme conjugate pathogen specific antibody is added. Again a colour development after the addition of enzyme substrate, suggests the presence of the pathogen in question (Dewey et al. 1990). As with LFIA's analysis these methods are fast and can be operated by non-scientifically personals, but quantitatively data are hard to obtain.

Multiplex Immunoassays

Promising new assays types, termed multiplex immunoassays have been developed in recent years. Multiplexing is defined as the detection of several pathogens simultaneously and is usually performed on the Luminex system. During this technology, based on use of beads (microspheres of 5.6 μm), which is internally stained with two flurochromes. A hundred different bead sets are available, with unique compositions of red and infrared fluorochromes. This makes detection of up to 100 different antigens possible. Both antibodies and nucleic probes can be covalently linked to the activated microspheres. In a Microsphere Immunoassay (MIA), sample are incubated with antibody coated beads. The pathogen/antigen are caught by the bead-antibody where after secondary antibodies with a reporter dye, are added, resulting in a sandwich type of assay. The samples are applied to a Luminex analyser, where one laser excites the internal dyes to identify each microsphere particle, and another laser excites the reporter dye on the secondary antibodies captured. This gives detailed data of the presence of pathogen. Several different beads (with different antibodies attached) can be added to one sample, which can thereby be scanned for the presence of many different pathogens (Vignali 2000). The technique has successfully been used for multiplex detection of three viral plant pathogens and four bacterial plant pathogens (Van der Wolf et al. 2005).

Automatic Diagnostic

Now a days Automatic detection of Plant disease within a short period of time is an essential research topic that can prove benefits in monitoring a large fields of crops, and thus automatically detect the symptoms of diseases as soon as they appear on plant leaves (Al-Bashish et al. 2011; Rumpf et al. 2010; Hillnhuetter and Mahlein 2008). Therefore, looking for fast, automatic, less expensive and accurate methods to detect plant disease cases is of great realistic significance (Prasad Babu and Srinivasa Rao 2010). Machine learning based detection and recognition of diseases part of plant can provide clues to detect, identify and treat the diseases in its early stages (Rumpf et al. 2010; Hillnhuetter and Mahlein 2008). Comparatively, visually identifying plant diseases is expensive, inefficient, and difficult. Which also, requires the expertise of trained personals in this area (Al-Bashish et al. 2011).

Now a days a fast and accurate new methods is/techniques are being developed which are based on computer image processing for differentiating/grading of plant disease, for that, leaf region was segmented by using Ostu method (Ostu 1979; Sezgin and Sankur 2003). After that the disease spot regions were segmented by using Sobel operator to detect the disease spot edges and Finally, that plant diseases are graded by calculating the quotient of disease spot and leaf areas.

LAMP Method

In contrast, to PCR, isothermal amplification methods avoid the use of thermal cycling equipment, allowing reactions to be carried out incubated in a simple water bath or simple heated block. It is performed in a single temperature. This Loop mediated isothermal amplification (LAMP) is an amplification method which uses two sets of primers (internal and external primers) and a DNA polymerase with strand-displacing activity to produce amplification products containing loop regions to which further primers can bind, allowing amplification to continue without thermal cycling. The Amplification reaction is accelerated by the use of an additional set of primes (loop primers) which binds to those loops which are of the in-corrected orientation for the internal primers to bind (Gill and Ghaemi 2008). This technique has previously been described for the detection of a different range of plant pathogen (Fukuta et al. 2003a, b, 2004; Tomlinson et al. 2007; Varga and James 2006). In case of this method the products can be detected by conventional agarose gel electrophoresis, by the use of spectrophotometric equipment to measure turbidity (Mori et al. 2004), in real time using intercalating fluorescent dyes (Lane et al. 2007), or by visual inspection through naked eyes) of turbidity or colour changes (Iwamtoo et al. 2003; Mori et al. 2001). Although this technique has advantages of requiring no use of any costly equipment, on the basis of colour or turbidity with the unaided eye. So, Equipment-free methods for unambiguous detection of LAMP products would increase the feasibility of using LAMP for diagnosis of phytopathogens outside the laboratory.

Advantages of LAMP method:

1. Amplification of DNA takes place at an isothermal condition (63–65 °C) with greater efficiency.
2. Thermal denaturation of double stranded DNA is not required.
3. This technique helps in specific amplification as it designs four primers to recognize six distinct regions on the target gene.
4. It is cost effective as it does not require special reagents or sophisticated equipment.
5. This technology can be used for the amplification of RNA templates in presence of reverse transcriptase.
6. This assay takes less time for amplification and detection.

Integrated Diseases Management

In recent era we have observed that a single method for the management of late blight is not sufficient. Therefore, it is demand of time, that we should follow integrated diseases management. Although late blight can be managed by growing resistant varieties, but sooner or later all the varieties develop some blight. Breeding new varieties is a key to success.

Cultural Control

They are employed to reduces/eliminates the initial inoculum of late blight form the seed tubers. Fields and to check the spread of the diseases. Important cultural methods include:

1. Selection of well drained soils for healthy cultivation of Potato
2. High ridging (about 15 cm) to avoid exposure of infected seed tubers which can act as source of the diseases.
3. Scouting of the field for identifying primary infection foci and their destruction by removal of the infected plants after drenching them with recommended fungicides.
4. As soon as the weather condition becomes congenial for late blight, irrigation should be stopped wherever applicable. Only light irrigation may be given later if required.
5. Destroy and remove the haulms form the fields when the diseases severity reaches >80 % (to avoid tuber infection).
6. **Use of healthy Seed:** only disease free seed should be used. Avoid seed from a field which has been infected by late blight in the previous year. In case the seed from the infected fields has to be used, the seed tubers should be thoroughly checked for late blight infection. The infected tubers should be removed and buried in the soil. This practice of sorting out late blight symptoms are east to be identified in cut-pieces where bronzing of the flesh can be seen.

Resistant Varieties

Northern Indian Plains: Kufri Anand, Kufri Pukhraj, K. Badshah, K. Chipsona I and II, K. Jawahar, Kufri Satluj, Kufri Surya, Kufri Khayati, Kufri Chipsona III, Kufri Arun, Kufri Lalima

HP Hills: K. Jyoti, K. Giriraj; K. Megha; K. Kanchan, Kufri Girdhari, Kufri Himalini, Kufri Shailja, Kufri Sawrna, Kufri Neelema

Chemical Control

Internationally, to protect the standing crops Protective sprays with a contact fungicide, viz., Mancozeb (0.2 %) before appearance of the disease is effective. Subsequent sprays if necessary should be repeated at 8–10 days interval. However if severe blight attack, one more sprays of Metalaxyl (0.25 %) are given to check the further spread of the disease. Similarly again Mancozeb is applied at an interval of 15 days after the Metalaxyl application.

A spray schedule of minimum of four fungicides sprays is recommended for managing late blight. However, the number sprays may b increased or decreased depending on diseases pressure.

1st spray as prophylactic measure, spray the crop with contact fungicides like macozeb 75 % WP (0.2 %), propioneb 70 % WP (0.2 %) or chlorothalonil (0.2 %) as soon as the weather conditions become congenial for late blight. do not wait or allow late blight to appear and establish in the field. Always use a sticker @ 0.1 % for proper sticking and uniform spread of fungicides on leaf surface.

2nd spray as soon as the diseases is noticed in the field, apply any of the systemic fungicides viz. Metalaxyl based (0.25 %) or cymoxanil based (0.3 %) fungicides.

3rd spray apply contact fungicides viz. Mancozeb (0.2 %), propioneb (0.2 %) or chlorothalonil (0.2 %) after 8–10 days of second application of fungicides. However, if weather is highly congenial, repeat application of cymoxanil based fungicides may be restored to.

4th spray apply systemic fungicides or contact fungicides as mentioned above depending on diseases severity and weather conditions.

A minimum of 400 litre water (during early growth period) to 1000 litre water would be required to spray 1 ha crop area. Ensure through coverage of plants to bottom with fungicides. Special attention should be given to lower leaves which need to be covered with fungicides. Biological Control is emerging as a new ray of hope for future control of harmful plant pathogens (Tomar et al. 2010, 2014).

Late Blight Forecasting

Development of late blight mainly depends on moisture, temperature and cloudiness. In India, the rains are heavy and the weather is cool and cloudy/foggy during summer in the hills but in plains the weather is generally clear with scanty rains (during autumn or spring) and therefore, the disease epidemic is not a regular feature. The monsoon moves from East to West in the Himalayas. Therefore, the blight occurs early in the eastern Himalayas. Taking weather parameters in account, Bhattacharya et al. (1983) developed forecasting models for Shimla, Shillong and Ootacamund i.e. (i) if the 7-day moving precipitation (30 mm for Shimla, 28.9 mm for Ootacamund and 38.5 mm for Shillong observed to be critical rainfall lines) associated with mean temperature of 23.9 °C or less continues for seven consecutive days, late blight would appear within 3 weeks and (ii) if hourly temperature ranges from 10 to 20 °C associated with high RH (80 % or more) for continuous 18 h for two consecutive days, the blight would appear within a week. Based on these criteria a late blight warning service was started since 1978 for Shimla hills and successful warnings are issued through All India Radio, Shimla every year.

Late blight forecasting in the sub-tropical plains is different to that of temperate highlands. In the hills, environmental conditions (temperature, RH, rainfall) favourable for late blight appearance are assured. There are plenty of rains during the crop season which led to high RH (>80 %) throughout the crop season. Temperature remains moderate and congenial throughout. It is therefore, possible to rely on weather parameters like, rainfall, RH and temperature for making disease forecasts. Such situations, however, do not exist in the sub-tropical plains, where there are scanty rains during the crop season. In such a situation, role of micro-climate, fog dew and sunshine becomes critical for the appearance of the disease. Besides, weather data for substantial period is required to develop reliable empirical models.

A late blight forecasting system has been developed for western Uttar Pradesh using temperature, RH and rainfall data. It consists of two models, one each for rainy and non-rainy years.

Future Requirements

Of late, the awareness of the diseases diagnostic and quality concern for planting material has crept into the minds of powers; more reliable kits are required to be developed for managing crop diseases. Fine tuning of detection techniques developed for hotriculture as well as agriculture crop pathogens is needed. Sampling is one of the major criteria for quality testing and certification which needs to be optimized for each pathogen attacking all fruit crops. Along with all these steps their will be a need of a specializes trained personal who are Crop-pathogen expert in identification, detection, for better understanding and helps in monitoring of the pathogen. Detection of pathogenic prologues in soil before establishing the planting will be helpful to keep the area free of diseases.

Conclusion

The utilization of a novel biomarker, a gene coding for green fluorescent protein (GFP) isolated from *Aequorea Victoria*, a jelly fish that live in the Pacific Ocean, has offered new possibilities in the area of host pathogen interaction, permitting histopathological studies, pathogen monitoring, during pathogenesis and ramification and quantification of pathogen in the host tissues like never before. Nevertheless, there are instances where nucleic acid based tests may be better value for money, for example, greater sensitivity may allow for samples to be bulked together, resulting in fewer assays being needed. This approach makes sense where negative result is the norm in a screening programme. Testing could also be carried out directly on tuber material, resulting in savings of time as well as costs. The nucleic acid based detection of fungal pathogens has emerged as a supplement to overcome these bottlenecks. The recent advancement in the area of PCR based approaches further extended its versatility. Assays like real-time PCR, multiplex PCR, nested PCR, Bio-PCR, repetitive PCR, LAMP are among the detection options that provides rapid data analysis with specificity. However, one has to choose the best or combination of options depending upon the needs. For example, when multiple pathogens are to be detected in a minimum time multiplex-PCR would be the best options. While, the pathogen detection limit in a sample is at zero tolerance level, nested PCR, bio-PCR should be carried out. Lastly a more sensitive test gives greater security. Detection of *P. infestans* is very essential in developing countries. Because there are seed production systems is semi-organise. After rigorous scrutinized of potato bags should issue the phytosanitry certificate for transportation of potato seed from one state to another state, so that movement of inoculums can be restricted.

References

Agindotan B, Perry KL (2007) Macroarray detection of plant RNA viruses using randomly primed and amplified complementary DNAs from infected plants. Phytopathology 97:119–127

Agindotan B, Perry KL (2008) Macroarray detection of eleven potato-infecting viruses and potato spindle tuber viroid. Plant Dis 92:730–740

Al-Bashish D, Braik M, Bani-Ahmed S (2011) Detection and classification of leaf diseases using K-means based segmentation and neural-network based classification. Inform Technol J 10:267–275

Angela GS, Evelyn MM, Hartwig HG (1996) Polymerase chain reaction based assays for species specific detection of *Fusariumculmorum*, *F. graminearum* and *F. avenaceaum*. Phytopathology 10:120–130

Bailey AM, Mitchell DJ, Manjunath KL, Nolasco G, Niblert CL (2002) Identification of the species level of the plant pathogens phytophthora and pythium by using unique sequences of the ITS1 region of ribosomal DNA as capture probes for PCR-ELISA. FEMS Microbiol Lett 207:153–158

Bhat RG, Browne GT (2007) Development of an improved PCR based techniques for detection of Phytophthora cactorum in strawberry plants. Phytopathology 97:S10, Abstract

Bhattacharyya SK, Phadatare SG, Khanna RN, Srivastava DS, Prasad B (1983) Efficacy of some fungicides in controlling late blight of potato in India. Indian J Agric Sci 53:153–157

Bhattacharyya SK, Shekhwat GS, Singh BP (1990) Potato late blight. Tech Bull. No. 27. Central Potato Research Institute, Shimla, 40 p

Bodrossy L, Sessitsch A (2004) Oligonucleotide microarrays in microbial diagnostics. Curr Opin Microbiol 7:245–254

Bonants PJM, Hagenaar, De-Veerdt M, Van Gent-Pelzer MP, Lacourt I, Cooke DEL, Duncan JM (1997) Detection and identification of *Phytophthora fragariae* Hickman by the polymerase chain reaction. Eur J Plant Pathol 103:345–355

Bonants PJM, Van Gent-Pelzer MPE, Hooftman R, Cooke DEL, Guy DC, Duncan JM (2004) A combination of baiting and different PCR formats, including measurement of real time quantitative fluorescence, for the detection of Phytophthora fragariae in strawberry plants. Eur J Plant Pathol 110:689–702

Browine J, Shawcross S, Theaker J, Whitecombe D, Ferrie R, Newton C, Litte S (1997) The elimination of primer-dimer accumulation in PCR. Nucleic Acids Res 25:3235–3241

Chahal SS, Pannu PPS (1997) Detection of *Tilletiaindica* in wheat and *T. Barclayana* in rice samples and its implication for seed certification. In: Hutchins JD, Reeves JC (eds) Seed health testing. CAB International. pp 153–158

Chamberlian JS, Gibbs RA, Ranier JE, Nguyen PN, Caskey CT (1988) Deletion screening of the Duchenne muscular dystrophy locus via multiplex DNA amplification. Nucleic Acids 16:11141–11156

Chehab FF, Jeff W (1992) Detection of multiple cystic fibrosis mutations by reverse dot blot hybridization: a technology for carrier screening. Hum Genet 89:163–168

Chen Y, Roxby R (1996) Characterization of a *Phytophthora infestans* gene involved in the vesicle transport. Gene 181:89–94

Chen W, Seifert K, Lévesque CA (2009) A high density COX1 barcode oligonucleotide array for identification and detection of species of *Penicillium* subgenus *Penicillium*. Mol Ecol Resour 9:114–129

Chowdappa P, Brayford D, Smith J, Flood J (2003) Molecular discrimination of *Phytophthora* isolates on cocoa and their relationship with coconut, black pepper and bell pepper isolates based on rDNA repeat and AFLP fingerprinting. Curr Sci 84:1235–1238

Clark MF, Adam AN (1977) Characteristics of the microplate method of enzyme-linked immunosorbent assay for the detection of plant viruses. J Gen Virol 34:475–483

Coleman WD, Tsongalis GJ (1997) The polymerase chain reaction. In: Coleman WB, Tsongalis GJ (eds) Molecular diagnostics for the clinical laboratorian. Human Press, Totowa, pp 47–55

Cook H (1947) Forecasting late blight epiphytotics of potatoes and tomatoes. J Agric Res 78:54–56

Cooke HT (1947) Forecasting tomato late blight. Food Packer. Apr 1947. 69–70 p

Cooke DEL, Drenth A, Dincan JM, Wagels G, Brasier CM (2000) A molecular phylogeny of *Phytophthora* and related oomycetes. Fungal Genet Biol 30:17–32

Cox AE, Large EC (1960) Potato blight epidemics throughout the world. Agricultural Research Service, United States Department of Agriculture, Agricultural Handbook No. 174. 230 p

Cullens DW, Hirsch PR (1998) Simple and rapid method for the direct extraction of microbial DNA from soil for PCR. Soil Biol Biochem 30:983–993

Danks C, Barker I (2000) On-site detection of plant pathogens using lateral flow devices. EPPO Bull 30:421–426

Dewey FM, Macdoland M, Phillips SI, Priestley RA (1990) Development of monoclonal antibody-ELISA and monoclonal antibody–dipstick immunoassays for penicillium-islandicum in rice grains. J Gen Microbiol 136:753–760

Dobrowolski MP (1998) Microsatellites in the mitochondrial genome of *Phytophthora cinnamomi* failed to provide highly polymorphic markers for population genetics. FEMS Microbiol Lett 163:243–248

Dorrance AE, Inglis DA (1997) Assessment of greenhouse and laboratory screening methods for evaluating potato foliage for resistance to late blight. Plant Dis 81:1206–1213

Fessehaie A, De Boer SH, Lévesque CA (2003) An oligonucleotide array for the identification and differentiation of bacteria pathogenic on potato. Phytopathology 93:262–269

Fry WE (1977) Integrated control of potato late blight: effects of polygenic resistance and techniques of timing fungicides applications. Phytopathology 67:415–420

Fry WE, Goodwin SB, Dyer AT, Matuszak JM, Drenth A, Tooley PW, Sujkowski LS, Koh YJ, Cohen BA, Spielman LJ, Deahl KL, Inglis DA, Sandlan KP (1993) Historical recent migrations of *Phytophthora infestans:* chronology, pathways and implications. Plant Dis 77:653–661

Fukuta S, Kato S, Yoshida K, Mizukami Y, Ishida A, Ueda J, Kanbe M, Ishimoto Y (2003a) Detection of *Tomato yellow leaf curl virus* by loop mediated isothermal amplification reaction. J Virol Methods 112:35–40

Fukuta S, Iida T, Mizumkami Y, Ishida A, Ueda J, Kanbe M, Ishimoto Y (2003b) Detection of Japanese *yam mosaic viruses* by RT-LAMP. Arch Virol 148:1713–1720

Fukuta S, Ohishi K, Yoshida K, Mizukami Y, Ishida A, Kanbe M (2004) Development of immunocapture reverse transcriptase loop mediated isothermal amplification for the development of tomato spotted wilt form chrysanthemum. J Virol Methods 121:49–55

Gill P, Ghaemi A (2008) Nucleic acid isothermal amplification technologies- a review. Nucleosides Nucleotides Nucleic Acids 27:224–243

Gitaitis R, Walcott R (2007) The epidemiology and management of seedborne bacterial diseases. Annu Rev Phytopathol 45:71–97
Glen M, Smith AH, Langrell SRH, Mohammed CL (2007) Development of nested polymerase chain reaction detection of Mycosphaerella spp. and its application to the study of leaf disease in eucalyptus plantations. Phytopathology 97:132–144
Goldsby RA, Kindt TJ, Osborne BA, Kuby J (2003) Immunology, 5th edn. W.H. Freeman and Company, New York
Goodwin SB (1997) The population genetics of *Phytophthora*. Phytopathology 87:462–473
Goodwin SB, Drenth A (1997) Origin of the A2 mating type of *Phytophthora infestans* outside Mexico. Phytopathology 87:993–999
Goodwin PH, Kirkpatrick BC, Duniways JM (1989) Cloned DNA probes for identification of *Phytophthora parasitica*. Phytopathology 79:716–721
Goodwin SB, Cohen BA, Fry WE (1994a) Panglobal distribution of single clonal lineage of the irish famine fungus. Proc Natl Acad Sci U S A 91:11591–11595
Goodwin SB, Cohen BA, Deahl KL, Fry WE (1994b) Migration from northern Mexico as the probable cause of recent genetic changes in populations of *Phytophthora infestans* in the United States and Canada. Phytopathology 84:553–558
Gottschling M, Plötner J (2004) Secondary, structure models of the nuclear internal transcribed spacer regions and 5.8 s r RNA in *Calciodinelloideae* (Peridiniaceae) and other dinoflagellates. Nucleic Acids Res 32:307–315
Haas BJ et al (2009) Genome sequence and analysis of the Irish potato famine pathogen Phytophthora infestans. Nature 461:393–398
Harrison JG, Barker H, Lowe R, Rees EA (1990) Estimation of amounts of *Phytophthora infestans* mycelium in leaf tissue by enzyme linked immunosorbent assay. Plant Pathol 39:274–277
Hermosa MR, Grondona I, Diaz-Minguez JM, Iturriaga EA, Monte E (2001) Development of a strain-specific SCAR marker for the detection of *Trichoderma atroviride* 11, a biological agent against soil borne fungal plant pathogens. Curr Genet 38:343–350
Hillnhuetter C, Mahlein AK (2008) Early detection and localisation of sugar beet diseases: new approaches. Gesunde Pfianzen 60(4):143–149
Holland PM, Abramson RD, Watson R, Gelfand DH (1991) Detection of specific polymerase chain reaction product by utilizing the 5′-3′ exonuclease activity of *Thermus aquaticus* DNA polymerase. Proc Natl Acad Sci U S A 88:7276–7280
Hussain T, Singh BP, Anwar F (in press) Development of specific marker for PCR diagnostic of Late blight of potato caused by Phytophthora infestans using RAPD based SCAR methodology. J Saudi Soc Agric Sci. Manuscript ID. JSSAS-D-13-00120R1 Accepted
Hussain S, Lees AK, Duncan JM, Cooke DEL (2005) Development of a species-specific and sensitive detection assay for *Phytophthora infestans* and its application for monitoring of inoculum in tubers and soil. Plant Pathol 54:373–382
Hussain T, Anwar F, Tomar S (2010) Rapid and sensitive detection of *Phytophthora infestans* in infected tubers by ITS region based Primers. In: 6th JK science congress, University of Kashmir, BT-7, p 225
Hussain T, Sharma S, Singh BP, Jeevalatha A, Sagar V, Sharma NN, Kaushik SK, Chakrabarti SK, Anwar F (2013) Detection of latent infection of *Phytophthora infestans* in potato seed tubers. Potato Journal 40(2):142–148.
Hussain T, Sharma S, Singh BP, Anwar F (2013) Fast, easier and sensitive detection of late blight of potato caused by *Phytophthora infestans* through Internal Transcribed Spacer (ITS) region based marker. In: UGC –DRS-1 sponsored national seminar on plant sciences: new technologies, conservation and environment. 23–24th Feb 2013, Department of Botany, AMU, Aligarh, p 14
Hussain T, Singh BP, Anwar F (2013a) A noval ITS region based marker for early detection of *Phytophthora infestans*, causing Late Blight of Potato in India. In: Proceedings of international conference on health, environment and industrial biotechnology (BioSangam-2013), MNNIT, Tata McGraw Publishers, Allahabad, UP, Nov 21st-23r 2013, pp 147–153
Hussain T, Singh BP, Anwar F (2013b) A Quantitative Real Time PCR of *Phytophthora infestans* in different Indian potato cultivars. IOSR J Agric Vet Sci 4(4):17–26

Hussain T, Singh BP, Anwar F (2013c) Real Time PCR based determination of *Phytophthora infestans* colonization in potato tubers at cold storage in India. Online J Biosci inform 4(2):159–178

Hussain T, Singh BP, Kaushik SK, Lal M, Gupta A (2014) PCR protocol for quick and combined detection of Early and Late Blights of Potato. In: National seminar on emerging problems of potato, 1st and 2nd Nov 2014, CPRI and IPA, Shimla, HP, p 169

Hussain T, Singh BP, Firoz A (2014) A quantitative real time PCR based method for the detection of *P. infestans* causing late blight of potato, in infested soil. Saudi J Biol Sci 21:380–386

Hussain T, Singh B, Anwar F, Tomar S (2015) A simple method for diagnostic of Phytophthora infestans (Mont.) de Bary from potato agricultural fields of potato. Turk J Agricult – Food Sci Tech 3(12):904–907

Iwamtoo T, Sonobe T, Hayashi L (2003) Loop mediated isothermal amplification for the direct detection of *Mycobacterium tuberculosis* complex, M. Avium, and M. Intracellular in sputum samples. J Clin Methods 41:2616–2622

Jackson GW, McNichols RJ, Fox GE, Willson RC (2007) Universal bacterial identification by mass spectrometry of 16S ribosomal RNA cleavage products. Int J Mass Spectrom 261:218–226

Jayan MH, Huang LC, Ann PJ, Liou RF (2002) Rapid detection of *Phytophthora infestans* by PCR. Plant Pathol Bull 11:25–32

Jeffers SN, Aldwinkle HS (1987) Enhancing detection of *Phytophthora cactorum* in naturally infested soil. Phytopathology 77:1475–1482

Judelson HS, Tooley PW (2000) Enhanced polymerase chain reaction methods for detecting and quantifying *Phytophthora infestans* in plants. Phytopathology 90:1112–1119

Karolev AA (1978) Slightly infected tubers as the source of Phytophthora. Zasbechita Rastenii 8:49

Kim K, Lee YS, Choi HS (1998) Analyses of genetic relationships of *rhizoctonia solani* from various crop species and rapid identification of anatomises group with RAPD method. Kor J Mycol 26:373–379

Kim YJ, Kim SO, Chung HJ, Jee MS, Kim BG, Kim KM, Yoon JH, Lee HS, Kim CY, Kim S, Yoo W, Hong SP (2005) Population genotyping of hepatitis C virus by matrix-assisted lase desorption/ionization time of flight mass spectrometry analysis of short DNA fragments. Clin Chem 51:1123–1131

Kong P, Hong CX, Tooley PW, Ivors K, Garbelotto M, Richardson PA (2004) Rapid identification of *Phytophthora ramorum* using PCR-SSCP analysis of ribosomal DNA ITS-1. Lett Appl Microbiol 38:433–439

Lane CR, Hobden E, Walker L, Barton VC, Inman AJ, Hughes KJD, Swan H, Colyer A, Barker I (2007) Evaluation of a rapid diagnostic filed test kit for identification of Phytophthora species including *P. ramorum* and *P. kernoviae* at the point of inspection. Plant Pathol 56:828–835

Lau A, Sorrell TC, Chen S, Stanley K, Iredell J, Halliday C (2009) Multiplex tandem PCR: a novel platform for rapid detection and identification of fungal pathogens from blood culture specimens. J Clin Microbiol 46:3021–3027

Le Floch G, Tambong J, Vallance J, Tirilly Y, Lévesque A, Rey P (2007) Rhizosphere persistence of three *Pythium oligandrum* strains in tomato soilless culture assessed by DNA macroarray and real-time PCR. FEMS Microbiol Ecol 61:317–326

Leushner J, Chiu NHL (2000) Automated mass spectrometry: a revolutionary technology for clinical diagnostics. Mol Diagn 5:341–348

Lévesque CA, Harlton CE, de Cock AWAM (1998) Identification of some oomycetes by reverse dot blot hybridization. Phytopathology 88:213–222

Lievens B, Thomma BPHJ (2005) Recent developments in pathogen detection arrays: implications for fungal plant pathogens and use in practice. Phytopathology 95:1374–1380

Lievens B, Brouwer M, Vanachter ACRC, Lévesque CA, Cammue BPA, Thomma BPHJ (2003) Design and development of a DNA array for rapid detection and identification of multiple tomato vascular wilt pathogens. FEMS Microbiol Lett 223:113–122

Liew ECY, Maclean DJ, Irwin JAG (1998) PCR based detection of *Phyophthora medicaginis* using the intergenic spacer region of the ribosomal DNA. Mycol Res 102:73–80

Manicon BQ, Bar-Joseph M, Rosmer A, Vigodsky-Haas H, Kotze JM (1987) Potential application of random DNA probes and restriction fragments length polymorphism in the taxonomy of the fusarial. Phytopathology 77:669–672

Mercado-Blanco J, Rodriguez-Jurado D, Jiménez-Diaz RM (2001) Detection of the nondefoliating pathotype of *Verticilium daliae* in infected olive plants by nested PCR. Plant Pathol 50:609–619

Mori Y, Nagamine K, Tomita N, Notomi T (2001) Detection of loop-mediated isothermal amplification reaction by turbidity derived from magnesium pyrophosphate formation. Biochem Biophys Methods Res Commun 289:150–154

Mori Y, Kitao M, Tomita N, Notomi T (2004) Real time turbidimetry of LAMP reaction for quantifying template DNA. J Biochem Biophys Methods 59:145–157

Morrison TB, Weis JJ, Wittwer CT (1998) Quantification of low-copy transcripts by continuous SYBR green I monitoring during amplification. Biotechniques 24:954–958

Mumford R, Skelton A, Metcalfe E, Walsh K, Boonham N (2004) The reliable detection of barley yellow and mild mosaic viruses using real time PCR (Taq Man). J Virol Methods 117:153–159

Niederhauser JS (1991) *Phytophthora infestans*: the Mexican connection. In: Lucas JA, Shattock RC, Shaw DS, Cooke LR (eds) Phytophthora. Cambridge University Press, Cambridge, UK, pp 25–45

Ostu N (1979) A threshold selection method from gray-level histograms. IEEE Trans Syst Man Cybern 9(1):62–66

Prasad Babu MS, Srinivasa Rao B (2010) Leaves recognition using back propogation neural network-advice for pest and disease control on crops. Technical report, Department of computer science and systems engineering. Andhra University, Visakhapatnam

Punja ZK, Wan A, Goswami RS, Verma N, Rahman M, Barasubiye T, Seifert KA, Lévesque CA (2007) Diversity of *Fusarium* species associated with discolored ginseng roots in British Columbia. Can J Plant Pathol 29:340–353

Ratti C, Budge G, Ward I, Clove G, Rubies-Autonell C, Henry C (2004) Detection and relative quantification of soil-borne cereal mosaic virus(SBCMV) and Polymyxa graminis in winter wheat using real time PCR (Taqman). J Virol Methods 122:95–103

Robertson NF (1991) The challenge of *Phytophthora infestans*. Adv Plant Pathol 7:1–30

Robideau GP, Caruso FL, Oudemans PV, McManus PS, Renaud MA, Auclair ME, Bilodeau GJ, Yee D, Désaulniers NL, DeVerna JW, Lévesque CA (2008) Detection of cranberry fruit rot fungi using DNA array hybridization. Can J Plant Pathol 30:226–240

Rollo F, Amici A, Francesca F, Di Silvestro I (1987) Construction and characterization of a cloned probe for the detection of *Phomatracheiphila* in plant tissues. Appl Microbiol Biotechnol 26:352–357

Rumpf T, Mahlein AK, Steiner U, Oerke EC, Dehne HW, Plumer L (2010) Early detection and classification of plant diseases with support vector machines based on hyperspectral reflectance. Comput Electron aAgric 74(1):91–99

Schena L, Cooke DEL (2006) Assessing the potential of regions of the nuclear and mitochondrial genome to develop a molecular tool box for the detection and characterization of phytophthora species. J Microb Methods 67:70–85

Schubert R, Bahnweng G, Nechwall GJ, Jung T, Cooke DEL, Duncan JM, Müller-Stark G, Langebartels C, Sandermann H Jr, Oßwald W (1999) Detection and quantification of *Phytophthora* species, which are associated withroot-rot disease in European deciduous forests by speciesspecific polymerase chain reaction. Eur J For Pathol 29:169–188

Sezgin M, Sankur B (2003) Survey over image threshold techniques and quantitative performances evaluation. J Electron Imaging 13(1):146–165

Sholberg PL, O'Gorman D, Bedford K, Lévesque CA (2005) Development of a DNA macroarray for detection and monitoring of economically important apple diseases. Plant Dis 89:1143–1150

Sholberg PL, O'Gorman D, Bedford KE (2006) Monitoring *Erwiniaamylovora* in pome fruit orchards using reverse dot-blot hybridization. Acta Horticult 704:91–98

Siricord C, O'Brien PA (2008) MALDI-TOF mass spectroscopy can be used for detection of pathogenic microorganisms in soil. Aust Plant Pathol 37:543–545

Singh BP, Roy S, Bhattacharyya SK (1994) Occurrence of the A_2 mating type of *Phytophthora infestans* in India. Potato Research 37:227–231

Stanley KK, Szewczuk E (2005) Multiplexed tandem PCR: gene profiling from small amounts of RNA using SYBR Green detection. Nucleic Acids Res 33:e180. doi:10.1093/nar/gni182

Summerbell RC, Lévesque CA, Seifert KA, Bovers M, Fell JW, Diaz MR, Boekhout T, de Hoog GS, Stalpers J, Crous PW (2005) Microcoding: the second step in DNA barcoding. Philos Trans R Soc B 360.1897–1903

Tambong JT, de Cock AWAM, Tinker NA, Lévesque CA (2006) Oligonucleotide array for identification and detection of *Pythium* species. Appl Environ Microbiol 72:2691–2706

Taylor EJA, Stevens EA, Bates JA, Morreal G, Lee D, Kenyon DM, Thomas JE (2001) Rapid-cycle PCR detection of *Pyrenophora graminea* from barley seed. Plant Pathol 50:347–355

Tomar S, Kumar S, Tyagi S and Hussain T (2010) Safe and environment friendly method for control of late blight disease of potato. In: 6th J&K science congress, University of Kashmir. BO-11, p 137

Tomar S, Singh BP, Lal M, Khan MA, Hussain T, Sharma S, Kaushik SK, Kumar S (2014) Screening of noval microorganism for biosurfactant and biocontrol activity against *Phytophthora infestans*. J Env Biol 35:893–899

Tomlinson JA, Barker I, Boonham (2007) Faster, simpler, more specific methods for improved molecular detection of *Phytophthora ramorum* in the field. Appl Environ Microbiol 73:4040–4047

Trout CL, Ristaino JB, Madritch M, Wangsomboondee T (1997) Rapid detection of *P. infestans* in late blight –infected potato and tomato using PCR. Plant Dis 81:1042–1048

Tsao P, Guy S (1977) Inhibition of *Mortierella* and *Pythium* in a *Phytophthora* isolation medium containing hymexazol. Phytopatholgy 67:796–801

Tyagi S, Kramer FR (1996) Molecular beacons – probes that fluoresce upon hybridization. Nat Biotechnol 14:303–308

Uehara T, Kushida A, Momota Y (1999) Rapid and sensitive identification of *Pratylenchus* spp. using reverse dot blot hybridization. Nematology 1:549–555

Van der Plank (1984) Diseases resistance in plants. Academic, Orlando, 194p

Van der Wolf JM, Peters J, Bergervoet JHW (2005) Multiplex detection of viral and bacterial pathogens with the luminex technology, a fluid microarray system. Cost action 853, Gdansk

Varga A, James D (2006) Use of reverse transcriptase loop-mediated isothermal amplification for the detection of *Plum pox virus*. J Virol Methods 138:184–190

Vignali DA (2000) Multiplexed particle based flow cytometric assays. J Immunol Methods 21:243–255

Wang Y, Ren Z, Zheng XB, Wang YC (2007) Detection of *Phytophthora melonis* in samples of soil, water and plant tissue with PCR. Can J Plant Pathol 29:172–181

Wen K, Segin P, St Arnaud M, Jabaji-Hare S (2005) Real time quantitative TR-PCR of defence-associated gene transcripts of Rhizoctoniasolani-infected bean seedlings in response to inoculation with a non pathogenic binucleate. Rhizoctonia isolate. Phytopathology 95:345–353

Whitcombe D, Theaker J, Guy SP, Brown T, Little S (1999) Detection of PCR products using self-probing amplicons and fluorescence. Nat Biotechnol 17:804–807

White T, Burns T, Lees S, Taylor J (1990) Amplification and direct sequencing of fungal ribosomal RNA genes for phylogenetics. In: Inms M, Gelfand D, Sninsky J, White T (eds) PCR protocols: a guide to methods and applications. Academic, San Diego, pp 315–322

Widmark AK (2010) The Late blight Pathogen: *Phytophthora infestans*: Interaction with potato plants and Inoculum source. Phd Thesis. Swedish University of Agriculture Science, Uppsala, 4 p

Williams R, Ward E, McCartney HA (2001a) Methods for integrated air sampling and DNA analysis for the detection of air borne fungal spores. Appl Environ Microbiol 67:2453–2459

Williams JG, Kubelik AR, Livak KJ, Rafalski JA, Tingey SV (2001b) DNA polymorphisms amplified by arbitrary primers are useful as genetic markers. Nucleic Acids Res 18: 6531–6535

Wittwer CT, Herrmann MG, Moss AA, Rasmussen RP (1997) Continuous fluorescence monitoring of rapid cycle DNA amplification. Biotechniques 22:130–138

Zhang Y, Coyne MY, Will SG, Levenson CH, Kawasaki ES (1991) Single-base mutational analysis of cancer and genetic diseases using membrane bound modified oligonucleotides. Nucleic Acids Res 19:3929–3933

Zhang N, Geiser MD, Smart CD (2007) Macroarray detection of solanaceous plant pathogens in the *Fusarium solani* species complex. Plant Dis 91:1612–1620

Zhang N, McCarthy ML, Smart CD (2008) A macroarray system for the detection of fungal and oomycete pathogens of solanaceous crops. Plant Dis 92:953–960

Biocontrol Mechanism of *Bacillus* for *Fusarium* Wilt Management in Cumin (*Cuminum cyminum* L.)

H.P. Gajera, Disha D. Savaliya, Darshana G. Hirapara, S.V. Patel, and B.A. Golakiya

Origin of Cumin

Cumin (*Cuminum cyminum* L.) is cultivated and grown in many countries including Malta, India, Sicily, Iran, Saudi Arabia, Mexico, and China. The crop is quite easy to grow and adapts well in many climates (Azeez 2008). Earlier, cumin was noticed in the Bible, particularly in the Old Testament (Isaiah 28:27) and the New Testament (Matthew 23:23) (Edison et al. 1991). The crop was initiated to cultivate in Iran and the Mediterranean region. The use of cumin bring up to the second millennium BC as indicated by the Syrian site Tell ed-Der. Spanish and Portuguese colonists were commenced the crop to start in Americas. The cultivation of cumin also spread in Southern England with turned down the frequency of its occurrence (Chattopadhyay and Maiti 1990). This spice has became popular in the middle ages, probably due to the renewed interest in ethnic dishes and spicy foods (Thamaraikannan and Sengottuvel 2012).

H.P. Gajera (✉) • D.D. Savaliya • D.G. Hirapara • S.V. Patel • B.A. Golakiya
Department of Biotechnology, College of Agriculture, Junagadh Agricultural University, Junagadh 362 001, India
e-mail: harsukhgajera@yahoo.com

P. Kumar et al. (eds.), *Current Trends in Plant Disease Diagnostics and Management Practices*, Fungal Biology, DOI 10.1007/978-3-319-27312-9_2

Taxonomic classification of cumin

Kingdom	Plantae
(Unranked)	Angiosperms
(Unranked)	Eudicots
(Unranked)	Asterids
Order	Apiales
Family	Apiaceae
Genus	*Cuminum*
Species	*C. cyminum*

Uses of Cumin

The distinctive flavour and aroma of cumin seeds make it used as a spice. The cumin seeds used as essential flavouring in many countries like South Asian, Northern African and Latin American cuisines. Cumin can be utilized in some cheeses and breads to make the product palatable. Cumin seeds contained dietary fibre and minerals like iron, copper, calcium, potassium, manganese, selenium, zinc and magnesium. It also be full of phyto-chemicals having antioxidant, carminative and anti-flatulent properties. Cumin seeds also restrain good amounts of vitamins A, C, E and B-complex like thiamin, pyridoxal, niacin, riboflavin. Cumin is wealthy in iron for pregnant and lactating mothers. Presence of many phenolics including flavonoids, anti-oxidants such as carotenes zeaxanthin and lutein are formulate the cumin seeds for therapeutic uses (Thamaraikannan and Sengottuvel 2012).

Production Constraints

Cumin requires a dry and cool environment for better growth, with a temperature between 25 and 30 °C. Cumin grows best on well drained sandy loam to loamy soils with a pH range of 6.8–8.3. Acidic soils and alkaline soils reduce yield unless soil acidity is lowered to pH 7.5 (Weiss 2002). The average yield of this crop is affected due to unawareness scientific cultural practices, short of superior varieties, and susceptibility to diseases like wilt, blight and powdery mildew incited by *Fusarium oxysporum* f. sp. *cumini*, *Alternaria burnsii* and *Erysiphe polygoni*, respectively. Among the diseases, *Fusarium* wilt causes yield losses up to 35 % in cumin (Vyas and Mathur 2002). The first described *Fusarium* wilt in pulse crops was reported by Butler (1910). The disease appeared in naturally infected soils and confirmed the causal agent as *Fusarium* spp.

Fusarium Wilt in Cumin

The soil borne fungus *Fusarium oxysporum* have a many specialized forms and races that causes *Fusarium* wilts in crop plants. The pathogen *F. oxysporum* can be found in a variety of soil types and also in many any host plants, once the fungus is introduced into a garden, nursery, greenhouse, or field. Thus, a broad range of economically imperative crops are infected by *F. oxysporum* to cause wilt disease (Booth 1984). The *Fusarium* is an anamorphic species confined by macro and micro morphological descriptors like colony colour, conidiophore structure, presence or absence of microconidia and chlamydospores, size and shape of the macroconidium (Nelson et al. 1981; Windels 1993).

The most of *Fusarium* species are virulence and cause disease to plants. At least one *Fusarium*-associated disease is found on many crop plants (Leslie et al. 2006). The plant diseases such as crown rots, head blights, scabs, vascular wilts, root rots, and cankers were caused by this fungi. The mycotoxin produced by *Fusarium* species affects 25 % of the world food crops and pose a severe threat to plant, animal and human health (Nik 2008).

Fusarium oxysporum is asymptomatic fungi and isolated mostly from roots of crop plants (Gordon and Martyn 1997). *Fusarium oxysporum* has an aptitude to stick without choice to pathogenesis. This fungial strains are found in agricultural soils throughout the world and having nature of pathogenic and non-pathogenic. Mostly, strains of *Fusarium oxysporum* are found in cultivated soils and wild plant systems (Gordon and Martyn 1997).

Tawfik and Allam (2004) isolated the causal pathogen *Fusarium oxysporum* f. sp. *cumini* from infected cumin plants by using hyphal tip technique. The pathogen was separated by placing surface sterilized root and stem parts in Petri-dishes containing acidified potato dextrose agar (PDA) medium to prevent bacterial contamination. The fungus was purified by transferring the culture and identified as *Fusarium oxysporum* f. sp. *cumini* based on their macro and micro-morphological characteristics of mycelia and spores as described by Booth (1977).

Taxonomy of *Fusarium oxysporum* (Booth 1971)

Kingdom	Fungi	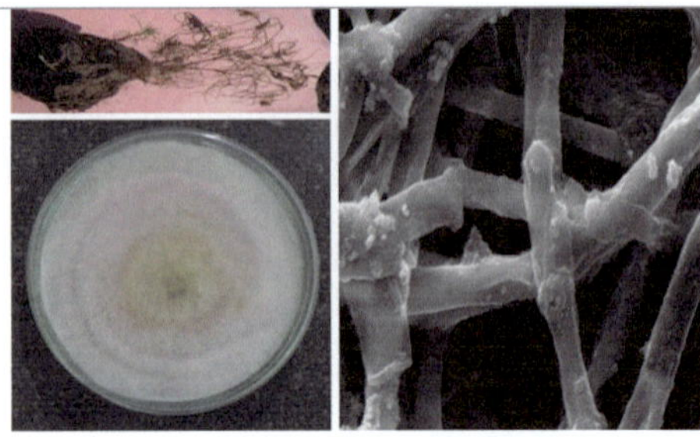
Phylum	Ascomycota	
Class	Sordariomycetes	
Order	Hypocreales	
Family	Nectriaceae	
Genus	*Fusarium*	
Species	*Fusarium oxysporum* f. sp. *cumini* (Pure culture and SEM image)	

Distribution and Diversity of Fusarium *Species*

Fusarium species is extensively dispensed in soils, plants, and air and well distributed across many geographical regions and substrates (Booth 1971; Burgess et al. 1994; Nelson et al. 1994; Summerell et al. 2003). Species occur predominantly in tropical, subtropical regions and cool to warm temperate regions while some species distributes in cosmopolitan geographic region (Burgess et al. 1994). The *Fusarium* species are also examined near to offices and hospitals buildings.

Temperature in different climatic regions also affects the species distribution and virulence. The *Fusarium* infection to cause disease in alfalfa was increased when the temperature is low (Saremi et al. 1999; Richard et al. 1982). Total 43 species were identified and isolated from various sources of crop plants like tobacco, rice, asparagus, banana, sugarcane, grass, soil, in Malaysia. Besides, Lim (1967) isolated five species of *Fusarium* from rice field soil in California which includes *F. moniliforme* (now known as *F. fujikuroi*) The stain was first reported of its species to be isolated from soil. However, ten *Fusarium* species were found with high diversity isolated from infected rice causing bakanae disease in Malaysia (Nur et al. 2011).

The distribution of *Fusarium* species in climate regions

Most climatic regions	Temperate regions	Subtropical and tropical regions
F. chlamydosporum	*F. acmuminatum*	*F. beomiforme*
F. equiseti	*F. avenaceum*	*F. compactum*
F. proliferatum	*F. crookwellense*	*F. decemcellulare*
F. oxysporum	*F. culmorum*	*F. longipes*
F. poae	*F. graminearum*	
F. semitectum	*F. sambucinum*	
F. tricinctum	*F. subglutinans*	

Disease Cycle in Crop Plants

The fungus may be seed borne and also survive in plant debris in soil. The fungus grows once within the plant, it multiplies in the vascular system (water and food conducting tissues) of the roots (Anonymous 1998).

Symptomatology

Typical symptoms of *Fusarium* wilt include the leaves started to become yellow and dropped, often starting on one side, and stunting of the plant. Disease symptoms initiated at the bottom of the stem and progress upwards, causing the leaves and flower heads to wilt, wither, and die. The lower stem parts are became dark and discolored, constantly on the inside and sometimes on the outside. The brown to black streaks are evident when infected stems are splitted in the vascular system (Fig. 1) (Anonymous 1998).

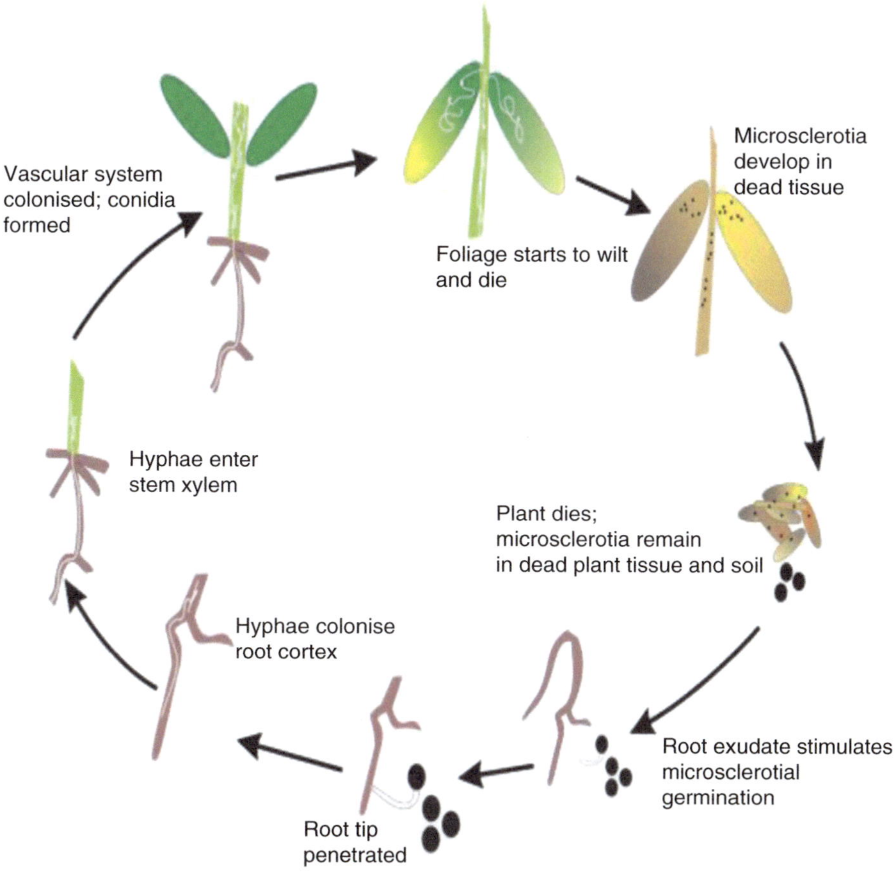

Fig. 1 Disease cycle of *Fusarium oxysporum* f. sp. *cumini*

Fusarium Wilt Management

The control of *Fusarium* fungi are difficult. The ability of the fungi to survive in soil for long periods, with or without a host plant, and the colonization of the vascular tissues within a plant are limited to eradicate the pathogen. Management of *Fusarium* wilt is mainly use of resistant cultivars and soil fumigation with fungicides. The broad spectrum of fungicides like methyl bromide used to fumigate soil before planting to control the soil borne pathogen which damage the soil environment particularly soil flora and fauna. Several measures can be taken, as listed below, to reduce the effects of the disease once *Fusarium* has been identified as producing the symptoms on the host plants (Anonymous 1998).

- Fumigate the soil with methyl bromide, chloropicrin, vorlex, or vapam and disinfest greenhouse, seedbed, and potting soil before planting.
- Crop rotation for 5–10 years may help to reduce the amount of infection and avoid to grow susceptible plants in *Fusarium*-infested soil where the same or closely related plants have grown previously.
- Plant materials like seeds cuttings, transplants, bulbs, corms should be healthy, top-quality and disease-free.
- Crop should be avoided with overwatering, deep planting, over fertilizing of nitrogen or phosphorus, and injuring plants when planting, cultivating, harvesting, grading, or sorting of the crop.

Extensive use of pesticides creates health hazards for humans and adverse effects on other non-target organisms. So, development of eco-friendly and safer plant disease control measure like use of biological control becomes a top priority for modern agriculture. For biological control of disease, an antagonism among microorganisms can be utilized to control plant pathogens. Biological control was relatively less effective in combating pathogens than insect pests (Butt et al. 2001). However, biological control could be included as one of the components of the integrated disease management strategies for wilt of cumin.

Biocontrol Antagonists on Plant Pathogen

Biological controllers like antagonistic microorganisms are an environmentally friendly and becoming more and more attentive to cope with problems associated with chemical control (Whipps 2001). Among the various bio-controllers, *Bacillus subtilis* has been frequently reported for inhibitory effect on plant pathogenic fungi in laboratory, greenhouse, and field studies (Pusey and Wilson 1984).

Soil-borne disease reduced yield and quality of agricultural products and a major problem of soil pollution. In this context, an effective way to solve this problem is a ecological remediation of soil. *Bacillus subtilis* SY1 was successfully utilized to antagonist soil borne fungal pathogens in eggplant (Yang et al. 2009). Tawfik and

Allam (2004) screened three fungal isolates (*Trichoderma harzianum*, *T. humatum*, and *T. viride*) and one bacterial (*Bacillus subtilis*) isolate on the basis of a preliminary in vitro antagonism test. Results showed that *Bacillus subtilis* (B7) was found with the most antagonistic bacterial isolate. *Trichoderma* spp. and *Bacillus subtilis* as an effective antagonistic reaction of the *Fusarium* wilt pathogen in different plant species were extensively documented (Larkin and Fravel 1998).

Landa et al. (2001) isolated rhizobacteria from the chickpea rhizospheres and utilized for suppression of *Fusarium* wilt in chickpea (*Cicer arietinum*) cv. PV 61 by seed and soil treatments. They studied the effects of temperature and inoculum density of *F. oxysporum* f. sp. *ciceris* race 5 to infect the disease and found that disease development was better at 25 °C compared with 20 and 30 °C. Total 23 bacterial isolates were tested, out of which 19 *Bacillus*, *Paenibacillus*, *Pseudomonas*, and *Stenotrophomonas* spp. were found to inhibit *F. oxysporum* f. sp. *ciceris* in vitro. Among the bacterial isolates, two strains – *Bacillus megaterium* RGAF 51 and *Paenibacillus macerans* RGAF 101 were the antagonists that significantly decreased the pathogen hyphal growth during in vitro conditions but did not suppress the disease.

Evaluation of antagonistic activity of three bacteria and fungi with direct confrontation method was carried out by Karkachi et al. (2010) and the filtrates culture against the growth of *F. oxysporum* f. sp. *lycopersici* showed the inhibition of the mycelia growth of test pathogen with *Bacillus cereus* energized the low activity.

Khan et al. (2011) used soil samples from different crop fields of paddy to isolate *Bacillus* species. Culture broth of the identified bacterial strain was tested for antifungal potential by disc diffusion technique against selected fungi such as *Candida albicans*, *Aspergillus niger*, *Rhizoctonia solani* and *Fusarium oxysporum*. Among the test fungi, *R. solani* was comparatively more sensitive to antifungal compound produced from the soil isolate of *B. subtilis*.

Gajbhiye et al. (2010) isolated *Bacillus subtilis* from cotton rhizosphere and evaluated it as biocontrol agent against *Fusarium oxysporum*. The performance of the pathogenic fungi in presence of the screened isolates was detected by poison food technique using antifungal assay medium. Twenty-one *Bacillus* isolates were obtained with antagonistic activity in the primary screening.

Antagonistic Activity of **Bacillus** *spp. Against Phytopathogens*

Biological control is the control of plant diseases, pest and weeds by other living organisms (Trigiano et al. 2004). Bio-control of plant diseases usually involves interactions between the antagonist, pathogen and host. *Bacillus subtilis* is one of the important bio-controller. The *Bacillus* commonly resident of soil is widely recognized as a powerful bio-controller. *Bacilli* including *B. subtilis* are potentially useful for biologic control of plant diseases due to ability to form endospores and produce different biologically active compounds with a broad spectrum of activity and a broad host range (Nagorska et al. 2007)

Taxonomic classification of antagonists (*Bacillus*)

Domain	Bacteria	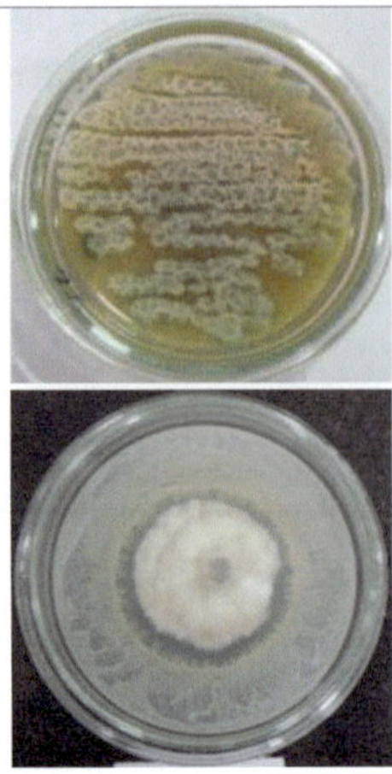	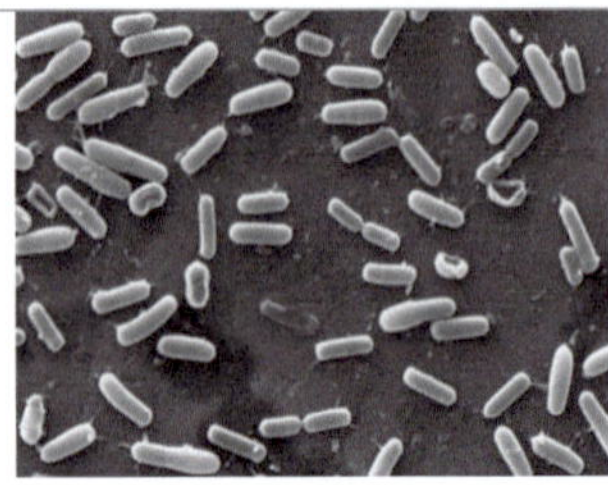
Division	Firmicutes	Bacillus isolates in specific HiChrome media and antagonism with *Fusarium oxysporum* f. sp. *ciceri*	SEM image of *Bacillus*
Class	Bacilli		
Order	Bacillales		
Family	Bacillaceae		
Genus	***Bacillus***		

Biocontrol Mechanism of Antagonist to Inhibit Fungal Pathogen

The *Bacillus* has been known for many years with the most profuse genus in the rhizosphere. This bacteria have antifungal and PGPR activities with a broad knowledge of the mechanisms involved (Probanza et al. 2002; Gutiérrez Mañero et al. 2003). The bacterial strain has a capacity to release number of antibiotics and metabolites which strongly inhibit the growth of fungal pathogen and affect the environment by increasing nutrient availability of the plants.

B. subtilis is naturally present in the instant vicinity of plant roots and able to maintain stable contact with higher plants to promote their growth. Besides that, bacterial inoculation at the beginning of the acclimatisation phase can be observed from the perspective of the establishment of the soil microbiota rhizosphere in a micropropagated plant system. *Bacillus licheniformis* can be used as a biofertilizer without changing normal management in greenhouses when inoculated on tomato and pepper with showing considerable colonisation (Garcia et al. 2004).

Jaizme-Vega et al. (2004) evaluated two micropropagated bananas on the first developmental stages for the effect of a rhizobacteria consortium of *Bacillus* spp. and summarized that this bacterial consortium can be useful to protect the plant against diseases and increases survival rates in commercial nurseries. *Bacillus* is examined for potential to increase the yield, growth and nutrition of raspberry plant under organic growing conditions (Orhan et al. 2006).

Competition for Nutrients and Siderophore Production

Classical' mode of action of biological control involves competition between biocontrol antagonist and phytopathogen for space and nutrient (Kaur et al. 2007; Bull et al. 1991). Siderophores are produced by microorganisms act as iron chelating molecules resulted to make low environmental iron concentrations. These molecules create iron limiting conditions for pathogenic microorganisms and help the bio-controller Bacillus to acquire iron, which is frequently biologically unavailable in the environment (Meyer and Stintzi 1998).

Many bacteria are able to produce several different antibiotics that have a broad range and sometimes overlap in their function (Leifert et al. 1995; Raaijmakers et al. 2002; Yu et al. 2002; Risoen et al. 2004). Some bacteria are genetically improved to produce new antibiotics to provide better protection (Bainton et al. 2004). Gupta and Gopal (2008) determined that under iron-limiting conditions, many bacteria secrete ferric iron-specific ligands, generically termed siderophores, which aid in sequestering and transport of iron. Siderophores produced PGPR bacteria was assayed by chrome azurol S method which is independent of siderophore structure and a general test.

Bacillus strains isolated from rhizosphere of soybean plants were characterized for their use as potential plant growth promoting rhizobacteria (Wahyudi et al. 2011). Out of 12, 3 isolates were able to inhibit the growth of *Fusarium oxysporum*, 9 isolates inhibited the growth of *Rhizoctonia solani*, and 1 isolate of *Bacillus* sp. inhibited the growth of *Sclerotium rolfsii*. All 12 isolates produced antibiotics siderophore to antagonize fungal pathogen and out of 12, 11 isolates possess extra PGPR activity and able to solubilize phosphate.

Production of Defence Related Substances

Synthesis of salicylic acid by bacteria can make the plant more tolerant to pests and pathogens by stimulating systemic acquired resistance (SAR), a common defense program induced in plants to combat pathogens (Bostock 2005). Induced systemic resistance (ISR) in the plant is another way that bacteria can protect plants (Van Loon et al. 1998).

Zhang et al. (2002) determine the role of salicylic acid (SA) in ISR against blue mold disease of tobacco elicited by PGPR. When plants treated with *Bacillus pumilus* strain SE34 were challenged with *P. tabacina*, levels of SA increased markedly 1 day after challenge, compared to the non-bacterized and challenged control. However, a similar increase in SA 1 day after pathogen challenge did not occur in plants treated with PGPR strains 90-166 or *Pseudomonas fluorescens* strain 89B-61.

Characterization of *Bacillus* sp. strains isolated from rhizosphere of soybean plants for their use as potential plant growth promoting rhizobacteria was carried out by Wahyudi et al. (2011). Out of 118 *Bacillus* isolates, 90 isolates able to produced phytohormone indole acetic acid (IAA). Ahmad et al. (2008) monitored

PGPR activities of free-living rhizospheric bacteria. These isolates were screened in vitro for production of indoleacetic acid (IAA), ammonia (NH3), hydrogen cyanide (HCN), siderophore, phosphate solubilization and antifungal activity. The IAA production was highest in the *Pseudomonas* followed by *Azotobacter* and *Bacillus* isolates at 50–500 mg ml^{-1} concentration of tryptophan.

Production of Mycolytic Enzymes (Fungal Cell Wall Degrading Enzymes)

Plant resistance to pathogenic fungi involves multiple reaction pathways like accumulation of hydrolytic chitinases and β-1,3 glucanase enzymes (Boller 1985). Bacteria are able to synthesize enzymes like chitinases, proteases, lipases and ß-1, 3-glucanases that are harmful for phytopathogens and further improve the biocontrol efficiency (Whipps 2001).

Chitinases have been implicated more in the plant defense mechanism (Schlumbaum et al. 1986; Linthrost et al. 1991). It is known that chitinases degrade chitin in fungal cell wall (Schlumbaum et al. 1986) and also act as synergistically with the β-1, 3 glucanase for inhibiting fungal growth (Mauch et al. 1984).

Chitinases are grouped in to 11 groups based on their primary structures and immunological properties. The enzyme can either be induced by pathogens (Van et al. 1991) or elicitors/chemicals (Boller and Mauch 1988). They are constitutively expressed at low levels in leaves and high level in roots and seeds. Increased levels of gene expression or enzymatic activity have been observed after elicitation.

Chitinases are extractable in acidic buffer, have low molecular weights (25–36 KDa), resistant to proteases and secreted extracellularly (Bol et al. 1990; Linthrost et al. 1991). Chen et al. (2009) identified the best antagonist *Bacillus subtilis* B579 against *F. oxysporum* f. sp. *cucumerinum* causing the disease in cucumber (*Cucumis sativus* L.) by dual plate assay. The strain B579 were evaluated for production of chitinase, ß -1, 3-glucanase, siderophores, indole-3-acetic acid (IAA), hydrogen cyanide (HCN), and phosphate solubilization, with the selected medium by in vitro tests. The vacuolation, swelling and lysis of fungal hyphae were found by the cell-free culture filtrate of B579 (20 % v/v).

Pleban et al. (1997) observed that *Bacillus cereus* 65 produce and excrete a chitinase with an apparent molecular mass of 36 kDa. The enzyme was classified as a chitobiosidase and application of *B. cereus* 65 directly to soil significantly protected cotton seedlings from root rot disease caused by *Rhizoctonia solani*. Melent'ev et al. (2001) determined the antifungal role of chitinase produced by *Bacillus* sp. 739. They noticed that both crude and purified chitinases were capable to lyse the cell walls of fungal mycelia. Besides, the role of chitinase produced by antagonists Bacillus sp. 739 is to utilize chitin present on cell wall of pathogenic fungi, it may also be inferred the antagonistic activity against micromycetes which is largely determined by low-molecular-weight nonenzymatic substances.

β- 1,3-glucanases belong to family of pathogenesis-related proteins. These enzymes catalyze the cleavage of β-1,3-glucosidic bonds of β-1,3-glucan, another constituent of the fungal cell wall. Unlike chitinases, the substrate for β-1,3-glucanases is widespread in plants and therefore these proteins are implicated in diverse physiological functions as well as in plant defense (Pan et al. 1991; Saikia et al. 2005; Ahmadzadeh et al. 2006). The enzyme can either be induced by infection with the pathogens (Van et al. 1991) or treatment with elicitors/chemicals (Boller and Mauch 1988) and wounding (Hedrick et al. 1988).

Leelasuphakul et al. (2005) studied an antagonistic activity of *Bacillus subtilis* NSRS 89-24 against rice blast and sheath blight. They purified and characterized the ß -1,3-glucanase activity and found that glucanase activity in the culture medium of *B. subtilis* NSRS 89-24 was inducible in the presence of 0.3 % chitin. The activity was evident maximum at 5 days incubations.

Plant resistance to pathogenic fungi involves multiple reaction pathways including the accumulation of hydrolytic enzymes such as chitinases and β-1, 3-glucanases (Boller 1985). Both the enzymes act synergistically and their optimal functions may be important in plant defense. β-1,3 glucan is an important component of fungal cell wall polysaccharide. Some β-1, 3- glucanase like the many other pathogenesis related proteins are acidic buffer extractable, have low molecular weights (25–36 KDa), resistant to proteases and secreted extra cellularly (Bol et al. 1990; Linthrost et al. 1991). Mandavia et al. (1999) reported the rise in level of cell wall degrading enzymes (Polymethyl galacturonase (PMG) and cellulase in cumin seedlings infected with *Fusarium oxysporum* f. sp. *cumini*. The increase in level of enzymes was varied according to degree of the infection. Our study (Gajera et al. 2012) reported in vitro inhibition *Macrophomina phaseolina* by *Trichoderma* to control root rot disease in castor. *T. koningii* inhibited maximum growth inhibition (74.3 %) of test pathogen followed by *T. harzianum* (61.4 %). The antagonistic effect of *Trichoderma* was due to induction of pathogenesis related enzymes (chitinase, ß-1, 3 glucanase) during antagonism which were positively correlated with growth inhibition of test pathogen and coiling pattern of antagonists.

Trichoderma is well documented for biologic control of fungal pathogen. Our study (Gajera et al. 2014) indicates *T. viride* JAU60 mediated systemic induction of phenolics for biologic control and their probable role in protecting groundnut against *A. niger* infection. Results confirmed that *T. viride* JAU60 treatment induces phenolics like gallic, ferulic and salicylic acids in groundnut seedlings challenged with rot pathogen. These phenolics may be synthesized upon activation of PAL by *Trichoderma* in groundnut seedlings under pathogen infestation.

It was observed that when *T. viride* JAU60 interact with *A. niger* and in vitro antagonism revealed the highest production of cell wall degrading enzymes. Study exhibited in vitro percent growth inhibition of *A. niger* and production of cell wall degrading enzymes chitinase, β-1,3 glucanase and protease in the culture medium of antagonist treatment are positively correlated and established a relationship to inhibit the growth of fungal pathogen by increasing the levels of these lytic enzymes (Gajera and Vakharia 2010). *T. viride* JAU60 also signifies induction of a lipoxygenase (LOX) related defense response to combat the collar rot disease incidence in groundnut seedlings (Gajera et al. 2015). The results showed that pathogen infection is necessary for LOX activation in tolerant (J-11, GG-2, GAUG-10) groundnut varieties but it is significantly enhanced upon seed treatment with strain *T. viride* JAU60 in susceptible varieties (GG-13, GG-20). Our work on groundnut treated with *T. viride* JAU60 suggests a pathogen-dependant systemic activation of the defense reaction.

Molecular Characterization of *Bacillus* Associated with Antagonism

Molecular characterization provides an immense source of data for identity, relatedness, diversity and selection of proper candidates for biological control. Molecular markers also offer a means of constructing quality control tests that are essential throughout the developmental processes (Avis et al. 2001). Polymerase chain reaction (PCR) refers to in vitro amplification of particular DNA sequences using arbitrary or specific primers and a thermostable DNA polymerase enzyme (Joshi 1999).

Randomly Amplified Polymorphic DNA (RAPD)

The RAPD is an inexpensive yet powerful typing method for many bacterial species. This technique is based on the PCR and has been one of the most commonly used molecular techniques to develop DNA markers (Fevzi 2001). Gajbhiye et al. (2010) performed RAPD profiling which revealed the diversity in the *Bacillus subtilis* group, ranging from 10 % to 32 %. Total 700 bands were scored in the analysis of nine isolates using 60 RAPD primers. All the nine isolates characterized on the basis of the RAPD molecular markers produced highly polymorphic patterns.

Aiming to develop a DNA marker specific for *Bacillus anthracis* and able to discriminate this species from *Bacillus cereus*, *Bacillus thuringiensis* and *Bacillus mycoides*, Daffonchio et al. (1999) applied RAPD fingerprinting technique to a collection of 101 strains of the genus *Bacillus*, including 61 strains of the *B. cereus* group. An 838-bp RAPD marker (SG-850) specific for *B. cereus*, *B. thuringiensis*, *B. anthracis*, and *B. mycoides* was identified. This fragment included a putative (366-nucleotide) open reading frame highly homologous to the ypuA gene of *Bacillus subtilis*.

RAPD-PCR technique was utilized for molecular characterization of the *Bacillus megaterium* isolates (Reddy et al. 2010). Selected two primers were used for fingerprinting and for estimation of genetic diversity among ten isolates of *B. megaterium* which generated a total of 20 RAPD bands. Out of 20 amplicons, 06 bands were unique and 11 bands were shared polymorphic, which were informative in revealing the relationship among the genotypes while 3 bands were monomorphic.

Hamshary et al. (2008) determined the genetic relationships of the six *Bacillus* isolates by RAPD-PCR method with seven random primers. Primers shaped multiple band profiles with a number of amplified DNA fragments ranging from 1 to 11. These primers had amplified 255 PCR product bands among 77 amplified bands, out of which 63 were polymorphic bands and 14 were monomorphic bands. Nilsson et al. (1998) suggested that RAPD-PCR can be used for large scale typing of *Bacillus cereus*. More than 3,000 strains of *B. cereus* were isolated from farms and dairies. Many different RAPD bands were observed and the GelcomparTM programme was used to analyse similarities between the strains. Jiyeon et al. (2011) obtained RAPD-PCR profiles of *B. cereus* strains using a 10-mer primer (S30) and a *B. cereus* specific 910 kb band was produced from all tested strains.

16s rDNA and 23s ITS Region Characterization

The *Bacillus* possess 16S–23S internal transcribed spacer (ITS) region which has been widely studied for the presence of functional motifs (Pfeiffer and Hartmann 1997), specific processing sites (Apirion and Miczak 1993) and secondary structures (Nour 1998; Liiv et al. 1998). The ITS-PCR fingerprints have been used to reveal length polymorphisms between *Bacillus* species (Daffonchio et al. 1998a) and at the intra-specific level (Daffonchio et al. 1998b). Specific part of the ITS region has been amplified by PCR and used for designing probe for the detection, identification and phylo-genetic analysis of Bacillus species (de Silva et al. 1998).

Beric et al. (2009) examined genetic diversity analysis of 205 *Bacillus* isolates collected from different geographical and ecological niches in Serbia and correlated the antagonists with production of hydrolytic enzymes. They determined 13 different groups of RAPD profiles within four species: *B. subtilis*, *B. cereus*/*B. thuringiensis*, *B. pumilus*, and *B. firmus* based on combining RAPD analysis and 16S rDNA sequencing,. Haque and Russell (2005) reported phenotypic and genotypic characterization of *Bacillus* cereus isolates from Bangladeshi rice. The sequence analysis of variable regions in the 16S rRNA gene gave four different groups by base differences at two positions.

Li et al. (2012) screened 400 *Bacillus* using a modified genotobiotic system for their capability in controlling *Fusarium* wilt of cucumber which were isolated from surface-sterilized roots of cucumber plants grown in greenhouses and fields. The molecular identifications of *Bacillus* by 16S rRNA gene and gyrA gene illustrated phylogenetic map indicating B068150 strain exhibits high levels of similarity to known *Bacillus* species. The B068150 strain was identified as *B. subtilis* B068150. *Bacillus* strains were recognized by conventional biochemical methods, fatty acid methyl ester (FAME) analysis and partial 16S rDNA sequencing which were isolated from the coastal environment of cochin, India (Parvathi et al. 2009). Biochemical and molecular data revealed that *Bacillus pumilus*, *B. cereus* and *B. sphaericus* were the predominant species in the region of coastal environment of cochin. The *B. pumilus* isolates were also characterized for antibiotic sensitivity profiling, arbitrarily primed PCR (AP-PCR), and PCR screening for known toxin genes associated with *Bacillus* spp.

Xu and Cote (2003) compared 3′ end 16S rDNA and 5′ end 16S–23S ITS nucleotide sequences and established the phylogenetic relationships between *Bacillus* species and related genera. The forward primer designed from about 200 nt upstream from the 3′ end of the 16S rRNA gene, and reverse primer located about 80 nt downstream from the 5′ end of the 23S rRNA gene, were amplified last 200 bp of the 16S rRNA gene and the entire 16S–23S ITS region from 40 Bacillaceae species. The amplified product was found in the range of 450–850 bp to identify the Bacillus.

Conclusions

The yield of cumin is affected by lack of superior varieties, scientific crop production technology and vulnerability to diseases like wilt, blight and powdery mildew incited by *Fusarium oxysporum* f. sp. *cumini*, *Alternaria burnsii* and *Erysiphe polygoni*, respectively. In these diseases, *Fusarium* wilt is most common, results in yield losses up to 35 % in cumin. *Fusarium oxysporum* is a causative agent of wilt disease in a wide range of economically important crops. *Fusarium* species is well distributed across many geographical regions and substrates, and also widely distributed in soils, plants, and air. The *Fusarium* species were distributed in different climatic regions according to temperature and virulence.

Cope with problems associated with chemical control, an environmentally friendly way of biological control using antagonistic microorganisms is becoming more and more attentive, and much research has been carried out in recent years. Soil-borne disease like *Fusarium* is a major problem of soil pollution, which affects yield and quality of agricultural products. Ecological remediation of soil is an effective way to solve this problem. *Bacillus* is the most abundant genus in the rhizosphere, and the PGPR activity of some of these strains has been known for many years, resulting in a broad knowledge of the mechanisms involved. The mechanism adopted by *Bacillus* for biologic control of *Fusarium* wilt viz., competition for nutrients and siderophore production, production of defense related substances, production of fungal cell wall degrading enzymes. Molecular characterization of antagonists and fungal pathogen provide an immense source of data for identity, relatedness, diversity and selection of proper candidates for biological control.

Acknowledgements The review chapter is a part of research project entitled "Biochemical and molecular characterization of *Bacillus* spp. isolated from rhizosphere of plants and their biocontrol potentials against *Fusarium oxysporum* f. sp. *cumini*" financial sponsored by Gujarat State Biotechnology Mission, Gandhinagar (India).

References

Ahmad F, Ahmad I, Khan MS (2008) Screening of free-living rhizospheric bacteria for their multiple plant growth promoting activities. Microbiol Res 163:173–181

Ahmadzadeh MH, Afsharmanesh M, Javan-Nikkhah A, Sharifi T (2006) Identification of some molecular traits in fluorescent *Pseudomonads* with antifungal activity. Iran J Biotech 4:245–253

Anonymous (1998) Report on plant diseases. Department of crop sciences, University of Illinois at Urbana-champaign

Apirion D, Miczak A (1993) RNA processing in prokaryotic cells. Bioessays 15:113–120

Avis TY, Hamelin RC, Belanger RR (2001) Approaches to molecular characterization of fungal biocontrol agents: some case studies. Can J Plant Pathol 23:8–12

Azeez S (2008) Cumin. In: Parthasarathy VA, Champakam B, Zachariah TJ (eds) Chemistry of spices. CABI International, Wallingford

Bainton NJ, Lynch JM, Naseby D, Way JA (2004) Survival and ecological fitness of *Pseudomonas fluorescens*, genetically engineered with dual biocontrol mechanisms. Microb Ecol 48:349–357

Beric T, Urdaci MC, Stankovic S, Jelena KV (2009) RAPD analysis of genetic diversity and qualitative assessment of hydrolytic activities in a collection of *Bacillus* spp. isolate. Arch Biol Sci 61:645–652

Bol JF, Linthrost HJM, Cornelissen BJC (1990) Plant pathogenesis related proteins induced by virus infection. Ann Rev Phytopathol 28:113–138

Boller T (1985) Induction of hydrolases as defense reaction against pathogens. In: Key JL, Kosuge TA, Liss R (eds) Cellular and molecular biology of plant stress. Alan R. Liss. New York, pp 247–262

Boller T, Mauch F (1988) Colorimetric assay of Chitinase. Methods Enzymol 161:431–435

Booth C (1971) The genus *Fusarium*. Common wealth Mycological Institute, Kew, p 237

Booth C (1977) *Fusarium* laboratory guide to the identification of major species. Commonwealth Mycological Institute, Kew, p 527

Booth C (1984) The *Fusarium* problem: historical, economic, and taxonomic aspects. pp 1–13, In: MO Moss, JE Smith (eds) The applied mycology of Fusarium. Cambridge University Press, Cambridge, 264pp

Bostock RM (2005) Signal crosstalk and induced resistance: straddling the line between cost and benefit. Annu Rev Phytopathol 43:545–580

Bull CT, Weller DM, Thomashow LS (1991) Relationship between root colonization and suppression of *Gaeumann omyces graminis* var. *tritici* by *Pseudomonas fluorescens* strain 2-79. Phytopathology 81:950–959

Burgess LW, Summerell BA, Bullock S, Gott KP, Backhouse D (1994) Laboratory manual for *Fusarium* research. 3rd edn. University of Sydney, Sydney, pp 116–117

Butler EJ (1910) Mem dep. Agric India (Bot.Ser) 2. No. 9. pp 1–64. Caesar-TonThat TC, Epstein L (1991) Adhesion reduced mutants and the wild type Nectria haematococea: an ultrastructural component of the macroconidia walls. Exp Mycol 15:193–205

Butt T, Jackson C, Magan N (eds) (2001) Fungi as biocontrol agents: progress, problems and potential. CAB International, Wallingford

Chattopadhyay SB, Maiti S (1990) Diseases of betelvine and spices, 2nd edn. Indian Council of Agricultural Research, New Delhi, p 160

Chen F, Wang M, Zheng Y, Luo J, Yang X, Wang X (2009) Quantitative changes of plant defense enzymes and phytohormones in biocontrol of cucumber *Fusarium* wilt by *Bacillus subtilis* B579. World J Microbiol Biotechnol 26:675–684

Daffonchio D, Borin S, Consolandi A, Mora D, Manachini PL, Sorlini C (1998a) 16S–23S rRNA internal transcribed spacers as molecular markers for the species of the 16S rRNA group I of the genus *Bacillus*. FEMS Microbiol Lett 163:229–236

Daffonchio D, Borin S, Frova G, Manachini PL, Sorlini C (1998b) PCR fingerprinting of whole genomes: the spacers between the 16S and 23S rRNA genes and of intergenic tRNA gene regions reveal a different intraspecific genomic variability of *Bacillus cereus* and *Bacillus licheniformis*. Int J Syst Bacteriol 48:107–116

Daffonchio D, Borin S, Frova G, Gallo R, Mori E, Fani R, Sorlini C (1999) A randomly amplified polymorphic DNA marker specific for the *Bacillus cereus* group is diagnostic for *Bacillus anthracis*. Appl Environ Microbiol 65:1298–1303

de Silva S, Petterson B, de Muro MA, Priest FG (1998) A DNA probe for the detection and identification of *Bacillus sporothermodurans* using the 16S–23S rDNA spacer region and phylogenetic analysis of some field isolates of *Bacillus* which form highly heat resistant spores. Syst Appl Microbiol 21:398–407

Edison S, Johny AK, Nirmal Babu K, Ramadasan A (1991) Spices varieties. In: A compendium of morphological and agronomic characters of improved varieties of spices in India. National Research Centre for Spices (ICAR), Kerala, p 63

Fevzi B (2001) Random Amplified Polymorphic DNA (RAPD) markers. Turk J Biol 25:185–196

Gajbhiye A, Alok R, Sudhir U, Dongre AB (2010) Isolation, evaluation and characterization of *Bacillus subtilis* from cotton rhizospheric soil with biocontrol activity against *Fusarium oxysporum*. World J Microbiol Biotechnol 26:1187–1194

Gajera HP, Vakharia DN (2010) Molecular and biochemical characterization of *Trichoderma* isolates inhibiting a phytopathogenic fungi *Aspergillus niger* Van Tieghem. Physiol Mol Plant Pathol 74:274–282

Gajera HP, Bambharolia RP, Patel SV, Khatrani TJ, Golakiya BA (2012) Antagonism of *Trichoderma* spp. against *Macrophomina phaseolina*: evaluation of coiling and cell wall degrading enzymatic activities. J Plant Pathol Microb 3:149. doi:10.4172/2157-7471.1000149

Gajera HP, Jadav JK, Patel SV, Golakiya BA (2014) *Trichoderma viride* induces phenolics in groundnut (*Arachis hypogaea* L.) seedlings challenged with rot pathogen (*Aspergillus niger* Van Tieghem). Phytoparasitica 42:703–712

Gajera HP, Savaliya DD, Patel SV, Golakiya BA (2015) Lipoxygenase-related defense response induced by *Trichoderma viride* against *Aspergillus niger* Van Tieghem, inciting collar rot in groundnut (*Arachis hypogaea* L.). Phytoparasitica 43:229–240

Garcia JAL, Probanza A, Ramos B, Palomino MR, Manero FJG (2004) Effect of inoculation of *Bacillus licheniformis* on tomato and pepper. Agronomie 24:169–176

Gordon TR, Martyn RD (1997) The evolutionary biology of *Fusarium oxysporum*. Ann Rev Phytopathol 35:11–128

Gupta A, Gopal M (2008) Siderophore production by plant growth promoting rhizobacteria. Indian J Agric Res 42:153–156

Gutiérrez Mañero FJ, Probanza A, Ramos B, Colon Flores JJ, Lucas Garcia JA (2003) Effects of culture filtrates of rhizobacteria isolated from wild lupine on germination, growth, and biological nitrogen fixation of lupine seedlings. J Plant Nutr 26:1101–1115

Hamshary E, Abo-Aba AN, Gomaa AM (2008) Genetic profile of *Bacillus cereus* and *Bacillus subtilis* indigenous isolates and their performance as bio-control agent against the plant pathogen *Fusarium oxysporum*. Res J Cell Mol Biol 2:30–38

Haque A, Russell NJ (2005) Phenotypic and genotypic characterisation of *Bacillus cereus* isolates from Bangladeshi rice. Int J Food Microbiol 98:23–34

Hedrick SA, Bell JN, Boller T, Lamb CJ (1988) Chitinase cDNA clonning and mRNA induction by fungal elicitor, wounding and infection. Plant Physiol 86:182–186

Jaizme-Vega MDC, Rodriguez-Romero AS, Guerra MSP (2004) Potential use of rhizobacteria from the *Bacillus* genus to stimulate the plant growth of micropropagated bananas. Fruits 59:83–90

Jiyeon L, Kwon G, Park J, Park C, Kwon D, Lim J, Kim J, Kim JH (2011) A RAPD-PCR method for the rapid detection of *Bacillus cereus*. J Microb Biotechnol 21:274–276

Joshi DH (1999) Studies on the seed microflora of groundnut, cotton, bajra, wheat and sesame under Gujarat condition. MSc (Agri) thesis, Gujarat Agricultural University, Sardar Krishinagar, Dantiwada

Karkachi NE, Gharbi S, Kihal M, Henni JE (2010) Biological control of *Fusarium oxysporium* f. sp. *lycopersici* isolated from Algerian tomato by *Pseudomonas fluorescens*, *Bacillus cereus*, *Serratia marcescens* and *Trichoderma harzianum*. Res J Agron 4:31–34
Kaur R, Kaur J, Singh RS, Alabouvette C (2007) Biological control of *Fusarium oxysporum* f. sp. *ciceri* by non pathogenic *Fusarium* and Fluorescent *Pseudomonas*. Int J Bot 3:114–117
Khan AAH, Naseem RL, Prathibha B (2011) Screening and potency evaluation of antifungal from soil isolates of *Bacillus subtilis* on selected fungi. Adv Biotechnol 10:35–37
Landa BB, Juan A, Ana H, Rafael M (2001) Influence of temperature and inoculum density of *Fusarium oxysporum* f. sp. *ciceris* on suppression of *Fusarium* wilt of chickpea by rhizosphere bacteria. Am Phytopathol Soc 91:807–816
Larkin RP, Fravel R (1998) Efficacy of various fungal and bacterial biocontrol organisms for control of *Fusarium* wilt of tomato. Plant Dis 82:1022–1028
Leelasuphakul W, Sivanunsakul P, Phongpaichit S (2005) Purification, characterization and synergistic activity of ß -1, 3-glucanase and antibiotic extract from an antagonistic *Bacillus subtilis* NSRS 89-24 against rice blast and sheath blight. Enzym Microb Technol 38:990–997
Leifert C, Li H, Chidburee S, Hampson S, Workman S, Sigee D, Epton HAS, Harbour A (1995) Antibiotic production and biocontrol activity by *Bacillus subtilis* CL27 and *Bacillus pumilus* CL45. J Appl Bacteriol 78:97–108
Leslie JF, Summerell BA, Bullock S (2006) The fusarium laboratory manual. Blackwell Publishing Ltd., Oxford
Li L, Mab J, Li Y, Wang Z, Tantan G, Wang Q (2012) Screening and partial characterization of *Bacillus* with potential applications in biocontrol of cucumber *Fusarium* wilt. Crop Prot 35:29–35
Liiv A, Tenson T, Margus T, Remme J (1998) Multiple functions of the transcribed spacers in ribosomal RNA operons. Biol Chem 379:783–793
Lim G (1967) *Fusarium* populations in rice field soils. Phytopathology 57:1152–1153
Linthrost HJM, Danhash N, Brederode FT, Van Kan JAL, De Wit PJGM, Bol JF (1991) Tobacco and tomato PR proteins homologus to win and pro Hevein Lack the "Hevein" domain. Mol Plant Microb Interact 4:586–592
Mandavia MK, Khan NA, Gajera HP, Andharia JH, Parmeswaran M (1999) Enzymes in host–pathogen interaction in *Fusarium* wilt. In: Parameswaran M (ed) Research and education in agricultural biochemistry. Proceedings of National Symposium, Junagadh
Mauch F, Mauch-Mani B, Boller T (1984) Antifungal hydrolases in pea tissue. II. Inhibition of fungal growth by combination of chitinase and β-1,3-glucanase. Plant Physiol 88:936–942
Melent'ev AI, Aktuganov GE, Galimzyanova NF (2001) The role of chitinase in the antifungal activity of *Bacillus* sp. 739. Microbiology 70:548–552
Meyer JM, Stintzi A (1998) Iron metabolism and siderophore production in Pseudomonas and related species. In: Montie (ed) Pseudomonas. Plenum Press, New York, pp 201–243
Nagorska K, Bikowski M, Obuchowskji M (2007) Multicellular behaviour and production of a wide variety of toxic substances support usage of *Bacillus subtilis* as a powerful biocontrol agent. Acta Biochim Pol 54:495–508
Nelson PE, Horst RK, Woltz SS (1981) *Fusarium* diseases of ornamental plants. In: Nelson PE, Toussoun TA, Cook RI (eds) *Fusarium*: diseases, biology and taxonomy. The Pennsylvania State University Press, University Park, pp 121–141, 193
Nelson PE, Dignani MC, Anaissie EJ (1994) Taxonomy, biology and clinical aspect of *Fusarium* species. Clin Microbiol Rev 7:479–504
Nik M (2008) Morphological characteristics, distribution, and mycotoxin profiles of *Fusarium* species from soils in Peninsular Malaysia. Universiti Sains Malaysia
Nilsson J, Svensson B, Ekelund K, Christiansson A (1998) A RAPD-PCR method for large scale typing of *Bacillus cereus*. Lett Appl Microbiol 27:168–172
Nour M (1998) 16S–23S and 23S–5S intergenic spacer regions of lactobacilli: nucleotide sequence, secondary structure and comparative analysis. Res Microbiol 149:433–448
Nur AIMZ, Siti NMS, Nor AJ, Azmi AR, Baharuddin S (2011) Isolation and identification of *Fusarium* species associated with *Fusarium* ear rot disease of corn. Pertanika J Trop Agric Sci 34:325–330

Orhan E, Esitken A, Ercisli S, Turan M, Sahin F (2006) Effects of plant growth promoting rhizobacteria (PGPR) on yield, growth and nutrient contents in organically growing raspberry. Sci Hortic 111:38–43

Pan SQ, Ye XS, Kuc J (1991) Association of β-1,3 glucanase activity and isoform pattern with systemic resistance to blue mold in tobacco induced by stem injection with *Peronospora tabacina* or leaf inoculation with tobacco mosaic virus. Physiol Mol Plant Pathol 39:25–39

Parvathi A, Krishna K, Jose J, Joseph N, Nair S (2009) Biochemical and molecular characterization of *Bacillus pumilus* isolated from coastal environment in Cochin. Indian Braz J Microbiol 40:269–275

Pfeiffer T, Hartmann RK (1997) Role of the spacer boxA of *Escherichia coli* ribosomal RNA operons in efficient 23S rRNA synthesis *in vivo*. J Mol Biol 265:385–393

Pleban S, Chernin L, Chet I (1997) Chitinolytic activity of an endophytic strain of *Bacillus cereus*. Appl Environ Microbiol 25:284–288

Probanza A, Lucas Garcia JA, Ruiz Palomino M, Ramos B, Gutierrez Manero FJ (2002) *Pinus pinea* L. seedling growth and bacterial rhizosphere structure after inoculation with PGPR *Bacillus* (*B. licheniformis* CECT 5106 and *B. pumilus* CECT5105). Appl Soil Ecol 20:75–84

Pusey PL, Wilson CL (1984) Post harvest biological-control of stone fruit brown rot by *Bacillus subtilis*. Int J Food Microbiol 68:753–756

Raaijmakers J, Vlami M, de Souza J (2002) Antibiotic production by bacterial biocontrol agents. Antonie Van Leeuwenhoek 81:537–547

Reddy DM, Mohan BK, Nataraja S, Krishnappa M, Abhilash M (2010) Isolation and molecular characterization of *Bacillus megaterium* isolated from different agro climatic zones of Karnataka and its effect on seed germination and plant growth of *Sesamum indicum*. Res J Pharm Biol Chem Sci 1:614

Richard C, Willemot C, Michaud R, Bernier-Cardou M, Gagnon C (1982) Low-temperature interactions in *Fusarium* wilt and root rot of alfalfa. Phytopathology 72:293–297

Risoen PA, Ronning P, Hegna IK, Kolsto AB (2004) Characterization of a broad range antimicrobial substance from *Bacillus cereus*. J Appl Microbiol 96:648–655

Saikia R, Singh BP, Kumar R, Arora DK (2005) Detection of pathogenesis-related proteins – chitinase and β-1,3-glucanase in induced chickpea. Curr Sci 89:659–663

Saremi H, Burgess LW, Backhouse D (1999) Temperature effects on the relative abundance of *Fusarium* species in a model plant–soil ecosystem. Soil Biol Biochem 31:941–947

Schlumbaum A, Mauch F, Vogeli U, Boller T (1986) Plant chitinase are potent inhibitors of fungal growth. Nature 324:365–367

Summerell BA, Salleh B, Leslie JF (2003) A utilitarian approach to *Fusarium* identification. Am Phytopathol Soc Plant Dis 87:117–128

Tawfik AA, Allam ADA (2004) Improving cumin production under soil infestation with *Fusarium* wilt pathogen: II. Field trial of different landraces and seed treatments. Asst Univ Bull Environ Res 7:35–45

Thamaraikannan M, Sengottuvel C (2012) Market survey on Cumin: can India maintain a monopoly? Facts for you (New Delhi) 32(8):11–14

Trigiano RL, Windham MT, Windham AS (2004) Plant pathology: concepts and laboratory exercises. CRC Press, Boca Raton, p 413

Van Loon LC, Bakker PAHM, Pieterse CMJ (1998) Systemic resistance induced by rhizosphere bacteria. Annu Rev Phytopathol 36:453–483

Van K, Mikkelsen JD, Collinge DB (1991) Induction purification and characterization of chitinase isolated from pea leaves inoculated with *Ascochyta pisi*. Planta 184:24–29

Vyas RK, Mathur K (2002) Distribution of *Trichoderma* species in cumin rhizosphere and their potential in suppression of wilt. Indian Phytopathol 55:451–457

Wahyudi AT, Astuti RP, Widyawati A, Meryandini A, Nawangsih AA (2011) Characterization of *Bacillus* sp. strains isolated from rhizosphere of soybean plants for their use as potential plant growth promoting rhizobacteria. J Microbiol Antimicrob 3:34–40

Weiss EA (2002) Spice crops. CABI International, Wallingford, p 299

Whipps JM (2001) Microbial interactions and biocontrol in the rhizosphere. J Exp Bot 52: 487–511
Windels A (1993) *Fusarium* species stored on silica gel and soil for ten years. Mycologia 85:21–23
Xu D, Cote JC (2003) Phylogenetic relationships between *Bacillus* species and related genera inferred from comparison of 3′ end 16S rDNA and 5′ end 16S–23S ITS nucleotide sequences. Int J Syst Evol Microbiol 53:695–704
Yang Z, Liu X, Liu Z, Pang J, Qiu J, Yang W (2009) Effect of *Bacillus Subtilis* SY1 on antifungal activity and plant growth. Int J Agric Biol Eng 2:55–61
Yu GY, Sinclair JB, Hartman GL, Bertagnolli BL (2002) Production of iturin A by *Bacillus amyloliquefaciens* suppressing *Rhizoctonia solani*. Soil Biol Biochem 34:955–963
Zhang S, Moyne AL, Reddy MS, Kloepper JW (2002) The role of salicylic acid in induced systemic resistance elicited by plant growth-promoting rhizobacteria against blue mold of tobacco. Biol Control 25:288–296

Eco Friendly Management of Damping-off of Solanaceous Crops Caused by *Pythium* Species

A. Muthukumar, R. Udhayakumar, and R. Naveenkumar

Introduction

Tomato (*Lycopersicon esculentum* Miller.), chilli (*Capsicum annuum* L.) and brinjal (*Solanum melongena* L.) are the major solanaceous vegetable crop and is native to sub tropics and tropics. Tomato is called as king of vegetables and it is cultivated for its fleshy fruits and has a high nutritional value. It is also believed that it gives protection from or reduces the risk of contracting chronic degenerative diseases. India has been the second largest exporter of chilli in the international market, exporting products ranging from dried form of chilli to chilli powder. It is an economically important tropical and sub-tropical crop because of its pungency and its nutritional value. Brinjal occupies second position among the vegetable. It is known to have ayurvedic medicinal properties and good for diabetic patients. It has also been recommended as an good retrieve for those suffering from liver disorder. Although India has a large growing area, their productivity is relatively low when compared to other countries. The prominent reason for this is the high incidence of fungal and viral diseases. Of these, damping-off incited by *Pythium* species caused more than 60 % death of seedlings in both nurseries and main field (Manoranjitham and Prakasam 2000; Jadhav and Ambadkar 2007).

The infected tissue become soft and water soaked, the collar region rots and the seedlings ultimately collapse and die. The guaranteed supply of disease free seedlings in required quantities is a major pre requisite for stabilized production of tomato, chilli and brinjal. While raising seedlings in nursery beds, the farmers face major problem of damping off incited by *Pythium* spp. The damping off in solanaceous crop is caused by *Pythium* spp including *P. aphanidermatum*,

A. Muthukumar (✉) • R. Udhayakumar • R. Naveenkumar
Department of Plant Pathology, Faculty of Agriculture, Annamalai University, Annamalai Nagar, Chidambaram, Tamil Nadu 608 002, India
e-mail: muthu78ap@yahoo.co.in

P. Kumar et al. (eds.), *Current Trends in Plant Disease Diagnostics and Management Practices*, Fungal Biology, DOI 10.1007/978-3-319-27312-9_3

P. irregulare, *P. debaryanum* and *P. ultimum* Trow, which can cause pre -emergence damping off and results in seed decay before the plants comes out from the soil. The post emergence damping off is designated by infection of the young tissues at the collar of the stem at above ground level.

The *Pythium* species are fungal-like organisms, commonly referred to as moisture loving fungi, (Domain Eukaryo; Kingdom Chromista; phylum Oomycota; class Oomycetes; order Pythiales and family Pythiaceae) are present all over the world and related with a wide variety of habitats ranging from terrestrial or aquatic environments, in cultivated or fallow soils, in plants or animals, in saline or fresh water. Among the oomycete class of fungi, the genus *Pythium* is one of the largest genus. It consists of more than 130 accepted species which are isolated from different crops and from different regions of the world (Paul et al. 2006; Bala et al. 2010; Robideau et al. 2011).

Almost all the *Pythium* species are known to affect the crop plants and ultimately cause severe damage (Hendrix and Campbell 1973; Kucharek and Mitchell 2000). In vegetables pre- and post-emergence damping-off caused by *Pythium* spp. is economically very important around the world (Whipps and Lumsden 1991). After exposure of sporangia (*Pythium*) to exudates or volatiles from seeds or roots (Osburn et al. 1989) followed by immediate infection make management of *Pythium* is very difficult (Whipps and Lumsden 1991). Damping-off disease in vegetables and field crops considered as important limiting factor in successful cultivation of crop plants worldwide. Yield loss due to *Pythium* species in different crops has been estimated approximately of multibillion dollar worldwide (Van West et al. 2003). Among these, *P. aphanidermatum* (Edson) Fitzp. is one of the most pathogenic and problematic species with host range and cause severe damage in many economically important crops.

Source of Inoculum

The spores of *P. aphanidermatum* may enter in to the crop plants through a number of ways viz., contaminated potting mixture (Moorman et al. 2002) and contaminated irrigation water (Hong and Moorman 2005). *P. aphanidermatum* can also survive in the form of oospores in soil, sporangia or mycelium in infected crop debris and farm implements (Moorman et al. 2002).

Host Range

Damping-off is a common nursery disease in a number of vegetable crop plants and it cause reduced germination percentage, vigour, quality and yield of crops. It also causes seedling rots, root rot, cottony-leak, cottony blight, stalk rot etc. It is known to cause infection on a wide range of plant species, belonging to different families viz., *Amaranthaceae*, *Amaryllidaceae*, *Araceae*, *Basellaceae*, *Bromeliaceae*, *Cactaceae*, *Chenopodiaceae*, *Compositae*, *Coniferae*, *Convolvulaceae*, *Cruciferae*, *Cucurbitaceae*, *Euphorbiaceae*, *Gramineae*, *Leguminosae*, *Linaceae*, *Malvaceae*, *Moraceae*, *Passifloraceae*, *Rosaceae*, *Solanaceae*, *Umbelliferae*, *Violaceae*,

Vitaceae, *Zingiberaceae* (Waterhouse and Waterston 1964). In greenhouse-grown crops damping-off and root rot caused by *P. aphanidermatum* is one of the most devastating diseases among them. Lot of research have been carried out on *P. aphanidermatum* in different parts of the world (Plaats-Niterink and Van Der 1981).

Distribution

Pythium species has also been isolated from different regions of the world and from different diseases such as damping-off, vascular wilt and root rot of groundnuts in Australia (Bellgard and Ham 2004), damping-off of cauliflower in India (Elliott 2003), damping-off of fenugreek in India (Mehra 2005), ginger soft rot in India (Kavitha and Thomas 2007), damping-off of cucumber in Oman (Deadman et al. 2007), *Pythium* rot of fig marigold in Japan (Kawarazaki et al. 2008), rhizome rot of ginger in India (Sagar et al. 2008), damping-off and root rot of soybean in USA (Rosso et al. 2008), damping-off of chilli in India (Manoranjitham and Prakasam 2000; Muthukumar et al. 2010a; Zagade et al. 2012), damping-off of Cucumis melon China (Juan et al. 2009), damping-off disease of tobacco in India (Subhashini and Padmaja 2009), damping-off of mustard in India (Khare et al. 2010), rhizome rot of turmeric in India (Radhakrishnan and Balasubramanian 2009) and damping-off of cabbage seedlings in Japan (Kubota 2010) etc.

The management of *Pythium* damping-off relies on the use of fungicides. However, fungicide application leads to phytotoxicity and fungicide residues are major problems leading to environmental pollution altered the biological balance in the soil by destroying non-target and beneficial microorganisms. Development of fungicide resistance in the pathogen has also been reported (Bharathi et al. 2004). Recent efforts have focused on developing environmentally safe, long lasting and effective biocontrol agents and plant products and plant essential oils for the management of damping-off disease.

Trichoderma Species

Introduction

Trichoderma spp. are very useful filamentous fungi. By producing beneficial effects on crops, they have actually sustained the agricultural yields that have supported the human population over the millennia. Together with other beneficial microbes, they help to maintain the general disease suppressiveness and fertility of soils, and aid in the maturation of compost for natural fertilizer production (Harman et al. 2004). *Trichoderma* spp. are ubiquitous and often predominant components of the mycoflora in native and agricultural soils throughout all climatic zones.

They colonize aboveground and belowground plant organs and grow intercellularly (endophytes), and they appear in plant litter, soil organic matter (saprophytes), and mammalian tissues (human pathogens). However, the ability of these fungi to recognition, invade, and destroy other fungi has been the major driving force behind their commercial success as biopesticides. These fungi not only protect plants by killing other fungi and certain nematodes but induce resistance against plant pathogens, impart abiotic stress tolerance, improve plant growth and vigour, solubilize plant nutrients, and bioremediate heavy metals and environmental pollutants.

Better understanding of how *Trichoderma* evolved to interact with other fungi and with plants will improve and expand their applications. The ability to attack other fungi, most importantly soil borne plant dominated the interest in *Trichoderma* for many years. Recent years have witnessed a wave of interest in plant disease resistance namely induced systemic resistance (ISR); systemic acquired resistance (SAR) induced by the *Trichoderma*-root symbiosis. These plant-centered mechanisms have rivaled mycoparasitism as an explanation for how *Trichoderma* controls plant diseases. The genome sequencing of *Trichoderma* species has stimulated the development of systems biological approaches, initiated and enhanced whole-genome expression studies, and provided unique data for phylogenetic and bioinformatic analyses toward understanding the roles of these opportunists in ecosystems (Mukherjee et al. 2013).

At present the, genome sequences of seven species: *Trichoderma reesei*, *T. virens*, *T. atroviride*, *T. harzianum*, *T. asperellum*, *T. longibrachiatum*, and *T. citrinoviride* are available (Grigoriev et al. 2012). The genome of *Trichoderma* spp. has been extensively investigated and has proven to contain many useful genes, along with the ability to produce a great variety of expression patterns, which allows these fungi to adapt to many different environments (soil, water, dead tissues, inside the plant, etc.). Several laboratories have recently started or planned to use proteomic and functional genomic analysis in the attempt to obtain an overall picture of the changes that occur in the *Trichoderma*, plant, and pathogen expressomes when they "talk" to each other, especially when an increase in disease resistance is generated. *Trichoderma* spp. are the most successful bio-fungicides used in today's agriculture with more than 60 % of the registered biofungicides world-wide being *Trichoderma*-based (Verma et al. 2007). In India alone, about 250 products are available for field applications (Singh et al. 2009). Despite this remarkable success, the share of bio-fungicides is only a fraction of the fungicides market, dominated by synthetic chemicals.

The major limitations of microbe-based fungicides are their restricted efficacy and their inconsistency under field conditions. The origin of these difficulties is that microbes are slow to act, compared to chemicals, and are influenced by environmental factors.

Trichoderma-Plant Interactions

Many *Trichoderma* spp. grow in the rhizosphere and are capable of penetrating and internally colonizing plant roots (Harman et al. 2004). This opportunistic/facultative symbiosis is driven by the ability of *Trichoderma* to derive sucrose or other nutrients from plants, in return for boosting plant immunity against invading pathogens. The presence of *Trichoderma* in the rhizosphere evokes a coordinated transcriptomic, proteomic and metabolomic response in the plants (Shoresh and Harman 2008). This reprogramming of the plant is often beneficial, improving growth, yield and resistance to pathogens. The combined ability to attack soil-borne pathogens while priming plant defenses, however, is what promotes *Trichoderma* as such a promising partner for sustainable management of plant diseases.

Mechanism of Action of Trichoderma Species

Of the bio-control agents, *Trichoderma* spp. has demonstrated effective and selective enough against most of the fungal diseases. *Trichoderma* spp. has developed numerous mechanisms by which they are attacking other fungi. These mechanisms include mycoparasitism (Haran et al. 1996), production of inhibitory compounds (Sivasithamparam and Ghisalberti 1998), competition for space and nutrients (Elad et al. 1999), inactivation of the pathogen's enzymes and induced resistance. Today, more than 50 different *Trichoderma*-formulated agriculture products can be found and are sold and applied to protect and improve the yield of fruits, vegetables and ornamentals (Lorito 2005). *Trichoderma* is completely safe to the humans, animals and environment.

Attachment to Host

Attachment and infection of host fungi by mycoparasitic *Trichoderma* is accompanied by the formation of appressoria and/or mycoparasitism (Druzhinina et al. 2011). The genetic studies underlying attachment of pathogen to the host are not well understood, although proteins like structures are possibly involved. Though experimental evidence is lacking, indirect support for the involvement of hydrophobins comes from the finding that *T. virens* mutants in the transcriptional regulator of secondary metabolism and morphogenesis, which have decreased hydrophobin expression, were defective in both hydrophobicity and mycoparasitism (Mukherjee and Kenerley 2010).

Killing the Host

Trichoderma species produces lytic enzymes (chitinase and glucanase) and antibiotics there by kill other fungi. Not surprisingly, the genomes of the mycoparasitic *Trichoderma* spp. are rich in genes encoding enzymes like chitinases and glucanases, and those for secondary metabolism like NRPSs (Kubicek et al. 2011). Glucanases are another group of cell wall-lytic enzymes with roles in mycoparasitism/biocontrol. Deletion of tvbgn3 (β-1, 6-glucanase-encoding) reduced the mycoparasitic and biocontrol potential of *T. virens* against *P. ultimum* (Viterbo and Horwotz 2010). In addition to chitinases and glucanases, proteases like Prb1/Sp1 are induced during mycoparasitism and play definitive roles in biocontrol (Djonovic et al. 2006).

Antibiosis

Antibiosis occurs during interactions involving low-molecular-weight diffusible compounds or antibiotics produced by *Trichoderma* strains that inhibit the growth of other microorganisms. Most *Trichoderma* strains produce volatile and nonvolatile toxic metabolites that impede colonization by antagonized microorganisms; among these metabolites, the production of harzianic acid, alamethicins, tricholin, peptaibols, antibiotics, 6-penthyl-α-pyrone, massoilactone, viridin, gliovirin, glisoprenins, heptelidic acid and others have been described (Vey et al. 2001). In some cases, antibiotic production correlates with biocontrol ability, and purified antibiotics mimic the effect of the whole agent. However, there are also examples of antibiotic-overproducing strains, such as gliovirin overproducing mutants of *T. virens*, which provide control similar to that of the wild-type, and of gliovirin-deficient mutants which failed to protect cotton seedlings from *P. ultimum*, whereas the parental strain did (Chet et al. 1997). In general, strains of *T. virens* with the best efficiency as biocontrol agents are able to produce gliovirin (Howell 1998). Also, the most effective isolates of *T. harzianum* against *Gaeumannomyces graminis var. tritici* produce pyrone antibiotics, and the success of the strains was clearly related to the pyrones they produced.

Peptaibols-a class of linear peptides that generally have strong antimicrobial activity against gram-positive bacteria and fungi-act synergistically with cell-wall-degrading enzymes (CWDEs) to inhibit the growth of fungal pathogens and elicit plant resistance to pathogens (Wiest et al. 2002). In tobacco plants, exogenous applications of peptaibols trigger a defense response and reduce susceptibility to tobacco mosaic virus (Wiest et al. 2002). A peptaibol synthetase from *T. virens* has recently been purified, and the corresponding gene, which has been cloned, will facilitate studies of this compound and its contribution to biocontrol. An extensive review on antibiosis and production of *Trichoderma* secondary metabolites is provided in Howell (Howell 1998, 2003).

Competition

Trichoderma species are generally considered to be aggressive competitors, grow very fast and rapidly colonize substrates to exclude pathogens such as *Fusarium* spp. (Papavizas 1985). Rhizosphere competence, following seed treatment is an important strategy to create a zone of protection against plant pathogens (Howell 2003). *Trichoderma* species, either added to the soil or applied as seed treatments, grow readily along with the developing root system of the treated plants (Ahmad and Baker 1987). Soil application with *T. harzianum* spores inhibited the infestations of *Fusarium oxysporum* f. sp. *vasinfectum* and *F. oxysporum* f. sp. *melonis*. Competition was a proposed mechanism, although it was not proven to be the main activity.

Parasitism

In contrast to studies on hyphal parasitism, very little research has been done on the molecular mechanisms of parasitism of resting structures. *Trichoderma* spp. are prolific producers of secondary metabolites and the genomes of the mycoparasitic

Trichoderma spp. are especially enriched in genes for secondary metabolism (Reino et al. 2008). Roles of antimicrobial secondary metabolites such as gliotoxin and gliovirin in suppression of *R. solani* and *P. ultimum* have been suggested, although contradictory reports exist (Viterbo and Horwotz 2010). Certain species like *T. atroviride* produce the volatile metabolite which plays an important role in *Trichoderma*-plant and *Trichoderma*-fungal interactions.

Trichoderma as Biological Control Agents (BCAs)

Trichoderma is one of a small group of beneficial fungi, which have been shown to act, and are commercially applied as BCAs against fungal pathogens. *Trichoderma* thrives in the leaf litter or mulch in orchard situations and it requires a minimum organic carbon level of 1 % to ensure proliferation in cropping locations. Weindling (1932) was first to show the antagonistic effects of a soil fungus to *Sclerotium rolfsii*. He observed that, the hyphae of *Trichoderma* spp. secreted some substance which was lethal to *S. rolfsii*. Later it was identified as gliotoxin. *Trichoderma* spp. have been successfully used for management of diseases caused by *R. solani* in bean, tomato, peanut, rice, lettuce, by *S. rolfsii* in lupine, tomato, peanut, sugarbeet and *Pythium* spp. in pea, tomato and brinjal, tobacco and sugar beet, and by *Macrophomina phaseolina* in sesame, and in okra (Dubey et al. 1996).

Commercial Formulation of Trichoderma

Many species of *Trichoderma* are now registered as bio-fungicides in several countries. Some of commercially available products of *Trichoderma* spp. formulated for the biocontrol of plant pathogens and/or plant growth promotion are Bio Fungus, Binab T, Root Pro, Root Shield/Plant Shield, T-22G, T-22 Planter Box, Trichodex, Trichopel, *Trichoderma* 2000, Tusal, Supresivit, Trichoject, SoilGard and *Trieco*.

Problems in Using Trichoderma Spp.

Although *Trichoderma* spp. is an effective biocontrol agent against several soil-borne fungal pathogens, possible adverse effects of this fungus on arbuscular mycorrhizal (AM) fungi might be a drawback in its use in plant protection. AM fungi are obligate biotrophic endosymbionts in roots of most of the herbaceous plants. These fungi grow from the roots out into the surrounding soil, forming an external hyphal network, which increases uptake of mineral nutrients and consequently promotes plant growth. The presence of *T. harzianum* in soil reduced root colonization by *G. intraradices*. The external hyphal length and density of *G. intraradices* was reduced by the presence of *T. harzianum*. Another problem has been low field performance of *Trichoderma* as biocontrol agent. Understanding the mechanisms by which *Trichoderma* controls fungal diseases assume importance. Further recent molecular biological techniques can be used to exploit these mechanisms, so that, fungal diseases can be effectively controlled in the field.

Trichoderma as a biocontrol agent against *Pythium* species

S. no	Antagonist(s)	Pathogen(s)	Crops(s)	References
1	*Trichoderma* sp.	*P. debaryanum*	Sugar beet	Dumitars and Fratilescu-Sesan (1979)
2	*T. virens*	*P. ultimum*	Cotton	Howell and Stipanovic (1983)
3	*T. viride*	*P. graminicolum*	Tomato	Padmanaban and Alexander (1984)
4	*T. harzianum*	*Pythium* spp.	Pea	Lifshitz et al. (1986)
5	*T. viride*	*P. indicum*	Tomato	Krishnamoorthy (1987)
6	*T. viride*	*P. myriotylum*	Ginger	Rathore et al. (1990)
7	*T. harzianum* and *T. virens*	*P. aphanidermatum* and *P. ultimum*	Cucumber	Woleffhechel and Jensen (1992)
8	*T. viride*	*P. aphanidermatum*	Ginger	Rathore et al. (1992)
9	*T. koningii*	*P. ultimum*	Peas	Nelson et al. (1992)
10	*T. viride*	*P. aphanidermatum*	Ginger	Shanmugam and Varma (1999)
11	*T. viride*	*P. aphanidermatum*	Tobacco	Jackisch-Matsuura and Menezes (2000)
12	*T. harzianum*	*P. aphanidermatum*	Tomato	Hazarika et al. (2000)
13	*T. viride*	*P. aphanidermatum*	Tomato and Chilli	Manoranjitham et al. (2000)
14	*T. viride*	*P. aphanidermatum*	Bhendi	Anitha and Tripathi (2001)
15	*T. harzianum*	*P. aphanidermatum*	Tomato	Pratibha Sharma et al. (2003)
16	*T. viride*	*P. aphanidermatum*	Brinjal	Ramesh (2004)
17	*T. viride*	*P. indicum*	Tomato	Neelamegam (2004)
18	*T. viride*	*P. aphanidermatum*	Tobacco	Loganathan et al. (2004)
19	*T. harzianum*	*P. aphanidermatum*	Tomato	Jayaraj et al. (2006)
20	*T. viride*	*P. aphanidermatum*	Tomato	Rakesh Kumar and Indra Hooda (2007a)
21	*Trichoderma* sp.	*P. aphanidermatum*	Chilli	Muthukumar et al. (2008)
22	*T. viride*	*P. aphanidermatum*	Tomato	Bhuvaneshwari (2008)
23	*T. viride*	*P. aphanidermatum*	Turmeric	Ushamalini et al. (2008)
24	*T. viride*	*P. aphanidermatum*	Mustard	Khare et al. (2010)
25	*T. viride*	*P. aphanidermatum*	Chilli	Muthukumar et al. (2011)
26	*T. viride* and *T. harzianum*	*P. aphanidermatum*	Tomato	Jeyaseelan et al. (2012)
27	*T. harzianum* and *T. longibrachatum*	*P. aphanidermatum*	Sugarbeet	Abdollahi et al. (2012)
28	*T. harzianum*	*P. aphanidermatum*	Tomato	Marzano et al. (2013)
29	*T. hamatum*	*P. ultimum*	Soybean	Hudge (2014)

Pseudomonas fluorescens and *Bacillus subtilis*

Introduction

Pseudomonas fluorescens is a common gram negative, rod-shaped bacterium. The name itself, it secretes a soluble fluorescent pigment is called as fluorescein. It is an obligate aerobe, except for some strains that can utilize NO_3 as an electron acceptor in place of O_2. It is motile by means of one/more polar flagella.

P. fluorescens has simple nutritional requirements and grows well in mineral salts media supplemented with any of a large number of carbon sources (Palleroni 1984). Because they are well adapted in soil, *P. fluorescens* strains are being investigated extensively for use in applications that require the release and survival of bacteria in the soil. Chief among these are biocontrol of pathogens in agriculture and bioremediation of various organic compounds. Certain members of the *P. fluorescens* have been shown to be potential agents for the biocontrol which suppress plant diseases by protecting the seeds and roots from fungal infection. They are known to enhance plant growth promotion and reduce severity of many fungal diseases (Hoffland et al. 1996; Wei et al. 1996). This effect is the result of the production of a number of secondary metabolites including antibiotics, siderophores and hydrogen cyanide (O'Sullivan and O'Gara 1992). Competitive exclusion of pathogens as the result of rapid colonization of the rhizosphere by *P. fluorescens* may also be an important factor in disease control.

Mode of Action

Direct Mechanisms

Facilitating Resource Acquisition

The best-studied mechanisms of bacterial plant growth promotion include providing plants with resources/nutrients that they lack such as fixed nitrogen, iron, and phosphorus. Many agricultural soils lack a sufficient amount of one or more of these compounds so that plant growth is suboptimal. To overcome this problem and obtain higher yields, farmers have become increasingly dependent on chemical sources of nitrogen and phosphorus. Besides being costly, the production of chemical fertilizers depletes non-renewable resources, the oil and natural gas used to produce these fertilizers, and poses human and environmental hazards. It would obviously be advantageous if efficient biological means of providing nitrogen and phosphorus to plants could be used to substitute for at least a portion of the chemical nitrogen and phosphorus that is currently used.

Nitrogen Fixation

Nitrogen (N) is the most vital nutrient for plant growth and productivity. Although, there is about 78 % N_2 in the atmosphere, it is unavailable to the growing plants. The atmospheric N_2 is converted into plant-utilizable forms by biological N_2 fixation (BNF) which changes nitrogen to ammonia by nitrogen fixing microorganisms using a complex enzyme system known as nitrogenase (Kim and Rees 1994). In fact, BNF accounts for approximately two-thirds of the nitrogen fixed globally, while the rest of the nitrogen is industrially synthesized by the Haber-Bosch process (Rubio and Ludden 2008). Biological nitrogen fixation occurs, generally at mild temperatures, by nitrogen fixing microorganisms, which are widely distributed in nature (Raymond et al. 2004). Furthermore, BNF represents an economically beneficial and environmentally sound alternative to chemical fertilizers (Ladha et al. 1997). Since the process of nitrogen fixation requires a large amount of energy in the form of ATP, it would be advantageous if bacterial carbon resources were directed toward oxidative phosphorylation, which results in the synthesis of ATP, rather than glycogen synthesis, which results in the storage of energy in the form of glycogen. In one experiment, a strain of *Rhizobium tropici* was constructed with a deletion in the gene for glycogen synthase (Marroquí et al. 2001).

Treatment of bean plants with this engineered bacterium resulted in a significant increase in both the number of nodules that formed and an increase in the plant dry weight in comparison with treatment with the wild-type strain. This is one of the very few examples of scientists genetically modifying the nitrogen fixation apparatus of a bacterium and obtaining increased levels of fixed nitrogen. Unfortunately, while this mutant increased nodule number and plant biomass in the field, it does not survive well in the soil environment.

The genes for nitrogen fixation, called nif genes are found in both symbiotic and free living systems (Kim and Rees 1994). Nitrogenase (nif) genes include structural genes, genes involved in activation of the Fe protein, iron molybdenum cofactor biosynthesis, electron donation, and regulatory genes required for the synthesis and function of the enzyme. The symbiotic activation of nif-genes in the Rhizobium is dependent on low oxygen concentration, which in turn is regulated by another set of genes called fix-genes which are common for both symbiotic and free living nitrogen fixation systems (Dean and Jacobson 1992; Kim and Rees 1994). Since nitrogen fixation is a very energy demanding process, requiring at least 16 mol of ATP for each mole of reduced nitrogen, it would be advantageous if bacterial carbon resources were directed toward oxidative phosphorylation, which results in the synthesis of ATP, rather than glycogen synthesis, which results in the storage of energy in the form of glycogen (Glick 2012). For instance, treatment of legume plants with rhizobia having a deleted gene for glycogen synthase resulted in a considerable augmentation in both the nodule number and plant dry weight with reference to treatment with the wild-type strain (Marroqui et al. 2001).

Phosphate Solubilization

Phosphorus is necessary for plant growth and is taken by the plants from soil as phosphate anions. Even so, phosphate anions are highly reactive and may be trapped via precipitation with cations such as Mg^{2+}, Ca^{2+}, Al^{3+} and Fe^{3+} depending on the quality of the soil. Phosphorus is extremely insoluble and unavailable to plants in these forms. As a result, the amount available to plants is usually a small proportion of this total. To overcome the P deficiency in soils, there are frequent applications of phosphatic fertilizers in agricultural fields. Plants absorb fewer amounts of applied phosphatic fertilizers and the rest is rapidly converted into insoluble complexes in the soil (Mckenzie and Roberts 1990). But regular application of phosphate fertilizers is not only costly but is also environmentally undesirable. This has led to search for an ecologically safe and economically reasonable option for improving crop production in low P soils. In this context, organisms coupled with phosphate solubilizing activity, often termed as phosphate solubilizing microorganisms (PSM), may provide the available forms of P to the plants and hence a viable substitute to chemical phosphatic fertilizers (Khan et al. 2006).

Of the various PSM(s) inhabiting the rhizosphere, phosphate-solubilizing bacteria (PSB) are considered as promising biofertilizers since they can supply plants with P from sources otherwise poorly available by various mechanisms (Zaidi et al. 2009). Bacterial genera like *Azotobacter*, *Bacillus*, *Beijerinckia*, *Burkholderia*, *Enterobacter*, *Erwinia*, *Flavobacterium*, *Microbacterium*, *Pseudomonas*, *Rhizobium* and *Serratia* are reported as the most significant phosphate solubilizing bacteria (Bhattacharyya and Jha 2012). Typically, the solubilization of inorganic phosphorus occurs as a consequence of the action of low molecular weight organic acids which are synthesized by various soil bacteria (Zaidi et al. 2009).

Conversely, the mineralization of organic phosphorus occurs through the synthesis of a variety of different phosphatases, catalyzing the hydrolysis of phosphoric esters (Glick 2012). Importantly, phosphate solubilization and mineralization can coexist in the same bacterial strain (Tao et al. 2008). Though, PSB are commonly found in most soils; their establishment and performances are severely affected by environmental factors especially under stress conditions (Ahemad and Khan 2010a, b, 2012a, b). However, the beneficial effects of the inoculation with PSB used alone (Ahemad and Khan 2011, 2012b) or in combination with other rhizospheric microbes have been reported (Zaidi and Khan 2005; Vikram and Hamzehzarghani 2008). Besides providing P to the plants, the phosphate solubilizing bacteria also augment the growth of plants by stimulating the efficiency of BNF, enhancing the availability of other trace elements by synthesizing important plant growth promoting substances (Suman et al. 2001; Ahmad et al. 2008; Zaidi et al. 2009).

Sideropore Production

Sequestering Iron Iron is one of the bulk mineral present in plenteous amount on earth, yet it is unavailable in the soil for the plants. This is because Fe^{3+} (ferric ion) is common form of iron found in nature and is meagerly soluble. To overcome this problem, PGPR's secrete siderophores. Siderophores are iron binding protein of low molecular mass and have a high binding affinity with ferric ion. Siderophores secreted by PGPR's improves plant growth and development by increasing the accessibility of iron in the soil surrounding the roots (Kloepper et al. 1980). Marschner and Romheld (1994) described that plants utilize siderophores secreted by PGPR for sequestering iron. Plants such as Oats, Sorghum, Cotton, Peanut, Sunflower and Cucumber demonstrate the ability to use microbial siderophores as sole source of iron than their own siderophores (Bar-Ness et al. 1991; Wang et al. 1993). Microbial siderophores are also reported to increase the chlorophyll content and plant biomass in plants of cucumber (Kloepper and Schroth 1978).

Phytohormone Production

Plant hormones play key roles in plant growth and development and in the response of plants to their environment (Davies 2004). Moreover, during its lifetime, a plant is often subjected to a number of nonlethal stresses that can limit its growth until either the stress is removed or the plant is able to adjust its metabolism to overcome the effects of the stress (Glick et al. 2007). When plants encounter growth limiting environmental conditions, they often attempt to adjust the levels of their endogenous phytohormones in order to decrease the negative effects of the environmental stressors (Salamone et al. 2005). While this strategy is sometimes successful, rhizosphere microorganisms may also produce or modulate phytohormones under *in vitro* conditions (Salamone et al. 2005) so that many PGPB can alter phytohormone levels and thereby affect the plant's hormonal balance and its response to stress (Glick et al. 2007).

(a) **Indoleacetic Acid**: Microbial synthesis of the phytohormone auxin (Indole-3-acetic acid/indole acetic acid/IAA) has been known for a long time. The most studied auxin, and much of the scientific literature considers auxin and IAA to be interchangeable terms (Spaepen et al. 2007) IAA affects plant cell division, extension and differentiation; stimulates seed and tuber germination; increases the rate of xylem and root development; control processes of vegetative growth; initiates lateral and adventitious root formation; mediates responses to light, gravity and florescence; affects photosynthesis, pigment formation, biosynthesis of various metabolites, and resistance to stressful conditions (Tsavkelova et al. 2006; Spaepen and Vanderleyden 2011). IAA produced by rhizobacteria likely, interfere the above physiological processes of plants by changing the plant auxin pool. Moreover, bacterial IAA increases root surface area and length, and thereby provides the plant greater access to soil nutrients. Also, rhizobacterial IAA loosens plant cell walls and as a result facilitates an increasing amount of root exudation that provides additional nutrients to support the growth of rhizosphere bacteria (Glick 2012). Thus, rhizobacterial IAA is identified as an effector molecule in plant-microbe interactions, both in pathogenesis and phytostimulation (Spaepen and Vanderleyden 2011). Starting with tryptophan, at least five different pathways have been described for the synthesis of IAA, and most pathways show similarity to those described in plants, although some intermediates can differ (Patten and Glick 1996; Spaepen and Vanderleyden 2011).

1. IAA formation via indole-3-pyruvic acid and indole-3-acetic aldehyde is found in a majority of bacteria like, *Erwinia herbicola*; saprophytic species of the genera *Agrobacterium* and *Pseudomonas*; certain representatives of *Bradyrhizobium*, *Rhizobium*, *Azospirillum*, *Klebsiella*, and *Enterobacter*
2. The conversion of tryptophan into indole-3-acetic aldehyde may involve an alternative pathway in which tryptamine is formed as in pseudomonads and azospirilla
3. IAA biosynthesis via indole-3-acetamide formation is reported for phytopathogenic bacteria *Agrobacterium tumefaciens*, *Pseudomonas syringae*, and *E. herbicola*; saprophytic pseudomonads like (e.g. *P. putida* and *P. fluorescens*)
4. IAA biosynthesis that involves tryptophan conversion into indole-3-acetonitrile is found in the cyanobacterium (*Synechocystis* sp.)
5. The tryptophan-independent pathway, more common in plants, is also found in azospirilla and cyanobacteria.

(b) **1-Aminocyclopropane-1-carboxylate (ACC) deaminase**: Generally, ethylene is an essential metabolite for the normal growth and development of plants (Khalid et al. 2006). This plant growth hormone is produced endogenously by approximately all plants and is also produced by different biotic and abiotic processes in soils and is important in inducing multifarious physiological changes in plants. Apart from being a plant growth regulator, ethylene has also been established as a stress hormone (Saleem et al. 2007). Under stress conditions like those generated by salinity, drought, water logging, heavy metals and pathogenicity, the endogenous level of ethylene is significantly increased which negatively affects the overall plant growth. For instance, the high concentration of ethylene induces defoliation and other cellular processes that may lead to reduced crop performance (Saleem et al. 2007; Bhattacharyya and Jha 2012). Currently, bacterial strains exhibiting ACC deaminase activity have been identified in a wide range of genera such as Acinetobacter, Achromobacter, Agrobacterium, Alcaligenes, *Azospirillum*, *Bacillus*, *Burkholderia*, *Enterobacter*, *Pseudomonas*, *Ralstonia*, *Serratia* and *Rhizobium* etc. (Nadeem et al. 2007; Zahir et al. 2008, 2009; Kang et al. 2010).

(c) **Cytokinins and Gibberellins**: Several studies have shown that many soil bacteria in general, and PGPB in particular, can produce either cytokinins or gibberellins or both (Taller and Wong 1989; Timmusk et al. 1999; Salamone et al. 2001). Thus, for example, cytokinins have been detected in the cell-free medium of some strains of *Azotobacter* spp., *Rhizobium* spp., *Pantoea agglo-*

merans, *Rhodospirillum rubrum*, *Pseudomonas fluorescens*, *Bacillus subtilis* and *Paenibacillus polymyxa*. Moreover, plant growth promotion by some cytokinin- or gibberellin-producing PGPB has been reported (Joo et al. 2005; Kang et al. 2009).

Indirect Mechanisms

The application of microorganisms to control plant diseases, which is a form of biological control, is an environment-friendly approach (Lugtenberg and Kamilova 2009). The major indirect mechanism of plant growth promotion in rhizobacteria is through acting as biocontrol agents (Glick 2012). They are

Siderophore

Siderophores (from the Greek: "iron carriers") are defined as relatively low molecular weight, ferric ion specific chelating agents elaborated by bacteria and fungi growing under low iron stress. Based on their iron-coordinating functional groups, structural features and types of ligands, bacterial siderophores have been classified into four main classes (carboxylate, hydroxamates, phenol catecholates and pyoverdines) (Crowley 2006). Hundreds of siderophores have been identified and reported for cultivable microorganisms, some of which are widely recognized and used by different microorganisms, while others are species-specific (Crowley 2006; Sandy and Butler 2009). The synthesis of siderophores in bacteria is induced by the low level of $Fe^{3+,}$ and, in acid soil, where solubility and availability grow, their protective effect comes down. In such conditions, the efficiency of iron fixation can be raised by gaining mutant strains capable of synthesizing siderophores. This process is independent of iron concentration in soil solution (Maksimov et al. 2011). Some rhizospheric bacteria can produce siderophores and there is evidence that a number of plant species can absorb bacterial Fe^{3+} siderophore complexes (Bar-Ness et al. 1991). However the significance of bacterial siderophore in the iron nutrition of plants is controversial (Vessey 2003). Some authors believe that the contribution of these siderophores to the overall iron requirements of plants is small (Glick 1995). The two bacterial siderophores (pseudobactin and ferrioxamine B) were inefficient as iron sources for plants and that rhizospheric siderophore-producing bacteria can be in competition with the plant for iron. Moreover microbial siderophores in the rhizosphere are frequently associated with biocontrol activities and not with plant nutrition (Vessey 2003).

Djibaoui and Ahmed (2005) tested the ability of *Pseudomonas* to grow and to produce siderophores is dependent on the iron content and the type of carbon sources in the medium. Under conditions of low-iron concentration the *Pseudomonas* isolates produced yellow-green fluorescent iron-binding peptide siderophores. Urszula (2006) tested the ability of six strains belonging to the genus *Pseudomonas* isolated from the rhizosphere of wheat to produce pyoverdin. The studied strains demonstrated a varied level of production of the siderophore, depending on the culture conditions. The highest level of pyoverdin was determined after 72 h of growth at 20–25 °C in iron-free medium supplemented with succinate.

The siderophores production by *Bacillus* and *Pseudomonas* when assessed both in the presence and in absence of technical grade of herbicides show that the metabolic activities of plant growth promoting rhizobacteria decline following herbicides application (Munees and Mohammad 2009).

Antibiotics and Lytic Enzymes

The synthesis of a range of different antibiotics is the PGPB trait that is most often associated with the ability of the bacterium to prevent the proliferation of plant pathogens (generally fungi) (Haas and Keel 2003; Mazurier et al. 2009). Some antagonistic bacteria produce enzymes including chitinases, β -1,3 glucanases, cellulases, proteases, and lipases that can lyse a portion of the cell walls of many pathogenic fungi. PGPB that synthesize one or more of these enzymes have been found to have biocontrol activity against a range of pathogenic fungi including *Botrytis cinerea*, *S. rolfsii*, *Fusarium oxysporum*, *R. solani*, and *P. ultimum* (Singh et al. 1999; Kim et al. 2008).

Phosphate Solubilization

Phosphorus is one of the most essential nutrient requirements in plants. Ironically, soils may have large reservoir of total phosphorous (P) but the amounts available to plants are usually a small proportion of this total. This low availability of phosphorous to plants is because of the vast majority of soil P is found in insoluble forms, while the plants can only absorb it in two soluble forms, the monobasic (H_2PO_4) and the diabasic ($HPO4_2$) ions (Glass 1989).

Several researchers consequently have isolated PSB from various soils and prove that inoculations of these bacteria increase the plant growth and yield (Sturz and Nowak 2000; Sudhakar et al. 2000; Mehnaz and Lazarovits 2006). The bacterial genera with phosphate solubilising capacity are *Alcaligenes*, *Actinobacter*, *Arthrobacter*, *Azospirillum*, *Burkholderia*, *Bacillus*, *Enterobacter*, *Erwinia*, *Flavobacterium*, *Pseudomonas*, *Rhizobium* and *Serratia*.

Importantly, phosphate solubilization and mineralization can coexist in the same bacterial strain (Tao et al. 2008). Besides providing P to the plants, the PS bacteria also augment the growth of plants by stimulating the efficiency of BNF, enhancing the availability of other trace elements (such as iron, zinc) and by synthesizing important plant growth promoting substances (Ponmurugan and Gopi 2006; Mittal et al. 2008). To make this form of P available for plant nutrition, it must be hydrolyzed to inorganic P by means of acid and alkaline phosphatase enzymes. Because the pH of most soils ranges from acidic to neutral values acid phosphatases should play the major role in this process (Rodríguez and Fraga 1999). The possibility of enhancing P uptake of crops by artificial inoculation with P-solubilising strains of rhizobacteria presents an immense interest to agricultural microbiologists.

Ethylene

Ethylene is a plant hormone that differs from other plant hormones in being a gas. Both abiotic and biotic factors affect the Agricultural crops and its yield under stress conditions. The PGPR's containing ACC deaminase activity are present in various soils and assures improvement of plant growth and development under stress conditions such as heavy metal stress, phytopathogens, flooding, drought and high salt concentration. Ethylene is a significant phytohormone, but excess secretion of ethylene will lead to root curling and shortening, even it can result in plant death under extreme conditions. ACC deaminase property/activity of PGPR helps plant to combat abiotic stress by hydrolyzing ACC, the precursor of ethylene, to alpha-ketobutyrate and ammonia, and encourages plant growth under stress environment. Use of biofertilizer containing PGPR with ACC deaminase activity may improve the plant growth and development by relieving harmful effects of salt stress ethylene (Belimov et al. 2001). Besides this, heavy metal stress can also be alleviated using PGPR's (Sharma et al. 2012).

Induced Systemic Resistance

Interaction of some rhizobacetria with the plant roots can result in plant resistance again some pathogenic bacteria, fungi and viruses. This phenomenon is called induced systemic resistance (ISR). Moreover, ISR involves jasmonate and ethylene signaling within the plant and these hormones stimulate the host plant's defense responses against a variety of plant pathogens (Glick 2012). Many individual bacterial components induce ISR, such as lipopolysaccharides (LPS), flagella, siderophores, cyclic lipopeptides 2,4-diacetylphloroglucinol, homoserine lactones, and volatiles like, acetoin and 2,3-butanediol (Lugtenberg and Kamilova 2009).

Moreover several researchers reported that *Bacillus* spp. produce a large number of antifungal metabolites such as bacitracin, gramicidin S, polymyxin, tyrotricidin, bacilysin, chlotetaine, iturin A, mycobacillin, bacilomycin, mycosubtilin, fungistatin and subsporin which are able to control plant disease (Collins and Jacobsen 2002; Trevor et al. 2003).

Antifungal Activity Against Pythium Species

Paulitz et al. (1992) reported that antifungal metabolites obtained from *Bacillus* spp. to control cucumber damping-off caused by *P. aphanidermatum*. Koomen and Jeffries (1993), who stated that most strains of *Bacillus* spp. have potential as bacterial antagonists against *P. aphanidermatum*. Chilli seeds treated with *P. fluorescens* reduced the damping-off incidence and increased the plant growth (shoot length and root length) and vigour (Harris et al. 1994). Pleban et al. (1995) reported that the bacterial antagonist like *Bacillus* and *Pseudomonas* spp. isolated from seeds and plants of different crops were tested for their antagonistic potential against various plant pathogenic fungi including *P. ultimum*, *R. solani* and *S. rolfsii*. The isolates of *P. fluorescens* were highly antagonistic to *P. ultimum* and *R. solani* by way of producing endochitinase and chitobiosidase (Nielson and Sorensen 1999).

Anitha and Tripathi (2001) recorded that *P. fluorescens* inhibited the mycelial growth of *P. aphanidermatum* and *R. solani* inciting seedling disease of okra. The phylloplane bacteria (*P. fluorescens*) isolated from the rambutan leaf surface can be effective against some soil-born pathogens, especially *P. aphanidermatum* (Yenjit et al. 2004). Kavitha et al. (2005) reported that the production of phenazine derivatives which was effective against *P. aphanidermatum* (chilli damping-off) which disorganized the hyphal morphology by inducing the formation of vacuolation in hyphal cells, degeneration of cell content followed by hyphal lysis. The bacterial antagonist *P. chlororaphis* strain PA23 and *B. subtilis* strain BSCBE4 showed the maximum inhibitory effect on the mycelial growth of *P. aphanidermatum* causing chilli damping-off (Nakkeeran et al. 2006). Muthukumar and Bhaskaran (2007) reported that 12 isolates of *P. fluorescens* were tested against the growth of *Pythium* sp. Of these, *P. fluorescens* three and four were highly effective in inhibiting the mycelial growth of *Pythium* sp. Tomato seeds treated with *P. fluorescens* increased the seed germination percentage, shoot length, root length and vigour of tomato seedlings (Valarmathi 2007).

Muthukumar (2008) reported that the endophytic bacterial isolate 5, 6 and 7 (isolated from stem and root) showed highest inhibition on the mycelial growth of *P. aphanidermatum* (51.4, 41.7 and 40.0 %) inciting damping-off of chillies. All the bacterial strains have the capacity to inhibit the mycelial growth of *P. aphanidermatum* causing tomato damping-off (Intana and Chamswarng 2007). Muthukumar et al. (2010a) reported that among the isolates, EBS (endophytic bacteria stem 20) produced largest inhibition zone and the least mycelia growth of *P. aphanidermatum*. The same isolate produced more amounts of salicylic acid, sideropore and HCN. Chilli seeds treated with *P. fluorescens*, Thiram 75WS and Captan 50WP and formulations reduce damping off incidence and enhanced yield in chilli (Saha et al. 2011). Dar et al. (2012) reported that seed treatment and soil application with talc based formulations of *P. fluorescens* reducing the damping-off of papaya caused by *P. debaryanum* and increased plant growth parameters of papaya seedlings.

Amaresan et al. (2014) reported that three bacterial isolates (BECS4, BECS5 and BECS7) showed highest antagonistic activity against *Pythium* sp. This type of antagonistic activity is might be due to the production of lytic enzymes (protease, cellulose, amylase and lipase). The three bacterial isolates (*P. fluorescens*-9A-14, *Pseudomonas* sp.-8D-45 and *Bacillus subtilis*-8B-1) were tested against the growth of *P. debaryanum* inciting damping-off and root rot of cucumber. The results revealed that all the three bacterial isolates inhibited the mycelia growth of *Pythium* sp. and increased the plant growth under *in vitro* condition. Further, cucumber seeds treated with peat and talc based formulations of these antagonistic bacteria effectively controlled the disease incidence and increased the plant growth (Khabbaz and Abbasi 2014).

Plant Extract

Introduction

Bioactive compounds or plant secondary metabolites (SMs) consist of low-molecular weight compounds that are regarded as not essential for sustaining life, but as crucial for the survival of the producing organism (Hadacek 2002). More than 50,000 structures have been identified in plants by NMR, MS and X-ray analysis. However, as only less than 20 % of all plants have been studied, it is very likely that the actual numbers of secondary metabolites (SMs) or bioactive compounds in the plant kingdom would exceed 100,000 structures (Wink 2006). The site of synthesis for SMs is not necessarily the site of accumulation. Hydrophilic compounds (such as alkaloids, flavonoids, tànnins, and saponins) are stored in the vacuole while the lipophilic SMs (such as terpenoids) are sequestered in resin ducts, laticifers, oil cells, trichomes, or in the cuticle. Bioactive compounds affect the fungi via interference with molecular targets in their organs, tissues and cells. The major targets include: Biomembrane, proteins and nucleic acids. Bioactive compounds are still regarded as a valuable pool for discovering novel mode of action (Engelmeier and Hadacek 2006).

Why We Consider Botanicals?

- Reduce crop losses
- Easily decomposed
- Low cost
- Environmentally friendly approach
- Sustainable solutions in agriculture
- Integrated Diseases Management (IDM)

Major Groups of Antimicrobial Compounds

Plants produce a array of bioactive metabolites which serve as plant defense mechanisms against pests and diseases. Some SMs give plants their odors (terpenoides), some are responsible for plant pigments (quinines and tannins) and others (e.g., some of terpenoids) are responsible for plant flavor. Cowan (1999) he classified the antimicrobial bioactive compounds into five classes. They are (i) terpenoids and essential oils, (ii) phenolics, (iii) polyphenols, (iv) alkaloids and (v) polypeptides and mixtures (crude extract). Plants have the capacity to synthesize secondary metabolites, like phenols, phenolic acids, quinones, flavones, flavonoids, lavonols, tannins and coumarins (Cowan 1999). The components with phenolic structures, like carvacrol, eugenol, and thymol shows antimicrobial effect and serves as plant defence mechanisms against pathogenic microorganisms (Das et al. 2010). The volatile antimicrobial substance allicin (diallyl thio sulphinate) is synthesized in garlic when the tissues are damaged and the substrate alliin (S-allyl-L-cysteine Sulphoxide) mixes with the enzyme alliin-lyase. Allicin is readily membrane-permeable and undergoes thiol-disulphide exchange reactions with free thiol groups in proteins. Lavanya et al. (2009) reported that the anti viral protein (AVP) extracted from *Bougainvillea spectabilis* and *Prosopis chilinesis* were found highly effective in reducing the sunflower necrosis virus (SFNV) infection both in cowpea and sunflower plants. At present, scientists are investigating for plant products of antimicrobial properties. It would be advantageous to standardize methods of extraction and in vitro antimicrobial efficacy testing so that the search for new biologically active plant products could be more systematic. Thousands of phyto chemicals which have inhibitory effects on all types of microorganisms in vitro should be subjected in vivo testing to evaluate the efficacy in controlling the incidence of diseases in crops, plants, and humans (Das et al. 2010).

Antimicrobial Secondary Metabolites

Plants have limitless ability to synthesize aromatic secondary metabolites, most of which are phenols or their oxygen-substituted derivatives. Important subclasses in this group of compounds include phenols, phenolic acids, quinones, flavones, flavonoids, flavonols, tannins and coumarins. These groups of compounds show antimicrobial effect and serves as plant defense mechanisms against pathogenic microorganisms.

Phenols Phenol, also known as carbolic acid, is an aromatic organic compound with the molecular formula C_6H_5OH. It is a white crystalline solid that is volatile. The molecule consists of a phenyl group ($-C_6H_5$) bonded to a hydroxyl group (–OH). Phenolic toxicity to microorganisms is due to the site(s) and number of hydroxyl groups present in the phenolic compound.

Quinone A quinone is a class of organic compounds that are formally derived from aromatic compounds [such as benzene or naphthalene] by conversion of an even number of –CH= groups into –C(=O)– groups with any necessary rearrangement of double bonds resulting in a fully conjugated cyclic dione structure.

Flavones, flavonoids and flavonols Flavones, flavonoids and flavonols are phenolic structure with one carbonyl group. They are synthesized by plants in response to microbial infection and are often found effective *in vitro* as antimicrobial substance against a wide array of microorganisms.

Tannins Tannin solutions are acid and have an astringent taste. It is responsible for the astringency, colour, and some of the flavour in tea. These compounds are soluble in water, alcohol and acetone and give precipitates with proteins.

Coumarins Coumarin is a phytochemical with a vanilla like flavour. It is a oxygen heterocycle phenolic substances made of fused benzene and α-pyrone rings.

The crude sap, volatile and essential oil extracted from whole plant or specialized plant parts like roots, stem, leaves, flowers, fruits and seeds are widely used in preparing the antimicrobial compounds which are significantly used against the different plant pathogens/diseases.

Mode of action of phytochemicals

Class	Sub-class	Mechanism
Phenolics	Simple -phenols	Membrane disruption, substrate deprivation
Phenolic acids	Phenolic acids	Bind to adhesins, complex with cell wall, inactivate enzymes
Terpenoids, essential oils	–	Membrane disruption
Alkaloids	–	Intercalate into cell wall
Tannins	–	Bind to proteins, enzyme inhibition, substrate deprivation
Flavonoids	–	Bind to adhesins, complex with cell wall, inactivate enzymes
Coumarins	–	Interaction with eukaryotic DNA
Lectins and polypeptides	–	Form disulfide bridges

Botanicals against plant diseases

Plant	Part used	Preparations	Disease/Pathogen	References
Turmeric (*Curcuma longa* Linn.), Ginger (*Zingiber officinale* Rosc.)	Rhizome	Crude extract	*Phytophthora infestens*, *Fusarium solani*, *Pyricularia oryzae*	Bandara et al. (1989)
Garden croton (*Codiaeum variegatum* Linn.)	Leaf	Phenolic compound	*Alternaria alternata*, *Fusarium oxysporum*	Naidu (1988)
Spearmint (*Mentha spicata* Linn.), Greek Sage (*Salvia fruticosa* Mill.), *Thymbra* spp.	Leaf	Essential oil	*Rhizoctonia solani*, *Sclerotium sclerotiorum*	Yegen et al. (1992)
Chloranthus japonicas, *Paulownia coreana*	Roots, stem	Crude extract	Rice blast, rice sheath blight and wheat leaf rust	Choi et al. (2004)
Neem/Margosa (*Azadirachta indica* A. Juss.),	Leaf	Achook formulations (azadirachtina)	Sheath blight of rice	Kandhari (2007)
Oregano (*Origanum hercleoticum*) (weed species)	Leaf	Essential oils	*Fusarium oxysporum*, *Phoma tracheiphila*	Salomone et al. (2008)
Neem/Margosa (*Azadirachta indica*), Black cumin (*Nigelia sativa* Linn.), Asfetida (*Ferula asafoetida* Linn.)	Seeds	Essential oils	*Fusarium oxysporum*, *A. niger*, *A. flavus*	Sitara et al. (2008)
Indian aloe (*Aloe barbadensis* Mill.), Neem/Margosa (*Azadirachta indica*), Tobacco Leaf (*Nicotiana tabacum* Linn.)	Leaf	Crude extract	Dry rot of yam *F. oxysporum*, *A. niger*	Taiga (2009)

Botanicals against plant diseases				
Plant	Part used	Preparations	Disease/Pathogen	References
Ginger (*Zingiber officinale* Rosc.), Aloe (*Aloe vera*), Bitter kola (*Garcinia cola*) and Neem (*Azadirachta indica*)	Leaf, fruit and seed	Crude	Root rot disease of cow pea (*Vigna unguiculata* L.)	Suleiman and Emusa (2009)
Eugenia aromatica, Piper betle, Alpinia galanga and *Sphaeranthus indicus*	Leaf	Crude	Stem rot disease of vanilla	Suprapta and khalimi (2009)
Metasequoia glyptostroboides	Leaf	Essential oil	*Fusarium oxysporum, F. solani, Phytophthora capsici, Colletotrichum capsici, Sclerotinia sclerotiorum, Botrytis cinerea* and *Rhizoctonia solani*	Bajpai and Kang (2010)

Antifungal Activity of Plant Extracts Against *Pythium* Species

Among the plant products tested, bulb extract of *Allium sativum* (garlic) recorded the minimum mycelial growth (176 mg) of *P. aphanidermatum* causing chilli damping-off. This was followed by *Lawsonia inermis* leaf extract. Chilli seeds treated with bulb extract of *Allium sativum* recorded maximum germination percentage, growth and vigour of chilli seedlings (Kurucheve and Padmavathi 1997). Sutthivaiyakit et al. (2000) recorded that the alcoholic leaf extract of *Chrysanthemum coronarium* has been found to be very active against *P. ultimum, Aspergillus* spp. and *Botrytis cinerea*.

Anti-*Pythium* activity of all tested *Centaurea* species in present study indicated that these plants are potent antifungal plants with possible potential for the control of damping-off diseases in cucumber. The antifungal activity of the plants in this genus was reported earlier (Karakenderes et al. 2006; Bahraminejad et al. 2011). The antifungal activity of plant extracts against *Pythium* has been previously reported by several workers (Sagar et al. 2007; Haouala et al. 2008; Suleiman and Emua 2009; Muthukumar et al. 2010a). Haouala et al. (2008) reported that water extract of fenugreek showed better inhibition on the mycelia growth of *P. aphanidermarum*.

The effect of plant species against rot causing fungi, *P. aphanidermatum*, has also been earlier investigated under *in vitro* and *in vivo* conditions (Sagar et al. 2007). The antimicrobial activity of some plant extracts against *P. aphanidermatum* inciting rhizome rot of ginger has been previously reported (Haouala et al.

2008; Suleiman and Emua 2009). Suleiman and Emua (2009) reported that PDA medium is poisoned with neem leaf extract found 55 % growth inhibition of *P. aphanidermatum* was observed. While ginger rhizome extract reduced 70 % damping-off infection on cowpea *in vivo*.

Ginger and aloe could completely inhibit the mycelial growth of *P. aphanidermatum* causing damping-off under *in vitro* condition (Suleiman and Emua 2009). Dana et al. (2010) reported that the aqueous extract of *Zygophyllum fabago* inhibits the mycelial growth of *P. aphanidermatum* causing damping-off in watermelon. The antifungal effects of 66 medicinal plants belonging to 41 families were evaluated against *P. aphanidermatum*, the causal agent of chilli damping-off. Of these, Zimmu (*Allium sativum* L. × *Allium cepa* L.) leaf extract at 10 % concentration had the highest inhibitory effect (13.7 mm) against mycelial growth of *P. aphanidermatum* (Muthukumar et al. 2010a).

Gomathi et al. (2011) reported that the methanolic extracts of various plants such as *Murraya Koenigii* (Karuveppilei), *Pithecellobium dulce* (Kodukkapuli), *Vitex negundo* (Karunocci), *Aleo vera* (kattalai) individually tested for antifungal activity against *P. debaryanum* by agar well diffusion method. The results revealed that the methanolic extracts of *vitex negundo* showed considerably highest antifungal activity against *P. debaryanum* than other plant extracts tested in the present study. In another study, the antifungal activity of five different medicinal plants namely *Lawsonia inermis* L, *Mimosa pudica* L, *Phyllanthus niruri* L., *Tephrosia purpurea* Pens., *Vinca rosea* L. were tested against *P. debaryanum* causing damping off of disease by agar well diffusion method. Among these, all the three solvents (n-butanol, methanol, aqueous) and the methanolic extracts of *Lawsonia inermis* showed maximum antifungal activity against *P. debaryanum in vitro* (Ambikapathy et al. 2011).

Recently, Tahira Parveen and Kanika Sharma (2014) stated that among the 20 plants tested, *Jacaranda mimosifolia*, *Moringa olifera*, exhibited 27.7 % inhibition of mycelial growth of *P. aphanidermatum*. Whereas, *Polyalthia longifolia* and *Terminallia arjuna* showed 22.2 % inhibition of *P. aphanidermatum*. Besides these, *Lawsonia inermis*, *Aegle marmelos*, *Nigella sativa*, *Azadirachta indica*, also exhibited maximum inhibitory activity against *P. aphanidermatum*.

Limitations of Botanicals for Plant Disease Management

- Extraction methods are not regulated
- Most studies are confined to laboratory condition
- Rapid humiliation
- Not easily formulated
- Some chemical compounds are harmful to human and plants
- Less effective

Plant Essential Oils

Introduction

Essential oils are volatile aromatic concentrated hydrophobic oily liquids which are obtained from various plant parts such as flowers, buds, seeds, leaves, twigs, bark, woods, fruits and roots. They are usually terpenoids responsible for the aroma and flavor associated with herbs, spices and perfumes, also called volatile oils because they easily diffuse into the air. The main constituents of essential oils are mono and sesquiterpenes including carbohydrates, phenols, alcohols, ethers, aldehydes and ketones are responsible for the biological activity as well as for their fragrance. Phenolic compounds present in essential oils have also been recognized as antimicrobial bioactive components (Sumonrat et al. 2008). Various plant materials are believed to have antifungal activity and many essential oils have been reported to have antifungal activities with no side effects on humans and animals (Sokmen et al. 1999).

In general, plant-derived essential oils and extracts are considered as non-phytotoxic compounds and potentially effective against several microorganisms including many fungal pathogens (Chuang et al. 2007). Therefore, they can be used as a natural therapy to inhibit fungal pathogens causing superficial infections. The active antimicrobial compounds of essential oils are generally terpenes, which are phenolic in nature and the site of action is through cell wall and cell membrane. Thus, active phenolic compounds might have several invasive targets which could lead to the inhibition of human infectious fungal pathogens.

The increasing resistance to antifungal compounds and the reduced number of available drugs led us to search for the new alternatives among aromatic plants and their essential oils, used for their antifungal properties. The antifungal activity can be attributed to the presence of some components such as carvacrol, α-terpinly acetate, cymene, thymol, pinene, linalool which are already known to exhibit antimicrobial activity against plant pathogens (Cimanga et al. 2002). A number of scientific investigations have highlighted the importance and the contribution of many plant families used as medicinal plants (Sheetal and Singh 2008). Some of the plant families, their antifungal activity of essential oil are summarized below.

Important Plant Families and Their Antifungal Activity

Asteraceae

The members comes under this family are herbaceous, but a very few belongs to shrubs, vines and trees. This family is distributed all over the world and is most common in the arid and semi-arid regions of subtropical and lowers temperate regions (Barkley et al. 2006). The essential oil obtained from the flower heads of garland chrysanthemum *Chrysanthemum coronarium* L. was evaluated against 12 plant pathogens (Alvarez -Castellanos et al. 2001). The results revealed that the oil was extremely effective against all the test pathogens. The main antifungal compounds present in the oil were camphor, α- and β-pinene and lyratyl acetate. Kordali et al. (2005) reported that the antimicrobial activity of essential oils from three *Artemisia* species i.e. *Artemisia absinthium*, *A. santonicum* and *A. spicigera*. The results clearly indicated that all the oils had same inhibitory effects over all the fungi tested. The antifungal activity of plant oil obtained from *Tagetes patula* L. exerted complete growth inhibition of *Botrytis cinerea* and *Penicillium digitatum* (Romagnoli et al. 2005). Further, the antimicrobial compounds were obtained and identified as piperitone and piperitenone. This was highly responsible for the inhibition of the pathogen. The essential oils from *Chrysactinia mexicana* Grag, caused complete mycelial growth inhibition of *Aspergillus flavus* (Cardenas et al. 2005) and *Helichrysum italicum* (Roth) Don showed antifungal activity against *Pythium ultimum* (Tundis et al. 2005).

Rutaceae

This family is commonly known as rue or citrus family. The antifungal activities of essential oils from *Citrus limon*, *C. paradise*, *C. sinesis* were highly effective against five phytopathogenic fungi (Sun et al. 2007). Pitipong et al. (2009) reported that essential oils from *Citrus hystrix* DC. showed strong antimicrobial activity against *Alternaria brassicicola*, *Aspergillus flavus*, *Bipolaris oryzae*, *Fusarium moniliforme*, *F. proliferatum*, *Pyricularia grisea* and *R. solani* the cause of rice seed-borne and sheath blight pathogen.

Liliaceae

Most of the ornamental plant are comes under this family and it is widely grown for their fragrance and beautiful flowers. It is native to temperate and subtropical regions. The activity of essential oils from *Allium fistulosum* L., *A. sativum* L. and *A. cepa* L. were investigated against human pathogen (Pyun et al. 2006). There is no report available on plant pathogens.

Lamiaceae

The plants comes under this family has world wide distribution. It is mainly used for medicinal, culinary, ornamental and various commercial utilizations. Earlier studies clearly indicated that essential oils were obtained from these families have a strong antifungal activity (Baratta et al. 1998). The essential oil of *Ocimum basilicum* L. has the potential for the treatment of fungal infections (Rios and Recio 2005). The main antifungal component of this oil was linalool.

Previous reports on the antimicrobial activity of the essential oils of some *Thymus* species showed strong antifungal activity against viruses, bacteria, food-derived microbial strains and fungi. The major antimicrobial compound of this oil was phenolic monoterpenes. These compounds were highly effective against *R. solani* and *Fusarium oxysporum*, and less effective against *A. flavus*. The essential oil extracted from *Rosmarinus officinalis* L. by hydro-distillation process (Angioni et al. 2004) and the chemical composition of this oil was identified as α-pinene, borneol, camphene, camphor, verbenone and bornyl acetate which has strong antimicrobial activity. Portillo et al. (2005) reported that essential oil obtained from the aerial parts of *Salvia mirzayanii* Rech. F. and Esfand. was tested against *F. solani*, and *Candida albicans*. The inhibitory effect was increased with increasing the concentration of essential oil. The major antifungal compounds present in this oils were linalool, linalyl acetate, α-terpinyl acetate, 1, 8-cineole, α-cadinol and δ-cadinene.

Verbenaceae

Tropical flowering plants mainly comes under this mainly. It constitutes trees, shrubs and herbs. Simic et al. (2004) reported that the essential oils were obtained from the aerial parts of *Lantana achyranthifolia* Desf. and *Lippia graveolens* Kunth. showed highest antifungal activity against *F. sporotrichum*, *A. niger*, *Trichophyton mentagrophytes* and *F. moniliforme*.

Lauraceae

It includes mostly flowering plants, which occur mainly in warm temperate and tropical regions, especially Southeast Asia and South America. Earlier studies revealed that the essential oil is extracted from the bark of *C. zeylanicum* has fungitoxic properties against *A. niger*, *A. fumigatus*, *A. nidulans* and *A. flavus* (Bullerman et al. 1977). The essential oils from cinnamon have strong inhibitory effect against the growth of *A. flavus* cause aflatoxin in ground nut and *Botrytis cinerea* (Montes-Belmont and Carvajal 1998; Lee et al. 2007; Sessou et al. 2012).

Apiaceae

This family is commonly called as carrot or parsley family. The oil extracted from carrot inhibited the mycelial growth of *A. niger*, *F. moniliforme* and *Curvularia lunata*. A complete mycelia growth inhibition was obtained at a 6 μl dose of the oil. The chemical composition of ajwain essential oil showed the presence of 26 components. Of these, Thymol was found to be a major component along with p-cymene, γ-terpinene, β-pinene and terpinen-4-ol (Sunita and Mahendra 2008).

Zingiberaceae

Plants come under this family are distributed in Tropical Africa, Asia and America. It is a flowering, aromatic perennial herbs with creeping horizontal or tuberous rhizomes. Pandey et al. (2010) reported that the essential oils from ginger (*Zingiber officinale* Roscoe) exhibited strong fungicidal activity against mycotoxin producers *Aspergillus flavus* and *A. parasiticus* and also effective against *F. oxysporum*. The major antifungal components present in zinger is zingiberene (Farag et al. 1989).

Theaceae

Theaceae is a family of flowering plants, composed of shrubs and trees. Pitipong et al. (2009) reported that essential oils from *Melaleuca alternifolia* (Maiden and Betche) Cheel showed strong antimicrobial activity against *Alternaria brassicicola*, *Aspergillus flavus*, *Bipolaris oryzae*, *Fusarium moniliforme*, *F. proliferatum*, *Pyricularia grisea* and *R. solani* the cause of rice seed-borne and sheath blight pathogen.

Important essential oils and their antifungal compounds

Common name	Scientific name	Antimicrobial compounds	References
Ajowan	*Trachyspermum ammi*	Thymol	Park et al. (2007)
American wild mint	*Mentha cervina*	Pulegone, isomenthone	Goncalves et al. (2007)
Basil	*Ocimum sativum*	Thymol	Soliman and Badea (2002)
	Ocimum basilicum	Thymol	
Brazilian rosewood	*Aniba rosaedora*	Linalool	Manjamalai and Berlin Grace (2012)
Bamboo piper	*Piper angustifolium*	Camphene	Tirillini et al. (1996)
Cascalote	*Caesalpinia cacalaco*	Gallic and tannic acids	Veloz-Garcia et al. (2010)
Celery	*Apium graveolens*	Angelicin, bergaptan, columbianetin, xanthotoxin	Afek and Carmeli (1995)
Chinese cinnamon	*Cinnamomum zeylanicum*	Trans-cinnamaldehyde	Pyun and Shin (2006)
Cinnamon	*Cinnamomum zeylanicum*	Cinnamaldehyde eugenol	Paranagama (1991) and Velluti et al. (2003)
Clove	*Syzygium aromatium*	Eugenol	Paranagama (1991)

Common name	Scientific name	Antimicrobial compounds	References
Clove	*Syzygium aromaticum*	Eugenol	Paranagama (1991) and Ranasinghe et al. (2002)
Calamint	*Calamintha nepeta* sub sp. *nepeta*	Sardinia: pulegone Portugal: isomenthone, 1,8-cineole	Marongiu et al. (2010)
Cumin	*Cuminum cyminum L.*	γ-Terpinene, cucumin aldehyde	Marjanlo et al. (2009)
Datura	*Datura metel*	Enzyme, peroxidase, β-1,3-glucanase and chitinase	Devaiah et al. (2009)
Eucaluptus	*Eucalyptus citriodora*	Essential oil	Pattnaik et al. (1996)
Fringed Rue	*Ruta chalepensis*	2-undecanone, 2-decanone and 2-do decanone	Mejri et al. (2010)
Garden thyme	*Thymus vulgaris*	Thymol, ρ-Cymene	Sun et al. (2007)
Geranium	*Pelargonium graveolens*	Essential oil	Pattnaik et al. (1996)
Golden shower tree	*Cassia* sp	Cassia oil	Feng et al. (2008)
Himalayan May Apple	*Podophllum hexandrum*	4′-O-demethyldehdropodophllotoxin and picropodophllone	Rahman et al. (1995)
Indian Aloe	*Aloe vera*	Crude extracts	Jasso de Rodriguez et al. (2005)
Kattukurumulagu	*Piper barberi*	1,8 ceneole, α-pinene, eugenol isomer, camphor	Raju and Maridass (2011)
Long leaf arnica	*Arnica longifolia*	Camphor, 1,8-cineole	Nurhayat et al. (2007)
Long pepper	*Piper longum*	Eugenol, piperine, piperlongumine and piperettine	Lee et al. (2009)
Lemon *Eucalyptus*	*Eucalyptus citriodora*	Citronellal, isopulegol	Chuang et al. (2007)
Lemon grass	*Cymbopogon citratus*	Citral, geranial neral	Pranagama et al. (2003) and Velluti et al. (2003)
Mustard	*Brassica* spp.	Allyl iso thiocyanate	Dhingra et al. (2009)
Neem	*Azadiracta indica*	Oleic acid, hexadecanoic acid	Somda et al. (2007)
Parupukkirai	*Chenopodium ambrosioides*	M-cymene, myrtenol	Prasad et al. (2009)
Periwinkle	*Catharanthus roseus*	5-hdrox flavones	Roy and Chatterjee (2010)
Peppermint	*Menthe piperita*	Essential oil	Pattnaik et al. (1996)
Pine	*Pinus* spp.	γ-terpineol	Knobloch et al. (1989)
Quinoa	*Chenopodium quinoa*	Titerpenoid saponins	Stuardo and San Martin (2008)

(continued)

(continued)

Common name	Scientific name	Antimicrobial compounds	References
Sage	*Saliva officinalis*	Essential oil	Pinto et al.(2007)
Sodom apple	*Calotropis procera*	Crude extract	Abdel-Monaim et al. (2011)
Rosebay	*Neerium oleander*		
Thulasi	*Osmium basilicum*		
Spearmint	*Mentha spicata*	1-carvone	Knobloch et al. (1989)
Tea tree oil	*Melaleuca aternifolin*	Terpinen-4-ol	Knobloch et al. (1989)
Thymus	*Thymus mastichina* sub sp. *matichina*	Linalool	Faleiro et al. (2003)
Thymus	*Thymus vulgaris*	Thymol, carvacrol, linalool	Soliman and Badea (2002)
Wild basil	*Ocimum gratissimum*	Thymol	Adekunle and Uma (2005)
Wild carrot	*Daucus carota* subsp. *halophilus*	Flowering umbels: sabinene, α-pinene, limonene; ripe umbels elemicin, sabinene	Tavares et al. (2008)
Yellow lavender	*Lavandula viridis*	1,8-cineole, camphor, α-pinene, linalool	Zuzarte et al. (2011)
Zinger	*Zingiber officinale*	Gingerenone A	Endoi et al. (1990)

Meliaceae

Mostly trees and shrubs of flowering plant come under this family. It is used for vegetable oil, soap-making and insecticides. Several reports have been made on the antifungal properties of *Azadirachta indica* (L.)Adelb. (neem) oil (Sunita et al. 2008). The essential oils from *A. indica* showed complete mycelial growth inhibition of *Alternaria alternata*, *A. niger* and *F. oxysporum* at 2–10 % neem oil (Sukatta et al. 2008).

Antifungal Activity of Essential Oils Against *Pythium* Species

Pandey and Dube (1994) extracted essential oil from the leaves of 30 angiospermic plants for their antifungal activity against *P. aphanidermatum* and *P. debaryanum*. Of these, essential oil from *Hyptis suaveolens*, *Murraa koenigii* and *Ocimum sanctum* showed strong antifungal activity against *P. aphanidermatum* and *P. debaryanum* causing damping-off disease of tomato. In *in vivo* condition, tomato seeds were soaked with essential oils of *H. suaveolens*, *M. koenigii* and *O. sanctum* exhibited 83.67 % and 50 %, respectively, in soil infected with *P. aphanidermatum* and 86 %, 71 %, and 43 %, respectively in soil infected with *P. debaryanum*.

Kishore and Dubey (2002) reported that seed treatment with Lippia oil exhibited 88.9 % and 71.3 % reduction of tomato damping-off disease when challenge inoculated with *P. aphanidermatum* and *P. debaryanum*, respectively. Among the plant oils tested, *Eucalyptus* oil recorded the maximum mycelia growth inhibition of *P. aphanidermatum* causing chilli damping-off. This was followed by Citrodara oil and Plamarosa oil in the decreasing order of merit (Muthukumar and Sangeetha 2008).

Nafiseh et al. (2012) reported that the antimicrobial activity of essential oil from *Eucalyptus* (*Eucalyptus camaldulensis* Dehnh.) was evaluated against three post-harvest pathogenic fungi (*Penicillium digitatum*, *A. flavus* and *Colletotrichum gloeosporioides*) and three soil borne pathogenic fungi (*P. ultimum*, *R. solani* and *Bipolaris sorokiniana*) under *in vitro*. The result showed that complete mycelial growth inhibition of *P. ultimum*, *R. solani* was obtained from *Eucalyptus* oil in all concentrations tested and it is not effective against post-harvest pathogens. The essential oil extracted from *T. vulgaris* was the most effective and caused complete mycelial growth inhibition of *P. aphanidermatum* and *R. solani*. The main antifungal compound present in *T. vulgaris* is thymol (Amini et al. 2012).

Recently, Fonseca et al. (2015) reported that the antifungal activity of *Origanum vulgare*, *O. majorana*, *Mentha piperita* and *R. officinalis* against *P. insidiosum*. The results revealed that the essential oil obtained from *O. vulgare* showed highest efficacy on *P. insidiosum*.

Conclusion

The current status of research suggests that there are really alternatives to replace the synthetic fungicides for management of this notorious soil as well as seed borne fungi: *Pythium aphanidermatum* which causes loss of multimillion dollars. However the farmers uses the common synthetic fungicides which leads into ill effects as well as many of the commonly used synthetic fungicides are unable to control *P. aphanidermatum* as it has got resistant against these synthetic fungicides. Hence, there is need to replace the chemical fungicides by bio-fungicides, prepared from plant extracts and essential oils and antagonistic microorganisms. Bio-fungicides will also be economical to the farmers and besides this, the use of biofungicides will not leave any ill effect in the soil, water as well as in the environment. It is possible that by combining these approaches, (use of plant extracts, essential oils and antagonistic micro-organisms) an economically viable alternative for crop production system can be developed. For the effective production of crops, formulation protocols as well as its using methods should be provided to the farmers. Formulation must have adequate shelf life, stability, and concentration. Before any formulated product is marketed, it must first be thoroughly tested by growers, whose comments, analysis, and suggestions for improvement, must consider. In future, the integration of all the three components will be tested *in vitro* and *in vivo* conditions and these can serve as natural fungicides.

References

Abdel-Monaim MF, Abo-Elyousr KAM, Morsy KM (2011) Effectiveness of plant extracts on suppression of damping- off and wilt diseases of lupine (*Lupinus termis* Forsik). Crop Prot 30(2):185–191

Abdollahi M, Rahnama K, Marabadi M, Ommati F, Zaker M (2012) The *in vitro* efficacy of *Trichoderma* isolates against *Pythium aphanidermatum*, the causal agent of sugar beet root rot. J Res Agric Sci 8:79–87

Adekunle AA, Uma NU (2005) Effect of benlate solution, crude leaf extracts of *Azadirachta indica* and *Ocimum gratissimum* on growth of fungi and preservation of melon seeds. Plant Pathol J 4:29–34

Afek U, Carmeli S (1995) Increasing celery resistance to pathogens during storage and reducing high-risk psoralen concentration by treatment with GA3. J Am Soc Hortic Sci 120:562–565

Ahemad M, Khan MS (2010a) Influence of selective herbicides on plant growth promoting traits of phosphate solubilizing *Enterobacter asburiae* strain PS2. Res J Microbiol 5:849–857

Ahemad M, Khan MS (2010b) Plant growth promoting activities of phosphate-solubilizing *Enterobacter asburiae* as influenced by fungicides. Asian J Biol Sci 4:88–95

Ahemad M, Khan MS (2011) *Pseudomonas aeruginosa* strain PS1 enhances growth parameters of green gram (*Vigna radiata* (L.) Wilczek) in insecticide-stressed soils. J Pest Sci 84:123–131

Ahemad M, Khan MS (2012a) Effect of fungicides on plant growth promoting activities of phosphate solubilizing *Pseudomonas putida* isolated from mustard (*Brassica campestris*) rhizosphere. Chemosphere 86:945–950

Ahemad M, Khan MS (2012b) Alleviation of fungicide-induced phytotoxicity in green gram (*Vigna radiata* (L.) Wilczek) using fungicide-tolerant and plant growth promoting *Pseudomonas* strain. Saudi J Biol Sci 19:451–459

Ahmad JS, Baker R (1987) Competitive saprophytic ability and cellulolytic activity ofrhizosphere-competent mutants of *Trichoderma harzianum*. Phytopathology 77:358–362

Ahmad F, Ahmad I, Khan MS (2008) Screening of free-living rhizospheric bacteria for their multiple plant growth promoting activities. Microbiol Res 163:173–181

Alvarez-Castellanos PP, Bishop CD, Pascual M (2001) Antifungal activity of the essential oil of flower heads of garland chrysanthemum (*Chrysanthemum coronarium*) against agricultural pathogens. J Phytochem 57(1):99–102

Alvarez-Castellanos PP, Bishop CD, Pascual M (2009) Antifungal activity of the essential oil of flower heads of garland chrysanthemum (*Chrysanthemum coronarium*) against agricultural pathogens. J Phytochem 57(1):99–102

Amaresan N, Jayakumar V, Thajuddin N (2014) Isolation and characterization of endophytic bacteria associated with chilli (*Capsicum annuum*) grown in coastal agricultural ecosystem. Indian J Biotechnol 13:247–255

Ambikapathy V, Gomathi S, Panneerselvam A (2011) Effect of antifungal activity of some medicinal plants against *Pythium debaryanum* (Hesse). Asian J Plant Sci Res 1:131–134

Amini M, Safaie N, Salmani MJ, Shams-Bakhsh M (2012) Antifungal activity of three medicinal plant essential oils against some phytopathogenic fungi. Trakia J Sci 10(1):1–8

Angioni A, Barra A, Cereti E, Barile D, Coisson JD, Arlorio M (2004) Chemical composition, plant genetic differences, antimicrobial and antifungal activity investigation of the essential of *Rosmarinus officinalis* L. J Agric Food Chem 52:3530–3535

Anitha K, Tripathi NN (2001) Laboratory screening of fungal and bacterial antagonists against *Rhizoctonia solani* khun. and *Pythium aphanidermatum* (Edson) Fitz. inciting seedling diseases of okra. Indian J Plant Prot 29:146–148

Bahraminejad S, Abbasi S, Fazlali M (2011) *In vitro* antifungal activity of 63 Iranian plant species against three different plant pathogenic fungi. Afr J Biotechnol 10:16193–16201

Bailey BA, Lumsden RD (1998) Direct effects of *Trichoderma* and *Gliocladium* on Plant growth and resistance to pathogens. In: Kubicek CP, Harman GE, Ondik KL (eds) *Trichoderma* and *Gliocladium*: enzymes, biological control and commercial applications. Taylor and Francis, London, pp 185–204

Bajpai VK, Kang SC (2010) Antifungal activity of leaf essential oil and extract of *Metaseqoia glystostroboides* Miki ex Hu. J Am Oil Chem Soc 87:327–336

Bala K, Robideau G, de Cock AWAM, Abad ZG, Lodhi AM, Shahzad S, Ghaffar A, Coffey MD, Lévesque CA (2010) Phytopythium gen. nov. Persoonia 24:136–137

Bandara BMR, Kumar NS, Samaranayake KMS (1989) An antifungal constituent from the stem-bark of *Butea monosperma*. J Ethnopharmacol 25:73–75

Baratta MT, Dorman HJD, Deans SG, Figueiredo AC, Barroso JG, Ruberto G (1998) Antibacterial and anti-oxidant properties of some commercial essential oils. Flavour Fragr J 13:235–244

Barkley TM, Brouillet L, Strother JL (2006) Magnoliophyta: Asteridae: Asteraceae, part 1 vol. 19, part 2 vol. 20, and part 3 vol. 21. In: Flora of North America Editorial Committee (eds) 1993+. Flora of North America North of Mexico, New York.

Bar-Ness E, Chen Y, Hadar Y, Marschner H, Römheld V (1991) Siderophores of *Pseudomonas putida* as an iron source for dicot and monocot plants. Plant and Soil 130:231–241

Belimov AA, Safronova VI, Sergeyeva TA, Egorova TN, Matveyeva VA, Tsyganov VE (2001) Characterization of plant growth promoting rhizobacteria isolated from polluted soils and containing 1-aminocyclopropane-1-carboxylate deaminase. Can J Microbiol 47:242–252

Bellgard S, Ham C (2004) Common diseases of peanuts in the top end of the NT *Arachis hypogaea* L. Agdex No: 141/633. Agnote Northern Territory of Australia (162), 7 p

Bharathi R, Vivekananthan R, Harish S, Ramanathan A, Samiyappan R (2004) Rhizobacteria-based bio-formulations for the management of fruit rot infection in chillies. Crop Prot 23:835–843

Bhattacharyya PN, Jha DK (2012) Plant growth-promoting rhizobacteria (PGPR): emergence in agriculture. World J Microbiol Biotechnol 28:1327–1350

Bhuvaneshwari S (2008) Biological control of damping-off disease of tomato caused by *Pythium aphanidermatum* (Edson) Fitz. MSc (Ag.) thesis, Annamalai University, Tamil Nadu

Bullerman LB, Lieu FY, Seire AS (1977) Inhibition of growth and alfatoxin production by cinnamon and clove oils, cinnamic aldehyde and eugenol. J Food Sci 42:1107–1116

Cardenas NC, Zavala MA, Aguirre JR, Perez C, Perez S (2005) Chemical composition and antifungal activity of essential oil of *Chrysactinia mexicana* Gray. J Agric Food Chem 53:4347–4349

Chet I, Inbar J, Hadar I (1997) Fungal antagonists and mycoparasites. In: Wicklow DT, Söderström B (eds) The mycota IV: environmental and microbial relationships. Springer-Verlag, Berlin, pp 165–184

Choi GJ, Jang KS, Kim JS, Lee SW, Cho JY, Cho KY, Kim JC (2004) *In vivo* fungal activity of 57 plant extracts against six plant pathogenic fungi. Plant Pathol J 20:184–191

Chuang PH, Lee CW, Chou JY, Murugan M, Shieh BJ, Chen HM (2007) Antifungal activity of crude extracts and essential oil of *Moringa oleifera* Lam. Bioresour Technol 98:232–236

Cimanga K, Kambu K, Tona L, Apers S, De Bruyne T, Hermans N (2002) Correlation between chemical composition and antibacterial activity of essential oils of some aromatic medicinal plants growing in the Democratic Republic of Congo. J Ethnopharmacol 79:213–220

Collins DP, Jacobsen BJ (2002) Optimizing a *Bacillus subtilis* isolate for biological control of sugar beet *Cercospora* leaf spot. Biol Control 26:153–161

Cowan MM (1999) Plant products as antimicrobial agents. Clin Microbiol Rev 12:564–581

Crowley DA (2006) Microbial siderophores in the plant rhizosphere. In: Barton LL, Abadia J (eds) Iron nutrition in plants and rhizospheric micro organisms. Springer, Netherlands, pp 169–189

Dana ED, Delomas JG, Sanchez J (2010) Effects of the aqueous extracts of *Zygophyllum fabago* on the growth of *Fusarium oxyosporum* f. sp. *melonis* and *Pythium aphanidermatum*. Weed Biol Manag 10:170–175

Dar WA, Bhat JA, Rashid R, Rehman S, Bhat ZA (2012) Bioefficacy of *Pseudomonas fluorescens* in management of damping-off disease in papaya (*Carica papaya* L.). J Agric Technol 8:693–697

Das K, Tiwari RKS, Shrivastava DK (2010) Techniques for evaluation of medicinal plant products as antimicrobial agent: current methods and future trends. J Med Plants Res 4:104–111

Davies PJ (2004) Plant hormones: biosynthesis, signal transduction, action. Kluwer Academic, Dordrecht

Deadman M, Perret J, Al-Jabri S, Al-Maqbali Y, Al-Sa'di A, Al-Kiyoomi K, Al-Hasani H (2007) Epidemiology of damping-off disease in greenhouse cucumber crops in the Sultanate of Oman. Acta Hortic 731:319–326

Dean DR, Jacobson MR (1992) Biochemical genetics of nitrogenase. In: Stacey G, Burris RH, Evans HJ (eds) Biological nitrogen fixation. Chapman and Hall, New York, pp 763–834

Devaiah SP, Mahadevappa GH, Shetty HS (2009) Induction of systemic resistance in pearl millet (*Pennisetum glaucum*) against downy mildew (*Sclerospora graminicola*) by *Datura metel* extract. Crop Prot 28:783–791

Dhingra OD, Jham GN, Rodrigues FA, Silva GJ, Costa MCN (2009) Retardation of fungal deterioration of stored soybeans by fumigation with mustard essential oil. Aust Plant Pathol 38:540–545

Djibaoui R, Ahmed B (2005) Effect of iron and growth inhibitors on siderophores production by *Pseudomonas fluorescens*. Afr J Biotechnol 4:697–702

Djonovic S, Pozo MJ, Kenerley CM (2006) Tvbgn3, a beta-1, 6-glucanase from the biocontrol fungus *Trichoderma virens*, is involved in mycoparasitism and control of *Pythium ultimum*. Appl Environ Microbiol 72:7661–7670

Druzhinina IS, Seidl-Seiboth V, Herrera-Estrella A, Horwitz BA, Kenerley CM, Monte E, Mukherjee PK, Zeilinger S, Grigoriev IV, Kubicek CP (2011) *Trichoderma* the genomics of opportunistic success. Nat Rev Microbiol 9:749–759

Dubey SC, Jha AK, Jha DK (1996) Integrated management of collar rot of okra through chemical and fungal antagonists. In: Diseases of economically importance plant of eastern India and their management. Nationals Symposium of Indian Phytopathological Society. 20 p.

Dumitars L, Fratilescu-Sesan (1979) Aspects of the antagonism of *Trichoderma viride* pers. to *Pythium debaryanum* Hesse. Biol Vegetala 31:63–67

Elad Y, David DR, Levi T, Kapat A, Kirshner B (1999) *Trichoderma harzianum* T-39 mechanisms of biocontrol of foliar pathogens. In: Lyr H, Russell PE, Dehne HW, Sisler HD (eds) Modern fungicides and antifungal compounds II. Intercept, Andover, Hants, UK, pp 459–467

Elliott SM (2003) Rhizome rot disease of ginger. Ministry of Agriculture Research and Development Division. St. Catherine, Jamaica

Endoi K, Kanno E, Oshima Y (1990) Structures of antifungal diarylheptenones, gingerenones A, B, C and isogingerenone B, isolated from the rhizomes of *Zingiber officinale*. Phytochemistry 29:797–799

Engelmeier D, Hadacek F (2006) Antifungal natural products: Assays and Applications. In: Ashok K., (ed). Advances in Phytomedicine, Naturally occurring bioactive compounds. Vol. 3, Elsevier, Amsterdam, The Netherlands, pp 423–467

Faleiro ML, Miguel MG, Ladeiro F, Venancio F, Ravares R, Brito JC, Figueiredo AC, Barroso JG, Pedro LG (2003) Antimicrobial activity of essential oils isolated from Portuguese endemic species of *Thymus*. Lett Appl Microbiol 36:35–40

Farag RS, Daw ZY, Abo- Raya SH (1989) Influence of some spice essential oils on *Aspergillus paraciticus* growth and production of alfatoxins in synthetic medium. J Food Sci 54:74–76

Feng W, Zheng X, Chen J, Yang Y (2008) Combination of cassia oil with magnesium sulphate for control of postharvest storage rots of cherry tomatoes. Crop Prot 27:112–117

Fonesca AO, Pereiro DJ, Jacob RG, Maia Filho FS, Oliverra DH, Maroneze BP, Valents JS, Osorio LG, Botton SA, Meireles MC (2015) *In vitro* susceptibility of Brazilian *Pythium insidiosum* isolates to essential oils of some Lamiaceae family species. Mycopathologia 179:253–258

Glass ADM (1989) Plant nutrition: an introduction to current concepts. Jones and Barlett publishers, Boston, p 234

Glick BR (1995) The enhancement of plant growth by free-living bacteria. Can J Microbiol 41:109–115

Glick BR (2012) Plant growth-promoting bacteria: mechanisms and applications. Hindawi Publishing Corporation, Scientifica, University Avenue South, Waterloo, Canada

Glick BR, Cheng Z, Czarny J, Duan J (2007) Promotion of plant growth by ACC deaminase-producing soil bacteria. Eur J Plant Pathol 119:329–339
Gomathi S, Ambikapathy V, Panneerselvam A (2011) Antimicrobial activity of some medical plants against *Pythium debaryanum* (Hesse). J Microbiol Biotechnol Res 1:8–13
Goncalves M, Vicente A, Cavaleiro C, Salgueiro L (2007) Composition and antifungal activity of the essential oil of *Mentha cervina* from Portugal. Nat Prod Res 21:867–871
Grigoriev IV, Nordberg H, Shabalov I, Aerts A, Cantor M (2012) The genome portal of the Department of Energy Joint Genome Institute. Nucleic Acids Res 40:26–32
Guirado ML, Sáez Y, Serrano E, Gómez J (2009) Aetiology watermelon "sudden death" of greenhouses in the southeast of Spain. Bol Sanid Veg Plagas 35(4):617–628
Haas D, Keel C (2003) Regulation of antibiotic production in root-colonizing *Pseudomonas* spp. and relevance for biological control of plant disease. Annu Rev Phytopathol l41:117–153
Hadacek F (2002) Secondary metabolites as plant traits: current assessment and future perspectives. Crit Rev Plant Sci 21:273–322
Haouala R, Hawala S, ElA-yeb A, Khanfir R, Boughanmi N (2008) Aqueous and organic extracts of *Trigonella foenumgraecum* L. inhibit the mycelia growth of fungi. J Environ Sci 20:1453–1457
Haran S, Schickler H, Oppenheim A, Chet I (1996) Differential expression of *Trichoderma harzianum* chitinases during mycoparasitism. Phytopathology 86:980–985
Harman GE, Howell CR, Viterbo A, Chet I, Lorito M (2004) *Trichoderma* species opportunistic, avirulent plant symbionts. Nat Rev Microbiol 2:43–56
Harris AR, Schisler DA, Correl RL, Ryder MH (1994) Soil bacteria selected for suppression of *Rhizoctonia solani* and growth promotion in bedding plants. Soil Biol Biochem 26:1249–1255
Hazarika DK, Sarmah R, Paramanick T, Hazarika K (2000) Biological management of tomato damping-off caused by *Pythium aphanidermatum*. Indian J Plant Pathol 18:36–39
Hendrix FF, Campbell WA (1973) *Pythiums* as plant pathogens. Annu Rev Phytopathol 11:77–98
Hoffland E, Halilinen J, Van Pelt JA (1996) Comparison of systemic resistance induced by avirulant and non-pathogenic *Pseudomonas* species. Phytopathology 86:757–762
Hong CX, Moorman GW (2005) Plant pathogens in irrigation water: challenges and opportunities. Crit Rev Plant Sci 24(3):189–208
Howell CR (1998) The role of antibiosis in biocontrol. In: Harman GE, Kubicek CP (eds) *Trichoderma & Gliocladium*, vol 2. Taylor & Francis, Padstow, pp 173–184
Howell CR (2003) Mechanisms employed by *Trichoderma* species in the biological control of plant diseases: the history and evolution of current concepts. Plant Dis 87:4–10
Howell CR, Stipanovic RD (1983) Suppression of *Pythium ultimum* induced damping-off of cotton seedlings by *Pseudomonas fluorescens* and its antibiotic pyoluteorin. Phytopathology 70:712–715
Hudge BV (2014) Management of damping off disease of soybean caused by *Pythium ultimum* Trow. Int J Curr Microbiol Appl Sci 4:799–808
Intana W, Chamswarng C (2007) Control of Chinese-kale damping-off caused by *Pythium aphanidermatum* by antifungal metabolites of *Trichoderma virens*. Songklanakarin J Sci Technol 29:919–927
Jackisch-Matsuura AB, Menezes M (2000) Effect of *Trichoderma* spp. in the control of *Pythium aphanidermatum* in tobacco (*Nicotiana tabacum L.*). Bio-control News Inf 21:38
Jadhav VT, Ambadkar CV (2007) Effect of *Trichoderma* spp. on seedling emergence and seedling mortality of tomato, chilli and brinjal. J Pl Dis Sci 2:190–192
Jasso de Rodríguez D, Hernández-Castillo D, Rodríguez-García R, Angulo-Sánchez JL (2005) Antifungal activity *in vitro* of *Aloe vera* pulp and liquid fraction against plant pathogenic fungi. Ind Crop Prod 21:81–87
Jayaraj J, Radhakrishnan NV, Velazhahan R (2006) Development of formulations of *Trichoderma harzianum* strain M1 for control of damping of tomato caused by *Pythium aphanidermatum*. Arch Phytopathol Plant Protect 39:1–8
Jayaseelan EC, Tharmila S, Niranjan K (2012) Antagonistic activity of *Trichoderma* sp and *Bacillus* spp. against *Pythium aphanidermatum* isolated from tomato damping-off. Arch Appl Sci Technol 4:1623–1627

Joo GJ, Kim YM, Kim JT, Rhee IK, Kim JH, Lee IJ (2005) Gibberellins-producing rhizobacteria increase endogenous gibberellins content and promote growth of red peppers. J Microbiol 43:510–515

Juan Z, Xue QH, Tang M (2009) Screening of antagonistic actinomycetes against 'Jiashi' *Cucumis melo* L. damping-off. J Northwest A & F Univ Nat Sci Ed 37(5):144–148

Kandhari J (2007) Management of sheath blight of rice through fungicides and botanicals. Indian Phytopathol 60:214–217

Kang SM, Joo GJ, Hamayun M (2009) Gibberellin production and phosphate solubilization by newly isolated strain of *Acinetobacter calcoaceticus* and its effect on plant growth. Biotechnol Lett 31:277–281

Kang BG, Kim WT, Yun HS, Chang SC (2010) Use of plant growth-promoting rhizobacteria to control stress responses of plant roots. Plant Biotechnol Report 4:179–183

Karamenderes C, Khan S, Tekwani BL, Jacob MR, Khan IA (2006) Antiprotozoal and antimicrobial activities of *Centaurea* species. Pharm Biol 44:534–539

Kavitha PG, Thomas G (2007) Evaluation of Zingiberaceae for resistance to ginger soft rot caused by *Pythium aphanidermatum* (Edson) Fitzp. J Plant Genet Resour Newslett 152:54–57

Kavitha K, Mathiyazhahan S, Sendhilvel V, Nakkeeran S, Chandrasekar G, Fernando WGD (2005) Broad spectrum action of phenazine against active and dormant structures of fungal pathogens and root knot nematode. Arch Phytopathol Plant Prot 38:69–76

Kawarazaki H, Nara Y, Kijima T, Goto M (2008) *Pythium* rot of fig marigold (*Lampranthus spectabile*) caused by *Pythium aphanidermatum*. J Gen Plant Pathol 74(1):94–95

Kazmi SAR, Shahzad S, Niaz I (1995) Effect of neem oil on *in vitro* growth of root infecting fungi. Pak J Bot 27(1):217–220

Khabbaz SE, Abbasi PA (2014) Isolation, characterization, and formulation of antagonistic bacteria for the management of seedlings damping-off and root rot disease of cucumber. Can J Microbiol 60:25–33

Khalid A, Akhtar MJ, Mahmood MH, Arshad M (2006) Effect of substrate dependent microbial ethylene production on plant growth. Microbiology 75:231–236

Khan MS, Zaidi A, Wani PA (2006) Role of phosphate solubilising microorganisms in sustainable agriculture – a review. Agron Sustain Dev 27:29–43

Khare A, Singh BK, Upadhyay RS (2010) Biological control of *Pythium aphanidermatum* causing damping-off of mustard by mutants of *Trichoderma viride* 1433. J Agric Technol 6:231–243

Kim J, Rees DC (1994) Nitrogenase and biological nitrogen fixation. Biochemistry 33:389–397

Kim YC, Jung H, Kim KY, Park SK (2008) An effective biocontrol bioformulation against *Phytophthora* blight of pepper using growth mixtures of combined chitinolytic bacteria under different field conditions. Eur J Plant Pathol 120:373–382

Kishore N, Dubey NK (2002) Fungitoxic potency of some essential oils in management of damping-off diseases in soil infested with *Pythium aphanidermatum* and *Pythium debaryanum*. Indian J For 25:463–468

Kloepper JW, Schroth MN (1978) Plant growth-promoting rhizobacteria on radishes: proceedings of the 4th international conference on Plant Pathogenic Bacteria, Station de Pathologie Vegetable et Phytobacteriologie, INRA, Angers. pp 879–882

Kloepper JW, Leong J, Teintze M, Schroth MN (1980) *Pseudomonas* siderophores: a mechanism explaining disease suppressive soils. Curr Microbiol 4:317–320

Knobloch K, Pauli A, Iberl B, Weis N, Weigand H (1989) Antibacterial activity and antifungal properties of essential oil components. J Essent Oil Res 1:119–128

Koomen, Jeffries P (1993) Effects of antagonistic microorganisms on the postharvest development of *Colletotrichum gloeosporioides* on mango. Plant Pathol 42:230–237

Kordali S, Cakir A, Mavi A, Kilic H, Yildirim A (2005) Screening of chemical composition and antifungal and antioxidant activities of the essential oils from three Turkish artemisia species. J Agric Food Chem 53:1408–1416

Krishnamoorthy AS (1987) Biological control of damping-off disease of tomato caused by *Pythium indicum*, MSc (Ag.) thesis, Tamil Nadu Agricultural University, Coimbatore

Kubicek CP, Herrera EA, Seidl SV, Martinez DA (2011) Comparative genome sequence analysis underscores mycoparasitism as the ancestral life style of *Trichoderma*. Genome Biol 12:40–63
Kubota M (2010) Diseases of cabbage plug seedlings in Japan and control of the diseases. Bull Natl Inst Veg Tea Sci 9:57–112
Kucharek T, Mitchell D (2000) Diseases of agronomic and vegetable crops caused by *Pythium*. Plant pathology fact sheet, University of Florida; p 53, http://128.227.207.24/takextpub/FactSheet/pp53.pdf
Kumar R, Hooda I (2007) Evaluation of antagonistic properties of *Trichoderma* species against *Pythium aphanidermatum* causing damping-off of tomato. J Mycol Plant Patho 37:240–243
Kurucheve V, Padmavathi R (1997) Effect of seed treatment with plant products on seed germination, growth and vigour of chilli seedlings (K-1). Indian Pathol 50:529–530
Ladha JK, de Bruijn FJ, Malik KA (1997) Introduction: assessing opportunities for nitrogen fixation in rice-a frontier project. Plant Soil 124:1–10
Lavanya N, Saravanakumar D, Rajendran L, Ramiah M, Raguchander T, Samiyappan R (2009) Management of sunflower necrosis virus through anti-viral substances. Arch Phytopathol Plant Prot 42:265–276
Lee SO, Choi GJ, Jang KS, Lim HK, Cho KY, Kim J-C (2007) Antifungal activity of five plant essential oils as fumigant against postharvest and soil borne plant pathogenic fungi. Plant Pathol J 23:97–102
Lee SW, Han SW, Sririyanum M, Park C-J, Seo Y-S, Ronald PC (2009) A type I–secreted, sulfated peptide triggers XA21-mediated innate immunity. Science 326:850–853
Lifshitz R, Windham MT, Baker R (1986) Mechanism of biological control of pre-emergence damping-off of pea by seed treatment with *Trichoderma* spp. Phytopathology 76:720–725
Loganathan M, Sible GV, Prabakar K, Samiyappan R (2004) Antagonism of yeast *Saccharomyces cerevisiae* against *Pythium aphanidermatum* (Edson) Fitz. in tobacco. Madras Agric J 91:530–532
Lorito M (2005) Molecular biology of the interactions between *Trichoderma*, phytopathogenic fungi and plants: opportunities for developing novel disease control methods. Second Global Conference. "Plant Health-Global Wealth". Abstracts. pp 162–163
Lorito M, Scala F (1999) Microbial genes expressed in transgenic plants to improve disease resistance. J Plant Pathol 81:73–88
Lorito M, Mach RL, Sposato P, Strauss J, Peterbauer CK, Kubicek CP (1996) Mycoparasitic interaction relieves binding of Cre1 carbon catabolite repressor protein topromoter sequence of *ech-42* (endochitinase-encoding) gene of *Trichoderma harzianum*. Proc Natl Acad Sci USA 93:14868–14872
Lugtenberg B, Kamilova F (2009) Plant growth promoting rhizobacteria. Annu Rev Microbiol 63:541–556
Maksimov IV, Abizgil'dina RR, Pusenkova LI (2011) Plant growth promoting rhizobacteria as alternative to chemical crop protectors from pathogens. Appl Biochem Microbiol 47:333–345
Manjamalai A, Berlin Grace VM (2012) Antioxidant activity of essential oils from *Wedelia chinensis* (osbeck) *in vitro* and *in vivo* lung cancer bearing C57BL/6 mice. Asian Pac J Cancer Prev 13:3065–3071
Manoranjitham SK, Prakasam V (2000) Management of chilli damping off using biocontrol agents. Capsicum Eggplant News Lett 19:101–104
Manoranjitham SK, Prakasam V, Rajappan K, Amutha G (2000) Control of chilli damping-off using bioagents. J Mycol Plant Pathol 30:225–228
Marjanlo AA, Mostofi Y, Shoeibi S, Fattahi M (2009) Effect of cumin essential oil on postharvest decay and some quality factors of strawberry. J Med Plants 8:25–43
Marongiu B, Piras A, Porcedda S, Falconieri D, Maxia A, Goncalves MJ, Cavaleiro C, Salgueiro L (2010) Chemical composition and biological assays of essential oils of *Calamintha nepeta* (L.) Savi subsp. Nepeta (Lamiaceae). Nat Prod Res 24:1734–1742
Marroqui S, Zorreguieta A, Santamarı'a C (2001) Enhanced symbiotic performance by Rhizobium tropici glycogen synthase mutants. J Bacteriol 183:854–864
Marschner H, Rohmeld V (1994) Strategies of plants for acquisition of iron. Plant Soil 165:261–274

Marzano M, Gailo A, Altomare C (2013) Improvement of biocontrol efficacy of *Trichoderma harzianum* VS *Fusarium oxysporum* f. sp. *lycopersici* through UV induced tolerance to fusaric acid. Biol Control 67:397–408

Mazurier S, Corberand T, Lemanceau P, Raaijmakers JM (2009) Phenazine antibiotics produced by fuorescent pseudomonads contribute to natural soil suppressiveness to *Fusarium* wilt. ISME J 3:977–991

McKenzie RH, Roberts TL (1990) Soil and fertilizers phosphorus update. In: Proceedings of Alberta soil science workshop proceedings, 20–22 Feb, Edmonton. pp 84–104

Mehnaz S, Lazarovits G (2006) Inoculation effects of *Pseudomonas putida*, *Gluconacetobacter azotocaptans* and *Azospirillum lipoferum* on corn plant growth under greenhouse conditions. Microb Ecol 51:326–335

Mehra R (2005) Seed spices: diseases and their management. Indian J Areca Nut Spices Med Plants 7(4):134–143

Mejri J, Abderrabba M, Mejri M (2010) Chemical composition of the essential oil of *Ruta chalepensis* L: influence of drying, hydro-dist illation duration and plant parts. Ind Crop Prod 32:671–673

Mittal V, Singh O, Nayyar H, Kaur J, Tewari R (2008) Stimulatory effect of phosphate solubilizing fungal strains (*Aspergillus awamori* and *Penicillium citrinum*) on the yield of chickpea (*Cicer arietinum* L. cv. GPF2). Soil Biol Biochem 40:718–727

Montes-Belmont, Carvajal (1998) Control of *Aspergillus flavus* in maize with plant essential oils and their components. J Food Prot 61:616–619

Moorman GW, Kang S, Geiser DM, Kim SH (2002) Identification and characterization of *Pythium* species associated with greenhouse floral crops in Pennsylvania. Plant Dis 86:1227–1231

Mukherjee PK, Kenerley CM (2010) Regulation of morphogenesis and biocontrol properties in *Trichoderma virens* by a VELVET protein, Vel1. Appl Environ Microbiol 76:2345–2352

Mukherjee PK, Benjamin A, Horwitz Estrella AH, Schmoll M, Kenerley CM (2013) *Trichoderma* research in the genome era. Annu Rev Phytopathol 51:105–129

Munees A, Mohammad SK (2009) Effects of quizalafop-p-ethyl and clodinafop on plant growth promoting activities of rhizobacteria from mustard rhizosphere. Ann Plant Prot Sci 17:175–180

Muthukumar A (2008) Management of chilli damping-off caused by *Pythium aphanidermatum* (Edson) Fitz. with bacterial endophytes (*Pseudomonas fluorescens*) in glasshouse conditions. Adv Plant Sci 21:295–298

Muthukumar A, Bhaskaran R (2007) Efficacy of antimicrobial metabolites of *Pseudomonas fluorescens* (Trevisan) Migula. against *Rhizoctonia solani* Khun. and *Pythium* sp. J Biol Control 21:105–110

Muthukumar A, Sangeetha G (2008) Broad spectrum antifungal activity of plant oils against chilli damping-off incited by *Pythium aphanidermatum*. Plant Dis Res 23:99–101

Muthukumar A, Eswaran A, Sanjeevkumar K (2008) Biological control of *Pythium aphanidermatum* (Deson) Fitz. Mysore J Agric Sci 42:20–25

Muthukumar A, Eswaran A, Nakkeeran S, Sangeetha G (2010a) Efficacy of plant extracts and biocontrol agents against *Pythium aphanidermatum* inciting chilli damping-off. Crop Prot 29:1483–1488

Muthukumar A, Bhaskaran R, Sanjeevkumar K (2010b) Efficacy of endophytic *Pseudomonas fluorescens* (Trevisan) migula against chilli damping-off. J Biopestic 3:105–109

Muthukumar A, Eswaran A, Sanjeevkumar K (2011) Exploitation of *Trichoderma* species on the growth of *Pythium aphanidermatum* in chilli. Braz J Microbiol 42:1598–1607

Nadeem SM, Zahir ZA, Naveed M, Arshad M (2007) Preliminary investigations on inducing salt tolerance in maize through inoculation with rhizobacteria containing ACC deaminase activity. Can J Microbiol 53:1141–1149

Nafiseh N, Mehrded L, Ali G (2012) Accumulation of chromium and its effect on growth of *Allium cepa* cv. Hybrid. Eur J Exp Biol 2:969–979

Naidu GP (1988) Antifungal activity in *Codieum variegatum* leaf extract. Curr Sci 57:502–503

Nakkeeran S, Kavitha K, Chandrasekar G, Renukadevi P, Fernando WGD (2006) Induction of plant defence compounds by *Pseudomonas chlororaphis* PA 23 and *Bacillus subtilis* BSCBE 4 in controlling damping-off of hot pepper caused by *Pythium aphanidermatum*. Biocontrol Sci Technol 16:403–416

Neelamegam R (2004) Evaluation of fungal antagonists to control damping-off of tomato (*Lycopersicon esculentum* Mill.) caused by *Pythium ultimum*. J Biol Control 18:97–102

Nelson RJ, Heschl MF, Craig EA (1992) Isolation and characterization of estrogenic suppressors of mutations in the SSA hsp70 genes of *Saccharomyces cerevisiae*. Genetics 131(2):277–285

Nielsen MN, Sorensen J (1999) Chitinolytic activity of *Pseudomonas fluorescens* isolates from barley and sugar beet rhizosphere. FEMS Microbiol Ecol 30:217–227

Nurhayat T, Betul D, Sara LC, Kemal HB, David EW (2007) Chemical composition and antifungal activity of *Arnica longifolia*, *Aster hesperius*, and *Chrysothamnus nauseosus* essential oils. J Agric Food Chem 55:8430–8435

O'Sullivan DB, O'Gara F (1992) Traits of fluorescent Pseudomonas spp. involved in suppression of plant root pathogens. Microbiol Rev 56:662–676

Osburn RM, Schroth MN, Hancock JG, Hendson M (1989) Dynamics of sugarbeet colonization by *Pythium ultimum* and *Pseudomonas* species: effects on seed rot and damping-off. Phytopathology 79:709–716

Padmanaban P, Alexander KC (1984) Biological control of cane seedling root rot. Sugarcane 1:11

Palleroni NJ (1984) Pseudomonadaceae. In: Kreig NR, Holt JG (eds) Bergey's manual of systematic biology. Williams and Wilkins Co, Baltimore, pp 141–199

Pandey VN, Dubey NK (1994) Antifungal potential of leaves and essential oils from higher plants against soil phytopathogens. Soil Biol Biocbem 26:1417–1421

Pandey KP, Mishra RK, Kamran A, Mishra P, Bajaj AK, Dikshit A (2010) Studies on antidermatophytic activity of waste leaves of *Curcuma longa* L. Physiol Mol Biol 16:177–185

Papavizas GC (1985) *Trichoderma* and *Gliocladium*: biology, ecology, and potential for biocontrol. Annu Rev Phytopathol 23:23–54

Paranagama P (1991) Lemongrass. In: Senanaake UM (ed) In: analysis of Sri Lankan essential oils by gas chromatography and mass spectroscopy. Industrial Technology Institute, Colombo, pp 39–40

Paranagama PA, Abeysekara KHT, Abeywickrama K, Nugaliadde L (2003) Fungicidal and anti-aflatoxingenic effects of the essential oil of *Cymbopogan citratus* (DC.) Stapf. (lemongrass) against *Aspergillus flavus* Link. isolated from store rice. Lett Appl Microbiol 37:86–90

Park-Kwon, Kim J, Lee S-G, Shin S-C (2007) Nematicidal activity of plant essential oils and components from Ajowan (*Trachyspermum ammi*), Allspice (*Pimenta dioica*) and Litsea (*Litsea cubeba*) essential oils against Pine Wood nematode (*Bursaphelenchus xylophilus*). J Nematol 39:275–279

Parveen T, Sharma K (2014) Management of "Soft Rot" of ginger by botanicals. Int J Pharm Life Sci 5:3478–3484

Patten CL, Glick BR (1996) Bacterial biosynthesis of indole-3- acetic acid. Can J Microbiol 42:207–220

Pattnaik S, Subramanyam VR, Kole CR (1996) Antibacterial and antifungal activity of ten essential oils *in vitro*. Microbios 86:237–246

Paul B, Bala K, Belbahri L, Calmin G, Sanchez-Hernandez E, Lefort F (2006) A new species of *Pythium* with ornamented oogonia: morphology, taxonomy, ITS region of its rDNA, and its comparison with related species. FEMS Microbiol Lett 254:317–323

Paulitz TC, Zhou T, Rankin L (1992) Selection of rhizosphere bacteria for biological control of *Pythium aphanidermatum* on hydroponically grown cucumber. Biol Control 2:226–237

Pinto E, Salgueiro LR, Cavaleiro C, Palmeira A, Gonçalves MJ (2007) *In vitro* susceptibility of some species of yeasts and filamentous fungi to essential oils of *Salvia officinalis*. Ind Crop Prod 26:135–141

Pitipong T, Jularat U, Apinya P, Phirayot K (2009) Screening for the antifungal activity of essential oils from Bergamot oil (*Citrus hystrix* DC.) and Tea tree oil (*Melaleuca alternifolia*) against economically rice pathogenic fungi: a driving force of organic rice cv. KDML 105 production. Asian J Agric Food Sci (Special Issue):S374–S380

Pleban S, Ingel F, Chet I (1995) Control of *Rhizoctonia solani* and *Sclerotium rolfsii* in the green house using endophytic bacteria. Eur J Plant Pathol 101:665–672

Ponmurugan P, Gopi C (2006) *In vitro* production of growth regulators and phosphatase activity by phosphate solubilizing bacteria. Afr J Biotechnol 5:348–350

Portillo MC, Viramontes S, Muñoz LN, Gastelum MG, Nevarez GV (2005) Antifungal activity of *Mexican oregano* (*Lippia berlandieri* Shauer.). J Food Prot 68:2713–2717

Prasad CS, Shukla R, Kumar A, Dubey NK (2009) *In vitro* and *in vivo* antifungal activity of essential oils of *Cymbopogon martini* and *Chenopodium ambrosioides* and their synergism against dermatophytes. Mycoses 53:123–129

Pyun MS, Shin S (2006) Antifungal effects of the volatile oils from *Allium* plants against *Trichophyton* species and synergism of the oils with ketoconazole. Phytomedicine 13:394–400

Radhakrishnan N, Balasubramanian R (2009) Salicylic acid induced defence responses in *Curcuma longa* (L.) against *Pythium aphanidermatum* infection. J Crop Prot 28(11):974–979

Rahman AU, Ashraf M, Choudhary MI, Rehman HU, Kazmi MH (1995) Antifungal *Aryltetralin lignans* from leaves of *Podophyllum hexandrum*. Phytochemistry 40:427–431

Raju G, Maridas M (2011) Composition, antifungal and cytotoxic activities of essential oils of *Piper barberi* Fruits. Int J Biol Technol 2:100–105

Ramesh R (2004) Management of damping-off in brinjal using bio-control agent. J Mycol Plant Pathol 34:666–670

Ranasinghe L, Jayawardenam B, Abeywickrama K (2002) Fungicidal activity of essential oils of *Cinnamomum zeylanicum* (L.) and *Syzygium aromaticum* (L.) Merr et L.M. Perry against crown rot and anthracnose pathogens isolated from banana. Lett Appl Microbiol 35:208–211

Rathore VRS, Mathur K, Lodha B (1990) Activity of volatile and non-volatile substances produced by *Trichoderma viride* on ginger rhizome rot pathogens. Indian Phytopathol 19:253–254

Rathore VRS, Mathur K, Lodha BC (1992) Activity of volatile and non-volatile substances produced by *Trichoderma viride* on ginger rhizome rot pathogen. Indian Phytopathol 45:253–254

Raymond J, Siefert JL, Staples CR, Blankenship RE (2004) The natural history of nitrogen fixation. Mol Biol Evol 21:541–554

Reino JL, Guerrero RF, Hernandez-Galan R, Collado IG (2008) Secondary metabolites from species of the biocontrol agent *Trichoderma*. Phytochem Rev 7:89–123

Rios JL, Recio MC (2005) Medicinal plants and antimicrobial activity. J Ethnopharmacol 100:80–84

Robideau GP, De Cock AW, Coffey MD, Voglmayr H, Brouwer H, Bala K, Chitty DW, Désaulniers N, Eggertson QA, Gachon CM, Hu CH, Küpper FC, Rintoul TL, Sarhan E, Verstappen EC, Zhang Y, Bonants PJ, Ristaino JB, Lévesque CA (2011) DNA barcoding of oomycetes with cytochrome c oxidase subunit I and internal transcribed spacer. Mol Ecol Resour 11(6):1002–1011

Rodriguez H, Fraga R (1999) Phosphate solubilizing bacteria and their role in plant growth promotion. Biotechnol Adv 17:319–339

Romagnoli C, Bruni R, Andreotti E, Rai MK, Vicentini CB, Mares D (2005) Chemical characterization and antifungal activity of essential oil of capitula from wild Indian *Tagetes patula* L. Protoplasma 225:57–65

Rosso ML, Rupe JC, Chen PY, Mozzoni LA (2008) Inheritance and genetic mapping of resistance to *Pythium* damping-off caused by *Pythium aphanidermatum* in 'Archer' soybean. J Crop Sci 48(6):2215–2222

Roy S, Chattergiee P (2010) A non toxic antifungal compound from the leaves of *Catharaanthus roseus* characterized as 5-hydroxy flavones by UV spectroscopic analysis and evaluation of its antifungal property by agar cup method. Ind Crop Prod 32:375–380

Rubio LM, Ludden PW (2008) Biosynthesis of the iron-molybdenum cofactor of nitrogenase. Annu Rev Microbiol 62:93–111

Sagar SD, Kulkarni S, Hegde YR (2007) Management of rhizome rot of ginger by botanicals. Int J Plant Sci 2:155–158
Sagar SD, Kulkarni S, Hegde YR (2008) Survey, surveillance and etiology of rhizome rot of ginger in Karnataka. J Plant Dis Sci 3(1):37–39
Saha S, Rai AB, Pandey S (2011) Efficacy of seed dressing agents against damping-off disease of chilli (*Capsicum frutescens*). Indian J Agric Sci 81:92–93
Salamone EG, Hynes RK, Nelson LM (2001) Cytokinin production by plant growth promoting rhizobacteria and selected mutants. Can J Microbiol 47:404–411
Salamone EG, Hynes RK, Nelson LM (2005) Role of cytokinins in plant growth promotion by rhizosphere bacteria. In: Siddiqui ZA (ed) *PGPR*: biocontrol and biofertilization. Springer, Amsterdam, pp 173–195
Saleem M, Arshad M, Hussain S, Bhatti AS (2007) Perspective of plant growth promoting rhizobacteria (PGPR) containing ACC deaminase in stress agriculture. J Ind Microbiol Biotechnol 34:635–648
Salomone A, Scaritto G, Sacco A, Cabras G, Angioni A (2008) Inhibitory effects of the main compounds of oregano essential oil against some pathogenic fungi. Modern fungicide and antifungal compounds V. Deut Phytomedizinische Ges 44:345–360
Sandy M, Butler A (2009) Microbial iron acquisition: marine and terrestrial siderophores. Chem Rev 109:4580–4595
Sessou P, Farougou S, Alitonou G, Djenontin TS, Yèhouénou, Azokpota P (2012) Chemical composition and antifungal activity of essential oil of fresh leaves of *Ocimum gratissimum* from Benin against six mycotoxigenic fungi isolated from traditional cheese wagashi. Int Res J Biol Sci 1:22–27
Shanmugam V, Varma AS (1999) Effect of native antagonists against *Pythium aphanidermatum* the causal organism of rhizome rot of ginger. J Mycol Plant Pathol 29:375–379
Sharma P, Sain SK, James S (2003) Compatibility study of *Trichoderma* isolates with fungicides against damping-off of cauliflower and tomato caused by *Pythium aphanidermatum*. Pestic Res J 15:133–138
Sharma D, Chauhan S, Kumar G, Singh KP, Kumar R (2012) Cadmium stabilization by plant growth promotory fluorescent pseudomonas in combination with Indian mustard *var*. Kranti South Asian J Exp Biol 3:128–135
Sheetal V, Singh SP (2008) Current and future status of herbal medicines. Vet World 1:347–350
Shoresh M, Harman GE (2008) The molecular basis of shoot responses of maize seedlings to *Trichoderma harzianum* T22 inoculation of the root: a proteomic approach. Plant Physiol 147:2147–2163
Simic A, Sokovic MD, Ristic M, Grujic S, Vukojevic J, Marin PD (2004) The chemical composition of some Lauraceae essential oils and their antifungal activities. Phytother Res 18:713–718
Singh PP, Shin YC, Park CS, Chung YR (1999) Biological control of *Fusarium* wilt of cucumber by chitinolytic bacteria. Phytopathology 89:92–99
Singh HB, Singh BN, Singh SP, Singh SR, Sharma BK (2009) Biological control of plant diseases: current status and future prospects. In: Johri JK (ed) Recent advances in biopesticides: biotechnological applications. New India Pub, New Delhi, p 322
Sitara U, Naseem J, Sultana N (2008) Antifungal effect of essential oils on *in vitro* growth of pathogenic fungi. Pak J Bot 40:409–414
Sivasithamparam K, Ghisalberti FL (1998) Secondary metabolism in *Trichoderma* and *Gliocladium*. In: Kubicek CP, Harman GE (eds) *Trichoderma* and *Gliocladium*, vol I. Taylor and Francis Ltd, London, pp 139–191
Sokmen A, Jones BM, Erturk M (1999) The *in vitro* antibacterial activity of Turkish medicinal plants. J Ethnopharmacol 67:79–86
Soliman KM, Badea RI (2002) Effect of oil extracted from some medicinal plants on different mycotoxigenic fungi. Food Chem Toxicol 40:1669–1675
Somda I, Leth V, Sereme P (2007) Antifungal effect of *Cymbopogon citrates*, *Eucalyptus camaldulensis* and *Azadirachta indica*. Asian J Plant Sci 6(8):1182–1189
Spaepen S, Vanderleyden J (2011) Auxin and plant-microbe interactions. Cold Spring Harb Perspect Biol 3:1–4

Spaepen S, Vanderleyden J, Remans R (2007) Indole-3-acetic acid in microbial and microorganism-plant signalling. FEMS Microbiol Rev 31:425–448
Stuardo M, Sanmartin R (2008) Antifungal properties of quinoa (*Chenopodium quinoa* Willd) alkali treated saponins against *Botrytis cinerea*. Ind Crop Prod 27:296–302
Sturz AV, Nowak J (2000) Endophytic communities of rhizobacteria and the strategies required to create yield enhancing associations with crops. Appl Soil Ecol 15:183–190
Subhashini DV, Padmaja K (2009) Exploitation of *Pseudomonas fluorescens* for the management of damping-off disease of tobacco in seedbeds. Indian J Plant Protect 37:147–150
Sudhakar P, Chattopadhyay GN, Gangwar SK, Ghosh JK (2000) Effect of foliar application of Azotobacter; Azospirillum and Beijerinckia on leaf yield and quality of mulberry (*Morus alba*). J Agric Sci 134:227–234
Sukatta U, Haruthaithanasan V, Chantarapanont W, Dilokkunanant U, Suppakul P (2008) Antifungal activity of clove and cinnamon oil and their synergistic against post harvest decay fungi of grape *in vitro*. Kasetsart J 42:169–174
Suleiman MN, Emua SA (2009) Efficacy of four plant extract in the control of root rot disease of cow pea (*Vigna unguiculata* Linn. Walp). Afr J Biotechnol 8:3806–3808
Suman A, Shasany AK, Singh M, Shahi HN, Gaur A, Khanuja SPS (2001) Molecular assessment of diversity among endophytic diazotrophs isolated from subtropical Indian sugarcane. World J Microbiol Biotechnol 17:39–45
Sumonrat C, Suphitchaya C, Tipparat H (2008) Antimicrobial activities of essential oils and crude extracts from tropical *Citrus* spp. against food-related microorganisms. *Songklanakarin*. J Sci Technol 30:125–131
Sun OL, Gyung JC, Kyoung SJ, He KL, Kwang YC, Jin-Cheol K (2007) Antifungal activity of five plant essential oils as fumigants against postharvest and soil borne plant pathogenic fungi. Plant Path J 23:97–102
Sunita B, Mahendra R (2008) Antifungal activity of essential oils from Indian medicinal plants against human pathogenic *Aspergillus fumigatus* and *A. niger*. World J Med Sci 3:81–88
Suprapta DN, Khalimi K (2009) Efficacy of plant extract formulation to suppress stem rot disease on vanilla seedlings. ISSA J 15:34–41
Sutthivaiyakit S, Thapsut M, Prachayasittikul V (2000) Constituents and bioactivity of the tubers of *Euphorbia sessiliflora*. Phytochemistry 53:947–950
Taiga A (2009) Efficacy of selected plant extracts in the control of fungal dry rot of white yam (*Dioscorea Rotundata*) Tubers in Kogi State. Am Eurasian J Sustain Agric 3:310–313
Taller BJ, Wong TY (1989) Cytokinins in *Azotobacter vinelandii* culture medium. Appl Environ Microbiol 55:266–267
Tao GC, Tian SJ, Cai MY, Xie GH (2008) Phosphate solubilizing and mineralizing abilities of bacteria isolated from soils. Pedosphere 18:515–523
Tavares AC, Gonçalves MJ, Cavaleiro C, Cruz MT, Lopes MC, Canhoto J, Salgueiro L (2008) Essential oil of *Daucus carota* subsp halophilus: composition, antifungal activity and citotoxicity. J Ethnopharmacol 119:129–134
Timmusk S, Nicander B, Granhall U, Tillberg E (1999) Cytokinin production by *Paenibacillus polymyxa*. Soil Biol Biochem 31:1847–1852
Tirillini B, Velasquez ER, Pellegrino R (1996) Chemical composition and antimicrobial activity of essential oil of *Piper angustifolium*. Planta Med 62:372–373
Trevor GP, Daniel J, Sullivan O, Larry LM (2003) Identification of bacilysin, chlorotetaine and itirin A produced by *Bacillus* sp. strain CS93 isolated from pozol, a maxican fermented dough. Appl Environ Microbiol 70:631–634
Tsavkelova EA, Klimova SY, Cherdyntseva TA, Netrusov AI (2006) Microbial producers of plant growth stimulators and their practical use: a review. Appl Biochem Microbiol 42:117–126
Tundis R, Statti GA, Conforti F, Bianchi A, Agrimonti C, Sachetti G (2005) Influence of environmental factors on composition of volatile constituents and biological activity of *Helichrysum italicum* (Roth) Don (Asteraceae). Nat Prod Res 19:379–387
Urszula J (2006) Synthesis of siderophores by soil bacteria of the genus *Pseudomonas* under various culture conditions. Acta Sci Polonorum Agric 5:33–44

Ushamalini C, Nakkeeran S, Marimuthu T (2008) Development of biomanure for the management of turmeric rhizome rot caused by *Pythium aphanidermatum*. Arch Phytopathol Plant Protect 41:365–378

Valarmathi A (2007) Studies on biocontrol potential of combined application of *Trichoderma* spp. and *Psudomonas* spp. against damping-off of tomato caused by *Pythium aphanidermatum* (Edson) Fitz., M.Sc. (Ag.) thesis, Annamalai University, Tamil Nadu, p 94

Van Der Plaats-Niterink AJ (1981) Monograph of the genus *Pythium*, vol 21, Studies in Mycology. Central Bureau Voor Schimmel cultures, Baarn

Van West P, Appiah AA, Gow NAR (2003) Advances in research on oomycete root pathogens. Physiol Mol Plant Pathol 62:99–113

Velluti A, Sanchis V, Ramos AJ, Marin S (2003) Inhibitory effect of cinnamon, clove, lemongrass, oregano and palmarose essential oils on growth and fumonisin B1 production by *Fusarium proliferatum* in maize grain. Int J Food Microbiol 89:145–154

Veloz-García R, Marín-Martínez R, Veloz- Rodríguez R, Rodríguez-Guerra R, Torres- Pacheco I, González-Chavira MM, Anaya-López JL, Guevara-Olvera L, Feregrino- Pérez AA, Loarca-Piña G, Guevara-González RG (2010) Antimicrobial activities of cascalote (*Caesalpinia cacalaco*) phenolics-containing extract against fungus *Colletotrichum lindemuthianum*. Ind Crop Prod 31:134–138

Verma M, Brar SK, Tyagi RD, Surampalli RY, Valero JR (2007) Antagonistic fungi, Trichoderma spp.: panoply of biological control. Biochem Eng J 37:1–20

Vessey JK (2003) Plant growth promoting rhizobacteria as biofertilizers. Plant and Soil 255:571–586

Vey A, Hoagland RE, Butt TM (2001) Toxic metabolites of fungal bio-control agents. In: Butt TM, Jackson C, Magan N (eds) Fungi as bio-control agents: progress, problems and potential. CAB International, Bristol, pp 311–346

Vikram A, Hamzehzarghani H (2008) Effect of phosphate solubilising bacteria on nodulation and growth parameters of greengram (*Vigna radiate* L. Wilczec). Res J Microbiol 3:62–72

Viterbo A, Horwotz BA (2010) Mycoparasitism. In: Borkovich KA, Ebbole DJ (eds) Cellular and molecular biology of filamentous fungi, vol 56. ASM Press, Herndon, pp 676–694

Wang GM, Stribley DP, Walker C (1993) Effect of pH on arbuscular mycorrhiza I. Field observations on long-term liming experiments at rothamsted and Woburn. New Phytol 124:465–472

Waterhouse GM, Waterston JM (1964) *Pythium aphanidermatum*. CMI descriptions of pathogenic fungi and bacteria, No. 36. Commonwealth Mycological Institute: Surrey. Waterhouse

Wei G, Kloepper JW, Tuzun S (1996) Induced systemic resistance to cucumber diseases and increased plant growth by plant growth promoting rhizobacteria under field condition. Phytopathology 86:221–224

Weindling R (1932) *Trichoderma lignorum* as a parasite of other soil fungi. Phytopathology 22:837–845

Whipps JM, Lumsden DR (1991) Biological control of *Pythium* species. Biocontrol Sci Tech 1:75–90

Wiest A, Grzegorski D, Xu B, Goulard C, Rebuffat S, Ebbole DJ, Bodo B, Kenerley C (2002) Identification of peptaibols from *Trichoderma virens* and cloning of a peptaibol synthetase. J Biol Chem 277:20862–20868

Wink M (2006) Plant breeding: importance of plant secondary metabolites for protection against pathogens and herbivores. Theor Appl Genet 75:225–233

Woleffhechel H, Jensen DF (1992) Use of *Trichoderma harzianum* and *Gliocladium virens* for the biological control of post-emergence damping-off and root rot of cucumber caused by *Pythium ultimum*. J Phytopathol 136:221–230

Yegen O, Begger B, Heitefuss R (1992) Studies on the fungitoxic effect for extracts of six selected plants from Turkey on phytopathogenic fungi. Z Pflanzenkrankh Pflanzenschütz 99:349–359

Yenjit P, Intanoo W, Chamswarng C, Siripanich J, Intana W (2004) Use of promising bacterial strains for controlling antracnose on leaf and fruit of mango caused by *Colletotrichum gloeosporioides*. Walailak J Sci Technol 1:56–69

Zagade SN, Deshpande GD, Gawade DB, Atnoorkar AA, Pawar SV (2012) Biocontrol agents and fungicides for management of damping-off in chilli. World J Agrl Sci 8:590–597

Zahir ZA, Munir A, Asghar HN, Shaharoona B, Arshad M (2008) Effectiveness of rhizobacteria containing ACC-deaminase for growth promotion of pea (*Pisum sativum*) under drought conditions. J Microbiol Biotechnol 18:958–963

Zahir ZA, Ghani U, Naveed M, Nadeem SM, Asghar HN (2009) Comparative effectiveness of *Pseudomonas* and *Serratia* sp. containing ACC-deaminase for improving growth and yield of wheat (*Triticum aestivum* L.) under salt-stressed conditions. Arch Microbiol 191:415–424

Zaidi A, Khan MS (2005) Interactive effect of rhizospheric microorganisms on growth, yield and nutrient uptake of wheat. J Plant Nutr 28:2079–2092

Zaidi A, Khan MS, Ahemad M, Oves M (2009) Plant growth promotion by phosphate solubilizing bacteria. Acta Microbiol Immunol Hung 56:263–284

Zuzarte M, Gonçalves MJ, Cavaleiro C, Canhoto J, Vale-Silva L, Silva MJ, Pinto E, Salgueiro L (2011) Chemical composition and antifungal activity of the essential oils *Lavandula viridis* L'Hér. J Med Microbiol 14:249–258

Biological Agents in *Fusarium* Wilt (FW) Diagnostic for Sustainable Pigeon Pea Production, Opportunities and Challenges

Deepak Kumar Verma, Shikha Srivastava, Vipul Kumar, and Prem Prakash Srivastav

Introduction

Pulses constitute one of the richest sources including valuable but incompletely balanced protein, particularly in vegetarian's diet (Ghadge et al. 2008) and are consequently known as an important part of the diet in many regions on the earth (Arinathan et al. 2003, 2009). Among the pulses crops, Pigeon pea [*Cajanus cajan*, (L.) Millspaugh] is a diploid (2n = 22, 44, or 66 chromosomes), most widely produced and consumed food legume worldwide. It also known as arhar, congo pea, tur dhal, frijol de árbol, gandul, gandure, gungo pea, no eye pea, poiscajan and red gram (Long and Lakela 1976) and belongs to family *Leguminosae*. The fruit of the pigeon pea are classified as a pod and each pod have three to five seeds with round or lens like shape.

Pigeon pea is an important grain legume crop of rain-fed agriculture in the semi-arid tropics (Mallikarjuna et al. 2011), which have probably originated from India but some people believe that it may have come from Africa. It is evident that pigeon pea had been originated in India and Asia, and moved to African countries

D.K. Verma (✉) • P.P. Srivastav
Agricultural and Food Engineering Department, Indian Institute of Technology,
Kharagpur 721 302, West Bengal, India
e-mail: deepak.verma@agfe.iitkgp.ernet.in; rajadkv@rediffmail.com;
pps@agfe.iitkgp.ernet.in

S. Srivastava
Department of Botany, Deen Dyal Upadhyay Gorakhpur University,
Gorakhpur 273 009, Uttar Pradesh, India
e-mail: shikha.sriv13@gmail.com

V. Kumar, PhD (✉)
Department of Agriculture, Lovely Professional University,
Phagwara 144 411, Punjab, India
e-mail: vipulpathology@gmail.com

P. Kumar et al. (eds.), *Current Trends in Plant Disease Diagnostics and Management Practices*, Fungal Biology, DOI 10.1007/978-3-319-27312-9_4

(Onyebuashi 1986). India and Africa both have been the centres of diversity for the genus *Cajanus* (van der Maesen 1990). Nowadays pigeon pea has become prominently second most important pulse crop after chickpea in India and fifth in the world. This pulse crop is widely cultivated between 30° N and 30° S in all tropical and semi-tropical zone of both the old and the new world including about 50 countries of Asia, Africa, and the Americas for a variety of uses in addition of food and fodder. For instance, green manure, soil conservation, rearing lac insects, wind breaks, fuel wood, hedges, roofing, and so on (Long and Lakela 1976; Sharma et al. 2006; Mallikarjuna et al. 2011; Patel and Patel 2012). India is well known as the biggest producer and consumer of pigeon pea. India has been leading producer of pigeon pea since long decades producing about 265,000°MT followed by Myanmar (900,000°MT), Malawi (237,210°MT), United Republic of Tanzania (206,057°MT), Kenya (89,390°MT), Uganda (84,200°MT), Dominican Republic (27,998°MT) and Nepal (14,082°MT) (FAOSTAT 2012).

Pigeon pea is used as proteinaceous food crops as well as nutritional alternative for human consumption and animal feed along with cereals. It is also grown as forage/cover crop which symbiotically fixes about 90 kg nitrogen per hectare (Adu-Gyamfi et al. 1997). It is an economic crop which is considered as major source of proteins for poor communities in many tropical and subtropical parts of the world viz. India, Myanmar, Malawi, United Republic of Tanzania, Kenya, Uganda, Dominican Republic, Nepal etc. (Singh et al. 1984; ICRISAT 1986; FAOSTAT 2012). Many developing countries including India have inadequate availability of proteinaceous foods. This is a global concern because a large number of populations of these developing countries are suffering from protein malnutrition (Arinathan et al. 2009; Soris and Mohan 2011). Only 20–30 % of proteins are estimated to meet the demand of world's population by the total legume production which is similar to wheat and over 50 % more than the rice or corn crop (Rockland and Radke 1981; Gopalan et al. 1985). Researchers are searching the available substitute of proteins for human nutrition that can impart the nutritional demand of pigeon pea in daily diet of human as protein contribute immense health-related benefits and also possess the best alternative due to their high nutritional value. Pigeon pea contain a high level of crude protein ranges from 21 % to 30 % and mostly some important essential amino acid like, methionine, lysine and tryptophan with phenyl alanine+tyrosine found to be of higher in content (110.4 mg/g of protein) (Udedibie and Igwe 1989; Amaefule and Onwudike 2000). Starch store energy and also known as the major constituent of pigeon pea (Ihekoronye and Ngoddy 1985). In addition, pigeon pea contains considerable amount of vitamin-B complex viz. thiamine, riboflavin and niacin (Bressani and Elias 1974; Arora 1977). Thus, pigeon pea is a staple crop because of its nutritional potential.

Pigeon pea suffers by their natural enemies viz. fungi (83), bacteria (4), viruses and mycoplasma (19) and nematodes (104) over 210 pathogens, reported in 58 countries (Nene et al. 1989, 1996; Reddy et al. 1990). Several fungal pathogens are involved to infect pigeon pea crop such as *Alternaria* spp., *Colletotrichum* spp., *Cercosporaindica*, *Sclerotium rolfsii*, *Rhizoctonia* spp., *Fusarium*spp, *Phytophthora* spp., *Xanthomonas* spp., *Pseudomonas* spp. etc. A list of fungal, bacterial, viral as

Table 1 Major diseases of pigeon pea and their pathogenic agents

Disease	Pathogenic agent
Fungal diseases	
Alternaria blight	*Alternaria* spp., *A. alternata* (Fries) Keissler, *A. tenuissima* (Kunze ex Persoon) Wiltshire
Anthracnose	*Colletotrichum cajani* Rangel, *C. truncatum*, *C. graminicola* (Ces.) Wilson
Botrytis gray mold	*Botrytis cinerea* Persoon ex Fries
Cercospora leaf spot	*Mycovellosiella cajani* (*Cercospora cajani*), *Cercospora indica*, *C. instabilis*, *C. thirumalacharii*
Collar rot	*Sclerotium rolfsii* Saccardo, *Athelia rolfsii* [teleomorph] = *Corticium rolfsii*
Dry root rot	*Macrophomina phaseolina* (Tassi), Goidanich *Rhizoctonia bataticola* (Taub.) Butler
Fusarium leaf blight	*F. pallidoroseum* (*F. semitectum*)
Fusarium wilt	*F. udum* Butler, *Gibberella indica* [teleomorph]
Phoma stem canker	*Phoma cajani* (Rangel)
Phyllosticta leaf spot	*Phyllosticta cajani*Sydow
Phytophthora blight	*Phytophthora drechsleri* Tucker *f. sp. cajani*
Powdery mildew	*Leveillula taurica* [Teleomorph], *Oidiopsis taurica* [Anamorph], *Ovulariopsis ellipsospora*
Rust	*Uredo cajani* Sydow
Bacterial diseases	
Bacterial leaf spot and stem canker	*Xanthomonas campestris* pv. *Cajani*
Halo blight	*Pseudomonas amygdali* pv. *Phaseolicola*
Viral diseases	
Phyllody	Mycoplasma-like organism vector: not known
Sterility mosaic	Vector: Eriophyid mite *Aceria cajani* Channabasavanna
Witches' broom	Mycoplasma-like organism vector: leaf hopper *Empoasca* spp.
Yellow mosaic	Mung bean yellow mosaic virus (MBYMV)
	Vector: India, Jamaica, Nepal, Puerto Rico, and Sri Lanka
Nematode diseases	
Dirty root (reniform nematode)	*Rotylenchulus reniformis* Linford and Oliveira
Pearly root (cyst nematode)	*Heterodera cajani* Koshy
Root-knot (root-knot nematode)	*Meloidogyne acronea* Coetzee, *M. arenaria* (Neal) Chitwood, *M. incognita* (Kofoid and White) Chitwood, *M. javanica* (Treub) Chitwood

Sources: Reddy et al. (1993b)

well as nematode diseases is summarized. (Table 1) (Kannaiyan et al. 1984; Hillocks et al. 2000; Joshi et al. 2001; Maisuria et al. 2008). The diseases of pigeon pea have significant importance including Phytophthora blight (*Phytophthora drechsleri*) (Kannaiyan et al. 1984), powdery mildew (Reddy et al. 1993a), sterility mosaic (Pigeon pea sterility mosaic virus) and wilt. Among them, FW is considered as most destructive disease which is described in detail.

Biological Agents for Plant Health Diagnostic (PHD), Why?

Plant health is a big issue throughout the world to fulfil increasing food demand and balanced food supply. Outbrake of plant diseases and upsurge of insect-pests pose a serious threat to food security. Both of the agents affect plants health leading to significant crop loss and hence productivity worldwide. These agents are needed to be controlled and more emphasis should be given to maintain the quality and abundance of food to mitigate the food demand of world's population. Different approaches are used in disease diagnosis to maintain the plant health. Among them, biological agents have become most promising agents to secure plant health from their pathogens. Biological control is free from use of chemicals, so it is eco-friendly approach which can be helpful to discard some environmental problems like bio-accumulation, bio-magnification and bio-diversity loss.

Plant health diagnostic (PHD) through biological agents is propitious contribution in crop productivity reported over the few decades. Miller et al. (2009) reviewed "Plant Disease Diagnostic, Capabilities and Networks" and stated that diagnostic of plant disease and detection of their pathogen are central to our ability to protect crops and natural plant systems, and are the crucial prelude to undertake management and prevention measures of PHD. According to Miller et al. (2009), Plant disease diagnostic is the determination of the cause of a disease or syndrome in a plant or plant population, whereas detection refers to the identification of microorganisms or their products, e.g., toxins, in any number of substrates including plant tissues, soil, and water.

Recently, biological diagnostic of plant disease (BDPD) have been recognised as swift alternative to chemical fungicides (Fig. 1) and more focused by researches because of the sustainability and eco-friendly. Recently, BDPD has been emerged as a useful technique of organic, eco-friendly and sustainable agriculture involving the use of antagonistic microorganisms to combat the various diseases in most of the crops. BDPD can be proved as best tool to secure pigeon pea from pathogens and control the target organism without being harmful to humans or any beneficial organisms in natural eco-systems. Nowadays the use of promising microorganisms or their formulations have been attracted attention due to increased incidence of disease. These promising microorganisms belong to bacterial as well as fungal genera are registered and commercially available (Table 2).

Fusarium Wilt (FW): Major Disease of Pigeon Pea

Fusarium wilt (FW) caused by soil borne pathogenic fungus *Fusarium udum* (Butler) is one of the major diseases of pigeon pea severely affecting demand, economy, production and seed yield worldwide (Kannaiyan et al. 1984; Reddy et al. 1990; Ajay et al. 2013). The loss of crop begins from pre pod stage. Total loss in

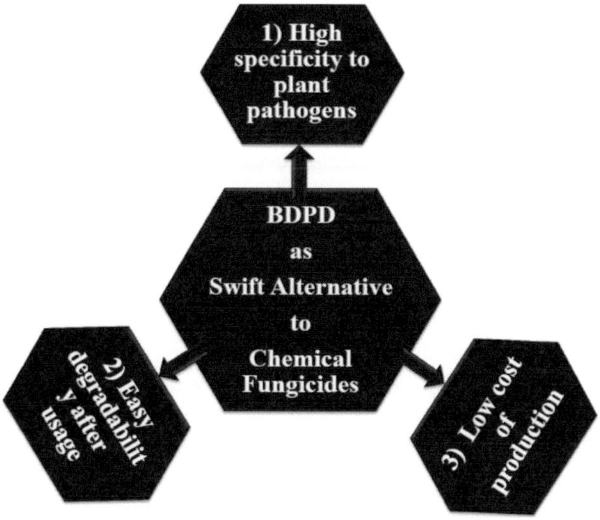

Fig. 1 Biological diagnostic of plant disease (BDPD) as swift alternative to chemical fungicides

Table 2 Promising microorganisms and their formulated products in plant disease diagnostics

Bacterial genera	Fungal genera
Agrobacterium spp.	*Ampelomyces* spp.
Bacillus spp.	*Candida* spp.
Pseudomonas spp.	*Coniothyrium* spp.
Streptomyces spp.	*Gliocladium* spp.
	Trichoderma spp.

Source: Vinale et al. (2008)

yield may ranges from 30 % to 100 % in pre-pod stage, about 67 % and 30 % at crop maturity and pre-harvest stage, respectively and almost 100 % yield losses in susceptible genotypes (Nene 1980; Upadhyay and Rai 1992; Kannaiyan and Nene 1981; Sheldrake et al. 1984; Reddy et al. 1990). The annual loss due to FW disease in India and eastern Africa is estimated to be approximately at US $71 and US $5 million (Reddy et al. 1993a, b; Saxena et al. 2002). The scenario of disease incidence in India is reported maximum in Maharashtra (22.6 %) and minimum in Rajasthan (0.1 %) (Kannaiyan et al. 1984; Upadhyay and Rai 1992). It is reported that the incidence of FW disease have to increased significantly over the time (Gwata et al. 2006) with an average of 10–15 % incidence and 16–47 % of crop loss (Prasad et al. 2003). The global crop loss due to FW disease is reported by Kannaiyan et al. (1984) and it was found to have 15.9 % (0–90 %), 20.4 % (0–60 %) and 36.6 % (0–90 %) in Kenya, Tanzania and Malawi respectively with annual loss estimated at 5 million US$ in each of the countries with 96 % of disease incidence in Tanzania (Mbwaga 1995). This disease poses annual loss by 470,000 and 30,000 ton of total grain production in India and Africa respectively which affects the economy by 71 million US$ (Reddy et al. 1993a; Joshi et al. 2001).

Fusarium udum: The Pathogenic Agent

Fusarium udum Butler. (Perfect stage: *Gibberella udum*) is causal organism of wilt disease of pigeon pea. In 1906, Butler firstly reported FW disease of pigeon pea in India (Butler 1906). The pathogenic agent was described as *F. udum* by Butler in 1990 (Butler 1910) and the fungus has subsequent multiple names *F. butleri*, *F. lateritium* f. spp. *cajani*, *F. lateritium* var. *uncinatum*, *F. oxysporum* f. spp. *udum*, *F. udum* f. spp. *cajani* and *F. uncinatum* (Dhar et al. 2005). The name *F. udum* suggests the presence of prominent hook shaped macro-conidia (Booth 1971). *F. udum* is a host specific (pigeon pea) pathogen with consistent pathogenic variability and morphological differences (Padwick 1940; Subramanian 1963; Booth 1971).

The mycelium of this fungus may be parasitic or saprophytic, hyaline, slender and branched. *F. udum* produces different type conidia (like macro and micro) and chlamydospores (Fig. 2). Macroconidia are 1–5 septate (predominantly 3 septate), curved to almost hooked and abundant in sporodochia (Fig. 2a–d) whereas microconidia are fusiform to reniform or oval and 0–1 septate (Fig. 2e–i). Chlamydospores are round or oval, thick walled, hyaline, sometimes in short chains, 5–10 μ in diameter (Fig. 2g–i). Perfect stage of pathogen (*G. udum*) needs further investigations. So far, five variants (strains) of *F. udum* have been identified and documented (Reddy et al. 1996; Mishra 2004).

In 2013, 14 isolates of *F. udum* from pigeon pea collected by Datta and Lal (2013) from major pulse growing parts of India and confirmed the genetic diversity between the races of FW in pigeon pea. A research paper was published in 1983 that revealed five category of *F. udum* on the basis of virulence differences (Pawar and Mayee 1983). Patil (1984) reported 9.4–12.0×3.1–3.3 μm size of conidia, 19.2×3.5–5.0 μm of macro conidia and it was mostly found to be whitish in the basal medium. Six isolates of *F. udum* described by Madhukeshwara and Sheshadri

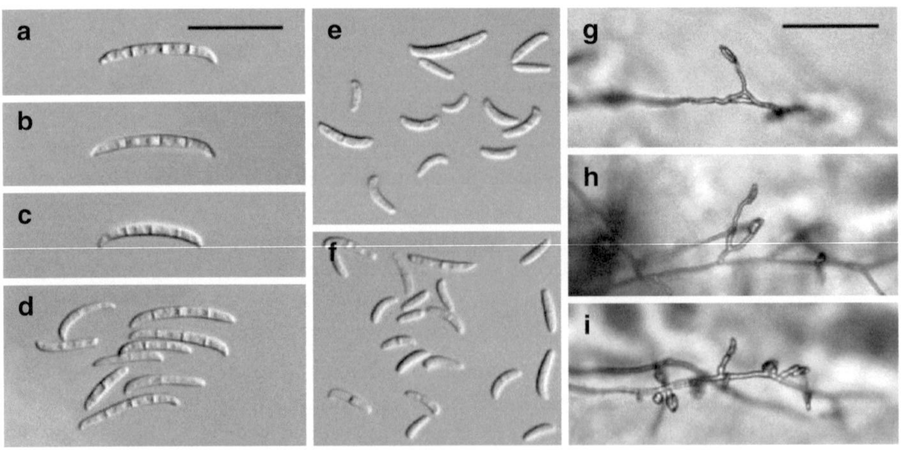

Fig. 2 Type of conidia produces by *F. udum*, they are macroconidia (**a–d**) and microconidia (**e–i**) (Adapted from Leslie and Summerell (2006))

(2001) with different colony characteristics, pigmentation and sporulation. One hundred ninety-five isolates of *F. udum* has been isolated (IIPR 2007–2008) and revealed that 135 were highly pathogenic (>50 % wilt), 33 moderately pathogenic (30–50 % wilt) and 32 were weak pathogenic (<30 % wilt) agent.

Distribution of Pathogen

Currently, FW diseases is considered as highly destructive (Nene et al. 1989) and distributed in form of both, seed borne as well as soil borne in several countries namely Bangladesh, Ghana, Grenada, Grenada, India, Indonesia, Kenya, Malawi, Mauritius, Myanmar (Burma), Nepal, Nevis, Tanzania, Thailand, Tobago, Trinidad, Uganda, Venezuela, Zambia etc. throughout the globe at where the field loss are widely prevalent (over 50 %) and more common in India, east Africa and Malawi (Kannaiyan and Nene 1981; Kannaiyan et al. 1984; Kimani 1991; Reddy et al. 1993a; Marley and Hillocks 1996; Ajay et al. 2013).

Disease Symptoms

The first symptoms of FW disease is usually seen in the field during early developmental stages (Fig. 3) when flowering and podding appears in the crop, sometimes may also be seen at seedling stage (Prasad et al. 2003) but never visible in later

Fig. 3 *Fusarium* wilt symptoms in the pigeon pea field appear during flowering and podding of early developmental stages but may be at seedling stage also

Fig. 4 (**a**) Prominent internal browning and blackening on 1–2 months old plants die from wilt (**b**), Development of *dark purple* bands on the stem surface extending upwards from the base, (**c**) Visible *black* streaks in xylem strands on the main stem or primary branches when it split open, and (**d**) Die-back symptoms with a purple band on branches extending from tip of the plant to downwards and starts drying (Pictures adapted from Reddy et al. (1993b))

developmental stages of pigeon pea (Reddy et al. 1990; Hillocks et al. 2000). The pathogen infects the host via vascularisation of injured root tips causing chlorosis of leaves and branches, wilting and collapsing of root system (Jain and Reddy 1995; Butler 1906). The pioneer symptom of FW is interveinal clearing and loss of turgidity in leaves with slight chlorosis. Leaves appear bright yellow before wilting (Reddy et al. 1990). FW infection is caused via tap root system and results into total wilt. There are many other factors that lead to partial wilting of plants like termite damage, drought and phytophthora blight (Nene 1980; Reddy et al. 1993a, b).

Diagnostic symptoms of FW appear as brown or black streaks on stem surface (Fig. 4a) which turns dark purple extending towards the tip of the main stem (Fig. 4b). The symptoms are more visible in interior section of the main stem or primary branch (Fig. 4c) (Reddy et al. 1990, 1993b). The severity of streaks reduces from base to the tip of the stem. Sometimes the streaks are not visible on main stem but lower branches start becoming non-viable due to die back symptoms which includes appearance of purple bands or streaks extending from upward to downward and blackening of xylem vessels (Fig. 4d) (Reddy et al. 1993b). It is also observed that young plants (1–2 months old) infested with FW may die due to partial wilt without showing characteristic purple bands (Fig. 4b).

Biological Agents in Diagnostic of *Fusarium* Wilt

Biological agents (including bacteria and fungi) have been listed as useful tool for disease diagnostic. Biological agents are devoid of chemical substances and can control target organism efficiently (Romero et al. 2007; Suarez-Estrella et al. 2007; Whipps and McQuilken 2009). These agents are environmental friendly, can be utilized experimentally for the control the enemies of crop plants without causing ill affect to human health or any beneficial organisms (Kaewchai et al. 2009). Some bacterial genera viz. *Bacillus*, *Pseudomonas* and *Rhizobium* and non-pathogenic, non-host *Fusarium* spp. are used as potent biological agent against the pigeon pea disease. Both in field and vitro study inferred significant reduction of disease incidence (Chérif et al. 2007). Biological agent mediated control has been a promising and attractive alternative for PHD and soil borne pathogens as it mitigate the adverse effects of fungicides and pesticides to the farmland.

There are many novel microorganism species viz. *Aspergillus* spp., Bacillus spp., *Pantoea* spp., *Pseudomonas* spp. and *Trichoderma* spp. were evaluated as potential alternative to replace the chemical such as thiram, bavistin and benomyl etc. against fungal pathogen *F. udum* (Upadhyay and Rai 1981; Bhatnagar 1996; Somasekhara et al. 1996, 1998; Gundappagal and Bidari 1997; Biswas and Das 1999; Prasad et al. 2002; Khan and Khan 2002; Anjaiah et al. 2003; Sawant et al. 2003; Roy and Sitansu 2005; Dhar et al. 2006; Maisuria et al. 2008; Ram and Pandey 2011). Many profitable rhizobacteria have been reported by many worker as bio-inoculants (Pusey 1989; Upadhyay and Rai 1992; Bapat and Shar 2000; Siddiqui et al. 2005; Siddiqui 2006; Siddiqui and Shakeel 2007). It has been shown that fungal or bacterial antagonists of pathogen inoculated to soil reduces FW and its pathogenesis (Bapat and Shar 2000; Singh et al. 2002; Anjaiah et al. 2003; Mandhare and Suryawanshi 2005; Maisuria et al. 2008). The mechanism for biological diagnostic of pigeon pea disease is shown in Fig. 5. The mechanisms of biological diagnostic of pigeon pea have different modes of action which are not pathogen specific and many of these mechanisms may be synergistically active and used by the same biological agent (Chérif et al. 2002; Mandeel and Baker 1991) which may not have efficacy to control the major diseases of pigeon pea.

Upadhyay and Rai (1981) reported many species of fungi viz. *A. niger*, *A. flavus*, and *A terreus* could be used for suppression of the population of *F. udum*. Soil antagonistic bacteria are well known to suppress the wilt through inducing resistance (Upadhyay and Rai 1981, 1992). Isolation of indigenous *Bacillus* spp. from the disease suppressive soil of the same environment may increase the probability of disease suppression (Cook and Baker 1983; Weller et al. 1985). Harman et al. (1989) studied combined effective strains of *T. harzianum* and solid matrix priming for biological seed treatment. The production of antibiotics by *P. cepacia* was used as biological control agent for soil borne plant pathogens (Homma et al. 1989). Bhatnagar (1996) studied the antifungal activity of three *Trichoderma* spp. as multiple action bio-inoculants and to control variable pathogenesis against wilt pathogen at different pH, temperatures and C/N ratios and found that all of them were equally efficient and showed maximum antagonistic properties at 35 ± 2 °C temperature and about of 6.5 pH.

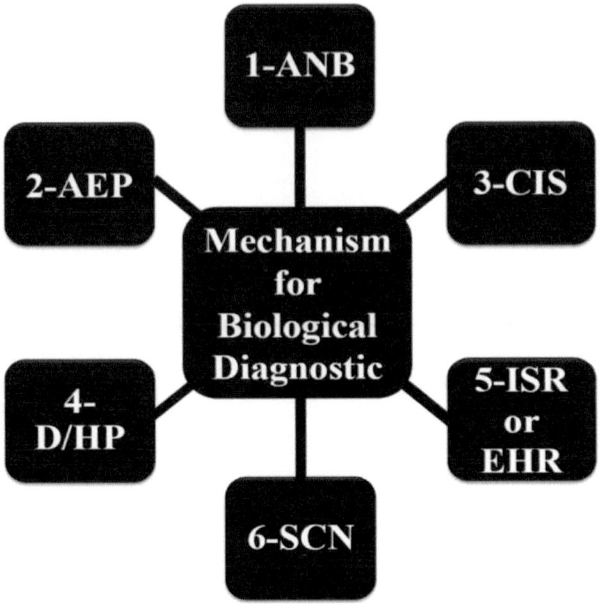

Fig. 5 Biological diagnostic mechanism for pigeon pea disease. (*1*) ANB (Antibiosis), (*2*) AEP (Antifungal enzyme production), (*3*) CIS (Competition for infection sites), (*4*) D/HP (Direct/ Hyper parasitism), (*5*) ISR or HER (Induced systemic resistance or Enhanced host resistance, and (*6*) SCN (Saprophytic competition for nutrients) (Concept adapted from Chérif and Benhamou (1990), Fuchs et al. (1997), and Chérif et al. (2002, 2003))

Apparently, Somasekhara et al. (1996) worked on two delivery systems (seed treatment and foliar application) by using six isolates of *Trichoderma* spp. and studied their efficacy which was found to be extreme on the 35 days of inoculation. As the plant is resistance to dry period, Gundappagal and Bidari (1997) used *T. viride* for seed treatment to resistant cultivar that can be effective in integrated disease management of pigeon pea under dry land cultivation. *Trichoderma* spp. are well known producer of extracellular volatile compound, which was found to be potent fungi toxic to wilt pathogen (Pandey and Upadhaya 1997). Somasekhara et al. (1998) evaluated *Trichoderma* isolates and their antifungal extracts as potential biocontrol agents against pigeon pea wilt pathogen, *F. udum*. Butler observed that non-volatile antibiotics of *T. viride* was highly toxic followed by *T. harzianum*, *T. harzianum* and *T. koningii*.

Biswas and Das (1999) performed in-vitro experiments to reduce pathogenesis and tested five *Trichoderma* spp. *T. harzianum* was found to be most effective antagonist followed by *T. hamatum*, *T. longiconis* and *T. koningii*. They also reported that by giving seed treatment of *T. harzianum* to pigeon pea, inoculants spores failed to reduce pathogen growth while soil amendment with *T. harzianum* in maize meal: sand applied at 40–60 g/kg soil resulted a significant reduction of wilt up to 90 %. Under field conditions, Prasad et al. (2002) studied the effect of soil and seed application of *T. harzianum* on pigeon pea wilt caused by *F. udum* and inoculation with *T. harzianum* controls the disease by 22–61.5 %. Khan and Khan (2002) confirmed differential behavior of multiple bio-control agents (*Trichoderma*, *Bacillus*, *Pseudomonas*) controlling FW and recorded 17–48 % of decrease disease incidence. Khan and Khan (2002) also observed that rhizospheric application of *B. subtilis*, *P. fluorescens*, *A. awamori*, *A. niger* and *Penicillium digitatum* resulted in significant decline of *F. oxysporum*.

Biological control of FW of pigeon pea had been reported by Vaidya et al. (2001, 2003) with chitinolytic activity of *Alcaligenes xylosoxydans*. Vaidya et al. (2003) conducted a pot experiment and field trials. *A. xylosoxydans* was used to treat pigeon pea seeds because it has antifungal activity due to chitinase production. The treated seeds were sown in *Fusarium* infested soil. He found that the incidence of wilt was reduced by 43.5 % and grain yield was increased by 58 %. Anjaiah et al. (2003) studied bio-control experiment to investigate the effect of genotype and root colonization in biological control of FW and reported that disease incidence of wilt was drastically reduced after inoculation of *P. aeruginosa* (PNA1) to both chickpea and pigeon pea in naturally infested soil. de Boer et al. (2003) experimented on combined *P. putida* strains to control of FW as it has different disease-suppressive mechanisms. Siddiqui et al. (2008) studied biological control of wilt disease of pigeon pea by fluorescent pseudomonads under pot and field conditions. He isolated a *Pseudomonas* strain Pa324, known as strong antagonist of *F. udum* and reported that this strain had an ability to produce hydrogen cyanide (HCN) and siderophore in excessive amount. Sometimes HCN is called as prussic acid (Gail et al. 2005). These bacterial origin volatile compounds produced by many fluorescent pseudomonads in the exponential growth phase in media containing $FeCl_3$ or inorganic phosphate may also influence plant root pathogen (Voisard et al. 1989) and suppresses the diseases (Glick 1995).

The efficacy and comparison of different biological control agents and their products studied by Sawant et al. (2003) against wilt of pigeon pea showed reduced wilt incidence by *Trichoderma* spp., and seed treatment with its formulated cell mass at 8 g/kg seed recorded the lowest wilt incidence. Many mutational and recombinant bio-inoculants have been experimented in this field to reduce the wilt incidence and found to be successful.

Roy and Sitansu (2005) published a research paper on biological control potential of some mutants of *T. harzianum* against wilt of pigeon pea and reported that recombinant *T. harzianum*, 50Th3II and 125Th4I reduced the wilt disease in non-sterilized soil, while 75Th4IV reduced the wilt disease in sterilized soil with a percentage of 36.51 %, 33.86 % and 33.33 % respectively. The application of *Trichoderma* spp. for managing FW of pigeon pea has been recommended by Mandhare and Suryawanshi (2005) as a seed treatment and soil application. The efficacy of *Trichoderma* spp. against pigeon pea wilt caused by *F. udum* was studied by Jayalakshmi et al. (2003). The observation of the study suggested that the seed of pigeon pea treated with *T. viride* followed by *T. harzianum* was found to be effective in reduction of the wilt disease by controlling *F. udum* effectively, when compared with individual treatments. In 2006, differential efficacy of bioagents namely *T. viride*, *T. harzianum* and *Gliocladium virens were* combined used by Dhar et al. (2006) against *F. udum* isolates and showed up to 35.5–57.3 % of reduction in disease incidence in FW of pigeon pea.

Burkholderia spp. reported as potential biological control agent (Heydari and Misaghi 1998; Zaki et al. 1998). Pandey and Maheshwari (2007) studied on bioformulation of *Burkholderia* spp. and reported antifungal properties against *F. udum*. These properties were due to an antibiotic 2-hydroxymethyl-chroman-4-one produced by *Burkholderia* spp.

Several *Bacillus* spp. have been proved to be used as bio-control agents for reduction of pathogen growth and disease incidence across the world (Siddiqui 2006). Many scientific evidences are available in literature, which have been reported that *Bacillus* species, most commonly found soil bacteria are excellent biocontrol agent (Dal-Soo et al. 1997; Bacon et al. 2001; Basha and Ulaganathan 2002; Chaurasia et al. 2005). Bapat and Shar (2000) used *B. brevis* as biological control agent of FW of pigeon pea as it produce antibiotic substance, which inhibit the growth of *F. oxysporum* and F. *udum* pathogen. Pandey et al. (2006) isolated HCN producing *Bacillus* spp. under in-vitro conditions. This inorganic compound reduces the radial growth of *F. udum*. Siddiqui and Shakeel (2007) screened *Bacillus* strains (B603, B613, B615) which had biological control potential against wilt disease of pigeon pea (*C. cajan*) under greenhouse and small-Scale field conditions. He found these agents can be used against *F. udum*, in both pot and field experiments and reported to be effective in terms of reduction in fungal growth and disease incidence. In 2008, Maisuria et al. (2008) reported *Pantoea dispersa* as biological control agent for FW of pigeon pea in field assessment.

Integrated management was recommended by Mahesh et al. (2010) in a combined way such as systemic fungicide, biological control agent and farmyard management as one of the most effective treatment of *F. udum* to control its infestation globally. The study showed considerable efficacy in controlling wilt incidence and increasing yield compared to untreated control with mean wilt incidence of 63.53 % and an yield of 362.72 kg/ha. Recent reports (Ram and Pandey 2011) suggested the combined use of *T. viride* and *P. fluorescens* for reduction of growth of *F. udum*. In 2011, by Gopalakrishnan et al. (2011) isolated five strains of *Streptomyces* spp. (CAI-24, CAI-121, CAI-127, KAI-32 and KAI-90) from herbal vermicomposting and reported that they have potential for biological control of FW.

Challenges Raised to Biological Agents in Disease Diagnostics

Field Application

Antagonistic microorganism and its formulation application influences the success of field trials, they are; (1) seed inoculation, (2) vegetative part inoculation, and (3) soil inoculation Several factors like, organic matter (%), pH, nutrient level, and moisture level of the soil influences the potential of antagonists from in vitro tests and efficacy of biological control agents and they often fail to work effectively (Lee et al. 1999).

Mixtures of Multiple Antagonists and Their Efficacy

Several microorganisms and its association are needed to control most pathogens in field. The appropriate combination of the microbial strains and their efficacy against pathogen can be significantly achieved with a higher level of protection (Becker et al. 1997; Raupach and Kloepper 1998; de Boer et al. 2003; Davelos et al. 2004).

Genetic Manipulation

The molecular techniques have been employed for strains modification to improve their ability against the soil borne pathogens. Advanced technologies in molecular genetics and genomics are been introduced to enhance new possibilities for improving the characterization, selection and management of biological control. The development in functional genomics-proteomics can give us the expression of crucial genes of biological control agents during mass production, application and mechanism of action. The major challenges in genetic manipulation of biological agents for disease diagnostic are the insertion of appropriate genes that express their antagonists to achieve the efficient control over plant pathogen (Baker 1989).

Whole-Genome Analysis

The revolutionary high throughput DNA sequencing of whole genomes have resulted tremendous success for understanding the mechanism of action of biological agents. The construction of artificial chromosome viz. bacterial artificial chromosome (BAC) and yeast artificial chromosome (YAC) libraries gene expression study and identification of genes of interest is of great value, especially in bacteria whose genome has not been sequenced, but having promising disease diagnostic potential (Rondon et al. 1999).

Formulation and Methods of Application

A correct formulation and right method of application of biological agents and its formulation are the major challenges. There is a lack of best alternative to come out of these challenges because formulations are being carried out without methodology. Greater efficacy, increased safety, lower production costs, ease of handling and compatibility with agricultural practices are major advantages of formulation.

Opportunities for Future Research

Significant efforts to broaden the genetic base and introduction of various traits for desirable biotic and abiotic stress are one of the important aspects of "Biological Agents in *Fusarium* Wilt (FW) Diagnostic for Sustainable Pigeon Pea Production, Opportunities and Challenges". Currently, fundamental knowledge in computing, molecular biology, biotechnology, statistics and chemistry have led to new research aimed at characterizing the functions of biological agents, pathogens, and host plants at sub-cellular and ecological levels. Biological agents in disease diagnostic are of supreme importance in the present crop production scenario, but its potential is still to be utilized and needs attention to produce the commercial formulations. Biological agents and their formulations are commercially available in market. But not getting adequate attention due to lack of information regarding its importance and use for sustainable production. Many research challenges are raised in this area to explore the biological agents for diagnostic of plant disease and have already been discussed above in five major points (see sections "Field Application", "Mixtures of Multiple Antagonists and Their Efficacy", "Genetic Manipulation", "Whole-Genome Analysis", and "Formulation and Methods of Application"). The challenges need to be addressed by the scientific community to solve the issue of use of multiple biological agents, their combined action on diagnostic of plan disease by controlling the plant pathogen.

References

Adu-Gyamfi JJ, Yoneyama IO, Devi TG, Katayama K (1997) Nitrogen management and biological nitrogen fixation in sorghum/pigeon pea intercropping on alfisols of the semi-arid tropics. Soil Sci Plant Nutr 43:1061–1066

Ajay BC, Prasad PS, Gowda MB, Ganapathy KN, Gnanesh BN, Fiyaz RA, Veerakumar GN, Babu HP, Venkatesha SC, Ramya KT (2013) Inheritance of resistance to Bangalore race of *Fusarium* wilt disease in pigeon pea (*Cajanus cajan* L.). Aust J Crop Sci 7(10):1520–1524

Amaefule KU, Onwudike OC (2000) Comparative evaluation of the processing methods of pigeon pea seeds (*Cajanus cajan*) as protein source for broilers. J Sustain Agric Environ 1:134–136

Anjaiah V, Cornelis P, Koedam N (2003) Effect of genotype and root colonization in biological control of Fusarium wilts in pigeon pea and chickpea by *Pseudomonas aeruginosa* PNA1. Can J Microbiol 49:85–91

Arinathan V, Mohan VR, John de Britto A (2003) Chemical composition of certain tribal pulses in South India. Int J Food Sci Nutr 54:209–217

Arinathan V, Mohan VR, Maruthupandian A, Athiperumalsami T (2009) Chemical evaluation of raw seeds of certain tribal pulses in Tamil Nadu. India Tropic Subtrop Agroecosyst 10:287–294

Arora SK (1977) Legume carbohydrates: in chemistry and biochemistry of legumes. Edited by Oxford and IBH, New Delhi

Bacon CW, Yates IE, Hinton DM, Meredith F (2001) Biocontrol of *Fusarium moniliforme* in maize. Environ Health Perspect 109:325–332

Baker R (1989) Some perspectives on the application of molecular approaches to biocontrol problems. In: Whipps M, Lumsden RD (eds) Biotechnology of fungi for improving plant growth. Cambridge University Press, Cambridge, pp 220–223

Bapat S, Shar AK (2000) Biological control of Fusarial wilt of pigeon pea by *Bacillus brevis*. Can J Microbiol 46(2):125–132
Basha S, Ulaganathan K (2002) Antagonism of *Bacillus* species (strain BC121) towards *Curvularia lunata*. Curr Sci 82:1457–1463
Becker DM, Kinkel LL, Schottel JL (1997) Evidence for interspecies communication and its potential role in pathogen suppression in a naturally occurring disease suppressive soil. Can J Microbiol 43:985–990
Bhatnagar H (1996) Influence of environmental conditions on antagonistic activity of *Trichoderma* spp. against *Fusarium udum*. Indian J Mycol Plant Pathol 26:58–63
Biswas KK, Das ND (1999) Biological control of pigeon pea wilt caused by *Fusarium udum* with *Trichoderma* spp. Ann Plant Prot Sci 7:46–50
Booth C (1971) The genus *Fusarium* (pp. Common Wealth Mycological Institute, Kew, p 114
Bressani R, Elias SG (1974) Nutritive value of legume crops for humans and animals. In: Summerfield RJ, Bunting AH (eds) Advances in legume science. Royal Botanical Gardens, London
Butler EJ (1906) The wilt disease of pigeon pea and pepper. Agric J India 1:25–26
Butler EJ (1910) The wilt disease of pigeonpea and parasitism of *Neocosmospora vasinfecta* Smith. Memories of the Department of Agriculture in India. Bot Ser 2:1–64
Chaurasia B, Pandey A, Palini LMS, Trivedi P, Kumar B, Colvin N (2005) Diffusible and volatile compounds produced by an antagonistic *Bacillus subtilis* strain cause structural deformations in pathogenic fungi in vitro. Microbiol Res 160:75–81
Chérif M, Benhamou N (1990) Cytochemical aspects of chitin breakdown during the parasitic action of a *Trichoderma* sp. On Fusarium oxysporum f. sp. radicislycopersici. Phytopathology 80:1406–1414
Chérif M, Sadafi N, Benhamou N, Hajlaoui MR, Boubaker A, Tirilly Y (2002) Ultrastructure and cytochemistry of *in vitro* interactions of the antagonistic bacteria *Bacillus cereus* X16 and *B. thuringiensis*55T with *Fusarium roseum* var. *sambucinum*. J Plant Pathol 84:83–93
Chérif M, Sadfi N, Ouellette GB (2003) Ultrastructure and cytochemistry of in vivo interactions of the antagonistic bacteria *Bacillus cereus* X16 and *B. thuringiensis*55T with *Fusariumroseum*v ar.*sambucinum*, the causal agent of potato dry rot. Phytopathol Mediterr 42:41–54
Chérif M, Arfaoui A, Rhaiem A (2007) Phenolic compounds and their role in bio-control and resistance of chickpea to fungal pathogenic attacks. Tunis J Plant Prot 2:7–21
Cook RJ, Baker KF (1983) The nature and practice of biological control of plant pathogens. Ame. Phytopat. Soci, St. Paul
Dal-Soo K, Cook RJ, Weller DM (1997) *Bacillus* sp. L324-92 for biological control of three root diseases of wheat grown with reduced tillage. Biol Control 87:551–558
Datta J, Lal N (2013) Genetic diversity of *Fusarium* wilt races of pigeon pea in major regions of India. Af Crop Sci J 21(3):201–211
Davelos AL, Kinkel LL, Samac DA (2004) Spatial variation in frequency and intensity of antibiotic interactions among Streptomycetes from prairie soil. Appl Environ Microbiol 70:1051–1058
de Boer M, Bom P, Kindt F, Keurentjes JJB, van der Sluis I, van Loon LC, Bajjer PAHM (2003) Control of Fusarium wilt of radish by combining Pseudomonas putida strains that have different disease-suppressive mechanisms. Phytopathology 93:626–632
Dhar V, Mishra S, Chaudhary RG (2005) Major diseases of pigeon pea and their management. In: Ali M, Kumar S (eds) Advances in pigeon pea research. Indian Institute of Pulses Research, Kanpur, pp 229–261
Dhar V, Mishra S, Chaudhary RG (2006) Differential efficacy of bioagents against *Fusarium udum* isolates. Indian Phytopathol 59:290–293
FAOSTAT (2012) Statistical Databases of Food and Agriculture Organization of the United Nations (http://www.faostat.org/)
Fuchs JC, Moënne-Loccoz Y, Défago G (1997) Nonpathogenic *Fusarium oxysporum* strain Fo47 induces resistance to Fusarium wilt in tomato. Plant Dis 81:492–496

Gail E, Gos S, Kulzer R, Lorösch J, Rubo A, Sauer M (2005) Cyano compounds, inorganic, Ullmann's encyclopedia of industrial chemistry. Wiley-VCH, Weinheim

Ghadge PN, Vairagar PR, Prasad K (2008) Physical Properties of Chick Pea Split (*Cicer arietinum* L.). Agricultural Engineering International: the CIGR Ejournal. Manuscript FP 07 039. Vol. X

Glick BR, Karatuprovic DM, Newwll PC (1995) A novel procedure for rapid isolation of plant growth promoting pseudomonads. Can J Microbiol 41:533–536

Gopalakrishnan S, Pande S, Sharma M, Humayun P, Kiran BK, Sandeep D, Vidya MS, Deepthi K, Om R (2011) Evaluation of actinomycete isolates obtained from herbal vermicompost for the biological control of Fusarium wilt of chickpea. Crop Protec 30:1–29

Gopalan C, Shastri BVR, Balsubramanian SC (1985) Nutritive value of Indian foods. National Institute of Nutrition (NIN), Indian Council of Medical Research, Hyderabad

Gundappagal RC, Bidari VB (1997) *Trichoderma viride* in the integrated management of pigeon pea wilt under dryland cultivation. Adv Agric Res India 7:65–69

Gwata ET, Silim SN, Mgonja M (2006) Impact of a new source of resistance to *Fusarium* wilt in pigeon pea. J Phytopathol 154:62–64

Harman GE, Taylor AG, Stasz TE (1989) Combining effective strains of Trichoderma harzianum and solid matrix priming to provide improved biological seed treatment systems. Plant Dis 73:631–637

Heydari A, Misaghi IJ (1998) Biocontrol activity of *Burkholderia cepacia* against *Rhizoctonia solani* in herbicide-treated soils. Plant Soil 202:109–116

Hillocks RJ, Minja E, Silim SN, Subrahmanyam P (2000) Diseases and pests of pigeon pea in eastern Africa. Inter J Pest Manag 46:7–18

Homma Y, Kato Z, Hirayama F, Konno K, Shirahama H, Suzui T (1989) Production of antibiotics by *Pseudomonas cepacia* as an agent for biological control of soilborne plant pathogens. Soil Biol Biochem 21:723–728

ICRISAT (1986) Uses of tropical grain legumes, International Crop Research Institute for Semiarid and Tropics, Patancheru, A.P. 502 324, India: ICRISAT. p 130

Ihekoronye AI, Ngoddy PO (1985) Integrated food science and technology for the tropics. Macmillan Publishers, London, pp 244–246

Jain KC, Reddy MV (1995) Inheritance of resistance to Fusarium wilt in pigeon pea (*Cajanus cajan* L.). Indian J Genet Plant Breed 55:434–437

Jayalakshmi SK, Sreermula K, Benig VI (2003) Efficacy of *Trichoderma* spp. against pigeon pea wilt caused by *Fusarium udum*. J Biol Control 17:75–78

Joshi PK, Parthasarathy RP, Gowda CLL, Jones RB, Silim SN, Saxena KB, Kumar J (2001) The world chickpea and pigeon pea economies: facts, trends, and outlook. ICRISAT, Patancheru

Kaewchai S, Soytong K, Hyde KD (2009) Mycofungicides and fungal biofertilizers. Fungal Divers 38:25–50

Kannaiyan J, Nene YL (1981) Influence of wilt at different growth stages on yield loss in pigeon pea. Trop Pest Manag 27:141

Kannaiyan J, Nene YL, Reddy MV, Ryan JG, Raju TN (1984) Prevalence of pigeon pea diseases and associated crop losses in Asia. Afr Am Trop Pest Manag 30:62–71

Khan MR, Khan SM (2002) Effect of root-dip treatment with certain phosphate solubilizing microorganisms on the *Fusarial* wilt of tomato. Bioresour Technol 85:213–215

Kimani PM (1991) Pigeon pea improvement research in Kenya: an overview. In: Singh L, Silim SN, Ariyanagam RP, Reddy MV (eds) Proceedings of the First Eastern and Southern Africa Regional Legumes (Pigeonpea) workshop, pp 108–117, 25–27 June 1990, Nairobi. Eastern Africa Regional Cereals and Legumes (EARCAL) Program, International Crops Research Institute for the Semi-Arid Tropics

Lee JY, Kim BS, Lim SW, Lee BK, Kim CH, Hwang BK (1999) Field control of *Phytophthora* blight of pepper plants with antagonistic rhizobacteria and DL-α-amino-*n*butyric acid. Plant Pathol J 15(4):217–222

Leslie JF, Summerell BA (2006) Species descriptions (chapter-13). In: Leslie JF, Summerell BA (eds) The *Fusarium* laboratory manual. Blackwell Publishing, USA, pp 270–271

Long RW, Lakela O (1976) A flora of tropical Florida. Banyan Books, Miami, p 962

Madhukeshwara SS, Sheshadri VS (2001) Studies on the variation in the growth phase of different isolates of *Fusarium udum* Butler. Curr Res 30:157–158

Mahesh M, Saifulla M, Sreenivasa S, Shashidhar RK (2010) Integrated management of pigeon pea wilt caused by *Fusarium udum* Butler. EJBS 2(1):1–7

Maisuria VB, Gohel V, Mehta AN, Patel RR, Chhatpar HS (2008) Biological control of *Fusarium* wilt of pigeon pea by *Pantoea dispersa*, a field assessment. Ann Microbiol 58(3):411–419

Mallikarjuna N, Saxena KB, Jadhav DR (2011) Cajanus. In: Kole C (ed) Wild crop relatives: genomic and breeding resources, legume crops and forages. Springer-Verlag, Berlin, pp 21–23

Mandeel Q, Baker R (1991) Mechanisms involved in biological control of *Fusarium* wilt of cucumber with strains of non-pathogenic *Fusarium oxysporum*. Phytopathology 81:462–469

Mandhare VK, Suryawanshi AV (2005) Application of *Trichoderma* species against pigeon pea wilt. JNKVV Res J 32(2):99–100

Marley PS, Hillocks RJ (1996) Effect of root-knot nematodes (*Meloidogyne* spp.) on fusarium wilt in pigeon pea (*Cajanus cajan*). Field Crops Res 46:15–20

Mbwaga AM (1995) *Fusarium* wilt screening in Tanzania. In: Silim SN, King SB, Tuwaje S (eds) Improvement of pigeon pea in Eastern and Southern Africa, pp. 101–102, Annual Research Planning Meeting 1994, 21–23 Sept 1994, Nairobi, Kenya. : International Crops Research Institute for the Semi-Arid Tropics, Patancheru

Miller SA, Beed FD, Harmon CL (2009) Plant disease diagnostic capabilities and networks. Annu Rev Phytopathol 47:15–38

Mishra S (2004) Studies on variability in *Fusarium udum* Butler, the pathogen of pigeon pea wilt disease and identification of resistant donors. PhD thesis, CSJM University, Kanpur

Nene YL (1980) Proc Consultants Group. Discussion on resistance to soil borne diseases in Legumes. ICRISAT, India

Nene YL, Kannaiyan J, Reddy MV, Zote KK, Mahmood M, Hiremath RV, Shukla P, Kotasthane SR, Sengupta K, Jha PK, Haque MF, Grewal JS (1989) Multilocational testing of Pigeon pea for broad based resistance to *Fusarium* wilt resistance in India. Indian Phytopathol 42:449–453

Nene YL, Sheeila VK, Sharma SB (1996) A world list of chickpea and pigeon pea pathogens, 5th edn. ICRISAT, Patancheru, pp 19–20

Onyebuashi COA (1986) Characteristic of water extraction protein from Cajanus Caj art: MSc thesis. Department of Food Science and Technology, University of Nigeria, Nsukka

Padwick GW (1940) Genus *Fusarium* 5 : *Fusarium udum* Butler, *F. vasinfectum* Atk. And *F. lateritium* var. *uncinatum* Wr. Indian J Agric Sci 10:863–878

Pandey P, Maheshwari DK (2007) Bioformulation of *Burkholderia* sp. MSSP with a multispecies consortium for growth promotion of Cajanus cajan. Can J Microbiol 53:213–222

Pandey KK, Upadhyay JP (1997) Selection of potential bio-control agents based on production of volatile and non-volatile antibiotics. Veg Sci 24:144–146

Pandey P, Saraf M, Dubey RC, Maheshwari DK (2006) Application of fusaria in agricultural and industrial biotechnology. In: Maheshwari DK, Dubey RC (eds) Biotechnological applications of microorganisms: a techno-commercial approach. I. K. International Publishing House, New Delhi, pp 199–212

Patel SI, Patel BM (2012) Pigeon pea wilt and its management: a review. AGRES Int J 1(4):400–413

Patil PY (1984) Study on pigeon pea wilt. PhD dissertation, Mahatma Phule Agricultural University, Rahuri

Pawar NB, Mayee CD (1983) Virulence differences in *Fusarium udum* isolates from Maharashtra State. ICPN 2:41–42

Prasad RD, Rangeshwaran R, Hegde SV, Anuroop CP (2002) Effect of soil and seed application of *Trichoderma haezianum* on pigeon pea wilt caused by *Fusarium udum* under field conditions. Crop Prot 21:293–297

Prasad P, Reddy NP, Anandam RJ, Lakshmikantha RG (2003) Isozymes variability among *Fusarium* udum resistant cultivars of pigeon pea (*Cajanus cajan* (L.) (Millsp). Acta Physiol Plant 25:221–228

Pusey PL (1989) Use of *Bacillus subtilis* and related organisms as biofungicides. Pestichem Sci 27:133–140

Ram H, Pandey RN (2011) Efficacy of bio-control agents and fungicides in the management of wilt of pigeon pea. Indian Phytopathol 64(3):269–271

Raupach GS, Kloepper JW (1998) Mixtures of plant growth promoting rhizobacteria enhance biological control of multiple cucumber pathogens. Phytopathology 88:1158–1164

Reddy MV, Sharma SB, Nene YL (1990) Pigeon pea: disease management. In: Nene YL, Hall SD, Sheila VK (eds) The pigeon pea. CAB International, Wallingford, pp 303–347

Reddy MV, Raju TN, Sharma SB, Nene YL, McDonald D (1993b) Handbook of pigeon pea diseases. In: En. Summaries in En., Fr. Information Bulletin No. 42. International Crops Research Institute for the Semi-Arid Tropics, Patancheru

Reddy MV, Raju TN, Sharma SB, Nene YL, McDonald D (1993a) Handbook of pigeobpea diseases. In: En. Summaries in En. Fr., Information Bulletin No. 14. International Crops Research Institute for the Semi-Arid Tropics, Patancheru

Reddy MV, Dhar V, Lene JM, Raju TN (1996) Proceedings of the Asian Pigeonpea Pathologists Group Meeting and Monitoring Tour, held during 20–25 Nov 1995, ICRISAT, Andhra Pradesh. International Crops Research Institute for the Semi-Arid Tropics, Patancheru

Rockland LB, Radke TM (1981) Legume protein quality. Food Technol 35(3):79

Romero D, de Vicente A, Zeriouh H, Cazorla FM, Fernandez-Ortuno D, Tores JA, Perez-Garcia A (2007) Evaluation of biological control agents for managing cucurbit powdery mildew on greenhouse-grown melon. Plant Pathol 56:976–986

Rondon MR, Raffel SJ, Goodman RM, Handelsman J (1999) Toward functional genomics in bacteria: analysis of gene expression in *Escherichia coli* from a bacterial artificial chromosome library of *Bacillus cereus*. Proc Natl Acad Sci U S A 96:6451–6455

Roy A, Sitansu P (2005) Biological control potential of some mutants of *Trichoderma harzianum* and *Gliocladium virens* against wilt of pigeon pea. J Interacademicia 9:494–497

Sawant DM, Kolase SV, Bachkar CB (2003) Efficacy of bio-control agents against wilt of pigeon pea. J Maharastra Agric Univ 28:303–304

Saxena KB, Kumar RV, Rao PV (2002) Pigeon pea nutrition and its improvement. J Crop Prod 5:227–260

Sharma KK, Sreelatha G, Dayal S (2006) Pigeonpea (*Cajanus cajn* (L.) Millsp.). In: Wang K (ed) *Agrobacterium* protocols. Methods in molecular biology, vol. 343. Humana Press, Totowa, New Jersey, pp 359–368

Sheldrake R, Narayanan A, Kannaiyan J (1984) Some effects of the physiological state of pigeon pea on the incidence of wilt disease. Trop Grain Legum Bull 11:24–25

Siddiqui ZA (2006) PGPR: prospective biocontrol agents of plant pathogens. In: Siddiqui ZA (ed) Biocontrol and biofertilization. Springer, Amsterdam, pp 111–142

Siddiqui ZA, Shakeel U (2007) Screening of *Bacillus* isolates for potential biocontrol of the wilt disease complex of pigeon pea (*Cajanus cajan*) under greenhouse and small-Scale field conditions. J Plant Pathol 89(2):179–183

Siddiqui S, Siddiqui ZA, Iqbal A (2005) Evaluation of fluorescent pseudomonads and *Bacillus* isolates for the biocontrol of wilt disease complex of pigeon pea. World J Microbiol Biotechnol 21:729–732

Siddiqui ZA, Shakeel U, Siddiqui S (2008) Biocontrol of wilt disease complex of pigeon pea by fluorescent pseudomonads and *Bacillus* spp. under pot and field conditions. Acta Phytopathol Entomol Hung 43:77–92

Singh U, Jain RC, Jambunathan RJ, Faris PG (1984) Nutritive value of vegetable pigeon pea, Mineral trace elements. J Food Sci 49:645–646

Singh R, Singh BK, Upadhyaya SK, Rai B, Lee YS (2002) Biological control of Fusarium wilt disease of pigeon pea. J Plant Pathol 18(5):279–283

Somasekhara YM, Anilkumar TB, Siddaramaiah AL (1996) Bio-control of pigeon pea [*Cajanus cajan* (L.) Millsp.] wilt (*Fusarium udum* Butler). Mysore J Agric Sci 30:159–163

Somasekhara YM, Siddaramaiah AL, Anilkumar TB (1998) Evaluation of *Trichoderma* isolates and their antifungal extracts as potential bio-control agents against pigeon pea wilt pathogen, *Fusarium udum* Butler. Curr Res 27:158–160

Soris PT, Mohan VR (2011) Chemical analysis and nutritional assessment of two less known pulses of Genus *Vigna*. Trop Subtrop Agroecosyst 14:473–484

Suarez-Estrella F, Vargas-Garcia C, Lopez MJ, Capel C, Moreno J (2007) Antagonistic activity of bacteria and fungi from horticultural compost against *Fusarium oxysporum* f. sp. *melonis*. Crop Prot 26:46–53

Subramanian S (1963) *Fusarium* wilt of pigeon pea I. Symptomatology and infection studies. Proc Indi Acad Sci B 57:134–138

Udedibie ABI, Igwe FO (1989) Dry matter yield and chemical composition of pigeon pea (*Cajans cajan*) leaf meal and the nutritive value of pigeon pea leaf meal grain meal for laying hens. Anim Feed Sci Technol 24:111–119

Upadhyay RS, Rai B (1981) Effect of cultural practices and soil treatments on incidence of wilt disease of pigeon pea. Plant and Soil 62:309–312

Upadhyay RS, Rai B (1992) Wilt of pigeon pea. In: Singh AN, Mukhopadhyay J, Kumar J, Chaube HS (eds) Plant diseases of international importance, vol I, Cereals and Pulses. Prentice-Hall, New Jersey, pp 388–414

Vaidya RJ, Shah IM, Vyas PR, Chhatpar HS (2001) Production of chitinase and its optimization from a novel isolate *Alcaligenes xylosoxydans*: potential in antifungal biocontrol. World J Microbiol Biotechnol 17:691–696

Vaidya RJ, Macmil SL, Vyas PR, Ghetiya LV, Thakor KJ, Chhatpar HS (2003) Biological control of *Fusarium* wilt of pigeon pea *Cajanus cajan* (L.) Millsp with chitinolytic Alcaligenes xylosoxydans. Indian J Exp Biol 41(12):1469–1472

van der Maesen LJG (1990) Pigeon pea: origin, history, evolution, and taxonomy. In: Nene YL, Hall SD, Sheila VK (eds) The pigeon pea. CAB International, Wallingford, pp 15–46

Vinale F, Sivasithamparamb K, Ghisalbertic EL, Marraa R, Wooa SL, Moritoa L (2008) *Trichoderma* plant pathogen interactions. Soil Biol Biochem 40:1–10

Voisard C, Keel C, Haas D, Defago G (1989) Cyanide production by *Pseudomonas fluorescens* helps suppress black root rot of tobacco under gnotobiotic conditions. EMBO J 8:351–358

Weller DM, Zhang BX, Cook RJ (1985) Application of a rapid screening test for selection of bacteria suppressive to take-all of wheat. Plant Dis 68:710–713

Whipps JM, McQuilken MP (2009) Biological control agents in plant disease control. In: Walters D (ed) Disease control in crops-biological and environmentally friendly approaches. Wiley Blackwell, Oxford, UK. pp 27–61

Zaki K, Misaghi IJ, Heydari A (1998) Control of cotton seedling damping-off in the field by *Burkholderia* (*Pseudomonas*) *cepacia*. Plant Dis 82:291–293

Recent Diagnostics and Detection Tools: Implications for Plant Pathogenic *Alternaria* and Their Disease Management

Dipak T. Nagrale, Lalan Sharma, Sunil Kumar, and Shailesh P. Gawande

Introduction

Plant diseases caused by *Alternaria* species are very common all over the world. They occurred on numerous field crops, vegetables, plantations, forest plants and in post harvest storage conditions. The genus *Alternaria* includes about 100 species which are reported worldwide over various agroclimtic zones and ecosystems and they are among the important phytopathogenic fungi. It is evidence that minimum 20 % of agricultural produce spoilage is caused by phytopathogenic *Alternaria* species. Many *Alternaria* species are known to produce a variety of secondary metabolite in fruits, vegetables and field crops; hence it plays important roles in quality of food items. Mycotoxins and secondary metabolites produced by *Alternaria* spp. mainly through fruits, vegetables and field crops is also a matter of concern (Andersen et al. 2015). Most *Alternaria* species are plant pathogenic and some are saprophytes has potential to produce a variety of secondary metabolites, which play important role in plant quarantine, plant pathology, food quality and safety of agro based products (EFSA 2011; Logrieco et al. 2009). Isolation, detection, identification and management of seed borne pathogens through various quality control practices from initial seeds harvest, purchase, marketing and applications in the agricultural fields are important for quality assurance of seeds, disease free material and genetic purity of seeds. The detection of these plant pathogens was generally done by conventional methods. However, with the advancement of technology, nucleic acid based

D.T. Nagrale (✉) • S.P. Gawande
ICAR- Central Institute for Cotton Research,
Panjari, Wardha Road, Nagpur 441108, Maharashtra, India
e-mail: dip29unique@gmail.com

L. Sharma • S. Kumar
ICAR- National Bureau of Agriculturally Important Microorganisms,
Kushmaur, Maunath Bhanjan 275103, Uttar Pradesh, India

P. Kumar et al. (eds.), *Current Trends in Plant Disease Diagnostics and Management Practices*, Fungal Biology, DOI 10.1007/978-3-319-27312-9_5

molecular methods (Chahal and Pannu 1997; Lopez et al. 2003), which are very specific, rapid and more reliable than conventional methods, were developed for rapid and accurate detection of seed borne plant pathogens. The molecular diagnostics approaches developed rapidly after the development of PCR based techniques in 1980s. In seed pathology, these diagnostic assays based on the PCR variants have been developed for seed borne diseases of wheat and other field food crops. With the rapid advancement of genome wide techniques for detection and characterization of plant pathogens in recent era had made the techniques more simple and rapid for pathogen detection and identification in disease complex. There are various species and strains of same genus, so it is important to have highly specific, rapid and reliable methods of identifying plant pathogens to develop strategies for effective plant disease management to minimize crop losses in agricultural produce and further prevention of disease spread.

Symptomatology

Various symptoms of diseases are caused by fungal plant pathogens on infected plants. They are localized (restricted to infected plant tissues/organs or/and systemic infections (away from court of infection or inside the plant system), which are produced by pathogenic fungi. Generally, foliar plant parts like leaves, stems and reproductive organs may show localized symptoms. Common symptoms like spots, blights in severe infections, shot holes appearance, rusts, and powdery mildews are observed mainly on foliar parts. Fruit spots and rots, fruit drops, head rots, grain discolourations are generally noticed when reproductive structures are infected with pathogens. Initially, the infection is restricted to tissues or organs at the point of infection. The systemic cause of pathogens infection leads to development of symptoms like damping off, smuts, downy mildews, root/stem rots and smuts as the pathogen spread from point of infection to other organs of host from site of infection. The recognition of systemic infection notices after long incubation period which are developed after initial stages of infection to the developed visible symptoms. The blights diseases are among major concern in crop production which cause heavy yield losses up to 32–57 %, despite several diseases by *Alternaria* spp. (Conn and Tewari 1990). The typical symptoms of this disease consist of small, circular to irregular brown to dark coloured necrotic spots on leaves, pods, fruits, twigs and tender plant parts with characteristics concentric rings of developed spots (Valkonen and Koponen 1990). As the disease progress, the spots coalesce to form large chlorotic patches, shot holes appearance on leaves in some cases, drying and blightening of leaves.

In case of apple infection by *Alternaria mali*, the infection caused through the lenticels but fruit does not rot in field stage or in storage conditions (Sawamura 1990). Ganie et al. (2013) reported that during earlier infection of early blight disease (*Alternaria solani*) of potato (*Solanum tuberosum* L.) in Kashmir valley, the spots are small, circular to irregular, light to dark brown spots appears on lower leaves, measuring up to 0.5 mm in size. As the disease progresses, the typically concentric ring are formed at the end of July as result of growth pattern by the pathogen in the foliar tissues. These lesions give 'target spot' like or 'bulls eye' appearance with disease development. The disease lesion enlarges maximum to 7.4 mm in first fortnight of August. Likewise, Kameniecki et al. (2013) observed small spots and lesions on the margins of *Terminalia australis* leaves after 20 days of inoculation of *Alternaria*. The diseased foliage mostly showed withering and necrosis within 2 weeks. In case of severe infection, leaves are dropped off from plants. However, it has been reported that symptoms do not showed any visible symptoms even for long period of time, mostly in case of perennial crops. The early blight symptoms are caused by both *A. solani* and *A. alternata*. Several researchers have simultaneously isolated these pathogens from diseased plants (Babler et al. 2004). Several ecotypes of the *Alternaria* blight pathogens are observed worldwide in various agro-ecosystems and forests. The early blight is major disease of tomatoes in northern USA which is caused by *Alternaria solani* (El 1. and G. Martin) Sor. The symptoms of the disease include dark spots with typical concentric rings, reported first on first matured lower leaves.

The host factors like level of resistance/susceptibility of germplasm or cultivars, edaphic factors, environmental conditions and degree of virulence of the pathogen decides the disease severity and induced symptoms caused by fungal pathogens. Various disease measurement scale/charts are available which represents level of disease severity by foliar fungal pathogens. Recently, the researchers have developed a digital image analysis method to measure disease severity for several foliar fungal diseases. The flatbed scanner or digital camera captures the image and the data is analyzed by available software package. Especially, scion image method is utilized to measure the foliar colour changes or tissues damage by fungal infections or their sporulation in diseased plants. Wijekoon et al. (2008) reported that Digital Image analysis using scion image method can be utilized to detect diseases early and rapidly it can quantify wide range of foliar fungal diseases. *Alternaria* species are known to attack all the aerial parts of its host plant. In succulent leaves like vegetables and ornamentals, symptoms of fungi infection initially appears as small, circular and light brown to dark spots. As the disease progresses, these spots enlarge in diameter usually 1 cm or more and turns brown to dark grey in colour. The disease on wheat plants appeared when plants are 7–8 weeks old and becomes severe when the crop is mature. Symptoms appeared as small spot, oval shape, discoloured irregular lesions and scattered on the leaves. Finally turns dark brown to black in colour, coalesce and cover entire leaves. The older leaves are more susceptible than younger leaves. Infection at the milking/dough stage of seed, the disease advances to glume, earhead and seed. The disease symptoms in carrots known as Alternariose may be develop on visible plant parts like leaves, stems, umbels and diseased seeds. Normally, *Alternaria* leaf spot symptoms first appear along the margin of the older leaves and later develops into irregular shape and turns dark brown to black in colour. However, the spots/lesions developed on the petioles and stems are dark brown in colour. As the disease progresses the whole leaves may wither, turns blighted and drop down. Akhtar et al. (2004) observed symptoms first on tomato leaves by tomato blight pathogen *Alternaria alternata* in Pakistan, initially started with yellowing and browning of older leaves and severity of disease increases with high humidity and congenial weather. The symptoms are often develop at leaf tip and get progressed towards leaf petiole margins. In case of favourable weather and heavy infection of pathogen, diseased lesions enlarge rapidly, coalesce and results in blightening of leaves and shoots. Under high humidity, the dark coloured sporulation can be seen from concentric ring layers of blighted leaves. In such cases, defoliation of flowers and shoots is observed with heavy yield losses. The environmental conditions plays major role in development of pathogens growth and subsequent disease developments. Thus, the growth of pathogen or spots appears as typical concentric rings.

Leaf spots caused by *Alternaria* species may kill plants also reduce aesthetic value of ornamentals and floricultural plants. These spots are the diagnostic for *Alternaria* diseases but there are few other pathogens having similar type of symptoms. In case of severe infections, the shot hole symptom develops with blightening of the aerial/ foliar parts. However, in high relative humidity, brown to grey colour mycelia fuzzy growth develops on lesion of host. This is due to sporulation of fungus and death of host tissues leading to starvation. Many *Alternaria* species are known to produce toxins which degrade the host cells and affects many normal physiological functions of host plants. Light to dark brown sunken spots caused by *Alternaria* species can be seen on various plant parts like leaves, fruits, stems, roots and rotting under storage conditions. Grayish to velvety growth of fungi may results on infected portion in humid conditions. Many foliar plant pathogens synthesized the metabolites which has role in development of disease symptoms (Strange 2007). In some cases, the host-pathogen interactions resulted in a favorable environment for the production of many different chemical compounds which has economical importance in various fields. Agostini et al. (2003) reported that presence of mycotoxins in citrus plants for *Alternaria* brown spot disease (*Alternaria alternata* pv. *citri*) have been a reason of serious concern for citrus growers all over the world. These mycotoxins impart lesions on younger and tender leaves, branches and fruits with effects on almost the whole citrus plant. On citrus fruits, the spots are minute initially which turns into deep depressed lesions (Agostini et al. 2003; Akimitsu et al. 2003; Peres and Timmer 2006). In case of severe infections, the infected spots or pustules, the eruptions are common which are dislodged/distorted, irregular in shape, reducing fruits quality and yields and eventually the market price of fruits (Akimitsu et al. 2003).

The Pathogen

The genus *Alternaria* Nees ex Fr. Classified into sub-division Pezizomycotina (Deuteromycotina) class Dothideomycetes (Hyphomycetes), order Pleosporales, family Pleosporaceae (Dematiaceae). They are cosmopolitan with several species and sub species, survives as necrotropic, saprophytes and in some cases as weak pathogens. The genus is well known for the formation of typical type of polymorphic conidiophores and conidia, produced either singly or in groups, shorter or in long chains. The conidia are multicelled, featured with characteristics longitudinal as well as transverse septa and may bear short or long beaks. *Alternaria* conidia have typical and unique appearance which makes them easy to recognize their different species under a microscope (Kirk et al. 2008). The fungi is having broad host range and spores/conidia can survives in the atmosphere, soil, water, plant debris and agricultural fields. The perfect/sexual stage (telemorph) is in rarely reported few species. Hence, they are placed in genus Pleospora class class Loculoascomycetes of sub-division Ascomycotina categorizing oval, muriform ascospores in bitunicate asci. The genus *Alternaria* was first time reported by Nees in 1817. However, Berkeley (1836) isolated and identified the pathogenic fungi from the plants of Brassicaceae, was named as *Macrosporium brassicae* Berk. But, it was then renamed as *Alternaria brassicae* (Berk.) Sacc (Saccardo 1886). The taxonomic description regarding Alternaria species were studied in details by Elliott (1917). Wiltshire (1933, 1938) did pioneer work in basic studies of *Alternaria* under class hyphomycetes. He worked on descriptive and fundamental studies of several fungi especially *Alternaria*, *Macrosporium* and *Stemphylium*. However, Neergaard (1945) studied extensively on the taxonomy, classification, parasitic nature and economic importance of the genus *Alternaria*. Later, the morphological differentiations of various *Alternaria* species were studied extensively by Joly (1959) and classified them into three groups and gave a synoptical key for the identification and classification of common species (Joly 1964). Then, Ellis classified the several species of *Alternaria* on the basis of their characteristics features and grouped them as "Dematiaceious Hyphomycetes" (Ellis 1971) as well as "More Dematiaceous Hyphomycetes" (Ellis 1976). The wide host range of phytopathogenic *Alternaria* species caused heavy crop losses and post harvest damages to agricultural produce (Rimmer and Buchwaldt 1995). The transmission or spread of the pathogen spores by winds, rain splashes and infected plant parts in agricultural as well as in storage. It was reported that a minimum leaf wetness duration of 13 h and temperatures ranging from 20 to 30 °C, ideal for sporulation and infection on host (Humpherson-Jones and Phelps 1989; Rotem 1994). Wild plants like *A. thaliana*, *C. sativa* and *C. bursa-pastoris* from cruciferaceae have ported to be resistance pathogen. But, no reliable source of resistance has been noticed from agronomicaly important Brassica species (Conn et al. 1988; Westman et al. 1999; Sigareva and Earle 1999a, b). The genetic strength of the germplasm were reported to have additive as well as dominant gene action (King 1994). Spores/conidia of Alternaria spp. are dark brown to olive in colour and borne singly or in chains. Some common *Alternaria* species are as follows.

Biology

The conidiogenus mycelia or conidiohores of most of the *Alternaria* species produce asexual spores or conidia and measurement ranging between 160 and 200 μm in length. In laboratory conditions, the sporulation occurs at 10–24 °C and maturity of the conidia develops after 14–24 h. The optimum temperature for sporulation ranged between 16 and 24 °C and time up to 12–14 h. The presences of moisture or high relative humidity are very important for sporulation and infection of fungal pathogen propagules. The minimum period of 9–18 h are crucial for most of the *Alternaria* species (Humperson-Jones and Phelps 1989). High relative humidity (91.5 %) and temperature of 20 °C or sometimes in higher range will produce full grown conidia

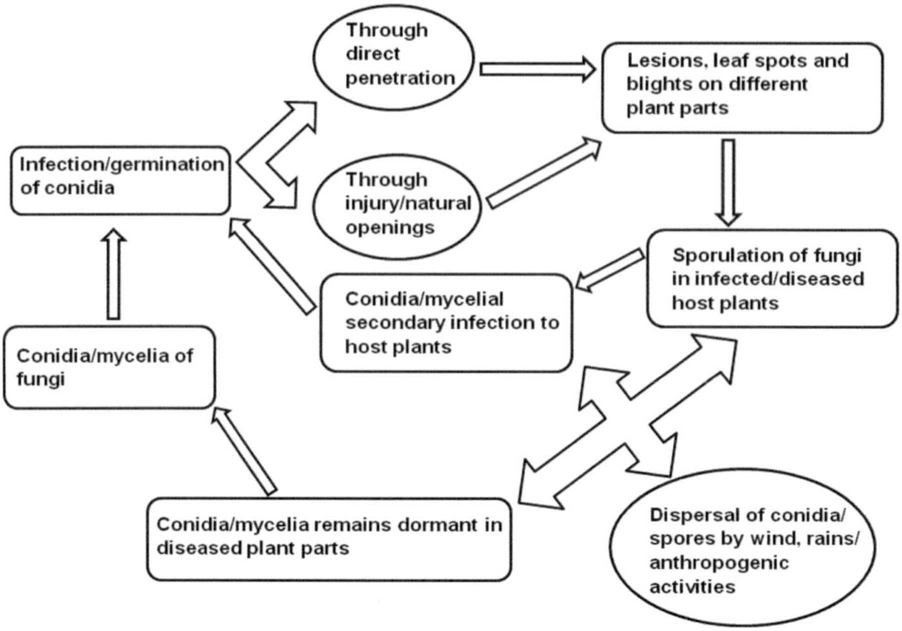

Fig. 1 Disease cycle of phytopathogenic *Alternaria* spp

in large quantities in period of 24 h (Humpherson-Jones and Phelps 1989). The plant pathogenic *Alternaria* spp. survives as spore and/or mycelium in the diseased crop debris, left over in fields and in infected seeds and/or fruits in storage conditions. The etiology of *Alternaria* diseases depends on mode of infection and carrier. In case of seed infection, they cause seedling damage, damping off, stem lesions or cankers, root or collar rot etc. However, soil moisture and relative humidity plays vital role in growth and sporulation of fungi in infected plant residues. Conidia are carried by winds, rain splashes, birds, anthropogenic activities onto healthy plant parts. The fungus can survives in alternate and/or collateral hosts, weeds or perennials in cropping systems. The free water is essential for germination, sporulation and infection of fungi. The active penetration can be occurred on host surface or passively through natural openings. The old and weak host tissues are more susceptible to infection than healthy and vigorous hosts. The *A. japonica* is carried or transmitted by seeds of radish was confirmed by seed germination tests. The pathogen propagules were noticed to present in the seed coat and can be seen in growing seedlings especially in cotyledons and epicotyl-hypocotyl. Generally, seed coats of most seeds are adhered to hypocotyls axes; hence in some cases the cotyledons may escape infection. The surface sterilization removes 90 % of *A. raphani* from the surface of seeds of radish. It was also reported that infected seeds by *Alternaria* spp. in *Brassica rapa* subsp. *Pekinensis* affected germination and produced symptoms like stripes or sharp-edged lesions on hypocotyls of developing seedling and killed them (Valkonen and Koponen 1990). The reduction in seed germination was up to 7 % where 35 % seeds may carry the fungus internally, externally or both (Fig. 1).

Physiological Disorders in Host Plant

The pathogen, *Alternaria* spp. mostly affect the host plant generally at all growth stages results in symptoms on all the aerial parts including leaves, petioles, inflorescence, stem, fruits, seeds and in some cases on roots. The pathogen mainly affects the foliar parts of plants thereby destructing the chlorophyll pigments, cell damages and normal metabolism thus reducing the photosynthetic potential of host plants. The symptoms initially appears on lower leaves or weak/older leaves as minute light brown to black spots, enlarges rapidly with round to irregular spots with concentric rings with variable shape, colour and intensity which are determined and depends on *Alternaria* spp., host attracted, genetic build up of host, nutritional level and climatic conditions. The rate of transpiration of the diseased plant increases exponentially and then declines results in malformation of leaves and inflorescence. With disease progression and severe infection, the foliar parts gets completely affected and results in foliage droppings. In case of infection to fruits, the marketing quality of fruits gets affected and cause unfavourable taste by released mycotoxins and metabolites. Similarly, in case of seeds, pods and fruits infections, the decaying is oftenly observed. Verma and Saharan (1994) noticed that in case of severe infection on the stem and pods of mustard results in premature ripening and seed shedding with heavy losses in yields. The diseases caused by various species of *Alternaria* significantly affects normal physiology of host plants and damages its cellular organization by altering the cell permeability, disturbing the cell organelles and nutrition conducting cells/tissues. The corky lesions were reported in tangerines and tangoles on surface of citrus fruits in post harvest storage by infections of *Alternaria alternata*. It may develop into scar or crater like injury and turn into large necrotic patches. Occasionally, these spots developed into decayed and rotten portion.

Agarwal et al. (1997) reported the occurrence of degenerated chloroplast or photosynthetic potential and damaged mitochondria, with less number of cristae in mustard leaves when they are severely infected by *A. brassicae*. Thus, it has been reported that *Alternaria* spp. causes several physiological disorders in plant systems infecting leaves, stems, fruits, reproductive organs, inner content of fruits etc. and severely affects normal metabolism of plant and physiology of hosts.

Epidemiology

The principal mode of carryover or transmission of pathogen are through infected plant propagating materials or seeds of infested lots where the spores/conidia/mycelia are present on seed surface or mycelia invaded in seed coat and seed surfaces. The dispersal of conidia/spores gets carried by wind, water, rain splashes, agronomic tools, farm animals and anthropogenic activities. It has also ben noticed that the *Alternaria* spp. can thrive on alternate host, susceptible weeds and/or some perennial crops or host (Chupp and Sherf 1960; Maude and Humpherson-Jones 1980b). The left over of crop residues after harvest of crops in agricultural fields are also act as major source of infection of many *Alternaria* spp. It was noticed that the diseased leaves of oil rapeseed and cabbage in storage on soil surface were known to produce viable spores till the diseased leaves remained attached to plants. The viability of oilseed rapseed remains up to 8 weeks and in case of cabbage up to 12 weeks after harvesting (Humpherson-Jones 1989). It is a well known fact that *Alternaria* has no perfect stage. Under unfavourable environmental conditions they are known to survive as mycelium/conidiospores or conidia on the left over debris, planting material, decaying plant material of previous crop (Humpherson-Jones 1989). In many cases, they are survived in alternate hosts, weeds or perennial crops (Maude and Humpherson-Jones 1980a, b). The fungal pathogen ovewinters in diseased plant parts, propagating material, fruits or/seeds of mustard for minimum of 1 year at room temperature (Shrestha et al. 2003; Ahmad and Sinha 2002). The *Alternaria* blight is major disease of rapeseed and mustard. Chattopadhyay et al. (2005) studied the time of appearance of *A. brassicae* on the host cultivar (*B. juncea* var. varuna) and the role of environmental factors on disease severity of *Alternaria* blight disease. They observed highest frequency of pathogen *A. brassicae* occurrence on leaves was between 67 and 84 DAS (days after sowing) whereas pods was between 67 and 142 DAS (days after sowing) respectively. They also reported that temperature ranging maximum from 18 to 27 °C and minimum 8–12 °C with an average RH >92 % reported as favour*able* for initiation and spread of *Alternaria* disease on the foliar parts. Likewise, temperature ranging from 20 to 30 °C along with an average RH >90 % supported with more than 9 h of sunlight and leaf wetness period of 10 h of favours the disease severity on pods. Singh and Verma (2009) studied several epidemiological parameters of *Alternaria* blight disease of *Adhatoda vasica* caused by fungi *Alternaria alternata*. The temperature plays vital role in disease development as it was noticed that disease severity was lowest in June when temperature exceeded than 34 °C and relative humidity was less than 44 %. The survival of the pathogen was longer in laboratory conditions nearly 10 months than in natural conditions (nearly 8 months). The plants were succumbing to *Alternaria alternata* infections ranging from 1 to 10 weeks of host age. The infection rate significantly increases as host becomes older up to eighth week when the disease severity on leaves recorded highest (38.57 %). In laboratory conditions, survival or viability of *Alternaria alternata* was studied at different temperatures, time periods and relative humidity on conidial/spore germination. Th result revealed that, the temperature 25 °C recorded highest germination percentage (92.32 %) and minimum of (10.58 %) at 40 °C. The germination of spores was completely inhibited at relative humidity less than 40 %. However, it was highest (95.66 %) with relative humidity of 100 %. The incremental increase in temperature from 30 to 40 °C with 30 min interval reduces spore germination. Similarly, *A. japonica* infection progresses rapidly in the temperature range of 22–26 °C in agricultural field conditions. Th severe disease intensity was recorded in boreal-temperate ecosystems where crops are grown on large scale in summer season. Seidle et al. (1995) reported that in Canada disease intensity of oil rapeseed shows variation from year to year and depends on weather parameters. The huge losses are noticed during frequent rains, greater leaf wetness period/and pod formation stage.

Detection and Diagnosis

Diagnosis of diseases caused by fungal pathogens induces certain characteristics and typical symptoms. On these basis the causative agents may predicted to some extent. However, the morphological characteristics of the fungal pathogens such as type, mycelia, shape, size and colour of sexual or asexual spore forms are generally used for their taxonomic studies and classification with certain limitations. Thus, considerable expertise is required for their identification and classification and to establish the phylogenetic relationship among those related fungi. Symptoms caused by *Alternaria* spp. are very similar to those caused by many other foliar fungal pathogens. Therefore, rapid and accurate identification of the fungal pathogen is important for effective disease management.

Morphological Characterization

Identification of plant pathogens through morphological studies are often more difficult, time consuming process and requires experience, extensive knowledge and expertise in the field of mycology. Although, it is the integral part of plant pathology for identification of fungal plant pathogens. The fungal pathogens, whose perfect stage has not been reported, are classified into Deuteromycetes (Fungi imperfecti). Most often techniques used is direct agar plating techniques (DAPI). This technique isolates the fungal plant pathogens from diseased tissues and/or infected plant samples. The agar medium commonly used for this technique is potato dextrose agar (PDA) (Lopes and Martins 2008). The inoculated petri plates are generally incubated at 27 ± 2 °C for 4–7 days. Different species of *Alternaria* grows rapidly ranging from 30 to 90 mm in size when incubated at 25 °C up to 7 days on potato glucose agar (PGA) media. The colonies reported as flat, woolly, brownish to grayish and sometimes short to aerial in growth. At initial stages the mycelia growth is whitish to grey in colour, which in turns becomes dark browns to olive brown in colour. When the reverse side of cultured petri plate is observed, it is noticed as brown to black due to pigment production(s) (Collier et al. 1998). Akhtar et al. (2004) and Shakir et al. (1997) recorded that the fungus *Alternaria alternata* was successfully isolated from the diseased plants and identified on the basis of morphological parameters. The fungus can grow well on PDA medium and produce grayish to dark brown colonies up to 90 mm in diameter when grown up to 7 days at incubation of 25 ± 2 °C. The conidia are formed in long chains, obclavate, muriform in shape, oftenly short, conical or cylindrical shape beak and generally one third lengths of the conidia. They have often three to seven transverse septa and more longitudinal or oblique arrangement of septa. Choi et al. (2014) observed that the colonies of fungus on PDA were dark brown and measured up to 47–54 mm in diameter with distinct margins and slightly raised mycelia. Whereas, on V8 juice agar medium, th colonies were profuse dark brown to black in colour and growth up to 52–60 mm in diameter with radial margins and slightly raised mycelia from the medium. It has been reported that the mycelia can grow on PDA as well as V8 juice agar medium between 5 and 35 °C with optimum temperature range of 25–30 °C. Corlett and Corlett (1999) and Malone and Muskett (1997) observed that the conidiophores of *Alternaria linicola* linseed/flax when grown on PDA are profuse in growth, occurs singly or in bunch, pale to olive brown in colour, septate mycelium and variable length ranging from 5 to 8 μm. On th other hand, conidia produce singly, smooth in appearance, olive brown in colour, long, obclavate, tapering at one end, short to long beak, muriform beak 4–16 μm, transverse septa, one to four longitudinal septa and may slightly constricted at septa.

El_Komy et al. (2012) reported that growth of *Alternaria* on potato dextrose agar (PDA) media plates when incubated at 25 °C for period of 7–10 days. When the cultures were grown at 25 °C for 6 days on PDA cultured plates near ultra violet radiation (UV) of 310–400 nm wavelength for 16 h per day, induces sporulation. Similarly, Ganie et al. (2013) reported that *A. solani* isolates of potato produces brown to dark colour mycelia, variable from grey, brown to olive brown in colour with tints. The colonies on media are cottony in texture, ranging from grey to brownish black in colour and velvety in appearance. The mycelium on PDA is septate, dark brown to black in coloured possessing grey to black tints. The conidiophores are septate, simple, short to elongated, erect, pale to olive brown in colour. They generally measured from 50 to 90 × 9 μm in size with an average conidiophores size of 60 × 7 μm. The conidia are dark coloured as light to dark brown, muriform, generally 9–11 transverse septa and 2–3 longitudinal septa. The shape of conidia is elliptical to oblong with short to long beak, occasionally as branched. The average size measured is 15–19 × 150–300 μm with mean average size of 17 × 163 μm of conidia. The beak of conidia is pale in colour and 2.0–5.0 μm wide. It is reported that infective pathogen fungus *Alternaria brassicola* of cruciferous plants, a semi selective medium (CW medium) favours better growth of fungus. It is noticed that the other fungal contaminates or saprophytic fungi growth was suppressed by CW medium (Wu and Chen 1999). Also, Strandberg (2002) developed a semi selective medium containing extracts of carrot leaves, glucose and minerals for the isolation of two fungal pathogens, *Alternaria dauci* and *A. radicina* from carrot seeds (Umbelliferae). The growth pattern of *A. dauci* conidiophores was reported by Lopes and Martins (2008) as single or in small groups. They are short to elongated, straight, three to seven longitudinal septa and their sizes ranging from 40 to 94 × 6–9 μm. Conidia may be straight or curved, obclavate, elongated beak of conidia, brown to gray in colour, maximum length size of 100–350 μm, 16–25 μm thickness at middle, 7–11 transverse septa, usually 1–4 longitudinal septa, mostly hyaline, occasionally branched and tapering at the apex. The growth of mycelium as immersed or semi-immersed, hyphae is branched, septate and light to brown in colour. The full growths of colony (7–8 days) on PDA are olivaceous to browinish in colour and cottony or velvety growth. The conidia generally are produced singly as short to moderate in length on for 7 days growth on PCA (potato carrot agar). The mycelia growth was spare, light brown to gray in the center of petri plates of MA. They observed conidia measurement of 12.5–37.5 (24.0 μm) × 7.5–12.5 (10.0 μm), mostly 0–3 longitudinal septa and beak length of 0–12.5 (4 μm). *Alternaria tenuissima* isolate had a sporulation pattern of Group 4, branching chains turns to short bushy groups branched spores chains. Conidial measurement was 0–40.0 (25.0 μm) × 5.0–17.0 (11.0 μm), mostly 0–4 longitudinal septa and a beak measurement of 0–7.5 (2.0 μm). These morphological features of conidia and conidiophores morpho-structures found silmilr with description of *A. alternata* (Fr.) Keissl. Guo-yin et al. (2013) observed that the fungus formed round grayish and aerial mycelial colonies of *Alternaria* and mycelia growth was dark green in colour with a clear zone at the backside of cultured plate. However, the colonies on potato carrot agar (PCA) were cottony white and dark grayish to olive black in colour. Their conidiophores were light brown, one or more septa and occasionally branched. Conidia were produced in long chains and occasionally as branched chains of 5–12 spores. These conidia were obclavate, dark brown in colour with 3–8 transverse septa and 0–2 longitudinal/oblique septa. The average measurement of conidia were 60.2 (20–98) × 10.4 (5.8–14.2) μm. Among all media, potato carrot agar (PCA) is best media (Roberts 2005) for growth and sporulation of plant pathogenic *Alternaraia* spp. under standard condition of incubation to observe morphological characters by extended keys by manual of Simmons (2007). These strains produced smooth, light brown conidia in branched chains, longer in shape, geniculate and conidiogenous conidiophores. However, on Dichloran Rose Bengal Yeast Extract Sucrose agar (DRYES; Samson et al. 2010), colonies were flat, woolly and pinkish white to greyish white in colour. All these strains belonged to A. infectoria sp.-grp from wheat samples. Andersen et al. (2015) studied 87 *Alternaria* strains from tomato, wheat, blueberries and walnuts. Out of these four strains belonged to the *A. infectoria* sp.-grp., 6 to the *A. arborescens* sp.-grp., 6 noticed similar sporulation pattern of "M" as per Simmons, 1 to that of *A. vaccinii*, and others 70 categorized into a diverse morphological groups "G" and "H" of Simmons (2007). In addition to these, six strains (AP015, AP016, AP018, AP022, AP073 and AP075) showed similar sporulation pattern of *A. arborescens* sp.-grp.(group "L" in Simmons (2007). The primary conidiophores were long, occasionally sub-terminal branched conidia in chains and secondary conidiophores originated from conidial apex. However, colonies growths on DRYES were sulcate and light to dark green. These six strains were obtained from tomato samples, matched the four strains for the *A. arborescens* sp.-grp. (BA0961, BA1343, BA1382 and BA1422) in morphological studies. In addition to these, six other strains from tomato (AP004, AP012, AP014, AP021, AP076 and AP095) reported a similar sporulation pattern to that of "M" in Simmons (2007). Their primary conidiophores were aerial and cobweb-like, but not Exactly arborescent. Two strains (AP004 and AP012) had similar morphological structure to *A. arborescens* sp.-grp. whereas other four had dark grey coloured colonies on DRYES. The colonies were flat, hairy texture and different from representative strains. A strain AP023 noticed a similar pattern of *A. vaccinii* E.G. Simmons (2007) with exceptionally long secondary conidiophores (up to 190 μm) with smooth conidia (16–46 × 10 μm) resembles to morphological group "G" in Simmons (2007). Colonies growth on DRYES were flat, light green with uneven edges.

The all other 70 Argentinean AP strains (Andersen et al. 2015) placed in distinct diverse group, without any identity representative strains, but these strains belonged to the morphological groups "G" and "H" in Simmons (2007). The *Alternaria* spp. collection was first done on the basis of mycelium morphology of fungal cultures, characteristics of the spores/conidia, fruiting bodies and reproductive structures (Gargouri-Kammoun et al. 2014). The fungus, *A. japonica* can be consistently isolated into pure form from the infected or diseased plant parts of raddish. The optimum growth of the fungus was 24–28 °C at pH ranging from 7.1 to 8.0 with profuse growth on PDA. However, the host decoction agar was reported as best medium for both and sporulation of fungus. The sporulation of fungus was highest at incubation temperature of 23–25 °C with treatment of 12 h alternating light and dark period of UV radiation (Verma and Saharan 1994).

Biochemical and Metabolic Profiling

Alternaria species are known to produce many toxins and metabolites like Alternariol (AOH), alternariol monomethyl ether (AME), tenuazonic acid (TeA), tentoxin (TEN) and altenuene (ALT) etc. These compound cause great risk to human health (EFSA 2011). The food inhabiting *Alternaria* species has potential to produce many other metabolites (Ostry 2008). The biochemical and toxicological data are very limited and incomplete. The compounds, AOH and AME are mutagenic (Prelle et al. 2013) and their presences in cereals are responsible high risk of human oesophageal cancer as reported in China (Liu et al. 1992). The Altertoxins (ALXs) are potent mutagens with acute toxicity to mice than AOH and AME (Scott 2004). Fleck et al. (2012) and Schwarz et al. (2012) noticed high genotoxic potency of ALX II in mammalian and human tissues. Tsuge et al. (2013) observed and identified that host selective toxins (HSTs) have supernumerary chromosomes which encodes for HST gene clusters in pathogenesis of *A. alternata* and discussed their role in evolution of pathotypes or strains. From the view plant pathology, many other *Alternaria* metabolites are phytotoxics and damage the plant tissues (Montemurro and Visconti 1992). But, there are many metabolites like infectopyrones, phomapyrones and novae-zelandins whose role is still unknown. Many *Alternaria* spp. which produces metabolites are unclear about toxicity and their mode of actions. Therefore, for accurate identification of phytopathogenic *Alternaria* spp., the biochemical as well as metabolic profiling is very important. Recently, polyphasic approaches along with secondary metabolite profiling, are being successfully used for the identification of plant pathogenic *Alternaria* spp. (Andersen et al. 2008; Brun et al. 2013).

Enzyme Analysis and Protein Profiling

The chemotaxonomic approaches have been used by several workers for the identification and classification some *Alternaria* species, was based on studies of known and unknown secondary metabolites with little success (Anderson et al.

2005, 2008). The multi locus enzyme analysis is one of the popular methods utilized for identification of unknown fungal species. Chowdappa and Lakshmi 2013 had identified 13 *Alternaria* spp., the studies based on multi-locus enzyme analysis. They obtained the zymograms for the 4 enzymes systems as well as for 13 different species of *Alternaria*. MDH and GDH resulted in seven different types of enzymatic phyllogeny. The studies with different enzymes revealed that combined SOD, GDH, MDH, GPI data may be utilized for the differentiation and identification of *Alternaria* species through combination of isoenzyme pattern studies of aconitase, malate dehydrogenase and isocitrate dehydrogenase. Similarly, Petrunak and Christ (1992) also reported that isolates of *Alternaria solani and Alternaria alternata* can be differentiated on the basis of enzymatic fingerprinting pattern. Hwang et al. (1987) studied *Alternaria* spp. apple isolates identification on the basis of esterase isozymes of *A. mali* within geographic location as more closely related isolates than geographically differentiated isolates. The interpretations of proteomic studies and isozymes analyses are comparatively easier than expensive molecular techniques and protocols. These techniques can be used to differentiate the fungal isolates which are morphologically similar in structures (Micales and Bonde 1995). In addition to this, Oudemans and Coffey (1991) also reported that isozymatic studies may be used as markers in molecular and genetic studies.

Immunological Assays and Serological Methods

Immunodiagnostics and serological techniques are rapid, sensitive, highly specific and more accurate than other biological approaches for taxinomic classification of fungal plant pathogens. Though several immunodiagnostics assays with their polyclonal antibodies (PAbs) are available but highly accurate and precise methods are not available with higher sensitivity. Different immunogen standards are made with whole fungal cells (Kraft and Boge 1994); crude fungal mycelia or conidia/spores extracts (Harrison et al. 1990); culture filtrates of extracellular metabolites (Brill et al. 1994), mycotoxins (Ward et al. 1990), less purified soluble proteins or crude proteins (Velichetti et al. 1993) and in some cases mycelia cell wall extracts have been studied which has varied degree of specificity and bioassays for detection of specific group of fungi. It has been reported that presence of high molecular weight immunodominant polysaccharides may have adverse effect on specificity functions of PAbs. This problem had tried to solve by some workers by the use of certain purified fungal components like ribosomes (Takenaka 1992), fungal mycelia proteins detected after the process of gel electrophoresis (Lind 1990) and in some instances, lectins (Kellens and Peumans 1991), to minimize these constraints of specificity. For the effective management of plant pathogens, it is an immediate need to identify and remove all the primary sources/inoculums of plant pathogens from the new introduced area or fields by early detection and identification of plant pathogens. This is very important to restrict the introduction and further spread of plant pathogens. One of the major drawbacks of conventional methods is that they can not differentiate closely related species, biotypes/ecotypes and races of most of the species as reliable one. Thus, there is an immediate need to have development of rapid and more accurate methods for the identification of plant pathogens. Generally, there are two methods are employed for the detection of phytopathogens. The specific methods for detection of certain species or group of after primary diagnostics, indicates the presence of specific pathogen. Another method is non specific which can be useful for the detection of non targeted or unknown phytopathogen (s). These techniques are very useful in quarantine and phytosanitory certification of consignment at seaport and airports (Chu et al. 1989). Among all the diagnostics methods, immunodiagnostics assays and nucleic acid based techniques are considered as more rapid, specific and reliable than other morphological and biochemical methods which are non specific and less reliability (Narayanasamy 2001) though they are easy to use.

The early diagnostic, detection and precise identification of fungal plant pathogens will enable timely and precise applications of fungicides for certain groups of plant pathogens or groups of pathogens at appropriate time (Narayanasamy 2001, 2005). The principles of immunological methods, its applications, advantages and limitations of several immunoassays for the diagnostics and identification of fungal pathogens are discussed further. It is reported that, one among the major fungal pathogen, *Alternaria alternata*, the melanins pigments obtained as derivative from 1, 8-dihydroxy naphthalene (DHN) has an important role in pathogenicity and survival in fields. It was demonstrated that Phage-displayed antibody (scFv) binds specific to 1,8 DHN present in the conidial septa and external primary walls of wild type of *A. alternata* conidia. Carzaniga et al. (2002) reported the importance in the utilization of M1 antibody to detect melanized fungal pathogens in infected host plant tissues has more scope for their detection in near future. It is well known fact that introduction of an antigen into a mammalian system results in activation of defense response. The lymphoid cells present in the mammalian body has receptors to recognize the antigen, results in secretion of antibodies specific to antigens known as humoral immunity. Similarly, cell mediated immunity involves activation of immune lymphocytes with specific antigen receptors without any concomitant liberation of circulating antibodies (Sissons and Oldstone 1980). The epitope may be continuous or discontinuous determinant sequence of amino acid residues depending on structures of protein (Atassi and Lee 1978). Another type is cryptotypes which are present in viral capsid proteins and becomes antigenically active only after depolymerization or denaturation of the antigen proteins (Jerne 1960). Polyclonal antisera among the plant pathogens for which polyclonal antibodies (PAbs) have been developed, consisting of viruses with simplest structure and bacteria, being single-celled with their complex structures. The fungi known to produce the various types of spores, fruiting bodies at different stages of their life cycle which again makes their antigens more complex in nature. Thus, the production of antisera by using the spores or mycelia in early growth stages may or may not be useful for detection of fungal pathogens in their early growth stages. Various mammals like rabbits, mouse, fowl and horses are being used for the production of antisera. However, among all, rabbits are extensively used as test animals for production of antisera against several groups of plant pathogens. There are specific protocols available for the production of PAbs for various fungal pathogens which are considerably different.

Monoclonal Antibody

Kohler and Milstein (1975) developed and introduced a hybridoma technology as an advanced process to cope with problems associated with production of antibodies associated with utilization of PAbs. The major advantage of hybridoma technology is the continuous production of monoclonal antibodies (MAbs) released by hybridomas which are obtained by fusion of antibody producing ß-lymphocytes cells and myeloma cells which have capacity of indefinite multiplication. They have the capacity to produce identical antibodies which are specific for each one of epitope present in the immunogen. Denis et al. (2012) worked on characterization *Alternaria alternata* which are specific monoclonal antibodies (mAbs) from rat. These mAbs were three in number and categorized into different isotypes of IgM. Among those, E5 types can recognise only the conidia/spores of *Alternaria* spp., however, they were very close to *Ulocladium botrytis*. These studies and types of E5 antibodies may be very useful for the immunodiagnostics assay for identification of *Alternaria* and *Ulocladium* species.

Enzyme-Linked Immunosorbent Assay (ELISA)

Apart from the known molecular techniques, the development in enzyme-linked immunosorbent assay (ELISA) is considered as one of the important achievement among several serological diagnostics methods for diseases caused by several plant pathogens and particularly for viral diseases (Clark and Adams 1977). Though the molecular techniques are more advanced, the ELISA based techniques and their variants are being extensively used for the diagnostics, detection, identification, classification and quantification of plant pathogens in host plant tissues and other residues. The main advantages of this techniques is that they are highly sensitive, more precise, availability of economical antiserum, quantifiable data and ease at large scale application, they are among the most popular and widely accepted technique. There are several variants of ELISA are available as per the need and suitability of experiments and researchers. The most common and popular method is double antibody sandwich (DAS)-ELSIA. The other widely adapted method is direct antigen coating (DAC)-ELISA method, Protein A coating (PAC)-ELISA as well as indirect method of ELISA by utilizing pathogen antibody along with labeled antiglobulin conjugate. Among all formats, DAS-ELISA is highly strain specific and having scope to use different antibody conjugate for specific strain or species of pathogen of interest and detection. Whereas, in PAC-ELISA method, the optimum concentration of protein A is very important as the higher concentrations may have non specific results and lower concentrations may results negative and false results. It is reported that the labeled methods of pathogen detection and identification, ELISA format have been used on large scale for the detection of several fungal pathogens.

Molecular Markers

For rapid and effective management of plant diseases, the proper identification of the phytopathogens, their variability in the pathogen population at ecological as well as genetic level is important. The conventional techniques are labourious, time consuming and mostly with non reproducible results. Thus, to reduce these drawbacks in conventional markers, identification of genetic variability in pathogen populations through DNA fingerprinting technology by the use of molecular markers has been developed by geneticians, plant breeders and plant pathologists. The DNA fingerprinting techniques do not require any prior knowledge of variable sequences available in the database, but deals with genome-wide biodiversity. Most commonly used molecular markers used for the studies and assessments on variability in fungal pathogens includes 18-23S rDNA spacer regions, internal transcribed spacer regions (ITS), simple sequence repeats (SSR), sequence characterization spacer regions (SCAR), restriction fragment length polymorphism (RFLP), randomly amplified polymorphic DNA (RAPD), amplified fragment length polymorphism (AFLP), inter-simple sequence repeats (ISSRs), sequence tagged repeats (STR), single nucleotide polymorphism (SNPs), Loop mediated isothermal amplification (LAMP) etc. The ITS regions are conserved parts in the fungal genome and most extensively used as preferred loci for DNA based studies in fungi at genera and species level. ITS regions of rDNA of *A. brassicae* and *A. brassicicola* were studied to evaluate the phylogenetic relatedness and lineage among the isolates. RAPD's is popular DNA fingerprinting techniques, have been effectively employed to analyze the genetic variations in several *Alternaria* species (Cooke et al. 1998; Gherbawy 2005). However, it was noticed that, the results obtained by RAPD were not reproducible which questions the authenticity of identification. Thus, the genetic variability studies in *Alternaria* species are being done with more sensitive techniques like AFLP (Bock et al. 2002) and with microsatellite markers (Avenot et al. 2005). Gargouri-Kammoun et al. (2014) practiced the use of molecular techniques based on the amplification of specific ITS1-5.8S-ITS2 region of rDNA and sequencing for several *Alternaria* spp. The detailed analysis of these data indicated that six species of *Alternaria* i.e. *A. alternata*, *A. tenuissima*, *A. mali*, *A. longipes*, *A. arborescens*, and *A. brassicae*. The dominant species isolated were *A. alternata* (36.1 %) and *A. tenuissima* (30.6 %). Mmbaga et al. (2011) developed specific diagnostics primers for the detection of *Alternaria* sp. and *A. alternata* causing blight of lilac leaves, designed on the basis of ITS region as well as four different kinds of genes which are involved in the pathogenesis of *A. alternata* namely glucanase precursor (mixed-linked), histone, endopolygalacturonase and hsp70.

A novel method, Random Amplified Microsatellites (RAMS) is comparatively cheaper, rapid and easier to use. This technique based on the principle of polymerase chain reaction (PCR), microsatellite sequences primers along with degenerate anchors at the 5′ end. It is more similar with RAPD technique (Random Amplified Polymorphic DNA) but uses longer 18 nucletides primers and annealing temperature during PCR were not dynamic. It was also observed that genomic regions with microsatellites evolve rapidly and with higher mutation than other genomic areas of isolate. It occurs due to mispairing of strands during replication and it is depend on length of repeat (Burgess et al. 2001). Hence, PCR optimization with higher annealing temperatures with longer nucleotides primers are more promising in RAMS markers techniques than RAPD markers (Roberts et al. 2000; Peever et al. 2002). Thus, it was concluded that the RAMS markers are better and highly specific than RFLP analysis resulted from amplification of rDNA (Aradhya et al. 2001) and ITS regions sequence analysis studying genetic variations among closely related fungal isolates (Kusaba and Tsuge 1995). It can be concluded that RAMS technique has advantages from RAPD and microsatellite markers analyses of isolates, for the genetic variability studies. Trivedi (2010), observed that in case of *Alternaria radicina* carrot seed application of quantitative PCR techniques were highly sensitive and rapid, detected even very low volume of DNA (100 fg) and atleast 10 conidia in soil for single reaction (60 conidia/g soil). Loop mediated isothermal amplification (LAMP) is one of the recent technique was employed by Amir et al. (2014) for rapid identification of plant pathogenic *Alternaria alternata* causing brown spot disease in citrus trees in Ramsar region of Iran. They utilized the specific pair of LAMP primers which were designed from ACT toxin gene. Though, the highly advanced molecular tools and techniques to study genetic variations among *Alternaria* at species, biotype and strains level, many areas are still remains unexplored and having great scope with extensive studies in different geographic locations and among variable hosts.

Nucleic Acid Based Identification

Extraction of DNA from Samples

Several methods are available for the extraction of DNA from the fungal mycelia as well as diseased plant samples. General method consists of DNA to be extracted from fungal mycelia by directly scraping the surface of petri plate fungal colonies as well as collected by using microcentrifuge filters when DNA extracted from seed infected samples. Various DNA extraction methods are exist for different group of fungi and application of particular method is depends on their accuracy, rapid extraction, quality of DNA obtained and environmental condition from which sample is obtained for DNA extraction. Klimyuk et al. (1993) demonstrated a simple technique for isolation of DNA directly from fungal colonies extracted when grown on media and processed with extraction buffer. Guo et al. (2000) developed a modified CTAB protocol for extraction of genomic DNA from fungi. This is very effective method as it allowed extraction of DNA from large fungal colonies from media within a short period of time and their diluted DNA extract is sufficient for PCR protocols. For the community analysis, the high quality of DNA is required when it is extracted and processed from environmental samples or requirement of large amount of DNA or for long term storage of DNA of fungal isolates. Also, the fungal DNA can be extracted from their colonies over media by several physical methods including grinding of mycelia in liquid nitrogen, bead beating method, cell wall degradation with enzymes and use of sodium dodecyl sulphate (SDS) detergents. The purification of DNA is done by phenol and phenol-chloroform washing to remove cell contaminants and enzymes degrading nucleic acids. The DNA precipitation is done by isopropanol and salts were removed by washing of DNA by ethanol and subsequent air drying. Lee and Taylor (1990) used RNase to remove RNA present in sample and minimizes the PCR troubleshooting in reaction. Extraction of DNA from diseased seeds involves lysis with a detergent-containing buffer at high temperature, extraction by utilizing organic solvents and nucleic acid precipitation with alcohol. Goodwin and Lee (1993) used the microwave miniprep procedure for the isolation of nucleic acids for PCR method. Several commercial kits are also available in market with their specific protocols. The universal primers ITS1/ITS2 are used as a positive control for the assessment of the quality of the extracted nucleic acids (White et al. 1990).

Among several molecular methods and tools available for plant pathogen detection and diagnostics, the nucleic acid based techniques are widely adapted and are very useful, rapid, efficient and more reliable detection methods. The nucleic acid based techniques mainly based on PCR variants are highly sensitive, rapid and more specific than most other methods of pathogen detection and identification. These nucleic acid-based diagnostic methods are widely used for detection, identification, quantification of propagules as well as classification of fungal pathogens in several crop plants and propagating plant materials. It is a well known fact that various species of *Alternaria* are carried by propagating plant materials and seeds of different plants. PCR technique is the selective amplification of specific segment of DNA in controlled condition to detect their specific DNA/RNA sequence. It consist of two oligonucleotide primers which amplifies the DNA fragment through repeated cycles of denaturation of DNA, primers annealing to complementary DNA sequences and primers extension with thermostable DNA polymerse (Mullis 1987). Hensen and French (1993) have made several attempt to develop specific PCR primers for the species specific fungal plant pathogens. The most widely used molecular markers for fungal evolutionary studies are fungal mitochondrial DNA analysis (Brunus et al. 1991). For the detection of two major fungal pathogens colonizing cereals grains i.e. *Alternaria alternata* and *A. solani*, PCR-based method was developed. The most commonly used conserved region in fungal DNA for amplification of pathogen DNA is ITS region. However, other functional as well as housekeeping genes are becoming popular especially β-tubulin gene (Fraaije et al. 1999; Hirsch et al. 2000), and in rare cases, mating type genes (Dyer et al. 2001; Foster et al. 2002). With the development of species specific primers, these functional genes are more reliable and very useful for the development of diagnostics probes and primers (McCartney et al. 2003). The most common primers designed and developed are based on the sequences of ITS1 and ITS2 regions of the 5.8S rDNA. The extensive studies of housekeeping genes with greater variability are being used to develop the diagnostics for fungal pathogens. The nuclear housekeeping genes such as β-tubulin are mostly used for detection and identification of fungi (Aroca et al. 2008; Fraaije et al. 2001; Mostert et al. 2006). Similarly, translation elongation factor 1 alpha or *TEF* 1α (Geiser et al. 2004; Knutsen et al. 2004; Kristensen et al. 2005) were used by the researchers. In addition to this, calmodulin gene (Mule et al. 2004), various avirulent genes (Lievens et al. 2009) and most recently the mitochondrial genes including multicopy *cox* I and *cox* II with their intergenic region (Martin and Tooley 2003; Seifert et al. 2007; Nguyen and Seifert 2008) are extensively used for better identification of fungal taxon. Peever et al. (2004) differentiated *A. limicola* isolates on phylogenetic basis of mitochondrial-LSU as well as β-tubulin. Their spores are larger in shape and size than others small size spored isolates from citrus which causes leaf spot disease of Mexican lime fruits. It was difficult to differentiate on the basis of evolutionary relationships than other sporulating fungi from citrus and associated hosts. Recently, Dube (2014) concluded that specific GAPDH and EF1α gene regions may be used to differentiate several isolates of *Alternaria alternata* which causes brown spot disease of potato in South Africa and results were found similar according to morphological features. The cluster analysis of the RFLP among several isolates of *Alternaria* showed a low average genetic difference (0.62). The separate clusters resulted among these isolates and grouped them into *A. alternata* and *A. bataticola*, despite of their geographical origin (Agnes 2009).

Zghair et al. (2014) observed that pathogen *A.alternata* can produce mycotoxin alternariol (AOH) govern by specific gene. The specific PKSJ primer was designed for target gene and amplified by PCR method for all isolates except for isolates no. 2 which is unable to produce AOH toxin. Finally, it was concluded that PKSJ primer is specific for polyketide synthase gene detection in *A. alternata*. Though, the results were similar, the PCR based diagnostic assay has an advantage over the conventional method of identification of pathogen. The results obtained by PCR based assay were reliable and rapid, thus are preferred over conventional methods (Konstantinova et al. 2002). The specific primers were designed for PCR by obtaining the sequences of ITS regions of rDNA for three fungal phytopathogens i.e. *A. brassicae*, *A. brassicola* and *A. japonica* causing black spot disease. All these pathogens then were detected in the macerated seeds (Lacomi-Vasilescu et al. 2002). The further study and analysis revealed that two clustered genes were involve in the pathogenicity and these genes were used to design two different sets of primers that were studied in conventional PCR and real time PCR methods. Recently, Guo-yin et al. (2013) demonstrated that *Alt a1* genes sequences are very useful than ITS sequences of rDNA for the identification of *Alternaria* spp. However, previous studies indicated that several species of *Alternaria* which were differentiated morphologically were significantly distinguished from each other by variation of ITS1 and ITS2 sequences (Kusaba and Tsuge 1995). Likewise, mostly ITS regions and 5.8S genes were employed to differentiate four different species of *Alternaria* i.e. *A. alternate*, *A. brassicae*, *A. brassicicola* and *A. raphani* and which were isolated from cruciferous crops (Jasalavich et al. 1995). Recently, Sharma et al. (2013) designed highly specific primer set for *A. brassicicola* which was designed from the sequences obtained from less conserved ITS region (100–110 bp 3′ end position and 520 bp 5′ end position), non specific to all other virulent and common *Alternaria* spp. Also, they were not specific to other seed mycoflora including *Fusarium oxysporum* and *Aspergillus niger*.

Molecular Techniques

Molecular method gives faster and more reliable results. The recent techniques and advanced molecular tool helped to answer complex questions on host-pathogens interaction and their biology. It will help for identification and quantification of inoculum sources, population behaviour and the epidemiology of *Alternaria* diseases.

Molecular methods mainly involve the polymerase chain reaction (PCR), which are used extensively as tools for the identification of several fungal plant pathogens (Fernandez et al. 1998; Niessen and Vogel 1998). The more common sensitive and rapid PCR based assay for the detection of fungal plant pathogens is based on amplification of conserved nucleotide sequences of the internal transcribed spacer (ITS) regions of rDNA. This method has been used to detect *A. brassicicola* or *A. japonica* infection in the cruciferous seeds (Lacomi-Vasilescu et al. 2002). However, the results were not specific to *A. brassicicola* and couldn't differentiate the *A. brassicae* contaminated seeds. The application of molecular techniques facilitates the identification of *Alternaria* spp. Comparison of Basic Local Alignment Search Tool (BLAST and analysis), it is easy to identify unknown fungus on the basis of DNA sequences of various isolates/strain, submitted previously in NCBI Gen-Bank in to find regions of similarity among these sequences (Pastor and Guarro 2008). Amplification of various ITS region specific primers, for multiple rRNA's and their sequencing of amplicon is very useful method for identification of *Alternaria* spp. (Pastor and Guarro 2008). Guillemette et al. (2004) studied two sets of primers for conventional PCR and real-time PCR and *A. brassicae* was detected specifically using DNA extracted from seed. They also reported the efficient quantification by using real time PCR for seed infection by *A. brassicae*. In case of *Alternaria* spp. host specificity typically depend on mycotoxin production. Therefore, for the efficient diagnostics and molecular probes for *Alternaria* spp. and their pathotypes differentiation, cloning of genes which are involved in virulence are necessary. The several researchers have studied and analysed the sequence data of ITS regions, mitochondrial LSU, β-tubulin gene, translation alpha-elongation factor, anonymous genomic regions (OPA1 and OPA2, actin, chitin synthase and many housekeeping genes (Pryor and Gilbertson 2000; Chou and Wu 2002; de Hoog and Horre 2002; Pryor and Bigelow 2003; Peever et al. 2004) have advanced the taxonomic levels of various *Alternaria* species into distinct species-groups and strains.

Real Time PCR

The techniques of real time PCR has open the new hope and avenues to detect, diagnose and study various plants pathogenic, antagonistic and mycotoxicant fungi. This method combines features of conventional PCR techniques by producing a specific fluorescent signal generation as real time analysis and reaction kinetics providing the quantification of DNA targets. The techniques of real time PCR detect the presence or absence of pathogen in plant sample as well as able to quantify the amount of DNA present in the sample(s), thus providing the quantitative assessment for distinct group of phytoathogens in the infected plant sample.

In case of real-time (RT) PCR, the fluorescent detection system has effectively reduced inoculum detection problem and better quantitative assessment and diagnostics of PCR amplicon. The variants of specific and non specific real time PCR have been proved to be highly sensitive, specific and rapid method as compared to conventional PCR techniques. Taylor et al. (2001) demonstrated that the technique of real-time PCR is highly sensitive, rapid and effective for the detection and quantification of DNA. Thus, applications of real time PCR would help in accurate assessment and quantification of several species of *Alternaria* in diseased sample. This method has prove a worth in early detection, quantification and assessment of two *Alternaria* species infecting potato leaves in early stages of pathogenesis and effective differentiation of two *Alternaria* spp. The application of real time PCR will help to know the disease within the host system by *Alternaria* spp. progression over the time, which will be help in epidemiology and forecasting of early blight disease. Thus, the technique of real time PCR was employed for the effective differentiation of *A. solani* and *A. alternata*. This advantage of real-time PCR will provides sensitive and rapid assessment of several *Alternaria* species infecting various fields, horticultural and plantation crop along with the etiology of *Alternaria* diseases and for effective disease management. Leiminger et al. (2014) stated that real time PCR is a very powerful tool for quantitative estimation of pathogen propagules during early stages of pathogenesis and further disease development in host plant. It was reported that higher amount of *A. alternata* DNA was reported as compared to other *Alternaria* spp. thus, indicating the dominant species in diseased samples. This technique also provided highly specific and sensitive detection of *Alternaria* spp. and the differentiation in radish as *A. brassicola* and *A. japonica*, also in radish and cabbage for phytopathogen *A. alternata*.

Reverse Transcription (RT)-PCR

One of the most important drawbacks in molecular techniques of pathogen detection is that they are unable to distinguish the living or dead fungal pathogens or propagules. Therefore, these results are to be cross checked for accurate detection and identification of fungal plant pathogens by routine pathogenicity tests. Sheridan et al. (1998) noticed that the mRNA of the dead cells of pathogens degrades very rapidly and therefore detection of mRNA by RT-PCR method gives better account of the cell viability. Similarly, Yang et al. (2010) revealed that RT-PCR techniques have very important application in the fungal plant pathology for the detection and analysis of fungal gene expression during pathogenesis and disease development in host plant. It is also very useful for the quantitative estimation of pathogen population in the fields. Several workers did gene expression studies by RT-PCR during disease development. McMaugh and Lyon (2003) and Schenk et al. (2003) worked out the application of RT-PCR for detection and quantification of expressed fungal pathogen genes during disease development expression of host plant genes and activation of disease resistance genes.

Nested-Polymerase Chain Reaction

Nested PCR approach is important when there is need for higher sensitivity and specificity of detection. Porter-Jordan et al. (1990) stated that nested PCR technique consist of two consecutive rounds of amplification where two external primers are applied to amplify large amplicon and thereby act as template for second round of amplification by applying two internal primers. These two reactions are carried in separate PCR tubes. Thus, it is time consuming, skillful work and increases the chances of faulty positive results by contamination. However, the method has improved over the time by employing relative concentrations of both external and internal primers for two reactions in a single closed tube resulting high output. Several workers have widely employed this technique for the detection, identification and subsequent characterization of many fungi (Aroca and Raposo 2007; Hong et al. 2010; Langrell et al. 2008; Meng and Wang 2010; Qin et al. 2011; Wu et al. 2011). The sensitivity, accuracy and detection of fungal pant pathogens in host plant tissues can be remarkably improved by the application of nested-PCR technique. It was reported that the sensitivity of nested PCR was 1,000 time more sensitive than conventional PCR techniques. This technique was applied in various samples from various field infection sites, crops, water and from different agro cultivation sites for analyzing and validation by nested PCR techniques (Grote et al. 2002).

Multiplex-PCR Assay

The multiplex PCR method involves the use of several PCR primers in a single reaction targeting highly sensitive detection of different DNA and their simultaneous amplification, thus reducing the time and cost than other methods. This method is very useful when the plants are infected by more than one pathogen and it is difficult to identify them by conventional methods. Thus, the specific fragment of target fungal DNA's are simultaneously amplified and can be identified on the basis of gel electrophoresis and detection on the basis of molecular sizes of amplicon. However, the efficiency of amplification is influenced to greater extent by the amplicon size (smaller amplicon amplify more than longer ones). Hence, proper designing of primers, their sensitivity and accuracy, optimization of PCR protocols are very important to solve these problems to have an enhanced and improved detection of fungi of interest. Chen et al. (2008) used this method has been used for simultaneous detection and distinguish two pathogenic fungi, i.e. *Podosphaera xanthii* and *Golovinomyces cichoracearum* in sunflower. The differentiation of two mating type pathogens i.e. *Tapesia yallundae* and *T. acuformis* was performed by Dyer et al. (2001). The two pathotypes of *Verticilliun albo-atrum* infecting hop plant was differentiated by the use of multiples PCR technique (Radisek et al. 2004). Recently, eleven taxa of wood decaying fungi infecting hardwood trees were distinguished by this technique (Guglielmo et al. 2007). This method does not employ the isolation of fungal plant pathogen from the diseased host plant tissues. Therefore, this method provides an extra advantage for the rapid detection and identification of fungal pathogens.

Amplified Fragment Length Polymorphism

AFLP technique literally is a combination of two molecular techniques i.e. RFLP and RAPD. This method consists of assisted specific amplification of higher number of restricted nucleotide fragment. Generally, 50–100 fragments are amplified and detected on gels by poly acrylamide gel electrophoresis. This is highly sensitive method to study polymorphism in genome wide studies and hence it is becoming very popular technique. This technique provides generation of large number of restriction fragment bands and therefore helps in detection of polymorphic bands giving better identification of pathogens (Skrede et al. 2012; Chen et al. 2012). Gannibal et al. (2007) studied the variation for 101 isolates of *A. tenuissima* by AFLP analysis of Russian isolates from obtained from wheat kernels and other associated hosts. The analysis of AFLP banding pattern obtained from the results showed that distinct genetic distance was present among isolates. The results also revealed that narrow level of gene flow indicating between eastern and two other group nearby to European parts of the country. They also indicated that similarity was present between North West regions and Caucasus isolates were wider in genetic distance. It was also stated that *Alternaria* isolates from wheat and Barley were not significantly distinct from each other, analysed on the basis of the Wright's fixation index. Similarly, Martınez et al. (2004) noticed the variability among *Alternaria solani* isolates from Cuban and other international strains isolated from different hosts and locations through AFLP method. The phylogenetic UPGMA clustering indicated the difference between *A. solani* isolates and others species. However, *A. porri* could be separated from group of *A. solani*. The major rapeseed pathogen, *Alternaria brassicicola* was analysed by AFLP genetic analysis and detected genetic variations among isolates (Bock et al. 2002). They analysed 18 isolates of *A. brassicicola* obtained from different locations noticed low level of genetic diversity among *A. brassicicola* populatins including New South Wales coast, Australia. Thus, it can be concluded that AFLP tools are very useful and reliable tool for the genetic studies and identification of *A. brassicicola* populations. Hong et al. (2006) evaluated genetic similarities among *Alternaria* species by AFLP fingerprinting and observed that *A. alternata* and *A. tenuissima* grouped into a single lineage. However, *A. arborescens* isolates placed into a distinct lineage. Oviedo et al. (2013) identified the isolates of *Alternaria* by using AFLP genetic markers to species level and also analysed the intra as well as inter specific level genetic variation among *A. alternata* and *A.infectoria* isolates from wheat. The distinct and specific polymorphic band pattern grouped *A. alternata* and *A. infectoria* species despite of primer combinations employed in PCR. Thus, the polymorphic banding patterns analysed as intra as well as inter species level of *A. alternata* and *A. infectoria* by AFLP would be very useful for identification. The intra specific variation and assessment in *A. alternata* and *A. infectoria* isolates forms a homogeneous group with greater similarity among them.

Random Amplified Polymorphic DNA (RAPD) Technique

This molecular method of genomic fingerprinting or marker system was given by Welsh and McClelland in 1991. This method is based on the principles of PCR techniques where genomic DNA from the fungal pathogen is used as template and short oligonuleotide primer(s) (about 10 mer). In this technique, the amplification of DNA is from a conserved region which is flanked by 10 bp priming site. The amplification of DNA for particular region of one individual but may not be for other individual thus indicating specific DNA polymorphism and can be used as genetic marker. The random amplified polymorphic DNA (RAPD) technique is based on PCR method employing arbitrary primers and hence it is very useful to distinguish species, races, biotype, strains and virulent or non virulent isolates of fungi. The lengths of the primers used in RAPD technique are very short in length (10 or more length bp) of nucleotides from known source. The PCR products obtained through RAPD technique when analyzed on gel electrophoresis, polymorphic bands are formed and some bands pattern may be very specific to particular group of fungal species, intraspecies, isolates, and/or stains. Thus, the particular banding pattern may be very useful for the detection, diagnosis and identification of certain plant pathogenic fungi. The highly specific bands may be cut down from the gel, sequence it and can produce highly specific primers for precise PCR amplification of individual fungi or in some cases may be useful for the development of probes in fungi detection protocols. Several workers have demonstrated the usefulness of RAPD analysis for detection, identification, confirmation and validation of taxonomic classes (Zheng and Ward 1998; Roberts et al. 2000; Pryor and Michailides 2002; Gherbawy 2005). Morris et al. (2000) used 29 different RAPD primers for the analysis of 69 *Alternaria alternata* isolates in California indicated high degree of genetic diversity and identified two major phylogenetic groups (Group 1 with 55 isolates and Group 2 with 14). The results also revealed that higher degree of variation or clustering among the isolates wasn't present in all over California. Whereas, Aneja et al. (2014) evaluated the banding pattern of RAPD fingerprinting and reported high degree of genetic variation among *Alternaria* isolates as *A. brassicae* (57–78 %), *A. brassicicola*

(78–92 %) and *A. alternata* (89–100 %). Pusz (2009) used OPA and OPB primers for the genetic variation of *A. alternata* by PCR amplification (RAPD-PCR), revealed that moderate level of genetic variation was present in *A. alternata* from Amaranthus host. Similarly, Kakvan et al. (2012) analysed the clusters fragments of *Alternaria* spp. of Iran obtained from RAPD-PCR of all applied primers and placed them into five groups at 85 % similarity. They indicated that genetic diversity of isolates was influenced by geographic locations, host plant affected and their virulence potential.

RFLP

Restriction fragment length polymorphism (RFLP) is based on restriction digestion of PCR amplicon or PCR products of amplified DNA by specific enzyme and thereby separation of these products by gel electrophoresis in agarose or polyacrilamide gels to obtain fragments and then analysis of these fragments to study differences in the banding pattern separated amplicon. The polymorphic banding pattern obtained by restriction digestion of PCR product, the cleavage sites are used for the differentiation of various fungi species. Several workers (Thies 2007; Kim et al. 2010; Martínez-García et al. 2011) have been used RFLP technique to study diversity of mycorrhizal fungi and soils inhabiting fungal populations. Likewise, Hyakumachi et al. (2005) employed this technique for the diversity studies and grouping of pathogenic fungi. This technique has been well supported by the other fingerprinting techniques based on PCR principles. The generation of polymorphic band pattern through hybridization indicates the sequence difference between the targeted pathogen populations. Abass (2013) did the characterization of *A. alternata* isolates obtained from date palm by analysis of ITS-RFLP. ITS-PCR amplicons were digested with restriction enzymes (*Eco*RI and *Sma*I). The banding profile pattern resulted from the restriction digestion of endonucleases, grouped into three categories. A type: Based on one undigested fragment, B type: Based on two fragments pattern (60–532 bp) and C type: Based on three fragments pattern. B type pattern resulted in ITS-RFLP analysis from *Eco*RI digestion of *A. alternata*, *A. niger*, *P. expansum* and *P. glabrum*. However, in acse of other fungal species, no detectable restriction digestion was observed. Similarly, restriction digestions through SmaI of PCR products were similar with previous results. The analysis of ITS-RFLP banding pattern indicated that *A. clavatus* fungal species were classified into C type. But, *A. alternata* and *A. citri* isolates categorized into A type.

Aradhya et al. (2001) evaluated the genetic variations in *Alternaria alternata* causing late blight in pistachio by RFLP pattern of rDNA through restriction digestion via *Eco*RI, *Hin*dIII, and *Xba*I restriction enzymes. Analysis of molecular variation revealed a significant amount of genetic diversity within populations (85 ± 8 %), with only marginal variation accounting for differentiation among populations. In addition to this, RFLP pattern analysis of rDNA did by many researchers (Shrama and Tewari 1998; Pryor and Michailides 2002).

Microsatellites Marker

Nowadays, the microsatellites markers are becoming very popular for genome wide fingerprinting of fungal pathogens, also known as simple sequence repeats (SSRs) or short tandem repeats (STRs), generally involves short motif of one to six nucleotides repeats in several times (especially in non-coding regions). These nucleotides regions may differ in repeat number among their communities and the distribution of these repeats is almost random genome wide. The utilization of flanking primers, variable regions PCR amplicon of different length sizes can obtain. Therefore, the microsatellites markers have potential to be use as dynamic genetic markers which are being exploited on large scale for DNA fingerprinting studies in fungi. One of the major benefit of these SSR or STR or microsatellites markers are that they involves multiallelic variation, co- dominance, highly polymorphic and huge amount of potentially polymorphic band pattern or markers are available. Similarly, the important advantage of these techniques is the analysis by low amount of DNA or degraded DNA is possible with repetitive results. However, there is requirement of large amount of microsatellite marker loci for the development of quality construction of phylogeny of related populations. But, with the advancement of next-generation sequencing technologies as well as multiplexing microsatellites, have reduced these problems. Recently, the cost of microsatellites development has reduced by new technique known as sequence tagged microsatellites (STMs). Hayden et al. (2002) reported that sequence tagged microsatellites (STMs) is amplified for conserved DNA sequence with single specific primer, flanking microsatellite repeat combinely with a universal primer attached to 5′-ends of the microsatellites.

Gel Electroforetic Profiling

The chemotaxonomic studies supported with molecular techniques for detection and identification is also very important in characterization of fungal pathogens. The stem canker of tomato is caused by fungus *Alternaria arborescens* (previously known as *A. alternata* f. sp. *lycopersici*) produces host specific AAL toxin which has particular role in pathogenesis (Yamagishi et al. 2006). The pulsed field gel electrophoresis (PFGE) studies and analysis revealed that fungus *A. arborescens* produced one CDC of 1.0-Mb (Akamatsu et al. 1999; Akamatsu 2004). Chowdappa and Lakshmi (2013) identified several species of *Alternaria* by studying and observing total proteins and isoenzyme system approaches (superoxide dismutase, malate dehydrogenase, glutamate dehydrogenase and glucose phosphate isomarase). The analysis was done for both native and dissociated proteins of *Alternaria* spp. The study revealed that the banding patterns for these native as well as dissociated proteins were recorded between 20 and 205 KDa. The unique protein mass absorption spectroscopy (MAS) was obtained for both native proteins. However, SDS-PAGE profile was noticed from several isolates of *Alternaria* species including *A.solani*, *A.porri*, *A.macrospora*, *A.dauci*, *A.alternata*, *A.brassicae*, *A.brassicicola*, *A.sesami*, *A.helianthi*, *A.ricini*, *A.carthami*, *A.brunsii*, and *A.mali*. The phylogenetic analyses which were developed on the basis of native as well as SDS-PAGE protein profiling patterns indicated that these isolates belonging to different groups related to 13 different *Alternaria* spp., thus, it can be concluded that these methods are very useful for detection and identification of *Alternaria* species on the basis of specific protein masses *of Alternaria* spp. In addition to this, the zygograms developed for the 4 isoenzymes and for 13 different species of *Alternaria* GPI resulted into 5 distinct zygograms with 4 zygograms for SOD. These zygograms were highly specific for their particular species and differentiated for 13 *Alternaria* species. But, it was also reported that the isozyme patterns of isolates within species were identical. Rotem (1994) reported that four *Alternaria* spp. including *A.solani*, *A.porri*, *A.dauci* and *A.macrospora* were grouped into same clade of *A.porri* group. Whereas, another *Alternaria* species i.e. *A.longipes*, *A.gaisen*, *A.citri* and *A.mali* were specifically grouped into *A. alternata* group (Anderson et al. 2001). Several workers have been successfully employed the techniques of electrophoretic profiling for native as well as dissociated proteins in differentiating the species, intraspecies, biotypes of fungal plant pathogens.

Fourier Transform Infrared Spectroscopy (FTIR)

The presently available methods of detection and identification of fungi such as molecular approaches, immunological, serological tests and several variants of PCR (polymerase chain reaction) may not be sensitive, rapid and specific. In many cases they are labourious and time consuming. Recently, Fourier-transform infrared (FTIR) attenuated total reflection (ATR) spectroscopy technique is highly advanced, comprehensive and sensitive method for detection of molecular changes even in intact cells. FTIR–ATR is highly sensitive, precise and effective method for the detection and differentiation of different groups of fungal genera (Naumann et al. 1991). Gupta et al. (2005) and Lamprell et al. (2006) have successfully used FTIR spectroscopy methods for detection and identification of several microorganism, particularly in food products. However, some studies also revealed that results are helpful not only for differentiating fungal genera but also for different species (Beattie et al. 1998; Lefier et al. 1997) and even at strain levels by Lamprell et al. (2006) and Udelhoven et al. (2000). The infrared spectra results in detail information on various cellular biomolecules like proteins, polysaccharides and lipids (Helm and Naumann 1995). However, these spectra are complex but consist of information from all other biomolecules. The technique of FTIR spectroscopy is a very important tool and have been successfully used in detection, identification, differentiation and the classification of specific group of microorganisms (Branan and Wells 2007; Erukhimovitch et al. 2005; Salman et al. 2002). It has been successfully employed in discrimination of several fungal species (Fischer et al. 2006; Naumann et al. 2005; Santos et al. 2010). FTIR technique also very useful for detection of changes in cellular components different stress conditions, detection and identification of fungal species when grown in optimum and/or stress conditions (Kaminskyj et al. 2008; Szeghalmi et al. 2007). Shoaib et al. (2013) observed the biochemical variation in canola plant when it was infected with *A. destruens* fungal blight pathogen by the application of FTIR technique to obtain specific pattern of changes in host to particular infection.

The infrared spectra of biochemical and/or biomolecules generate highly specific profile pattern and can act as efficient "finger print". The large amount of data is available about spectral bands produced from FTIR spectroscopy of living cells. Thus, FTIR spectroscopy method provided greater scope for detection, identification and classification of plant pathogens.

DNA Microarray

A recent DNA microarray technology is new and highly advanced tool for diagnostics of pathogens was developed originally to study gene expression studies through the development of single nucleotide polymorphism (SNP) profiles. The DNA microarray technology is a powerful tool which provides a strong platform with wide scope and multiple potential. This technology was first time introduced by Schena et al. in 1995 applying in various fields of biological sciences. The advancement of oligonucleotide microarrays technology, it is possible to customize and identify the presence and quantification of specific fungal plant pathogenic species or taxon group by ITS or rRNA database available and their functional genes specially virulence and resistance and/or tolerance genes. The community analysis of microorganisms from the environmental samples is possible by screening several thousands of oligonuleotide probes already available with microchip in DNA microarray machine. The DNA obtained from the environmental samples is labeled and the resulting profile patterns are analyzed as compared to database of reference known organisms. Upadhyay et al. (2014) worked on the differential gene expression pattern during pathogenesis (PR) related protein genes in tomato by inoculation with virulent fungus, *A. solani*. The analysis of the results indicated that 32 genes were classified in this group which showed promising changes in resistant and susceptible germplasm (EC-520061 and CO-3). Out of 32 genes, 22 genes showed up regulation in resistant germplasm, however non significant up regulation reported in fold change (FC) in susceptible germplasm.

Hence, the DNA microarray technology is very useful to study all fungal community in particular environmental for their detection as well as quantification. It is estimated that soil consists of over 30 millions different types of genes in 1 g of soil as compared to about 30,000 genes in human genome which might increase the complexity when we analyze environmental samples, may produce false positive results. Therefore, it is very important to specify certain techniques to isolate specific group of nucleic acids from these samples.

Genomics and Proteomics

The combination of all available approaches like comparative genomic and proteomics, pathogenesis (PR proteins) induced proteins, resistance/tolerance proteins when combine with other advanced techniques i.e. DNA microarray technology, functional genomics, biochemical and metabolomic profiling and host-pathogen interaction studies, provides better idea on account of protein function in fungal pathogens. Bouws et al. (2008) and Kim et al. (2008) reported that comprehensive work has been done on development of catalogue of several fungal structures including morphological studies of mycelia, spores/conidia and released proteins known as secretome, among all database of different group of fungal species. These approaches are well supported by protein identification following recent techniques of SDS-PAGE and/or 2D-PAGE separation of different molecular weight of proteins or proteomics studies by analyzing total digestion of fungal proteins through tandem LC-MS techniques and results in development of protein database (Carberry et al. 2006; Braaksma et al. 2010). Recently, Hu et al. (2012) reported that the virulence of one of the major pathogen of tomato i.e. *Alternaria arborescens* determined by availability of conditionally dispensable chromosome (CDC) consisting of hot specific toxin producing genes. In case of plant pathogenic *Alternaria* species, several species carries conditionally dispensable chromosomes (CDCs) (Johnson et al. 2001; Hatta et al. 2002). It was reported that these chromosomes normally less than 2 MB in size and more frequently may horizontally transferred among the sub species or strains of their population and may impart new pathogenic potential or features to new recipient strain(s) (Salamiah et al. 2001; Masunaka et al. 2005). The CDCs are known to carry several gene clusters coding genes for host specific toxins (HSTs). Similarly, Nakashima et al. (1985) reported AK-toxin from the pathotype of Japanese pear and tangerine pathotype producing ACT-toxin. Mukherjee et al. (2010) worked on proteomics of *Arabidopsis thaliana*, a model test plant when it was infected with fungal pathogen, *Alternaria brassicicola*. The study revealed that the host pathogen interaction of *Arabidopsis–A. brassicicola* can be developed into a model genetic system as this is a unique incompatible host plant-fungi interaction. It was resulted from the experiment that pathogenesis related protein PR4, glycosyl hydrolase and antifungal protein osmotin were highly up regulated. It was also reported that two members from Arabidopsis glutathione S-transferase (GST) family observed in higher quantity in diseased leaves. Similarly, Kannan et al. (2012) analysed the expression of MAP2K9 and MAPK6 during disease development when *Arabidopsis thaliana* (ecotype Columbia) infected with *Alternaria*. They concluded that MAP2K9 and MAPK6 pathway module may play an important role during infection of *Alternaria*. These genetic expression studies and modulation with test plant in Arabidopsis and/or Brassica will be help in developing suitable genetic modification of host plant to develop defense management practices against *Alternaria* blight.

Hatta et al. (2002), it was concluded and gave hypothesis that genomic constitution of CDCs in *Alternaria* species were transmitted through horizontal gene transfer (HGT) among those populations. Within the advancement of technology of whole genome sequencing and de-novo assembly strategies, Hu et al. (2012) noticed CDC sequences in fungal pathogen, *A. arborescens*. The comparison of nucleotide mapping between CDC and EC contigs, it was found that it is more likely the host specific toxin genes are transferred through HGT in *A. arborescens*.

Management of Diseases

A judicious use of various methods is very effective for management of fungal diseases.

Cultural Practices

For the effective management of *Alternaria* blight diseases several cultural practices can be adapted in the field as well as in storage condition. Meena et al. (2002) reported that early sowing of clean, certified and well stored seed lot, deep ploughing of agricultural fields, strategic weeding practices, optimum population of crop density in field can efficiently manage *Alternaria* blight disease. The irrigation practices should be avoided during flowering and pod formation stages for better disease control.

Source of Planting Material

Use of the disease resistant varieties is one of the major tools in integrated disease management for efficient management of *Alternaria* blight as it provides the resistance or tolerance to crop plant throughout life cycle of plant. The resistant or tolerant varieties released for the cultivation improve the cost benefit ratio to farmers as they reduce the cost of cultivation on disease and/pest management.

Host plant resistance is an integral part of integrated disease management and very important in host-pathogen interaction studies for effective diseases management. The development and release of resistant varieties with strong genetic constitution against pathogens provide economically and ecofriendly options to farmers in crop cultivation. Katiyar et al. (2001a, b) recorded three varieties of bottle gourd i.e. Azad Harit 7002 and 7003 resistant to *A. cucumerina* fungal pathogen. Matharu et al. (2006) reported several resistant germplasm of tomato like Arka Saurabh, Arka meghali, IIHR-305, IIHR-308, IIHR-2266, IIHR-2285 and IIHR-2288 resistant against *Alternaria* early blight of tomato. Recently, the researchers all over the world are developing varieties with expression of various genes which encodes proteins and induces the resistance in crops. Ethiopian mustard (*B. carinata*) observed to high degree of tolerance to *Alternaria* blight (Kolte 1996), but reported with undesirable agronomic characters and not cultivated on large scale in India. The inheritance of resistance in *Alternaria* blight of oilseed brassicas is goverened by additive genes; hence there are very limited sources of resistance genes (Krishnia et al. 2000). Good resistance sources are available in wild and related species of Brassicas. Also, it was reported that in case of commercial apple varieties, resistance is governed by a single recessive gene (Saito and Takeda 1984; Shin and Ko 1992). *A. mali* is major pathogen among all the apple growing areas and hence the development of resistant varieties is among the top priorities for Asian breeding programs in *Malus* spp. through conventional breeding programmes with resistant varieties (Saito and Takeda 1984). In present era, few *A. mali* resistant cultivar have been released in world market from by targeted disease resistant breeding programme.

Epidemiology and Forecasting Models

The epidemiology and forecasting of plant diseases is one of the effective strategies of plant diseases management. Most of leaf spot and blight pathogen of Alternaria spp. can be transmitted though infected seeds and infected planting materials and also survives in plant debris or left over of crop after harvesting. The spread of inoculums or propagule of the pathogenic fungus is depends on several parameters like primary amount of inoculums present in diseased material, temperature of atmosphere, leaf wetness period and susceptibility of hosts. As it is well known fact that *Alternaria* spp. are seed borne in nature and reported in most of the areas where there is cultivation of carrots. Since, it was reported that cool weather favours by *A. dauci* for the disease progression and this was evidenced in Portugal that the fungal pathogen o the symptoms of the disease observed in late season of September and October during harvest time of *D. carota* (Anonymous 2005). Filajdic and Sutton (1992b) developed the disease prediction model in North Carolina for *Alternaria* leaf blotch disease symptoms. This model was based on the similar pattern with earlier reported fungal plant pathogenic prediction models like in case of warning systems in Apple scab diseases (Black spot), it also considers the environmental conditions or parameters like air temperatures, rainfall, leaf wetness periods and level of disease severity in early stages. Madden et al. (1978) had developed a computerized forecasting system (FAST) for the prediction of *Alternaria solani disease* severity in tomato crop for the identification of critical periods when environmental parameters becomes favourable for the development of early blight of tomato and proper scheduling for efficient applications of fungicides. Generally, the forecasting systems is based on two empirical models considering day to day environmental conditions i.e. maximum and minimum air temperature, leaf wetness period in hours, maximum and minimum leaf wetness period, relative humidity period in hours (>90 %) and precipitation over period. The FAST forecasting model schedules resulted in lesser requirement of fungicides applications for the management of same level of disease as in regular weekly schedules of disease control. The efforts are being made to improve the forecasting models from collection of more data from similar type of experiments and etiology of the pathogen considering inoculums, spread of propagules and spore ecology as well as survival for possible disease spread based on climatic and environmental variables. It is the best strategy to employ the area and variety specific crop model for forecasting of diseases through regression analysis especially at early weeks after sowing, age of crop at which different diseases appears mainly on leaves and pods as well as highest disease severity at maturity of crop age. The predictions will help farmers to arrange the strategic application of effective fungicides spray schedule. Very few epidemiological data sets were reported with reasonable prediction accuracy with an independent datas.

Van der Waals et al. (2003); in South Africa stated that early blight of potato forecasting model (PLANT-Plus) predicted timely application of fungicides for effectively management of disease and reduced the spray schedules without affecting the yield as compared to traditional methods of fungicides applications.

Strategies to Impart Fungal Disease Resistance

Conventional as well as molecular approaches are being utilized either singly or in combination to insert desirable traits and to improve the genetic constitution into new cultivar(s). However, nowadays, many germplasm lines and cultivars are available as resistance sources for many diseases. Inter and intraspecific hybridization methods with tolerance/resistant lines with desirable agronomic characters are used to produce disease resistant cultivars.

Hot Water Treatment

It is also very effective for the management of diseases when the propagative materials are cane sets, tubers, fruits and any vegetative portion. Mohsan et al. (2011) reported that hot water treatments of mango fruits significantly reduced development of decay caused by *Alternaria alternaria*. These studies revealed that hot water treatment of 55–60 °C proved effective against *Alternaria alternate* black rot of mango fruits. Nega et al. (2003) used the hot water treatment for the management of seed borne pathogens of vegetable seeds produced in organic farming and noticed the significant reduction in seed borne pathogen without much germination losses at hot water treatment of 50 °C for 20–30 min and 53 °C for 10–30 min. The recorded hot water treatment against several *Alternaria* species (*A. dauci*, *A. radicina*, *A. alternata*, *A. brassicicola*) was more than 95 %. However, Tohamy et al. (2004) observed that in vitro suspensions of *A. alternata* spore when treated with hot water at 55 °C for 7 min, could not germinate. Similarly, fungal discs when exposed to hot air of 40 °C for 72 h, significantly suppress the growth. It is also reported that dip treatment of tomato fruits in hot water at 55 °C for 7 min or with hot air at 38 °C for 72 h reduced decay by *Alternaria alternata* for 15 days at 20 °C. In addition to these, Le et al. (2010) reported that optimizing the conditions of hot water at 55 °C for 3 min and vapour heat treatment at 46.5 °C for 40 min with special treatment of 3 °C at storage condition reduce incidence of *A. alternata* in native Taiwan mango fruits (Tuu Shien).

Crop Rotation

Crop rotation is one of the important strategies in management of plant diseases. The successive planting of crops from different family or groups in the same agricultural field from one season to another is widely adapted cultural practice in crop cultivation. The concept may also include a fallow period or land kept in rested condition without cultivation of land. These practices improves the soil organic matter, nutrition, overall soil fertility, moisture content, soil texture, management of weeds and disease control. It has been noticed that successful results are observed with greater fallow period to the crops than the survival period of pathogens in the field conditions.

The pulse crops should be followed by cereals in the successive season to avoid the carryovers of pathogen inocula from one season to another season. Guillemette et al. (2004) observed that crop rotations with non-cruciferous crops, destruction of crop residues and weed management also help to reduce the incidence of *Alternaria brassicae* disease in cruciferous seed. The deep ploughing and burying of diseased crop plants in soil decomposes the crop residues and lowers *Alternaria* viability and infection. Hampton et al. (2012) observed that *A. radicina* fungal pathogen of carrot reduced by 29 % (from 240 CFU/g) in pot culture trial experiment when grown with wheat plants for a period of 4.5 months. It was also noticed significant reduction in presence of barley and feba bean but not with perennial ryegrass or pea. The inhibition of fungal pathogens may results due to root exudates containing antifungal compounds.

Biological Control

In the era of organic farming and eco-friendly disease management, biological control is considered as one of the important and promising tool as alternative to fungicides where the applications of microorganisms to seeds, planting materials, crop plants, fruits and vegetables helps to protect the crops from virulent plant pathogens (Janisiewicz and Korsten 2002; Spadaro and Gullino 2004). There are several antagonists bacteria, fungi, actinomycetes are reported which can be used as effective bio-control agents (BCA). They are being increasingly popular and are nature friendly without any harmful effect on crop plants and human beings. Many of them have dual strategies of pathogen control ability and plant growth promoting traits. Babu et al. (2000) observed significant inhibition of fungal mycelia of *A. solani* causing blight disease of tomato by bioagents like *Bacillus subtilis* and *Trichoderma viridae*. Similarly, Zhao et al. (2008) reported that *Bacillus* and *Pantoea* bacteria have strong antifungal property under in vitro and in vivo conditions. However, *Curtobacterium* and *Sphingomonas* noticed antifungal properties only under in vitro against *A. solani* fungal pathogen of tomato.

Many commercialized technologies are used like Mycostop (formulation of Streptomyces) for seed treatment against *A. brassicicola* (White et al. 1990), soil isolates of *Trichoderma virdie* (Meena et al. 2004) and bulb extracts of *Allium sativum* (Meena et al. 2004; Patni and Kolte 2006) for *B. juncea* in oilseed brassicas. Kantwa et al. (2014) concluded that among botanicals, garlic clove extract had strong inhibition potential against growth of mycelium and sporulation of *Alternaria alternata* under in vitro. The next effective plant extracts were neem and datura leaf extract. Rivillas-Acevedo and Soriano Garcia (2007) isolated antifungal peptide, Ay-AMP from the seeds of *Amaranthus hypochondriacus*, found effective for the management of *A. alternata* at least doses. It was reported that Ay-AMP degrades chitin layer of fungal pathogen, resistant to proteases and high temperature. Cheng et al. (2008) observed that methanol extracts obtained from leaves and stems of *Myoporum bontioides* have antifungal property against *A. alternata* with inhibition more than 58 % after 12 h at 10 gL^{-1}. The analysis revealed that the antifungal compounds was (−) epingaione. It was also noticed that ethanol extracts *Glycyrrhiza glabra* and metannol extract of *Taverniera cunefolia* have inhibitory effect on *A. brassicicola* at 0.02 % concentrations (Zore et al. 2004). Similarly, Kumar et al. (2004) reported comparative effect of four neem based products on *A. brassicae*, achook and bioneem were effective than furpume and nimbicidine. The extensive screening of botanicals (20 plants) against *A. brassicae* was done by Bhardwaj and Laura (2007) and noticed significant antifungal activity by extracts of *Camellia sinensis* (leaves). The next superior botanicals against pathogen were *Asparagus racemosus*, *Aloe vera*, *Acacia nilotica* and *Anthocephalus cadamba* (all root extracts) with moderate effect by *Astercantha longifolia*.

Trivedi (2010) reported that two times applications of fungal biocontrol agent *Ulocladium oudemansii* with 2 × 1,011 CFUs/ha, significantly reduce *A. radicina* as reported in harvested seeds when applied at weekly interval before swathing. The early blight of tomato caused by A. *solani*, mycelium growth was significantly inhibited by *Bacillus subtilis* and *Trichoderma viride* (Babu et al. 2000).

Panwar et al. (2006) observed the bacterial strains *Brevibacillus brevis* KH-7 and *Bacillus firmus* M-10 showed significant inhibition against *A. solani* and resulted in improved growth and yield of potato. Recently, Waghe et al. (2015) recorded significant inhibition by botanicals with Neem extract (63.05 % and 68.88 %) followed by Karanj extract (56.38 % and 63.60 %) at 10 % and 20 %, respectively. Also, Amein et al. (2011) observed efficient strain of *P. fluorescens* L18 and *Trichoderma viride* TV6903 which reduced the disease incidence and increased healthy cabbage by 20 %.

The neem leaf extracts are very effective in inhibiting the mycelia growth of *A. solani* (43.3 %) at 0.1 % concentration (Sharma et al. 2007). Pandey (2010) used the culture plate method to study antagonistic effect of *Trichoderma harzianum* and *Trichoderma viride* against *Alternaria alternata* of *Capsicum frutescens* and reported reduction in growth of *Alternaria alternata* (67.07 %) by *Trichoderma harzianum* and *Trichoderma viridae* by 66.67 % reduction of pathogenic fungi. Ramjegathesh et al. (2011) evaluated bioefficacy of plant oils, botanicals extracts and microbial bioagents against *Alternaria alternata* causing leaf blight of onion. Field study indicated that two sprays of neem oil (3 %), first at disease appearance and second after 15 days later significantly reduced the percent disease index (22.22 %) of onion blight and enhances the yield. On the other hand, among botanicals, rhizome extract (10 %) of *Acorus calamus* was found effective in reducing disease incidence (34.78 %). Abdalla et al. (2014) noticed four Bacillus spp. (Bacillus B25, B35, B41, B45) as effective bioagents against early blight of tomato caused by *Alternaria alternata*. Thakur and Harsh (2014) assessed the efficacy of antagonist fungi against *A. alternata* causing disease of Spilanths oleraces. They reported several strains of Trichoderma spp. i.e. *Trichoderma harzianum* ISO-1,

T.harzianum ISO-2 and *T. piluliferum* with maximum inhibition of *A. alternata* by 90 %.

Siddiqui (2007) isolated fluorescent pseudomonads and *Bacillus* spp. were isolated from *Alternaria triticina* suppressive soils of wheat fields. It was found that strain *Pseudomonas* sp. B28 was best in enhancing growth of wheat in presence of *A. triticina*. Pane and Zaccardelli (2015) noticed that spore-forming bacteria which were isolated from phylloplane of solanaceous crops, were evaluated under in vitro condition for antagonistic activity against early blight pathogen of tomato i.e. *Alternaria alternata*. Out of total bacteria, four strains of *Bacillus* spp. were effective in reducing disease intensity of *Alternaria*. Ganie et al. (2013) evaluated the bioefficacy of botanicals and bioagents against *Alternaria solani* and recorded significant inhibition of *A. solani* mycelia by *T. harzianum* (71.85 %) and among botanical extract of *D. stramonium* reported significant inhibition (61.12 %) of *A. solani* mycelia growth. Hence, it can be concluded that there are several biocintrolagents, botanical extracts and essential oils which are reported to have significant potential of disease control ability caused by *Alternaria* spp. without any harmful effect on human beings and environment.

Fungicidal Management of Diseases

The applications of fungicides in the management of the diseases caused by *Alternaria* spp. are among the effective disease control strategies.

Seed/planting material treatment with broad spectrum fungicides, after the hot water treatment will reduce inocula associated with the seed/planting material and will reduce pre-emergence seedling blight/rots caused by many fungal pathogens. Dubey et al. (2000) worked on the fungicidal management of blight of broad bean caused by *Alternaria alternata* under in vitro condition and reported 100 % inhibition by contaf fungicide. Ahmad (2009) evaluated the systemic and non-systemic fungicides for the management of *Alternaria mali causing* blotch of apple. It was observed that, hexaconazole 5EC (0.05 %) was significant in reduction of disease among systemic fungicides (PDI 0.9 %). Whereas, copper oxychloride and ziram among non-systemic fungicides both at 0.4 % concentrations were effective in reducing disease intensity by (PDI 4.8 %) and PDI 4.6 %), respectively as compared to control (PDI 25.6 %). Sharma et al. (2013) evaluated the effect of fungicides for the management of blight of cumin caused by *Alternaria burnsii*. They recorded that propiconazole have at par results with carbendazim + iprodione and chlorothalonil with minimum percent disease index (PDI) as compared to other treatments. Arun Kumar et al. (2011) studied the effect of fungicides in field condition against *Alternaria alternata* causing chrysanthemum leaf blight. They reported hexaconazole (0.1 %) as most effective fungicide for disease management with least PDI (4.49) than Chlorothalonil (0.2 %) and Mancozeb (0.2 %). Venkataramanamma et al. (2014) studied the efficacy of *Alternaria helianthi* causing blight of sunflower and noticed that seed treatment with SAAF@3 g/kg+Propiconazole (0.1 %) resulted highest seed yield with lowest PDI. Similar results were obtained when experiment repeated for the 3 years. Recently, Ramdaswaghunde et al. (2014) studied the effect of fungicides under in vitro condition and noticed that at 1,000 and 1,500 ppm concentrations, the fungicides like hexaconazole, combination of thiophanate methyl and hexaconazole (5 %)+captan (70 %) had 100 % inhibition of fungal mycelia and sporulation. Whereas, Waghe et al. (2015) noticed that SAAF at 2,000 ppm (90.36 %) was recorded with significant inhibition of fungal growth than with Mancozeb at 2,500 ppm (88.88 %) under in vitro. The most significant fungicide in controlling fungal growth of *A. solani* was mancozeb (0.2 %) as reported by Choulwar et al. (1989) and confirmed by Singh et al. (2001). Verma and Verma (2010) reported that hexaconazole inhibit cent percent growth of fungal mycelium of *A. alternata* blight of tomato. Katiyar et al. (2001a, b) reported the best control of bottle gourd leaf spot disease caused by *Alternaria* with spray application of Indofil M-45 (0.2 %). Among the newer method of disease management, sulfanilamide derivatives (chitosan) noticed most effective against *A. solani* at different concentration ranging from 50 to 500 μg/ml (Mei et al. 2007). In addition to this, potassium bicarbonate, sodium bicarbonate and Nerol (essential oil fractions from citrus) had significant inhibition against *A. solani* causing early blight disease of potato. Abd-el-Kareem (2007) reported 100 % inhibition of *Alternaria* fungus by potassium or sodium bicarbonate (each @2 % and Nerol @0.5 %).

Mohsan et al. (2011) studied the effect of five different fungicides like prochloraz, deconil, carbendazim, TBZ (thiabendazol) and mancozeb through dip treatments for the management of black spot decay of mango acused by (*Alternaria alternata*) during storage at 20 °C. The fungicides, mancozeb and prochloraz were proved as the most effective fungicides with minimum development of lesion on mango fruits. Balai and Singh (2013) noticed that two way management of disease by seed treatment as well as fungicidal spray of Mancozeb along with biocontrol agent fungi *T. viride* was reported as most effective in reduction of disease severity. They reported management of disease (11.37 % and 72.69 %) thereby combination of Mancozeb with *T. harzianum* (11.43 % and 72.53 %) and Mancozeb (11.49 % and 72.40 %) respectively. Several researchers have reported that seed dressing with different types of fungicides are very effective in the management of many seed borne pathogens. Mondal et al. (1989) and Verma and Saharan (1994) reported significant control of *A. japonica* by seed dressings with different groups of fungicides.

The chemical control are effective but they may have residual toxicity problem and their indiscriminate use causes many environmental and health problems. They also reported to kill many beneficial micro flora and results in development of resistance in target pathogen. It has been reported to affect the quality of the produce (McCartney et al. 1999). Despite of many deleterious effects caused by the use of synthetic fungicides, recently the focus is on the development of novel biotechnological tools for the development of sustainable disease resistant/tolerant cultivars, adaptive in adverse environmental conditions and high degree of crop produce.

Biotechnological Approaches

Despite of conventional methods of cultivar improvement and pest management, nowadays biotechnological techniques like tissue culture and genetic transformation based recombinant DNA technologies have revolutionized research on desirable agronomic traits and pest resistant genes in germplasm. These tools have provided great access to transform traits from distantly related wild races/cultivars for development of improved and superior varieties of crops.

Tissue Culture Techniques

Embryo rescue one of the most successful techniques was employed for the production of inter-specific and inter-generic hybrids from naturally incompatible crosses. This technique has proved a very effective for transfer of *Alternaria* blight tolerance in oilseed brassicas (Aneja and Agnihotri 2013). The hybrids resulted from these crosses are then multiply using somatic embryogenesis techniques. Gupta et al. (2010) used inter-specific hybridization method with in vitro ovule culture to incorporate high tolerance trait to *Alternaria* blight. Somatic hybridization consists of isolation of protoplasts by removal of cell wall which then fused with desired germplasm protoplasts to produce hybrids. This technique has been reported in *Alternaria* blight resistant trait from *Moricandida arvensis* to *B. oleracea* (Toriyama et al. 1987), and from *Sinapis alba* to *B. napus* (Primard et al. 1988). Four highly resistant hybrids were developed after repeated back crosses by somatic hybrids between *S. alba* x *B. oleracea* by protoplast fusion followed by embryo rescue technique (Sigareva and Earle 1999a, b, c). Somaclonal variations refer to chromosomal rearrangements of species at different ploidy levels in outcrossing, inbreeding, vegetatively and seed propagated crops under in-vitro conditions. Sharma and Singh (1995) reported the use of somaclonal variations by incorporating disease resistance/tolerance traits against *Alternaria* blight. The chemical mutagenesis treatment with ethyl methanesulfonate (EMS) and ethyl nitrosourea (ENU) microspores/microspore derived embryos (Agnihotri et al. 2009).

Transgenic Plants

Production of transgenic plants through recombinant DNA technology is another biotechnological approach to incorporate disease resistance genes from resistant/tolerant genotypes to the agronomically and economically important superior varieties. Transgenic in *B. juncea* have been developed which imparts delayed development of the *Alternaria* blight. Low disease intensity of *Alternaria* blight was developed in *B. juncea* cv. RLM-198 by Kanrar et al. (2002), cDNA sequence encoding “hevein-rubber tree lectin” from *Hevea brasiliensis* by genetic transformation. Chitinase, which has major role in plant defenses, gene tagged with over expressing promoter 35S CaMV transformed in *B. juncea* (cv. RLM198)

(Mondal et al. 2003). These transformed plants reported to have delayed disease progression and 12–56 % approx. reduction in fungal colonization. A transformed *B. juncea with* Osmotin protein, which has role in signal transduction, also reported tolerance to *A. brassicae* (Taj et al. 2004). Likewise, Mondal et al. (2007) developed genetic transformed Indian mustard expressing "class I glucanase" gene under CaMV 35S promoter. Recently, Verma et al. (2012) reported the introgression of cysteine rich-antimicrobial peptide, PmAMP1 from *Pinus monticola* into *B. napus*, imparting enhanced protection against broad spectrum of fungal pathogens like *A. brassicae*, *Leptosphaeria maculans* and *Sclerotinia sclerotiorum*.

Induced Host Resistance

The plant disease management through induced systemic resistance by the application of biotic and abiotic agents is one of the important strategies under eco-friendly disease management. Under abiotic activators of inducing host resistance against plant pathogens, application salicylic acid, jasmonic acid and phosphatic salts are common. Klessig and Malamy (1994) reported that chemicals like salicylic acid, jasmonic acid, cow urine and microbial bioagents like *Pseudomonas* sp. induces systemic host resistance with challenge inoculation of these agents/ compounds in certain quantities. Ratnam et al. (2001) observed that sunflower seeds when treated with salicylic acid (5 mM) and bion (5 mM) induces systemic host plant resistance as evidenced from higher phenol content in host plant and recoreded reduced disease severity.

The avirulent strain or weak pathogens are inoculated on plants which activates the defense related genes in comparatively susceptible plants (Deverall 1995). The avirulent strain or weak pathogen triggers the cascade of defense related genes by bringing the metabolic shift through release of elicitors and results in production of PR proteins (pathogenesis related) and/or pathogen inhibiting compounds. Similarly, some abiotic agents like phosphate salts also activate the defense related pathways in plants. To have the benefits of durable host plant resistance by challenge inoculations of biotic and abiotic agents, it is important to understand the pattern and behavior of pathogens development, pathogenesis on various hosts, etiology of diseases in different ecological conditions. Surekha et al. (2014) noticed that defense related enzymes and phenolic content in the host plant (*Vigna mungo*) when challenge inoculated with *Trichoderma viride* in presence of phytopathogens, *Fusarium oxysporum* and *Alternaria alternata*. The defense related enzymes induced by *T. viride* in the host were peroxidase, polyphenol oxidase and phenyl alanine ammonia lyase and the phenolic contents. Hence, it was then concluded that induced plant defense enzymes play major role in reducing biotic stress in black gram by *T. viride* as biocontrol agent. Recently, Song et al. (2011) concluded that application of abscisic acid improves the host resistance against seedlings of *Alternaria solani*. There was rapid up-regulation of defense related genes i.e. PR1, b-1, 3-glucanase, PPO, POD, and superoxide dismutase by external application of

ABA. They suggested that application of ABA may impart resistance to *A. solani* infection in tomato by defense genes activation and mediate the production of defense related enzymes.

Induced host resistance was reported in susceptible *B. juncea* cv. PR-15 against highly virulent isolate *A. brassicae* A (AbA) and considerably virulent isolate C (AbC) from *B. carinata* cv. PPCS-1 by challenge inoculation with avirulent isolate D (AbD) which was isolated from same cultivar (Vishwanath et al. 1999). There was up to 60 % reduction in disease intensity against both AbA and AbC pathogenic isolates by induction of defense response. But, the mechanism of defense activation in otherwise susceptible host is still unclear with limited studies in molecular characterization and host pathogen interactions. The taxonomic position and etiological studies of pathogen causing diseases will help to plan the induction of stable resistance against pathogenic *Alternaria* spp. in diverse agro ecological zones and cropping systems. This will provide a path towards ecofriendly management of diseases, reduction of harmful chemical fungicides for sustainable crop production.

Conclusions

From the various experiments and studies, it can be concluded that diseases caused by *Alternaria* spp. are very devastating and widespread on several economically important field crops, vegetables, forests, horticultural, ornamentals and plantation crops.

However, with the advancement of molecular and technological developments, it has been significantly improved the accurate detection, characterization and identification of various *Alternaria* spp. and related strains and biotypes. Recently, significant progress has been made on host-pathogen interaction, biosynthesis of biochemical/toxins pathways, role in disease development and molecular basis of pathogenesis. The utilization of several variants of PCR, development of molecular and protein/enzymes markers, genetic studies, pathogenesis cascade will help to understand the multiple role of virulence factors in disease development and impact on normal physiology of host plants. The high end techniques like FTIR, DNA microarray, proteomics and genomics studies of host-pathogen interaction will provide new insight into disease development and to work out possible measurement of disease management. The most common method in the management of *Alternaria* diseases is through application of fungicides. However, due to indiscriminate use of chemicals for the disease management, they have caused serious health hazards and environmental problems. By considering these problems, efforts are being made to employ disease resistant/tolerant varieties, use of botanical extracts and oils, plant defense activators, biocontrol agents and improved cultural/agronomic practices etc. Also, these are important part of integrated disease management (IDM), organic farming and sustainable agriculture. They are safe, cost effective and eco-friendly. Due to continuous and higher dose of fungicides for disease control, resistant populations in *Alternaria* spp. have been reported rapidly in cultivated plants. Therefore, it is an immediate need to adapt anti resistant strategies for the disease management.

Resistance was noticed in pistachio orchards by *Alternaria* to azoxystrobin and to strobilurins by *Alternaria* causing late blight. The oilseed brassicas, saaflower and sunflower are heavily damaged by severe infection of *Alternaria* spp. therefore, immediate efforts to be made for effective management of *Alternaria* diseases, reduce losses, high yields and disease free production and development of improved cultivars in crops. With development in biotechnological field, the conventional plant breeding process gets highly benefited and enabled to incorporate and combine superior agronomic, disease and pest resistant and quality traits. Though, lots of studies and efforts have been made on incorporation of disease resistant/tolerant characters, gene pyramiding especially against blight in oilseed brassicas cause by *Alternaria* spp., needs some improved target specific and thorough research. The pathogen, *Alternaria* spp. is cosmopolitan and attacks to crop wide host range distributed over different geographic and agro-climatic zones with greater variability in inter and intra species of *Alternaria.* It has been noticed that the most of the *Alternaria* spp. have been common lineage of *A. alternata*, *A. brassicae* and *A. brassicicola* and their subspecies also grouped into these categories. Therefore, it is very important to correlate the *Alternaria* spp. on the basis of morphological, physiological, biochemical, aggressiveness and virulence factors with molecular diversity and lineage of the pathogen. The recent approaches is the extensive studies on pathogen associated molecular pattern (PAMPs) and the signal transduction pathways and mechanism of *Alternaria* spp. for better understanding of host-pathogen interaction and activation of defense related genes during pathogenesis. The detection and accurate identification of pathogenesis related proteins (PR-proteins) and defense related proteins/enzymes generated during infection of *Alternaria* diseases will help botanist, geneticist, molecular biologist and plant pathologist to develop *Alternaria* resistant/tolerant crops/varieties and/or plausible genetic modifications. This will eliminate environmental hazards by reducing application of toxic fungicides. It can be summarized that we have to combine both the traditional, biotechnological as well as advanced techniques to develop strategies for integrated disease management (IDM) including durable germplasm, disease resistant/tolerant varieties of crops/plants along with superior agronomic characters.

References

Abass MH (2013) A PCR ITS-RFLP method for identifying fungal contamination of date palm (*Phoenix dactylifera* L.) tissue cultures. Afr J Biotechnol 12(32):5054–5059

Abd-el-Kareem F (2007) Potassium or sodium bicarbonate in combination with Nerol for controlling early blight disease of potato plants under laboratory, greenhouse and field conditions. Egypt J Phytopathol 35(1):73–86

Abdalla SA, Algam SAA, Ibrahim EA, Naim AME (2014) In vitro screening of Bacillus isolates for biological control of early blight disease of tomato in Shambat soil. World J Agri Res 2(2):47–50

Agarwal A, Garg GK, Devi S, Mishra DP, Singh US (1997) Ultrastructural changes in Brassica leaves caused by *Alternaria brassicae* and destruxin B. Int J Plant Biochem and Biotechnol 6(1):25–28

Agostini JP, Bushong PM, Timmer LW (2003) Greenhouse evaluation of products that induce host resistance for control of scab, melanose, and *Alternaria* brown spot of citrus. Plant Dis 87:69–74

Agnes A (2009) Distribution and characterization of sweet potato *Alternaria* blight isolates in Uganda. A thesis submitted to the school of graduate studies in partial fullfilment of the requirement for the award of a master of science in crop science of Makerere University, Uganda

Agnihotri A, Gupta K, Prem D, Sarkar, Mehra VS, Zargar SM (2009) Genetic enhancement in rapeseed- mustard for quality and disease resistance through *in vitro* techniques. In: Proceedings 16th Australian Research Assembly on Brassicas, Ballarat, 14–16 Sept, p 28

Ahmad S (2009) Management of *Alternaria* blotch of apple using fungicides. Indian J Plant Prot 37(1 and 2):134–138

Ahmad MS, Sinha KK (2002) Mycotoxin contamination of mustard seeds during storage. Indian Phytopathol 55(3):299–302

Akamatsu H (2004) Molecular biological studies on the pathogenicity of *Alternaria alternata* tomato pathotype. J Gen Plant Pathol 70:389

Akamatsu H, Taga M, Kodama M, Johnson R, Otani H, Kohmoto K (1999) Molecular karyotypes for Alternaria plant pathogens known to produce hostspecific toxins. Curr Genet 35(6):647–656

Akimitsu K, Peever TL, Timmer LW (2003) Molecular, ecological and evolutionary approaches to understanding *Alternaria* diseases of citrus. Mol Plant Pathol 4:435–446

Akhtar KP, Saleem MY, Asghar M, Haq MA (2004) New report of *Alternaria alternata* causing leaf blight of tomato in Pakistan. New Dis Rep 9:43

Amein T, Wright SAI, Wikström M, Koch E, Schmitt A, Stephan D, Jahn M, Tinivella F, Gullino ML, Forsberg G, Werner S, der Wolf JV, Groot SPC (2011) Evaluation of non-chemical seed treatment methods for control of *Alternaria brassicicola* on cabbage seeds. J Plant Dis Prot 118(6):214–221

Amir M, Hamid M, Javad H (2014) Rapid identification of *Alternaria alternata* fungi, the cause of brown spot disease by loopmediated isothermal amplification. Iran J Public Health 43(2):51

Andersen B, Dongo A, Pryor BM (2008) Secondary metabolite profiling of *Alternaria dauci, A. porri, A. solani & A. tomatophila*. Mycol Res 112(2):241–250

Andersen B, Nielsen KF, Pinto VF, Patracia A (2015) Characterization of *Alternaria* strains from Argentinean blueberry, tomato, walnut and wheat. Int J Food Microbiol 196:1–10

Anderson B, Kroger E, Roberts RG (2001) Chemical and morphological segregation of *Alternaria alternata*, *Alternaria gaisen* and *Alternaria longipes*. Mycol Res 105:291–299

Anderson B, Hansen ME, Smedsgaard J (2005) Automated and unbiased image analysis as tool in phenotypic classification of small spored *Alternaria* spp. Phytopathology 95:1021–1029

Aneja JK, Agnihotri A (2013) *Alternaria* blight of oilseed Brassicas: epidemiology and disease control strategies with special reference to use of biotechnological approaches for attaining host resistance. J Oilseed Brassica 4(1):1–10

Aneja JK, Agarwal A, Agnihotri A (2014) Inter and intra-specific diversity in *Alternaria* species infecting oilseed Brassicas in India. J Oilseed Brassica 5(2):102–117

Anonymous (2005) Crop production yearbook. GPPAA- Ministério da Agricultura do Desenvolvimento Rural e das Pescas. Lisboa, Castel-Publicações e Edições AS. Appl Environ Microbiol 64(4):1313–1318. ISSN 0099-2240

Aradhya MK, Chan HM, Parfitt DE (2001) Genetic variability in the pistachio late blight fungus, *Alternaria alternata*. Mycol Res 105:300–306

Aroca A, Raposo R (2007) PCR-based strategy to detect and identify species of *Phaeoacremonium* causing grapevine diseases. Appl Environ Microbiol 73(9):2911–2918, ISSN 0099-2240

Aroca A, Raposo R, Lunello P (2008) A biomarker for the identification of four *Phaeoacremonium* species using the beta-tubulin gene as the target sequence. Appl Microbiol Biotechnol 80:1131–1140

Arun Kumar GS, Kamanna BC, Benagi VI (2011) Management of chrysanthemum leaf blight caused by *Alternaria alternata* (fr.) Keissler under field condition. Plant Arch 11(1):553–555

Atassi MZ, Lee C (1978) The precise and entire antigenic structure of native lysozyme. Biochem J 171:429–434

Avenot H, Dongo A, Simoneau NB, Iacomivasilescu B, Hamon B, Peltiers D, Simoneau P (2005) Isolation of 12 polymorphic microsatellite loci in the phytopathogenic fungus *Alternaria brassicicola*. Mol Ecol Notes 5:948–950

Babler E, Asensio N, Leiminger J, Hausladen H, Bahnweg G, Zinkernagel V (2004) Investigations on the appearance and control of *Alternaria solani* and *Alternaria alternate* on potato. Mitt Biol Bundesanst Land- Forstwirtschaft 396:100–101

Babu S, Seetharaman K, Nandakumar R, Johanson I (2000) Efficacy of fungal antagonists against leaf blight of tomato caused by *Alternaria solani* (Ell. and Mart.) Jones and Grout. J Biol Control 14(2):79–81

Balai LP, Singh RB (2013) Integration management of *Alternaria* blight of pigeonpea with some fungicides and antagonists in pot condition. Bioscan 8(3):881–886

Beattie SH, Holt C, Hirst D, Williams AG (1998) Discrimination among *Bacillus cereus B. mycoides* and *B. thuringiensis* and some other species of the genus *Bacillus* by Fourier transform infrared spectroscopy. FEMS Microbiol Lett 164:201–206

Berkeley MJ (1836) Fungi. In: Smith JE and Hooker JW (eds). The English Flora. Vol. V, Part II. Longman, London. pp 339–340

Bhardwaj SK, Laura JS (2007) Antifungal potential of some botano-extracts against *Alternaria brassicae*. J Plant Dis Sci 2(2):135–137

Bock CH, Peter HT, Burbaker CL, Burdon JJ (2002) Detection of genetic variation in *Alternaria brassicicola* using AFLP fingerprinting. Mycol Res 106(4):428–434

Bouws H, Wattenberg A, Zorn H (2008) Fungal secretomes – nature's toolbox for white biotechnology. Appl Microbiol Biotechnol 80:381–388

Braaksma M, Martens-Uzunova ES, Punt PJ, Schaap PJ (2010) An inventory of the *Aspergillus niger* secretome by combining *in silico* predictions with shotgun proteomics data. BMC Genomics 11:584

Branan N, Wells TA (2007) Microorganism characterization using ATR–FTIR on an ultrathin polystyrene layer. Vib Spectrosc 44(1):192–196, http://dx.doi.org/10.1016/j.vibspec.2006.10.002

Brill LM, McClary RD, Sinclair JD (1994) Analysis of two ELISA formats and antigen preparations using polyclonal antibodies to *Phomopsis longicolla*. Phytopathology 84:173–179

Brun S, Madrid H, Gerrits van den Ende B, Andersen B, Marinach-Patrice C, Mazier D, de Hoog GS (2013) Multilocus phylogeny and MALDI-TOF analysis of the plant pathogenic species *Alternaria dauchi* and relatives. Fungal Biol 117:32–40

Brunus TD, White TJ, Taylor JW (1991) Fungal molecular systematic. Ann Rev Ecol Syst 22:525–564

Burgess T, Wingfield MJ, Wingfield BW (2001) Simple sequence repeat markers distinguish among morphotypes of *Sphaeropsis sapinea*. Appl Environ Microbiol 67:354–362

Carberry S, Neville CM, Kavanagh KA, Doyle S (2006) Analysis of major intracellular proteins of *Aspergillus fumigatus* by MALDI mass spectrometry: identification and characterisation of an elongation factor 1B protein with glutathione transferase activity. Biochem Biophys Res Commun 314:1096–1104

Carzaniga R, Fiocca D, Bowyer P, O'Connell RJ (2002) Localization of melanin in conidia of *Alternaria alternata* using phage antibodies. Mol Plant-Microbe Interact 15:216–224

Chahal SS, Pannu PPS (1997) Detection of *Tilletia indica* in wheat and *T. Barclayana* in rice samples and its Implication for seed certification. In: Hutchins JD, Reeves JC (eds) Seed health testing: Progress towards 21st century. CAB Direct, pp 153–158

Chattopadhyay C, Agrawal R, Kumar A, Bhar LM, Meena PD, Meena RL, Khan SA, Chattopadhyay AK, Awasthi RP, Singh SN, Chakravarthy NVK, Kumar A, Singh RB, Bhunia CK (2005) Epidemiology and forecasting of *Alternaria* blight of oilseed Brassica in India – a case study. J Plant Dis Prot 112(4):351–365

Chen RS, Chu S, Cheng C-W, Chen W-Y, Tsay J-G (2008) Differentiation of two powdery mildews of sunflower (*Helianthus annuus*) by a PCR-mediated method based on ITS sequences. Eur J Plant Pathol 121:1–8

Chen M, Chang Y, Zhang J, Ding J, Sun Q (2012) A genetic linkage map of Japanese scallop *Mizuhopecten yessoensis* based on Amplified Fragment Length Polymorphism (AFLP) and Microsatellite (SSR) markers. Afr J Biotechnol 11(46):10517–10526

Cheng DY, Zhen Y, Zhen YY, Lian BX (2008) Inhibitory activity against plant pathogenic fungi of extracts from *Myoporum bontioides*. A Gray and indentification of active ingredients. Pest Manag Sci 64(2):203–207

Choi YP, Paul NC, Lee HB, Yu SH (2014) First record of *Alternaria simsimi* causing leaf spot on sesame (*Sesamum indicum* L.) in Korea. Mycobiology 42(4):405–408

Chou HH, Wu WS (2002) Phylogenetic analysis of internal transcribed spacer regions of the genus *Alternaria* and the significance of filament-beaked conidia. Mycol Res 106:164–169

Choulwar AB, Datar VV, Kurundkar BD (1989) Efficacy of fungitoxicants on the mycelial growth of *A. solani*. Pestology 13:17–19

Chowdappa P, Lakshmi MJ (2013) Identification of *Alternaria* species associated with leaf blights of fruit, vegetable and oil seed crops based on protein and multi-locus enzyme finger prints. Pest Manag Horticult Ecosyst 19(1):45–56

Chu PWG, Waterhouse PM, Martin RR, Gerlach WL (1989) New approaches to the detection of microbial plant pathogens. Biotechnol Genet Eng Rev 7:45–111

Chupp C, Sherf AF (1960) Crucifer diseases IK. In: Vegetable diseases and their control. The Ronald Press Company, New York, Chapter 8, pp 237–288

Clark MF, Adams AN (1977) Characteristics of the microplate method of enzyme linked immunosorbent assay for the detection of plant viruses. J Gen Virol 34:475–483

Collier L, Balows A, Sussman M (1998) Topley & Wilson's microbiology and microbial infections, vol 4, 9th edn. Arnold, London

Conn KL, Tewari JP (1990) Survey of *Alternaria* blackspot and *Sclerotinia* stem rot in central Alberta in 1989. Can Plant Dis Surv 70:66–67

Conn KL, Tewari JP, Dahiya JS (1988) Resistance to *Alternaria brassicae* and phytoalexin elicitation in rapeseed and other crucifers. Plant Sci 56:21–25

Cooke DEL, Forster JW, Jenkins PD, Gareth DJ, Lewis DM (1998) Analysis of intraspecific and interspecific variation in the genus *Alternaria* by the use of RAPD-PCR. Ann Appl Biol 132:197–209

Corlett M, Corlett M (1999) Fungi Canadensis No 341 Alternaria linicola. Can J Plant Pathol 21:55–57

de Hoog GS, Horre R (2002) Molecular taxonomy of the *Alternaria* and *Ulocladium* species from humans and their identification in the routine laboratory. Mycoses 45:259–276

Denis O, Van Cauwenberge A, Treutens G, Es Saadi B, Symoens F, Popovic N, Huygen K (2012) Characterization of new *Alternaria alternata* specific rat monoclonal antibodies. Mycopathologia 173(2–3):151–162

Deverall BJ (1995) Plant protection using natural defense systems of plants. Adv Plant Pathol 11:211–228

Dube JP (2014) Characterization of *Alternaria alternata* isolates causing brown spot of potatoes in South Africa. M.Sc. thesis submitted in partial fulfillment of the requirements for the master in science (Agriculture) Plant Pathology to University of Pretoria, Pretoria, South Africa

Dubey SC, Patel B, Jha OK (2000) Chemical management of *Alternaria* blight of broad bean. Indian Phytopathol 53(2):213–215

Dyer PS, Fumeaux PA, Douhan G, Murray TD (2001) A multiplex PCR test for determination of mating type applied to the plant pathogens *Tapesia yallundae* and *Tapesia acuformis*. Fungal Genetic Biol 33:173–180

EFSA (2011) EFSA J. 9, 2407. http://dx.doi.org/10.2903/j.efsa.2011.2407 (97 pp, www.efsa.europa.eu/efsajournal)

El_Komy MH, Saleh AA, Molan YY (2012) Molecular characterization of early blight disease resistant and susceptible potato cultivars using random amplified polymorphic DNA (RAPD) and simple sequence repeats (SSR) markers. Afr J Biotechnol 11(1):37–45

Elliott JA (1917) Taxonomic characters of the genera *Alternaria* and *Macrosporium*. Am J Bot 4:439–476

Ellis MB (1971) Dematiaceous hyphomycetes. Commonwealth Mycological Institute, Kew, pp 464–497

Ellis MB (1976) More dematiaceous hyphomycetes. Commonwealth Mycological Institute, Kew, pp 411–427

Erukhimovitch V, Tsror L, Hazanovsky M, Talyshinsky M, Mukmanov I, Souprun Y, Huleihel M (2005) Identification of fungal phyto-pathogens by Fourier transform infrared (FTIR) microscopy. J Agric Technol 1(1):145–152

Fernandez D, Ouinten M, Tantaoui A, Geiger JP, Daboussi MJ, Langin T (1998) Fot 1 insertions in the *Fusarium oxysporum* f. sp. *albedinis* genome provide diagnostic PCR targets for detection of the date palm pathogen. Appl Environ Microbiol 64:633–636

Filajdic N, Sutton TB (1992) Influence of temperature and wetness duration on infection of apple leaves and virulence of different isolates of *Alternaria mali*. Phytopathology 82:1279–1283

Fischer G, Braun S, Thissen R, Dott W (2006) FT-IR spectroscopy as a tool for rapid identification and intra-species characterization of airborne filamentous fungi. J Microbiol Meth 64:63–77

Fleck SC, Burkhardt B, Pfeiffer E, Metzler M (2012) *Alternaria* toxins: altertoxin II is a much stronger mutagen and DNA strand breaking mycotoxin than alternariol and its methyl ether in cultured mammalian cells. Toxicol Lett 214:27–32

Foster SJ, Ashby AM, Fitt BDL (2002) Improved PCR-based assays for presymptomatic diagnosis of light leaf spot and determination of mating type of *Pyrenopeziza brassicae* on winter oilseed rape. Eur J Plant Pathol 108:374–383

Fraaije B, Lovell DJ, Rohel EA, Hollomon DW (1999) Rapid detection and diagnosis of *Septoria tritci* epidemics in wheat using a polymerase chain reaction/Pico green assay. J Appl Microbiol 86:701–708

Fraaije BA, Lovell DJ, Coelho JM, Baldwin S, Hollomon DW (2001) PCR-based assays to assess wheat varietal resistance to blotch (*Septoria tritici* and *Stagonospora nodorum*) and rust (*Puccinia striiformis* and *Puccinia recondita*) diseases. Eur J Plant Pathol 107(9):905–917, ISSN 0929-1873

Ganie SA, Ghani MY, Nissar Q, Shabir-u-Rehman (2013) Bioefficacy of plant extracts and biocontrol agents against *Alternaria solani*. Afr J Microbiol Res 7(34):4397–4402

Gannibal PB, Klemsdal SS, Levitin MM (2007) AFLP analysis of Russian *Alternaria tenuissima* populations from wheat kernels and other hosts. Eur J Plant Pathol 119:175–182

Gargouri-Kammoun L, Bensassi F, Mnari-Hattab M, Rhouma A, Bacha H, Hajlaoui MR (2014) Identification of *Alternaria* species recovered from stored durum wheat kernels in Tunisia. Tunis J Plant Prot 9:119–129

Geiser DM, Jiménez-Gasco M, Kang S, Makalowska I, Veeraraghavan N, Ward TJ, Zhang N, Kuldau GA, O'Donnell K (2004) *Fusarium*-ID v.1.0: A DNA sequence database for identifying *Fusarium*. Eur J Plant Pathol 110(5–6):473–479, ISSN 0929-1873

Gherbawy YAMH (2005) Genetic variation among isolates of *Alternaria* spp. from selected Egyptian crops. Arch Phytopathol Plant Protect 38(2):77–89

Goodwin DC, Lee SB (1993) Microwave miniprep of total genomic DNA from fungi, plants, protists and animals for PCR. Biotechniques 15:438–444

Grote D, Olmos A, Kofoet JJ, Tuset E, Bertolini E, Cambra M (2002) Specific and sensitive detection of *Phytophthora nicotianae* by simple and nested-PCR. Eur J Plant Pathol 108(3):197–207, ISSN 0929-1873

Guglielmo F, Bergemann SE, Gonthier P, Nicolotti G, Garbelotto M (2007) A multiplex PCR-based method for the detection and early identification of wood rotting fungi in standing trees. J Appl Microbiol 103:1490–1507, ISSN 1364-5072

Guillemette T, Iacomi-Vasilescu B, Simoneau P (2004) Conventional and real-time PCR based assay for detecting pathogenic *Alternaria brassicae* in cruciferous seed. Plant Dis 88:490–496

Guo LD, Hyde KD, Liew ECY (2000) Identification of endophytic fungi from *Livistona chinensis* (Palmae) using morphological and molecular techniques. New Phytol 147:617–630

Guo-yin T, Zhi-ling Y, Zhi-lin Y, Shou-an Z (2013) Morphological, molecular and pathogenic characterization of *Alternaria longipes*, the fungal pathogen causing leaf spot on *Atractylodes macrocephala*. Afr J Microbiol Res 7(21):2589–2595

Gupta MJ, Irudayaraj JM, Debroy C, Schmilovitch Z, Mizrach A (2005) Differentiation of food pathogens using FTIR and artificial neural networks. Trans ASAE 48:1889–1892
Gupta K, Prem D, Agnihotri A (2010) Pyramiding white rust and *Alternaria* resistance in low erucic *Brassica juncea* using *B. carinata*. J Oilseed Brassica 1(2):55–65
Hampton JG, Trivedi RS, Townshend JM (2012) Can crop rotation reduce inoculum of the carrot pathogen *Alternaria radicina*? Agron N Z 42:155–162
Harrison JG, Barker H, Lowe R, Rees EA (1990) Estimation of amounts of *Phytophthora infestans* mycelium in leaf tissue by enzyme-linked immunsorbent assay. Plant Pathol 39:274–277
Hatta R, Ito K, Hosaki Y, Tanaka T, Tanaka A, Yamamoto M, Akimitsu K, Tsuge T (2002) A conditionally dispensable chromosome controls host-specific pathogenicity in the fungal plant pathogen *Alternaria alternata*. Genetics 161(1):59–70
Hayden MJ, Good G, Sharp PJ (2002) Sequence tagged microsatellite profiling (STMP): improved isolation of DNA sequence flanking target SSRs. Nucleic Acids Res 30:e129
Helm D, Naumann D (1995) Identification of some bacterial cell components by FT-IR spectroscopy. FEMS Microbiol Lett 126:75–80
Hensen JM, French R (1993) The polymerase chain reaction and plant disease diagnosis. Ann Rev Phytopathol 31:81–109
Hirsch PR, Mauchline TH, Mendum TA, Kerry BR (2000) Detection of the nematophagous fungus *Verticillium chlamydosporium* in nematode-infested plant roots using PCR. Mycol Res 104:435–439
Hong SG, Maccaroni M, Figuli PJ, Pryor BM, Belisario A (2006) Polyphasic classification of *Alternaria* isolated from hazelnut and walnut fruit in Europe. Mycol Res 110:1290–1300
Hong SY, Kang MR, Cho EJ, Kim HK, Yun SH (2010) Specific PCR detection of four quarantine *Fusarium Species* in Korea. Plant Pathol J 26(4):409–416, ISSN 1598-2254
Hu J, Chen C, Peever T, Dang H, Lawrence C, Mitchell T (2012) Genomic characterization of the conditionally dispensable chromosome in Alternaria arborescens provides evidence for horizontal gene transfer. BMC Genomics 13:171
Humpherson-Jones FM (1989) Survival of *Alternaria brassicae* and *Alternaria brassicicola* on crop debris of oilseed rape and cabbage. Ann Appl Biol 115:45–50
Humpherson-Jones FM, Phelps K (1989) Climatic factors influencing spore production in *Alternaria brassicae* and *Alternaria brassicicola*. Ann Appl Biol 114:449–458
Hwang BK, Yun JH, Kim ZS (1987) Geographic variation of esterase isozymes in populations of *Alternaria mali* on apple leaves. Phytopathol Z 119:225–231
Hyakumachi M, Priyatmojo A, Kubota M, Fukui H (2005) New anastomosis groups, AG-T and AG-U, of binucleate *Rhizoctonia* causing root and stem rot of cut-flower and miniature roses. Phytopathology 95:784–792
Janisiewicz WJ, Korsten L (2002) Biological control of postharvest diseases of fruits. Ann Rev Phytopathol 40:411–441
Jasalavich CA, Morales VM, Pelcher LE, Seguin-Swartz G (1995) Comparison of nuclear ribosomal DNA sequences from *Alternaria* species pathogenic to crucifers. Mycol Res 99:604–614
Jerne NK (1960) Immunological speculations. Annu Rev Microbiol 14:341–358
Johnson LJ, Johnson RD, Akamatsu H, Salamiah A, Otani H, Kohmoto K, Kodama M (2001) Spontaneous loss of a conditionally dispensable chromosome from the *Alternaria alternata* apple pathotype leads to loss of toxin production and pathogenicity. Curr Genet 40(1):65–72
Joly P (1959) Morphological variations and the idea of species in the genus *Alternaria*. Bull Soc Mycol Fr 75:149–158
Joly P (1964) In: Le Genre *Alternaria* (Encycl. Mycol.) 33:213
Kakvan N, Zamanizadeh H, Morid B, Taheri H, Hajmansor S (2012) Study on pathogenic and genetic diversity of *Alternaria alternata* isolated from citrus hybrids of Iran, based on RAPD-PCR technique. European J Exp Biol 2(3):570–576
Kameniecki M, Wright ER, Rivera MC (2013) Identification of *Alternaria* spp. as Pathogenic on the native species *Terminalia australis* and *Salvia guaranitica*. Am J Plant Sci 4:36–41

Kaminskyj S, Jilkine K, Szeghalmi A, Gough K (2008) High spatial resolution analysis of fungal cell biochemistry – bridging the analytical gap using synchrotron FTIR spectromicroscopy. FEMS Microbiol Lett 284(1):1–8, http://dx.doi.org/10.1111/j.1574-6968.2008.01162.x

Kannan P, Pandey D, Gupta AK, Punetha H, Taj G, Kumar A (2012) Expression analysis of MAP2K9 and MAPK6 during pathogenesis of *Alternaria* blight in *Arabidopsis thaliana* ecotype Columbia. Mol Biol Rep 39:4439–4444

Kantwa SL, Tetarwal JP, Shekhawat, K (2014) In vitro effect of fungicides and phyto-extracts against Alternaria alternata causing leaf blight of groundnut. 7(6) Ver. I: 28–31

Kanrar S, Venkateswari JC, Kirti PB, Chopra VL (2002) Transgenic expression of hevein, the rubber tree lectin in Indian mustard confers protection against *Alternaria brassicae*. Plant Sci 162(3):441–448

Katiyar A, Kant S, Chauhan SS, Alka S (2001a) Chemical control of *Alternaria* leaf spot of bottle gourd. Ann Plant Prot Sci 9(2):339–341

Katiyar A, Kant S, Chauhan SS, Alka S (2001b) Sources of resistance in bottle gourd to *Alternaria* leaf spot. Ann Plant Prot Sci 9(1):155–157

Kellens JTC, Peumans WJ (1991) Biochemical and serological comparison of lectins from different anastomosis groups of *Rhizoctonia solani*. Mycol Res 95:1235–1241

Kim Y, Nandakumar MP, Marten MR (2008) The state of proteome profiling in the fungal genus Aspergillus. Brief Funct Genomic Proteomic 7:87–94

Kim YT, Cho M, Jeong JY, Lee HB, Kim SB (2010) Application of terminal restriction fragment length polymorphism (T-RFLP) analysis to monitor effect of biocontrol agents on rhizosphere microbial community of hot paper (*Capsicum annum* L.). J Microbiol 48(5):566–572

King SR (1994) Screening, selection, and genetics of resistance to *Alternaria* diseases in *Brassica oleracea*. Ph.D. thesis, Cornell University, Ithaca

Kirk PM, Cannon PF, Minter DW, Stalpers JA (2008) Dictionary of the fungi, 10th edn. CABI, Wallingford, p 22

Klessig DF, Malamy J (1994) The salicylic acid signal in plants. Plant Mol Biol 26:1439–1458

Klimyuk VI, Carroll BJ, Thomas CM, Jones JDG (1993) Alkali treatment for rapid preparation of plant material for reliable PCR analysis. Plant J 3:493–494

Knutsen AK, Torp M, Holst-Jensen A (2004) Phylogenetic analyses of the *Fusarium poae*, *Fusarium sporotrichioides* and *Fusarium langsethiae* species complex based on partial sequences of the translation elongation factor-1 alpha gene. Int J Food Microbiol 95(3):287–295, ISSN 0168-1605

Köhler G, Milstein C (1975) Continuous cultures of fused cells secreting antibody of predetermined specificity. Nature (Lond) 256:495–497

Kolte SJ (1996) Diseases. In: Chopra VL, Prakash S (eds) Oilseeds and vegetable brassicas. An Indian perspective. Oxford & IBH Publishing Co. Pvt. Ltd, New Delhi, pp 184–207. 294 pages

Konstantinova P, Bonantis PJM, Gent-Pelzer MPE, Van Der Zouwen P, van Der Bulk R (2002) Development of specific primers for detection and identification of *Alternaria* spp. in carrot material by PCR and comparison with blotter and plating assays. Mycol Res 106:23–33

Kraft JM, Boge WL (1994) Development of an antiserum to quantify *Aphanomyces euteiches* in resistant pea lines. Plant Dis 78:179–183

Krishnia SK, Saharan GS, Singh D (2000) Genetics of *Alternaria* blight resistance in inter and intraspecific crosses of *Brassica juncea* and *B. carinata*. Ann Biol 16(2):212–216

Kristensen R, Torp M, Kosiak B, Holst-Jensen A (2005) Phylogeny and toxigenic potential is correlated in *Fusarium* species as revealed by partial translation elongation factor 1 alpha gene sequences. Mycol Res 109(2):173–186, ISSN 0953-7562

Kumar S, Mehta N, Sangwan MS, Kumar R (2004) Relative sensitivity of various isolates of *Alternaria brassicae* (Berk.) Sacc. to fungicides. J Mycol Plant Pathol 34(1):28–32

Kusaba M, Tsuge T (1995) Phylogeny of *Alternaria* fungi known to produce host-specific toxins on the basis of variation in internal transcribed spacers of ribosomal DNA. Curr Genet 28:491–498

Lacomi-Vasilescu B, Blancard D, Guénard M, Molinero-Demilly V, Laurent E, Simoneau P (2002) Development of a PCR based diagnostic assay for detecting pathogenic *Alternaria* species in cruciferous seeds. Seed Sci Technol 30:87–95

Lamprell H, Mazerolles G, Kodjo A, Chamba JF, Noel Y, Beuvier E (2006) Discrimination of *Staphylococcus aureus* strains from different species of *Staphylococcus* using Fourier transform infrared (FTIR) spectroscopy. Int J Food Microbiol 108:125–129

Langrell SRH, Glen M, Alfenas AC (2008) Molecular diagnosis of *Puccinia psidii* (guava rust) – a quarantine threat to Australian eucalypt and *Myrtaceae* biodiversity. Plant Pathol 57(4):687–701, ISSN 0032-0862

Le TN, Shiesh CC, Lin HL (2010) Effect of vapor heat and hot water treatments on disease incidence and quality of Taiwan native strain mango fruits. Int J Agric Biol 12:673–678

Lee SB, Taylor JW (1990) Isolation of DNA from fungal mycelia and single spores. In: Innis MA, Gelfand DH, Sninsky JS, White TJ (eds) PCR protocols: a guide to methods and applications. Academic, London, pp 282–287

Lefier D, Hirst D, Holt C, Williams AG (1997) Effect of sampling procedure and strain variation in *Listeria* monocytogenes on the discrimination of species in the genus *Listeria* by Fourier transform infrared spectroscopy and canonical variates analysis. FEMS Microbiol Lett 147:45–50

Leiminger J, Bahnweg G, Hausladen H (2014) Differentiation of *Alternaria* species and quantification of disease development using real-time PCR. In Fourteenth Euroblight Workshop PPO – Special report no. 16:189–194

Lievens B, Van Baarlen P, Verreth C, Van Kerckhove S, Rep M, Thomma BPHJ (2009) Evolutionary relationships between *Fusarium oxysporum* f. sp *lycopersici* and *F. oxysporum* f. sp *radicis-lycopersici* isolates inferred from mating type, elongation factor-1 alpha and exopolygalacturonase sequences. Mycol Res 113:1181–1191, Part: 10, ISSN 0953-7562

Lind V (1990) Isolation of antigens for serological identification of *Pseudocercosporella herpotrichoides* (Fron) Deighton. J Plant Dis Plant Prot 97:490–501

Liu GT, Qian YZ, Zang P et al (1992) Etiological role of *Alternaria alternata* in human esophageal cancer. China Med J 105(5):394–400

Logrieco A, Moretti A, Solfrizzo M (2009) *Alternaria* toxins and plant diseases: an overview of origin, occurrence and risks. World Mycotoxin J 2(2):129–140

Lopes MC, Martins VC (2008) Fungal plant pathogens in Portugal: *Alternaria dauci*. Rev Iberoam Micol 25:254–256

Lopez MM, Bertolini E, Olmos A, Caruso P, Gorris MT, Llop P, Penyalver R, Cambra M (2003) Innovative tools for detection of plant pathogenic viruses and bacteria. Int Microbiol 6:233–243

Madden L, Pennypacker SP, Macnab AA (1978) FAST, a forecast system for *Alternaria solani* on tomato. Phytopathology 68:1354–1358

Malone JP, Muskett AE (1997) In: Sheppard JW (ed) Seed-borne fungi. Description of 77 fungus species. International Seed Testing Association, Zurich, pp 19–20

Martin FN, Tooley PW (2003) Phylogenetic relationships among *Phytophthora* species inferred from sequence analysis of mitochondrially encoded cytochrome oxidase I and II genes. Mycologia 95(2):269–284, ISSN 0027-5514

Martínez SP, Snowdon R, Pons-Kühnemann J (2004) Variability of Cuban and international populations of *Alternaria solani* from different hosts and localities: AFLP Genetic Analysis. Eur J Plant Pathol 110(4):399–409

Martínez-García LB, Armas C, Padilla FM, Miranda JD, Pugnaie FI (2011) Shrubs influence arbuscular mycorrhizal fungi communities in semiarid environment. Soil Biol Biochem 43:682–689

Masunaka A, Ohtani K, Peever TL, Timmer LW, Tsuge T, Yamamoto M et al (2005) An isolate of *Alternaria alternate* that is pathogenic to both tangerines and rough lemon and produces two host selective toxins, ACT and ACR toxins. Phytopathology 95:241–247

Matharu BK, Sharma JR, Manrao MR (2006) Synthesis and antifungal potential of 2-chlorobenzal derivatives. Pestic Res J 18(2):113–115

Maude RB, Humpherson-Jones FM (1980a) The effect of iprodione on the seed-borne phase of *Alternaria brassicicola*. Ann Appl Biol 95:321–327

Maude RB, Humpherson-Jones FM (1980b) Studies on the seed-borne phases of dark leaf spot (*Alternaria brassicicola*) and grey leaf spot (*Alternaria brassicae*) of *brassicas*. Ann Appl Biol 95:311–319

McCartney HA, Doughty KS, Morton G, Booth EJ, Kightey SPJ, Landon G, West G, Walker KC, Thomas JE (1999) A study of the effect of disease on seed quality parameters of oilseed rape. In: Proceedings of 10th international rapeseed congress, Canberra, 26–29 Sept

McCartney HA, Foster SJ, Fraaije BA, Ward E (2003) Molecular diagnostics for fungal plant pathogens. Pestic Manag Sci 59:129–142

McMaugh SJ, Lyon BR (2003) Real-time quantitative RT-PCR assay of gene expression in plant roots during fungal pathogenesis. Biotechniques 34(5):982–986

Meena PD, Chattopadhyay C, Singh F, Singh B, Gupta A (2002) Yield loss in Indian mustard due to white rust and effect of some cultural practices on *Alternaria* blight and white rust severity. Brassica 4:18–24

Meena PD, Meena RL, Chattopadhyay C, Kumar A (2004) Identification of critical stage for disease development and biocontrol of *Alternaria* blight of Indian mustard (*Brassica juncea*). J Phytopathol 152(4):204–209

Mei ZZ, Rong C, Ronge X, Lin CX, Song L, Yong GZ, Xia J, Lin W, Cheng LP (2007) Synthesis and antifungal properties of sulfonilamide derivatives of chitosan. Carbohydr Res 342(16):2390–2395

Meng J, Wang Y (2010) Rapid detection of *Phytophthora nicotianae* in infected tobacco tissue and soil samples based on its *Ypt1* gene. J Phytopathol 158(1):1–7, ISSN 0931-1785

Micales AJ, Bonde RM (1995) In Isozymes: methods and applications, Molecular methods in plant pathology, chapter 9

Mmbaga MT, Shi A, Kim M-S (2011) Identification of *Alternaria alternata* as a causal agent for leaf blight in Syringa species. Plant Pathol J 27(2):120–127

Mohsan M, Intizar-ul-Hassan M, Ali L (2011) Chemotheraptic management of alternaria black spot (*Alternaria alternata*) in mango fruits. J Agric Res 49(4):499–506

Mondal SN, Rashid MA, Monowar SM, Abdullah AM (1989) Efficacy of fungicides in controlling *Alternaria* blight of radish seed crop. Thai J Agric Sci 22:191–196

Mondal KK, Chatterjee SC, Viswakarma N, Bhatacharya RC, Grover A (2003) Chitinase mediated inhibitory activity of Brassicas transgenic on growth of *Alternaria brassicae*. Curr Microbiol 47:171–173

Mondal KR, Bhattacharya RC, Koundal KR, Chatterjee SC (2007) Transgenic Indian mustard (*Brassica juncea*) expressing tomato glucanase leads to arrested growth of *Alternaria brassicae*. Plant Cell Rep 26:247–252. doi:10.1007/s00299-006-0241-3

Montemurro N, Visconti A (1992) *Alternaria* metabolites-chemical and biological data. In: Chełkowski J, Visconti A (eds) Alternaria: biology, plant diseases and metabolites. Elsevier Sciences Publisher, Amsterdam, pp 449–557

Morris PF, Connolly MS, St clair DA (2000) Genetic diversity of *Alternaria alternata* isolated from tomato in California assessed using RAPDs. Mycol Res 104(3):286–292

Mostert L, Groenewald JZ, Summerbell RC, Gams W, Crous PW (2006) Taxonomy and pathology of *Togninia* (Diaporthales) and its *Phaeoacremonium* anamorphs. Stud Mycol 54:1–113, ISSN 0166-0616

Mukherjee AK, Carpa M-J, Zuchmanc R, Zivc T, Horwitza BA, Gepstein S (2010) Proteomics of the response of *Arabidopsis thaliana* to infection with *Alternaria brassicicola*. J Proteomics 73:709–720

Mule G, Susca A, Stea G, Moretti A (2004) Specific detection of the toxigenic species *Fusarium proliferatum* and *F. oxysporum* from asparagus plants using primers based on calmodulin gene sequences. FEMS Microbiol Lett 230(2):235–240

Mullis K (1987) Specific synthesis of DNA *in vitro* via a polymerase-catalyzed chain reaction. Methods Enzymol 155:335–350

Nakashima T, Ueno T, Fukami H, Taga T, Masuda H, Osaki K et al (1985) Isolation and structures of AK-Toxin I and II, host-specific phytotoxic metabolites produced by *Alternaria alternata* Japanese pear pathotype. Agric Biol Chem 49:807–815

Narayanasamy P (2001) Plant pathogen detection and disease diagnosis, 2nd edn. Marcel Dekker, New York

Narayanasamy P (2005) Immunology in plant health and its impact on food safety. The Haworth Press, New York

Naumann D, Helm D, Labischinski H (1991) Microbiological characterizations by FT-IR spectroscopy. Nature 351:81–82

Naumann A, Navarro-Gonzalez M, Peddireddi S, Kues U, Polle A (2005) Fourier transform infrared microscopy and imaging: detection of fungi in wood. Fungal Genet Biol 42(10):829–835, http://dx.doi.org/10.1016/ j.fgb.2005.06.003

Neergard P (1945) Danish species of *Alternaria* and *Stemphylium*. Oxford University Press, London

Nega E, Ulrich R, Werner S, Jahn M (2003) Hot water treatment of vegetable seed-an alternative seed treatment method to control seed borne pathogens in organic farming. J Plant Dis Prot 110(3):220–234

Nguyen HDT, Seifert KA (2008) Description and DNA barcoding of three new species of *Leohumicola* from South Africa and the United States. Persoonia 21:57–69, ISSN 0031-5850

Niessen ML, Vogel RF (1998) Group specific PCR-detection of potential trichothecene- producing *Fusarium*-species in pure cultures and cereal samples. Syst Appl Microbiol 21:618–631

Ostry V (2008) *Alternaria* mycotoxins: an overview of chemical characterization, producers, toxicity, analysis and occurrence in foodstuffs. World Mycotoxin J 1:175–188

Oudemans P, Coffey MD (1991) A revised systematic of twelve papillate *Phytophthora* species based on isozymes analysis. Mycol Res 95:1025–1046

Oviedo MS, Sturm ME, Reynoso MM, Chulze SN, Ramirez ML (2013) Toxigenic profile and AFLP variability of *Alternaria alternata* and *Alternaria infectoria* occurring on wheat. Braz J Microbiol 44(2):447–455

Pandey A (2010) Antagonism of two *Trichoderma* Species against *Alternaria alternata* on *Capsicum frutescens*. J Exp Sci 1(5):18–19

Pane C, Zaccardelli M (2015) Evaluation of *Bacillus* strains isolated from solanaceous phylloplane for biocontrol of *Alternaria* early blight of tomato. Biol Control 84:11–18

Panwar S, Singh L, Sunaina V, Vats P (2006) Evaluation of *Brevibacillus brevis* and *Bacillus firmus* strains in management of fungal pathogens. Pestic Res J 18(2):159–161

Pastor FJ, Guarro J (2008) *Alternaria* infections: laboratory diagnosis and relevant clinical features. Clin Microbiol Infect 14:734–746

Patni CS, Kolte SJ (2006) Effect of some botanicals in management of *Alternaria* blight of rapeseed-mustard. Ann Plant Prot Sci 14:151–156

Peever TL, Ibanez A, Akimitsu K, Timmer LW (2002) Worldwide phylogeography of the citrus brown spot pathogen, *Alternaria alternata*. Phytopathology 92:794–802

Peever TL, Su G, Carpenter-Boggs L, Timmer LW (2004) Molecular systematics of citrus-associated *Alternaria* species. Mycologia 96:119–134

Peres NA, Timmer LW (2006) Evaluation of the Alter-Rater model for spray timing for control of *Alternaria* brown spot on murcott tangor in Brazil. Crop Protec 25:454–460

Petrunak DM, Christ BJ (1992) Isozyme variability in *Alternaria solani* and *A.alternata*. Phytopathology 82:1343–1347

Porter-Jordan K, Rosenberg EI, Keiser JF, Gross JD, Ross AM, Nasim S, Garrett CT (1990) Nested polymerase chain reaction assay for the detection of cytomegalovirus overcomes false positives caused by contamination with fragmented DNA. J Med Virol 30(4):85–91, ISSN 0146- 6615

Prelle A, Spadaro D, Garibaldi A, Gullino ML (2013) A new method for detection of five *Alternaria* toxins in food matrices based on LC–APCI-MS. Food Chem 140:161–167

Primard C, Vedel F, Mathieu C, Pelletier G, Chevre AM (1988) Interspecific somatic hybridization between *Brassica napus* and *Brassica hirta* (*Sinapis alba* L.). Theor Appl Genet 75:546–552

Pryor BM, Bigelow DM (2003) Molecular characterization of *Embellisia* and *Nimbya* species and their relationship to *Alternaria*, *Ulocladium*, and *Stemphylium*. Mycologia 95:1141–1154

Pryor BM, Gilbertson RL (2000) Phylogenetic relationships among *Alternaria* and related fungi based upon analysis of nuclear internal transcribed sequences and mitochondrial small subunit ribosomal DNA sequences. Mycol Res 104:1312–1321

Pryor BM, Michailides TJ (2002) Morphological, pathogenic, and molecular characterization of *Alternaria* isolates associated with *Alternaria* late blight of pistachio. Phytopathology 92:406–416

Pusz W (2009) Morpho-physiological and molecular analyses of *Alternaria alternata* isolated from seeds of *Amaranthus*. Phytopathologia 54:5–14

Qin L, Fu Y, Xie J, Cheng J, Jiang D, Li G, Huang J (2011) A nested-PCR method for rapid detection of *Sclerotinia sclerotiorum* on petals of oilseed rape (*Brassica napus*). Plant Pathol 60(2):271–277, ISSN 0032-0862

Radisek S, Jakse J, Javornik B (2004) Development of pathotype-specific SCAR markers for detection of *Verticillium albo-atrum* isolates from Hop. Plant Dis 88(10):1115–1122. doi:10.1094/PDIS.2004.88.10.1115

Ramdaswaghunde R, Patil RK, Shinde MS (2014) Occurrence and fungicidal management of *Alternaria* fruit rot (*Alternaria altarnata* (Fr.) Keissler) of Aonla. J Mycopathol Res 52(2):311–316

Ramjegathesh R, Ebenezar EG, Muthusamy M (2011) Management of onion leaf blight by *Alternaria alternata* (FR.) Keissler by botanicals and bio-control agents. Plant Pathol J 10(4):192–196

Ratnam SV, Reddy PN, Rao KC, Raju SK, Kumar JH (2001) Effect of salicylic acid and bion seed treatment in induction of systemic resistance against sunflower leaf blight. Alternaria helianthi (Hansf.). Tub Et Nish Indian J Plant Prot 29:79–81

Rimmer SR, Buchwaldt H (1995) Diseases. In: Kimber D, McGregor DI (eds) Brassica oilseeds-production and utilization. CAB International, Allingford, pp 111–140

Rivillas-Acevedo LA, Soriano Garcia M (2007) Isolation and biochemical characterization of an antifungal peptide from Amaranthus hypochondriacus seeds. J Agric Food Chem 55(25):10156–10161

Roberts RG (2005) *Alternaria yaliinficiens* sp. nov. on Ya Li pear fruit: from interception to identification. Plant Dis 89:134–145

Roberts RG, Reymond ST, Andersen B (2000) RAPD fragment pattern analysis and morphological segregation of small-spored *Alternaria* species and species groups. Mycol Res 104:151–160

Rotem J (1994) The genus *Alternaria*: biology, epidemiology and pathogenicity. American Phytopathological Society Press, St Paul

Saccardo PA (1886) *Sylloge fungorum omninum hucusque cognitorum*. Pavia, 4: pp. 1–807

Saito K, Takeda K (1984) Genetic analysis of resistance to *Alternaria* blotch (*Alternariamali* Roberts) in apple. (Studies on the breeding of the apple. VIII). Jpn J Breed 34:197–209

Salamiah H, Yukitaka F, Hiroshi O, Motoichiro K (2001) Construction and genetic analysis of hybrid strains between apple and tomato pathotypes of *Alternaria alternata* by protoplast fusion. J Gen Plant Pathol 67(2):97–105

Salman A, Erukhimovitch V, Talyshinsky M, Huleihil M (2002) FTIR spectroscopic method for detection of cells infected with herpes viruses. Biopolymers- Biospectroscopy Sect 67(6):406–412, http://dx.doi.org/10.1002/bip.10171

Samson RA, Houbraken J, Thrane U, Frisvad JC, Andersen B (2010) Food and indoor fungi. CBS-KNAW— Fungal Biodiversity Centre, Utrecht, pp 1–398

Santos C, Fraga ME, Kozakiewicz Z, Lima N (2010) Fourier transform infrared as a powerful technique for the identification and characterization of filamentous fungi and yeasts. Res Microbiol 161(2):168–175

Sawamura K (1990) *Alternaria* blotch. In: Aldwinckle HS (ed) Compendium of apple and pear diseases. APS Press, St Paul, p 100

Schena M, Shalon D, Davis RW, Brown PO (1995) Quantitative monitoring of gene-expression patterns with a complementary-DNA microarray. Science 270:467–470

Schenk PM, Kazan K, Manners JM, Anderson JP, Simpson RS, Wilson IW, Somerville SC, Maclean DJ (2003) Systemic gene expression in *Arabidopsis* during an incompatible interaction with *Alternaria brassicicola*. Plant Physiol 132:999–1010

Schwarz C, Tiessen C, Kreutzer M, Stark T, Hofmann T, Marko D (2012) Characterization of a genotoxic impact compound in *Alternaria alternata* infested rice as Altertoxin II. Arch Toxicol 86:1911–1925

Scott PM (2004) Other mycotoxins. In: Magan N, Olsen M (eds) Mycotoxins in food: detection and control. Woodhead Publishing, Cambridge, UK, pp 406–440

Seidle E, Rude S, Petrie A (1995) The effect of *Alternaria* black spot of canola on seed quality and seed yield and studies on disease control. Agriculture and Agri-Food Canada Saskatoon, Saskatoon, 41 pp

Seifert KA, Samson RA, deWaard JR, Houbraken J, Lévesque CA, Moncalvo J-M, Louis-Seize G, Hebert PDN (2007) Prospects for fungus identification using CO1 DNA barcodes, with Penicillium as a test case. Proc Natl Acad Sci USA 104(10):3901–3906, ISSN 0027-8424

Shakir AS, Mirza JH, Akhtar KP (1997) New records of *Alternaria* species from Pakistan. Pak J Phytopathol 9:102–104

Sharma TR, Singh BM (1995) Generation and evaluation of somaclones of Brassica juncea for resistance to *Albugo candida* and *Alternaria brassicae*. Proc Indian Natl Sci Acad B61(2):155–162

Sharma A, Dass A, Pau MS (2007) Antifungal effect of neem extract on some common phytopathogenic fungi. Adv Plant Sci 20(2):357–358

Sharma P, Deep S, Sharma M, Bhati DS, Chowdappa P (2013) PCR based assay for the detection of *Alternaria brassicicola* in Crucifers. Indian Phytopathol 66(3):263–268

Sheridan GEC, Masters CI, Shallcross JA, Mackey BM (1998) Detection of mRNA by reverse transcription-PCR as an indicator of viability in *Escherichia coli* cells. Appl Environ Microbiol 64(4):1313–1318, ISSN 0099-2240

Shin YU, Ko KC (1992) Studies on Alternaria leaf spot (A. mali Roberts) in apple trees (Malus domestica Borkh): epiphytotics, screening methods and inheritance of resistance. III. Inheritance of resistance. J Korean Soc Hortic Sci 33:372–378

Shoaib A, Akhtar N, Nafisa, Aftab A (2013) Fourier transform-infrared spectroscopy to monitor modifications in canola biochemistry caused by Alternaria destruens. Pak J Phytopathol 25(02):105–109

Shrama R, Tewari JP (1998) RAPD analysis of three *Alternaria* species pathogenic to crucifers. Mycol Res 102(7):807–814

Shrestha SK, Munk L, Mathur SB (2003) Survival of *Alternaria brassicae* in seeds and crop debris of rapeseed and mustard in Nepal. Seed Sci Technol 31(1):103–109

Siddiqui ZA (2007) Biocontrol of *Alternaria triticina* by plant growth promoting rhizobacteria on wheat. Arch Phytopathol Plant Protect 40(4):301–308

Sigareva MA, Earle ED (1999a) Production and characterization of somatic hybrids between *Camilina sativa* and rapid-cycling *B. oleracea*. Theor Appl Genet 98:164–170

Sigareva MA, Earle ED (1999b) Camalexin induction in intertribal somatic hybrids between *Camelina sativa* and rapid cycling *Brassica oleracea*. Theor Appl Genet 98:164–170

Sigareva MA, Earle ED (1999c) Regeneration of plants from protoplasts of *Capsella bursapastoris* and somatic hybridization with rapid cycling *Brassica oleracea*. Plant Cell Rep 18:412–417

Simmons EG (2007) *Alternaria*: an identification manual. Centraalbureau voor\Schimmelcultures, Utrecht

Singh N, Verma OP (2009) Epidemiology of *Alternaria* blight of *Adhatoda vasica* caused by *Alternaria alternata*. Indian J Agric Sci 79(11):945–948

Singh NK, Saxena RP, Pathak SP, Chauhan SKS (2001) Management of *Alternaria* leaf spot disease of tomato. Indian Phytopathol 54:508

Sissons JGP, Oldstone MBA (1980) Antibody-mediated destruction of virus-infected cells. Adv Immunol 29:209–260

Skrede I, Carlsen T, Stensrud O, Kauserud H (2012) Genome wide AFLP markers support cryptic species in Conphora (Boletales). Fungal Biol 116:778–784

Song W, Ma X, Tan H, Zhou J (2011) Abscisic acid enhances resistance to Alternaria solani in tomato seedlings. Plant Physiol Biochem 49:693–700

Spadaro D, Gullino ML (2004) State of the art and future prospects of the biological control of postharvest fruit diseases. Int J Food Microbiol 91:185–194

Strandberg JO (2002) A selective medium for the detection of *Alternaria dauci* and *Alternaria radicina*. Phytoparasitica 30:269–284

Strange RN (2007) Phytotoxins produced by microbial plant pathogens. Nat Prod Rep 24:127–144

Surekha C, Neelapu N, Prasad BS, Ganesh PS (2014) Induction of defense enzymes and phenolic content by *Trichoderma Viride* in *Vigna mungo* infested with *Fusarium oxysporum* and *Alternaria alternata*. Int J Agric Sci Res 4(4):31–40

Szeghalmi A, Kaminskyj S, Gough KM (2007) A synchrotron FTIR microspectroscopy investigation of fungal hyphae grown under optimal and stressed conditions (conference paper). Anal Bioanal Chem 387(5):1779–1789, http://dx.doi.org/10.1007/s00216-006-0850-2

Taj G, Kumar A, Bansal KC, Garg GK (2004) Introgression of osmotin gene for creation of resistance against *Alternaria* blight by perturbation of cell cycle machinery. Indian J Biotechnol 3:291–298

Takenaka S (1992) Use of immunological methods with antiribosome serums to detect snow mold fungi in wheat plants. Phytopathology 82:896–901

Taylor MJ, Hughes MS, Skuce RA, Neill SD (2001) Detection of Mycobacterium bovis in bovine clinical specimens using real-time fluorescence and fluorescence energy transfer probe rapid cycle PCR. J Clinic Microbiol 39:1272–1278

Thakur S, Harsh NSK (2014) Phylloplane fungi as biocontrol agent against *Alternaria* leaf spot disease of (Akarkara) *Spilanthes oleracea*. Biosci Discov 5(2):139–144

Thies JE (2007) Soil Microbial community analysis using terminal restriction fragment length polymorphisms. Soil Sci Soc Am J 71:579–591

Tohamy MRA, Helal GA, Ibrahim KI, Abd-El-Aziz SA (2004) Control of postharvest tomato fruit rots: II. Using heat treatments. Egypt J Phytopathol 32(1–2):129–138

Toriyama K, Hinata K, Kameya T (1987) Production of somatic hybrid plants "Brassicomoricandia", through protoplast fusion between *Moricandida arvensis* and *Brassica oleracea*. Plant Sci 48(2):123–128

Trivedi RKS (2010) Strategies to control *Alternaria radicina* in carrot seed production. A Ph.D thesis submitted to Lincoln University, Seed Research Centre, Bio-Protection Research Centre, Lincoln University Canterbury, Lincoln

Tsuge T, Harimoto Y, Akimitsu K, Ohtani K, Kodama M, Akagi Y, Egusa M, Yamamoto M, Otani H (2013) Host-selective toxins produced by the plant pathogenic fungus *Alternaria alternata*. FEMS Microbiol Rev 37:44–66

Udelhoven T, Naumann D, Schmitt J (2000) Development of a hierarchical classification system with artificial neural networks and FT-IR spectra for the identification of bacteria. Appl Spectrosc 54:1471–1479

Upadhyay P, Rai A, Kumar R, Singh M, Sinha B (2014) Differential expression of pathogenesis related protein genes in tomato during inoculation with *A. Solani*. J Plant Pathol Microbiol 5:217. doi:10.4172/2157-7471.1000217

Van der Waals JE, Denner FDN, van Rij N, Korsten L (2003) Evaluation of PLANT- Plus, a decision support system for control of early blight on potatoes in South Africa. Crop Protec 22:821–828

Valkonen JPT, Koponen H (1990) The seed-borne fungi of Chinese cabbage (*Brassica pekinensis*), their pathogenicity and control. Plant Pathol 39:510–516

Velichetti RK, Lamison C, Brill LM, Sinclair JB (1993) Immunodetection of *Phomopsis* species in asymptomatic plants. Plant Dis 77:70–73

Venkataramanamma K, Madhusudhan P, Neelima S, Narasimhudu Y (2014) Field evaluation of fungicides for the management of *Alternaria* leaf blight of sunflower. Indian J Plant Prot 42(2):165–168

Verma PR, Saharan GS (1994) Monograph on *Alternaria* diseases of crucifers. Agriculture and agri-food Canada, pp 1–146

Verma N, Verma S (2010) *Alternaria* diseases of vegetable crops and new approaches for its control. Asian J Exp Biol Sci 1(3):681–692

Verma SS, Yajima WR, Rahman MH, Shah S, Liu JJ, Ekramoddoullah AK, Kav NN (2012) A cysteine-rich antimicrobial peptide from *Pinus monticola* (PmAMP1) confers resistance to multiple fungal pathogens in canola (*Brassica napus*). Plant Mol Biol 79(1–2):61–74

Vishwanath, Kolte SJ, Singh MP, Awasthi RP (1999) Induction of resistance in mustard (*Brassica juncea*) against *Alternaria* black spot with an avirulent *Alternaria brassicae* isolate-D. Eur J Plant Pathol 105(2):217–220

Waghe KP, Wagh SS, Kuldhar DP, Pawar DV (2015) Evaluation of different fungicides, bioagents and botanicals against *Alternaria* blight caused by *Alternaria helianthi* (Hansf) of sunflower. Afr J Agric Res 10(5):351–358

Ward CM, Wilkinson AP, Bramham S, Lee HA, Chan WHS, Butcher GW, Hutchings A, Morgan MRA (1990) Production and characterization of polyclonal and monoclonal antibodies against aflatoxin B1 oxime-BSA in an enzyme-linked immunosorbent assay. Mycotoxin Res 6:73–83

Welsh J, McClelland M (1991) Genomic fingerprinting using arbitrarily primed PCR and a matrix of pairwise combinations of primers. Nucleic Acids Res 19(19):5275–5279

Westman AL, Kresovich S, Dickson MH (1999) Regional variation in *Brassica nigra* and other weedy crucifers for disease reaction to *Alternaria brassicicola* and *Xanthomonas campestris* pv. *campestris*. Euphytica 106:253–259

White TJ, Bruns T, Lee S, Taylor J (1990) Amplification and direct sequencing of fungal ribosomal genes for phylogenetics. In: PCR protocols: a guide to methods and applications. Academic, San Diego

Wijekoon CP, Goodwin PH, Hsiang T (2008) Quantifying fungal infection of plant leaves by digital image analysis using Scion Image software. J Microbiol Method 74:94–101

Wiltshire SP (1933) The foundation species of *Alternaria* and *Microsporium*. Trans Br Mycol Soc 18:135–160

Wiltshire SP (1938) The original and modern conceptions of *Stemphylium*. Trans Br Mycol Soc 21:211–239

Wu WS, Chen TW (1999) Development of a new selective medium for detecting *Alternaria brassicola* in cruciferous seeds. Seed Sci Technol 27:397–409

Wu CP, Chen GY, Li B, Su H, An YL, Zhen SZ, Ye JR (2011) Rapid and accurate detection of *Ceratocystis fagacearum* from stained wood and soil by nested and realtime PCR. Forest Pathol 41(1):15–21, ISSN 1437-478

Yamagishi D, Akamatsu H, Otani H, Kodama M (2006) Pathological evaluation of host-specific AAL-toxins and fumonisin mycotoxins produced by *Alternaria* and *Fusarium* species. J Gen Plant Pathol 72(5):323–327

Yang F, Jensen JD, Svensson B, Jorgensen HJL, Collinge DB, Finnie C (2010) Analysis of early events in the interaction between *Fusarium graminearum* and the susceptible barley (*Hordeum vulgare*) cultivar Scarlett. Proteomics 10(21):3748–3755, ISSN 1615-9853

Zghair FS, Mohamed BT, Neda SM (2014) Molecular assay of Polyketide Synthase gene of Alternariol (AOH) produce by *Alternaria alternata*. Int J Innov Appl Stud 9(3):1124–1127

Zhao Y, Tu K, Shao XF, Jing W, Yang W, Su ZP (2008) Biological control of the post-harvest pathogens *Alternaria solani*, *Rhizopus stolonifer* and *Botrytis cinerea* on tomato fruit by *Pichia guilliermondii*. J Hortic Sci Biotechnol 83(1):132–136

Zheng FC, Ward E (1998) Variation within and between *Phytophthora* species from rubber and citrus trees in China, determined by polymerase chain reaction using RAPDs. J Phytopathol 146(2-3):103–109

Zore GB, Surwase BS, Karuppayil SM (2004) Antifungal activity in two medicinal plants. J Mycol Plant Pathol 34(2):543–547

Molecular Prospecting: Advancement in Diagnosis and Control of *Rhizoctonia solani* Diseases in Plants

Hemant J. Patil and Manoj K. Solanki

Introduction

Agriculture is one of the most dependable and widely used resources to satisfy the food need of the living world, including human being. This tends us to pay prime attention to grow healthy plants and protect them from phytopathogens to achieve more quantity of good quality food. In the history of mankind, agriculture is one of the oldest trading sector and play major role in the world economy. The scope of agriculture is not limited to food production only, while it is contributing to fulfilling other needs such as fiber production, feed for animal farming, moreover attempts were also made for fuel production. In growing years, commercialization of food products through globalization as well as alarming population growth lead agricultural practices more evolved aiming more production with available resources. In one report by Tom (2009), the world population will jump from seven billion to nine billion by 2050, and food production will be required to be doubled by then to keep proper pace. The agriculture business is on the third revolution stage, also known as "Green Revolution", which corresponds to twentieth century and involves biotechnology, genetic engineering, chemical fertilizers, and many more to achieve mass agricultural production (web link as on June 30th, 2015: Growing a Nation). According to Lori (2013), it was analyzed that farmers are

H.J. Patil (✉)
Institute of Soil, Water and Environmental Sciences, Volcani Center, Agricultural Research Organization, PO Box 6, Bet Dagan 50250, Israel
e-mail: hemantpatil8311@gmail.com

M.K. Solanki
Guangxi Crop Genetic Improvement and Biotechnology Lab, Guangxi Academy of Agricultural Sciences, Nanning 530007, China

P. Kumar et al. (eds.), *Current Trends in Plant Disease Diagnostics and Management Practices*, Fungal Biology, DOI 10.1007/978-3-319-27312-9_6

producing 262 % more food at this stage with 2 % fewer inputs such as seeds, labor, fertilizers, etc., than they did in 1950s, which indicates evolving agricultural practices in view of sustainable agriculture.

Phytopathogens or plant pathogens are known as the organisms parasitic on plant host and causes disease in plants, which results in unusual plant growth and reduced production capacity compared to non-diseased plant, i.e. absence of typical phytopathogen. In modern agriculture, to overcome these harmful organisms, intensive use of fertilizers and pesticides is being practiced irrelevant to the crop type, land quality and more importantly the phytopathogen (Higa and Parr 1994). This ultimately results reduced land fertility, environmental degradation through saturation and resistance in phytopathogens for these chemical compounds and or pesticides (Tilman et al. 2002). In case of cotton, the world agricultural land use is around 2.4 %, while it accounts for 24 % and 11 % of global sale of insecticide and pesticide, respectively (Web link: Farming pollution). Based on crop and their diverse growth conditions, various organisms has pathogenic attributes such as fungi, bacteria, viruses, nematodes, oomycetes, phytoplasmas, viroids, virus like organisms, protozoa and algae with wide host range to cause infection and express disease symptoms (Lopez and Castano 2012). Among phytopathogens, *Rhizoctonia* holds attention being important phytopathogens, as it has a remarkable history almost as long as the history of plant pathology (Ogoshi 1996). *R. solani* is an ecologically diverse and globally occurred common soilborne pathogen with wide host range including soybean (*Glycine max* L.), cotton (*Gossypium hirsutum* L.), canola (*Brassica campestris* L.), wheat (*Triticum aestivum* L.), beet (*Beta vulgaris* L.), potato (*Solanum tuberosum* L. subsp., tuberosum), and rosemary (*Rosemarinus officinalis* L.), turfgrass, tobacco and many more (Baker 1970; Anderson 1982; Blazier and Conway 2004). It has the ability to cause various infections such as damping off, root rot, collar rot, stem canker, crown rot, bud and fruit rots and foliage blight on range of susceptible hosts (Baker 1970; Anderson 1982). The ability of infecting various plant parts including below ground level, such as the seeds, hypocotyls, and roots as well as above ground level, such as pods, fruits, leaves and stems, enable *R. solani* a potent and eye catching phytopathogen (Parmeter 1970). It has distinctive ability to cause infection at various stages of its life cycle such as mycelium, sclerotia as well as basidiospore, where later one can serve as a source for rapid and long distance dispersal of this pathogenic fungus (Parmeter 1970). The control of such widespread phytopathogen from being disseminating and spread of infection will certainly serve for the betterment of economy and ecology.

Plant pathology is usually defined as a scientific study of plant diseases caused due to phytopathogens. It has been well elaborated by Singh (2008) as a study deals with the cause, etiology, resulting losses and control or management of the plant diseases. Also the objectives were designated for plant pathology as study of living as well as non-living entities involved in plant disease or disorder, mechanism involved in plant disease, the host-pathogen interactions and the methods involved in plant disease management aiming to reduce losses (Singh 2008). Traditionally, plant pathology involves taxonomical tools for identification of pathogens from infected plant parts, which takes hours, that too results less specific diagnosis (Khiyami et al. 2014). It also involves recognizing disease symptoms, isolation of pathogens, followed by their preliminary identification based on biochemical and physiological methods such as microscopy, immunological techniques, etc., (Fox 1993). However, advancement in rapid and highly sensitive in-situ diagnosis of phytopathogens for immediate action towards restricting the epidemic spread of the causative agent is highly recommended. The molecular plant pathology includes diagnosis of pathogens through advanced tools such as PCR, qPCR, DNA sequencing, etc., and plant disease management for protecting crop yield (Sue et al. 2014). In recent years, very few reports were available as a step ahead, showing nano-diagnosis of phytopathogens to make disease control much faster and cheaper (Khiyami et al. 2014). This review sheds light on the emphasis of advancements in plant pathology modules during the agricultural revolution in the past few decades, over time consuming traditional methods considering *R. solani* as representative of high impact phytopathogens.

Rhizoctonia solani

Rhizoctonia was first described as a genus by DeCandolle in 1815 following *R. crocorum* (Pers.) as the type species (Sneh et al. 1998) while Kuhn in 1858 stated *R. solani* as the most important species of *Rhizoctonia* (Ogoshi 1996). *R. solani* is widely spread fungal plant pathogen belongs to the phylum basidiomycete means they does not produce any asexual spores (called conidia) and may sometimes produce sexual spores (basidiospores). In the early phase of *Rhizoctonia* classification, due to lack of genus specific characteristic features, many irrelevant fungi were grouped as *Rhizoctonia* spp. (Parmeter and Whitney 1970; Moore 1987), while in growing years, Ogoshi (1975) remarkably specify the genus concept of *Rhizoctonia* and was elaborated the characteristic features of *R. solani* at genus level. On this basis of revealed genus concept, *Rhizoctonia* species being classified using color of mycelia, nuclei per young vegetative hyphal cell as well as the morphology of their teleomorph. The teleomorph of *Rhizoctonia* spp., belongs to the sub-division Basidiomycota, class Hymenomycetes (Yang and Li 2012).

Taxonomic Grouping: Classical and Molecular

As mentioned above, the classification of genus *Rhizoctonia* founds tricky due to unavailability of specific characteristics, the classical intraspecies grouping of *R. solani* and other *Rhizoctonia* species was performed on the basis of affinity for hyphal fusion, i.e. anastomosis (Parmeter et al. 1969; Parmeter and Whitney 1970; Burpee et al. 1980; Ogoshi et al. 1983). Based on anastomosis grouping, *R. solani* have been grouped in 13 anastomosis groups (AG), while very recently AG-BI has been reported as 14th AG group (Carling 1996, 2002a, b). The interaction scale, such as perfect fusion, imperfect fusion, contact fusion and no fusion (Matsumoto et al. 1932; Yang and Li 2012), led anastomosis to categorize in four groups viz. C3–C0, as follows (Carling 1996).

C3: walls fuse; membranes fuse, accompanied with protoplasm connection; anastomosis point frequently is not obvious; diameter of anastomosis point is equal or nearly equal hyphal diameter; anastomosing cells and adjacent cells may die, but generally do not. This category occurs in the same anastomosis group, same vegetative compatibility populations (VCPs) and the same isolate.

C2: wall connection is obvious, but membrane contact is uncertain; anastomosing and adjacent cells always die. This category occurs in same AG, but not between different VCPs.

C1: wall contact between hyphae is apparent, but both wall penetration and membrane membrane contact do not occur; occasionally one or both anastomosing cells and adjacent cells die. This category occurs between different AGs or in the same AG.

C0: no reaction. This category occurs between different AGs.

The members of the same AG group represent ≥50 % frequency for hyphal fusion (i.e. From C3 to C1 reaction), except non-self-anastomosing isolates (Hyakumachi and Ui 1988), while there occurs low frequency ≤30 % or no fusion among members of different AG (i.e. C0 reaction).

On the other hand, the most reliable and widely accepted molecular biology has observed to be crucial and undetachable module towards determining the appropriate classification and grouping of organisms on the basis of genetic information and evolutionary base (Hebert and Gregory 2005). Though molecular biology involves advanced tools to determine taxonomical relatedness, it concurrently supports classical groupings of organisms and was observed in the case of *R. solani*, as molecular markers based clustering as well as percent sequence similarity found supportive of the AGs and subgroups based on hyphal fusion anastomosis (Sharon et al. 2006).

Host Range and Diseases Caused by *R. solani*

The genus *Rhizoctonia* known to have wide host range and hence is known to be an economically important plant pathogen all over the globe. It has a destructive lifestyle as a non-obligate parasite causes necrosis and damping-off on various host plant species and host range of *R. solani* is extensive and it affected the growth of various economic significant crop plants throughout the world (Table 1) including species in the *Asteraceae*, *Braccicaceae*, *Fabaceae*, *Poaceae*, and *Solanaceae* in addition to forest trees and ornamental plants (Ogoshi 1996). Disease symptoms contain damping-off, rots on roots, leaf spots, leaf blights, aerial blight, shoots and fruits, canker lesions on sprouts and stolons, and sclerotial diseases, while few of

Table 1 Host range and plant diseases associated with *Rhizoctonia* anastomosis groups

Host	Disease	AG	Reference
Barley	Root rot	AG5	Rush et al. (1994)
	Stunt	AG8	Burton et al. (1988)
Broan bean	Reduced growth	AG5	Valkonen et al. (1993)
Buckwheat	Damping-off	AG1	Herr and Fulton (1995)
Cabbage	Bottom rot	AG1	Tu et al. (1996)
	Web blight	AG4 HGI	Yang et al. (2007)
Carrot	Damping-off	AG1	Grisham and Anderson (1983)
	Crown and brace root rot	AG-2-4	Sumner and Phatak (2003)
Cereals	Bare patch	AG8	Mazzola et al. (1996)
Clover	Root rot	AG2	Wong and Sivasithamparam (1985)
	Damping-off	AG2	Wong and Sivasithamparam (1985)
Common bean	Leaf blight	AG1	Muyolo et al. (1993)
	Web blight	AG1	Muyolo et al. (1993)
	Root rot	AG2	Muyolo et al. (1993)
	Root rot	AG4	Muyolo et al. (1993)
	Reduced growth	AG5	Valkonen et al. (1993)
Corn	Leaf blight	AG1	Tomaso-Peterson and Trevathan (2007)
	Root rot	AG4	Mazzola et al. (1996)
Cotton	Root rot	AG4	Rothrock (1996)
	Root canker	AG7	Baird and Carling (1997)
	Minor pathogen	AG13	Carling et al. (2002a)
Eggplant	Brown spot	AG3	Kodama et al. (1982)
Flower bulbs	Root rot	AG2	Dijst and Schneider (1996)
Lettuce	Damping-off	AG1	Herr (1993)
Lupin	Late emergence	AG5	Valkonen et al. (1993)
	Minor pathogen	AG10	MacNish et al. (1995)
Narrow-leafed lupins	Damping-off and hypocotyl rot	AG11	Kumar et al. (2002)
Oilseed rape	Damping-off	AG2	Kataria et al. (1991)
	Basal rot	AG4	Verma (1996a)
Onion	Damping-off	AG4	Erper et al. (2005)
Orchid	Mycorrhizal	AG6	Perkins and Mcgee (1995)
	Mycorrhizal	AG12	Carling et al. (1999)
Pea	Stem rot	AG4	Hwang et al. (2007)
Pea	Root rot	AG4	Hwang et al. (2007)
Potato	Stem canker	AG2	Chand and Logan (1983)
	Stem canker	AG4	Anguiz and Martin (1989)
	Stem canker	AG5	Bandy et al. (1984)
	Black scurf	AG5	Bandy et al. (1984)
	Minor pathogen	AG9	Carling et al. (1994)

(continued)

Table 1 (continued)

Host	Disease	AG	Reference
Radish	Root rot	AG2	Grisham and Anderson (1983)
Rice	Sheath blight	AG1	Sayler and Yang (2007)
	Web blight	AG1	Hashiba and Kobayashi (1996)
	Sheath blight	AG2	Hashiba and Kobayashi (1996)
Snapbean	Web blight	AG4 HGI	Yang et al. (2007)
Soybean	Rot	AG1	Yang et al. (1990)
	Bud rot	AG1	Hwang et al. (1996)
	Damping-off	AG2	Nelson et al. (1996)
	Root rot	AG4	Liu and Sinclair (1991)
	Root rot	AG5	Nelson et al. (1996)
Sugarbeet	Root rot	AG2	Herr (1996)
	Damping-off	AG2	Herr (1996)
	Leaf blight	AG2	Herr (1996)
Tobacco	Target spot	AG3	Ogoshi (1987)
Tomato	Leaf blight	AG3	Date et al. (1984)
	Root rot	AG4	Montealegre et al. (2010)
	Fruit rot	AG4	Strashnov et al. (1985)
Turfgrass	Brown patch	AG1	Herr and Fulton (1995)
	Brown patch	AG2	Herr and Fulton (1995)
	Large patch	AG2	Burpee and Martin (1996)
Wheat	Root rot	AG5	Rush et al. (1994)

Lehtonen (2009)

them have symbiotic mycorrhizal relationships with orchid plants (Carling et al. 1999; Chang and Chou 2007). This fungus is able to persist as hyphae and sclerotia in both soil and plant parts. *R. solani* infection begins when sclerotium germinating mycelia or hyphae starts to grow towards a suitable host as a result of drawing chemical exudates, like amino acids, organic acids, sugars, protein substances and phenols, from the plants (Keijer 1996). After the primary contact, unfastened and still separated hypha starts to nurture over the plant and after few hours the hypha flattens and directional growth over the epidermal cells is initiated (Lehtonen 2009). Before actual active penetration of the host, T-shaped hyphal branches form thick infection cushions that fasten strongly to the host epidermis (Keijer 1996). Topological signaling and host recognition by its surface structure play the significant role in the establishment of the infection (Lehtonen 2009). The fungus enters actively in to the plant by weak spots, where it can rupture the defensive layer (Weinhold and Sinclair 1996), on the other hand the passive entry happens rarely and stay limited for leaf pathogenic isolates (Weinhold and Sinclair 1996) and the passive entry is not the usual infection mechanism for soilborne *R. solani* (Keijer 1996). Swollen hyphal tips on infection lesion parallely form infection pegs and then enters through the cuticle and epidermal cell walls into the host epidermal tissue and outer layer of the cortex (Demirci and Döken 1998). Lytic enzymes such as cutinases (Baker and Bateman 1978), pectinases (Bertagnolli et al. 1996; Jayasinghe et al. 2004) and xylanases (Peltonen 1995), are most probably involved in infection and penetration, where hydrostatic pressure also plays role (Demirci and Döken 1998). Inside the host, it grows and degrades inter- and intracellular tissues, which can be seen as necrotic lesions on epidermal tissue of shoot, root, stolon or as young seedlings damping-off (Demirci and Döken 1998). This pathogen is known for damping-off, root rot, stem rot and sometimes in leaf blight, cankers, yellow spots etc., (Lehtonen 2009). Moderately moist soil conditions and temperate soil favored this pathogen. The members of the different AGs may have different pathogenic attributes, physical characteristics and evolutionary relationship in past (Gonzalez et al. 2006). *R. solani* is known to be one of the most destructive pathogen for rice, as it causes sheath blight which is one of the most prevalent rice diseases (Willocquet and Savary 2011). Members of AG8 have ability to cause bare patch of cereals and legumes (MacNish and Neate 1996), and severe root rot in canola (Khangura et al. 1999). According to Cook et al. (2002) AG8 patches were associated with up to 30 % yield loss of rice in the US and around $59 m of annual losses of wheat in Australia (Murray and Brennan 2009). Moreover, AG2-1 is extremely pathogenic on canola, causes severe hypocotyl rot on mustard and mild symptoms of hypocotyl rot on narrow-leafed lupin and clover, but failed to infect cereals, such as wheat, oats, barley, and ryegrass (Khangura et al. 1999). Other *R. solani* isolates cause severe diseases for other crops including potato, (Banville 1989; Hide et al. 1992), canola (Verma 1996a, b), maize (Kluth and Varrelmann 2010) and sugar beet (Kiewnick et al. 2001; Kühn et al. 2009). Carling et al. (1989) and Read et al. (1989) have reported yield losses up to 10–30 % on marketable size potato tubers by *Rhizoctonia* disease. Similarly, Olaya and Abawi (1994) also reported that AG 2-2 caused 88 % loss in table beet in western New York and Galindo et al. (1982) observed that AG 4 caused 55 % loss in snap bean and AG-1 on cabbage (Abawi and Martin 1985). While members of the AG-5, AG-4, AG-2-1 and binucleate *Rhizoctonia* are known to be associated with table beets infection (Olaya and Abawi 1994). Those associated with snap beans belonged to AG-1 and AG-2 (Galindo et al. 1982).

Conventional Approach to Diagnose and Control *R. solani*

In the early years, by conventional means, plant infection or the presence of the pathogen was determined by visual observation of disease symptoms followed by isolation of plant pathogen from diseased plant part, such as in black scurf disease caused by *R. solani* (Narayanasamy 2011). These methods involve cultivation of diseased plant part using appropriate growth medium to grow phytopathogens in specific or non-specific manner. This further involves determining morphological characteristics of the pathogens through visual observation or light microscopy, which requires knowledge of taxonomy, time consuming and laborious too. The inability of conventional methods to discriminate among closely related species and/ or strains of the same species made it non-specific and less reliable (Narayanasamy 2001). In some cases, detection of pathogen becomes difficult, especially when pathogen levels are low, low/ no fruiting bodies, latent infection, indistinct disease symptoms, etc., (Agrios 1988). The isolation methods sometimes consumes much time, i.e. from 1 to 4 weeks (Errampalli et al. 2001; Kristensen et al. 2007), however delays cannot be affordable when deals with high cash crops and quick diagnosis and rapid measures are preferred to control disease spread in host plants (Chakrabarty et al. 2007). Based on target pathogen and single or multiple pathogens, detection method can be categorized as specific as well as non-specific for ease of isolation and characterization (Chu et al. 1989).

In the old era of agriculture, the *Rhizoctonia* infection was controlled through general or non-specific approaches such as soil fumigation, soil amendments, planting tricks, maintaining soil moisture, etc., either single or in combination. Where soil fumigation was carried out using methyl bromide (bromomethol), metam sodium in combination with seed coating using some fungicidal or pesticidal compounds such as pencycuron, thiram, imidacloprid, captafol, etc., of which, pesticidal activity has the ability to restrict few other pathogens also. However the high costs and harmful effects of these compounds on the environment made them out of competition in the growing years. The agricultural practices such as irrigation intervals and planting tricks, either alone or in combination found more effective than earlier mentioned approaches (Narayanasamy 2011).

Molecular Approaches

PCR Based Tools

In current timeframe, polymerase chain reaction (PCR) based tools have been used widely due to its high sensitivity (Bounoua et al. 1999). PCR amplifies the low copy number of DNA by millions of times with high sensitivity in view of easy diagnosis. The design of primers for use in PCR (Liu et al. 1995) that recognize unique DNA sequences can result in direct detection of very low levels of target DNA in plant material and soil. Such PCR methods have been used to detect *R. solani* AG-1-IB (Grosch et al. 2007), AG-2 and subgroups (Salazar et al. 2000), AG-3 in potatoes (Bounoua et al. 1999), AG-4 and AG-8 in wheat (Brisbane et al. 1995) and AG-8 in soil (Whisson et al. 1995). To develop species specific diagnostic kit for R. solani seems challenging because of its multiple AG's composition. However, Bounoua et al. (1999) reported a PCR-based restriction mapping method, using one restriction endonuclease, *Xho* I for the detection of *R. solani* AG-3 and this detection method is very specific and reliable and it can be applied on plant tissue and soil. Moreover, Toda et al. (2004) has also concluded that PCR-based technique using specific primers A091-F/R is useful for the rapid detection of *R. solani* AG 2-2 LP isolates from leaf sheaths of zoysia grass exhibiting large-patch symptoms. Likewise, Pannecoucque and Höfte (2009) used two enzymes *Ava*II and/or *Hinc*II for PCR RFLP analysis and they observed the same level of diversity within the strains of *R. solani* as they shares common cluster in sequence based grouping.

Analysis Based on *rRNA* Genes

Generally, taxonomic and phylogenetic studies based on the known conserved genes with enough sequence variant are selected for designing PCR diagnostic assays. The DNA sequences that encode ribosomal RNA genes have been extensively utilized for characterization of phylogenetic analysis of fungal genera (Cubeta et al. 1996). In case of fungi, these genes are usually positioned in either mitochondria or nuclei (Gardes and Bruns 1993). Fungal nuclear rRNA gene contains several hundreds of tandem repeats per genome and each unit enclosed with three genes, viz. small rRNA genes 18S and 5.8S and the large rRNA gene 28S (Capote et al. 2012). These genes are regularly used for fungal taxonomy and phylogeny of fungal species (Vilgalys et al. 1994). The conserved sequences exist in large subunits (LSU) and small subunits (SSU), while the internal transcribed spacer (ITS) regions between the subunits are variable and hence use to differentiate among closely related taxa (Gardes and Bruns 1993). In one report by Vilgalys and Gonzales (1990), it was determined that *Thanatephorus praticola* (anamorph: *R. solani* AG 4) have rDNA repeats of length about 8.8 kb and the predictable number of rDNA copies to be 59 per haploid genome. Sequence database of ITS region (ITS1-5.8S-ITS2) of *R. solani* is widely available in the gene banks and these ITS database facilitates expanded phylogenetic analysis of *R. solani*. Similarly, the sequences of the 18S subunit and the large (28S) subunit (LSU) regions are available for *R. solani* isolates. The rRNA gene sequences were utililized to investigate the genetic relatedness within an AG and sometimes within AG subgroups and confirmed that they are genetically distinct and it has concurrency with DNA/DNA hybridization, RFLP analysis, PCR fingerprinting and other methods (Capote et al. 2012). All available data have discovered a completely conserved 5.8S region, but show enormous deviation in internal transcribed spacers (ITS) regions (Boysen et al. 1996; Kuninaga et al. 1997; Salazar et al. 2000; Gonzalez et al. 2001; Carling et al. 2002a, b). Gonzalez et al. (2001) concluded that the ITS region was not easy to align and exhibited more homology than the large subunit region and ITS region were concentrated in six highly variable regions. The difference in ITS regions and rDNA nucleotide sequence correlate with differences in biological properties like pathogenicity and habitat (Boysen et al. 1996; Kuninaga et al. 1997; Salazar et al. 2000; Gonzalez et al. 2001; Carling et al. 2002b) indicates authenticity of the molecular tools to be used in diagnosis.

Real Time PCR

Now a day's Real-time PCR method is considered as one of the advanced diagnostic tool for detecting plant pathogens. The remarkable benefit of this technique is to detect, amplify and quantify even very less quantity of DNA in targets sample (Wittwer et al. 1997). Quantitative real-time PCR (qPCR) is appropriate for taxon-specific quantification of pathogen DNA in infected host tissue or soil samples. Apart from delivering quantitative data, qPCR is faster and more sensitive than conventional PCR. Chemistry of RT-PCR can be based on the use of doubled-stranded DNA binding dyes, such as SYBR Green, specific fluorescent labeled probes such as TaqMan, Molecular Beacons, or Scorpions, or dye-primer based systems, for example hairpin primers or Plexor system. Guo et al. (2012) has described the qPCR for quantification of *R. solani* AG2-2 IIIB in soil and plant samples with significant observations. The qPCR assays can also quantify *R. solani* AG-8 in natural soils at levels below population densities of 57–87 ppg associated with patches in the field (Paulitz and Schroeder 2005; Okubara et al. 2008). Budge et al. (2009) have developed and reported the sophisticated protocol for the DNA extraction from soil as well as real-time PCR assays for 11 AGs and subgroups of *R. solani*. Real-time PCR assay for AG-2-1 has demonstrated that in naturally infected field soils, *R. solani* target on the upper soil layers (Budge et al. 2009). Edel-Hermann et al. (2009) has also reported the quantification of the *R. solani* AG-2-2 in the soil originating from the diseased sugar beet field at different time points. Zhao et al. (2014) observed the *R. solani* AG-3 from infected tobacco tissue and soil by using a specific primer pair based on the internal transcribed spacer region of the fungal pathogen DNA sequence, *R. solani* AG-3 DNA at quantities as low as 100 fg of purified pathogen DNA could be successfully detect by qPCR.

DNA-DNA Hybridization

DNA-DNA hybridization is the earliest molecular methods for the detection and identification of *Rhizoctonia* species (Kuninaga 1996). Kuninaga and Yokosawa (1985a, b) and Vilgalys (1988) utilized the DNA comparison method to resolve the genetic relatedness within AGs of *R. solani* and each AG isolate showed the highly similar kind of GC content. But, the *R. solani* complex was showed wide dissimilarity in GC content which range from 43.2 to 49.5 % (Kuninaga 1996). A case study by Kurtzman in 1987 on fungal strains had accomplished that diverse isolates showed 2.0–2.5 % different by GC contents belonged to different species (Kuninaga 1996). Though a wide range of GC contents values among the different *R. solani* isolates showed the possibility of DNA-DNA hybridization technique based separations provide more conclusive evidence to support this concept. The isolates belongs the same AGs showed a very high association (≥90 %), and vice-versa (Lübeck 2004). In AGs 2-1, 3, 5, 7, 8 and BI, the members of the same AG hybridized with each other at a rate of ≥91 %. The hybridization studies also showed 11 considerable genetic differences among subgroups of the certain AGs, viz. 1, 2, 4, 6 and 9 (Lübeck 2004). AG subgroups with low relatedness were compatible with earlier identified subgroups based on cultural, morphology, pathogenicity and vitamin requirements (Kuninaga 1996). Between different AGs, the hybridization value is normally less than 15 %, while within an AG is mostly less than 60 % (Lübeck 2004). The genomic DNA-DNA hybridization experiments not only determine the genetic relatedness, but also may possibly reveal the former unknown heterogeneity. New subgroups of AG 4 and AG 6 were assessed by this method and between members of AG 4-HGI and HGII were 30–47 % hybridization values occurs and isolates of AG 6 subgroups HGI and GV were showed 47–62 % hybridization values which confirm their relationship to separate subgroups (Kuninaga 1996). AG of the *R. solani* complex representing distinct evolutionary units corresponding by DNA relatedness into separate species was termed as "genomic species" by Kuninaga (1996). On the darker side of the technique, only few things can be noted such as time consuming, need genomic DNA rather gene specific region, and hence in modern era it has been replaced by newer, easier and quicker methods such as polymerase chain reaction (PCR) based fingerprinting techniques (Lübeck 2004).

PCR Fingerprinting Techniques

PCR fingerprinting techniques allow the amplification of tandem repeats of random regions present on the genomic DNA, which helps identifying species-specific pattern when conserved genes have not enough difference to successfully identify the species (McCartney et al. 2003). Fingerprinting analyses are normally used to study the phylogenetic structure of microbial populations. In the present era, PCR fingerprinting techniques (RAPD and DAF) have been widely used to assess genetic differences among AG subgroups of *R. solani* (Ceresini et al. 2002; Priyatmojo et al. 2001; Stodart et al. 2007). Moreover, universally primed PCR (UP-PCR) method also very similar to RAPD for PCR fingerprinting and being used widely (Trigiano et al. 2004; Lübeck and Lübeck 2005). *R. solani* isolates obtained from diseased sugar beets and potatoes in many cases enabled to their identification by visual assessment of aligned UP-PCR banding profiles with reference isolates (Lübeck and Poulsen 2001). Various PCR fingerprinting techniques uses only based on the same principle of DNA polymerase mediated amplification of DNA fragments to generate multiple copies of target genome sites. The difference of these techniques depends on primarily in the design or choice of primers and level of stringency. However, amplified fragment length polymorphism (AFLP) is different than the above methods (Vos et al. 1995; Lübeck 2004; Lübeck and Lübeck 2005). Majer et al. (1996) have tested AFLP to detect inter- and intra-specific genetic variation of two fungal pathogens *Cladosporium fulvum* and *Pyrenopeziza brassicae*, proved that AFLP showed efficient polymorphism as compared to RFLP analysis. The number of bands generated from AFLP was in the range of 50–70. The advantages of this method are higher reproducibility and higher proportion of the genome being analyzed per reaction.

UP-PCR Based Cross-Hybridization

Cross-hybridization of universally primed polymerase chain reaction (UP-PCR) is a kind of the UP-PCR method, which results from a single UP-primer and can be used to examine the sequence similarity (homology) of unknown *Rhizoctonia* strains to that of UP-PCR hybridization groups (Lübeck and Lübeck 2005). The UP-PCR products from different strains are blotted on to a membrane first and then readily labeled UP-PCR products of a reference strain are used as a hybridization probe in each blot. The major advantage of this method is the ability to examine sequence homology of numerous strains simultaneously. Probe DNA can be labeled by radioactive phosphorus or a non-radioactive molecule, for example digoxigenin (DIG). The hybridization signal strength is observed to determine the relatedness of the strain to the unknown one. Radioactive probe shows strong, visible signals by autoradiograph after one hour and these signals indicate that the hybridized strains belong to the same UP-PCR hybridization group. The signal intensity is directly proportional to the degree of similarity, such as weak signals shows less degree of similarity, while no signal means no relationship among the tested strains. Antigen-antibody reactions are used on same working principle as non-radioactive detection method. Lübeck and Poulsen (2001) have reported to use a UP-PCR cross-hybridization assay for rapid detection and grouping of 21 *Rhizoctonia* isolates to 11 AGs using single UP primer. They have concluded the strong level cross hybridization within isolates of same AG subgroups, while modest or no cross-hybridization among different AGs. Moreover, 16 *Rhizoctonia* isolates were determined using this tool have found concurrently similar to that of total DNA-DNA hybridization.

SCAR Approach

A sequence-characterized amplified region (SCARs) is a PCR based approach, especially intended to amplify members of the same genus (Lübeck and Lübeck 2005). However, these markers have the ability to zoom in at genus level; the same is unable to discriminate the population at species level. This approach can be used, particularly in the cases where intra-species evolutionary relationship is strong and hence, members can be diagnosed from soil or infected plant parts at the genus level with high specificity (Lübeck and Lübeck 2005). Nevertheless, to detect the unique molecular markers present in the target organisms, the fingerprinting methods with UP-PCR can be used for rapid identification of organisms of interest. The sequence information of the SCAR markers can also be used to develop primers that selectively amplify the marker in identifying the target organisms in diagnostic assays. This tool leads the rapid detection of the disease causing agents in view of the rapid disease management. It is recommended that the SCAR markers need to be identified at an AG subgroup level in order to use this technique in a meaningful way. However, UP-PCR derived markers for strains of a subgroup may not be available. As an example, UP-PCR method has been used to selectively distinguish a single antagonistic strain of *Clonostachys rosea* (syn. *Gliocladium roseum*) which is a well known mycoparasite of many soil borne pathogens (Bulat et al. 2000). The antagonists were screened by UP-PCR and afterward, a strain specific marker was identified for the strain GR5. The marker was rehabilitated into a SCAR and a specific primer pair was designed for selective amplification of GR5. Eighty two strains and DNA from 31 soil samples, mostly of Danish origin, were tested for specificity of SCAR markers. Out of 82 strains two responded to SCAR primers which were similar to GR5 but not identical. The total DNA extracted from soil samples infested with and without GR5 demonstrated the SCAR primers could detect GR5 in a pool of mixed DNA and that this particular marker was not present in other microorganisms, indicating high specificity of the approach.

A DNA Sequence Based Approach

Morphological characters are not always giving the proper identification of *R. solani* like pathogen, especially when having members of the closely related groups such as AG. Sequence data analysis of PCR amplification of a target gene with universal primers amplifying a conserved region in the, followed by comparison with the reference databases is becoming an important tool for characterizing new fungal species. On the other hand, the use of sequence databases based on DNA similarity may have few drawbacks such as incomplete sequences, sequences associated with misidentified organisms, the inability to easily change or update data, and problems associated with defining species boundaries, all of them leading to erroneous interpretation of search results. To procure proper identification high quality data by the researcher community should be the remedy of this drawback (Kang et al. 2010).

The Sanger sequencing technique has been upgraded or replaced to several "next-generation" sequencing technologies able to generate a high number of short sequences from multiple organisms in short time (Capote et al. 2012). Massive sequencing technologies offer dramatic increases in commercial sequence throughput, having a tremendous impact on genomic research. For example, Hane et al. (2014) reported that a high quality genome of *R. solani* AG8 isolate WAC10335 was isolated from lupin was assembled and a manually curated set of 13,964 genes supported by RNA-seq. The heterozygous SNP mutation rate within single isolate of AG8 was scrutinized to be superior as compared to SNP mutation rates observed across the fungal populations. Comparative analyses were combined to predict biological processes relevant to AG8 and 308 proteins with effector-like characteristics, forming a valuable resource for further study of this pathosystem. Predicted effector-like proteins had elevated levels of non-synonymous point mutations relative to synonymous mutations (dN/dS), suggesting that they may be under diversifying selection pressures. Moreover, significant genomic resources for other AGs of *R. solani* have also recently become publicly available, formerly been limited to EST libraries of AG1-IA and AG4 (Lakshman et al. 2012; Hane et al. 2014).

Future Perspectives

Being an essential component of the plant disease control, an early diagnosis of phytopathogens plays a key role and is generally carried out through various means. The advancements in the diagnostic tools from time-consuming culture dependent techniques to the rapid and precise molecular techniques made it one of the widely accepted approach. The early diagnosis of phytopathogen helps in controlling the disease as well as its further spread. In the current time frame, various diagnostic tools have been reported, however, accuracy and reproducibility is equally essential along with rapidity. Molecular tools are usually facilitating qualitative as well as quantitative estimation of phytopathogens, which may help to determine the specific pathovar as well as infection scale. The qPCR and loop-mediated isothermal amplification (LAMP) are the most commonly used molecular tools for detecting phytopathogens in rapid and precise manner (Heard and West 2014). Along with molecular tools, many other techniques such as biosensor, lateral flow devices, immune based techniques such as fluorescence microscopy, resonance imaging, ELISA, bookmakers for detection of volatile or particular toxin, metabolites using electrochemical sensors, etc., have been implemented for the same (West et al. 2013). The use of air samplers has also been reported as single tool or in combination with other techniques for in situ determination of pathogen (West et al. 2013). The molecular plant pathology is seems to extend its horizon in growing years, where few reports can be seen about the use of nano-sensors for rapid diagnosis of multiple phytopathogens in the field itself. It indicates the future potential and scope of the molecular prospecting in the plant pathology for the betterment of the agriculture.

Conclusion

The advancements in the agricultural practices have marked as an important move for maintaining the agricultural economy through determining phytopathogens at an early stage of infection followed by plant disease control. The strength of molecular tools such as highly specific, rapid, and in situ diagnosis of phytopathogens, made them superior over other available tools. These tools can potentially diagnose and discriminate between closely related pathogenic strains belongs to same genus or species such as pathovars or biovars including multiple pathogens. Though, in situ detection of phytopathogen is highly appreciable in view of broadcasting of the field details such as disease information, infection scale, etc., in the public domain towards sustainable agriculture.

Acknowledgement HJP gratefully acknowledge Agricultural Research Organization, the Volcani Center, Israel.

References

Abawi GS, Martin SB (1985) *Rhizoctonia* foliar blight of cabbage in New York State. Plant Dis 69(2):158–161

Agrios GN (1988) Plant pathology, 3rd edn. Academic, New York, p 803

Anderson NA (1982) The genetics and pathology of *Rhizoctonia solani*. Ann Rev Phytopathol 20:329–347

Anguiz R, Martin C (1989) Anastomosis groups, pathogenicity, and other characteristics of *Rhizoctonia solani* isolated from potatoes in Peru. Plant Dis 77:199–201

Baird RE, Carling DE (1997) First report of *Rhizoctonia solani* AG–7 in Georgia. Plant Dis 81:832

Baker KF (1970) Types of *Rhizoctonia* diseases and their occurrence. In: Parmeter JR Jr (ed) *Rhizoctonia solani*: biology and pathology. California University Press, Berkeley, pp 124–148

Baker CJ, Bateman DF (1978) Cutin degradation by plant pathogenic fungi. Phytopathology 68:1577–1584

Bandy BP, Zanzinger DH, Tavantzis SM (1984) Isolation of anastomosis group 5 of *Rhizoctonia solani* from potato field soils in Maine. Phytopathology 74:1220–1224

Banville GJ (1989) Yield losses and damage to potato plants caused by *Rhizoctonia solani*. Kuhn Am Potato J 66:821–834

Bertagnolli BL, Dal Soglio FH, Sinclair JB (1996) Extracellular enzyme profiles of the fungal pathogen *Rhizoctonia solani* isolate 2B–12 and two antagonists, *Bacillus megaterium* strain B153–2–2 and *Thrichoderma harzianum* isolate Th008. I. Possible correlations with inhibition of growth and biocontrol. Physiol Mol Plant Pathol 48:145–160

Blazier SR, Conway KE (2004) Characterization of *Rhizoctonia solani* isolates associated with patch diseases on turfgrass. Proc Okla Acad Sci 84:41–51

Bounoua S, Jabaji-Harec SH, Hogueb R, Charesta PM (1999) Polymerase chain reaction-based assay for specific detection of *Rhizoctonia solani* AG-3 isolates. Mycol Res 103:1–8

Boysen M, Borja M, del Moral C, Salazar O, Rubio V (1996) Identification at strain level of *Rhizoctonia solani* AG4 isolates by direct sequence of asymmetric PCR products of the ITS region. Curr Genet 29:174–181

Brisbane PG, Neate SM, Pankhurst CE, Scott NS, Thomas MR (1995) Sequence-tagged site markers to identify *Rhizoctonia solani* AG4 or AG8 infecting wheat in South Australia. Phytopathology 85:1423–1427

Budge GE, Shaw MW, Colyer A, Pietravalle S, Boonham N (2009) Molecular tools to investigate *Rhizoctonia solani* distribution in soil. Plant Pathol 58:1071–1080

Bulat SA, Lübeck M, Alekhina IA, Jensen DF, Knudsen IMB, Lübeck PS (2000) Identification of a universally primed–PCR–derived sequence–characterized amplified region marker for an antagonistic strain of *Clonostachys rosea* and development of a strain–specific PCR detection assay. Appl Environ Microbiol 66:4758–4763

Burpee LL, Martin SB (1996) Biology of turfgrass diseases incited by *Rhizoctonia* species. In: Sneh B, Jabaji–Hare S, Neate S, Dijst G (eds) *Rhizoctonia* species: taxonomy, molecular biology, ecology, pathology and disease control. Kluwer Academic Publishers, Dordrecht, pp 359–368

Burpee LL, Sanders PL, Cole H Jr, Sherwood RT (1980) Anastomosis groups among isolates of Ceratobasidium cornigerum and related fungi. Mycologia 72:689–701

Burton RJ, Coley–Smith JR, Wareing PW, Gladders P (1988) *Rhizoctonia oryzae* and *R. solani* associated with barley stunt disease in the United Kingdom. Trans Br Mycol Soc 91(3):409–417

Capote N, Pastrana AM, Aguado A, Sánchez-Torres P (2012) Molecular tools for detection of plant pathogenic fungi and fungicide resistance. In: Dr. Cumagun CJ (ed) Plant pathology. InTech, Croatia, pp 152–202. ISBN 978-953-51-0489-6

Carling DE (1996) Grouping *Rhizoctonia solani* by hyphal anastomosis reaction. In: Sneh B, Jabaji–Hare S, Neate S, Dijst G (eds) *Rhizoctonia* species: taxonomy, molecular biology, ecology, pathology and disease control. Kluwer Academic Publishers, Dordrecht, pp 37–47

Carling DE, Leiner RH, Westphale PC (1989) Symptoms, signs and yield reduction associated with *Rhizoctonia* disease of potato induced by tuberborne inoculum of *Rhizoctonia solani* AG–3. Am Potato J 66:693–701

Carling DE, Rothrock CS, MacNish GC, Sweetingham MW, Brainard KA, Winters SW (1994) Characterization of anastomosis group 11 (AG–11) of *Rhizoctonia solani*. Phytopathology 84:1387–1393

Carling DE, Pope EJ, Brainard KA, Carter DA (1999) Characterization of mycorrhizal isolates of *Rhizoctonia solani* from an orchid, including AG–12, a new anastomosis group. Phytopathology 89:942–946

Carling DE, Baird RE, Gitaitis RD, Brainard KA, Kuninaga S (2002a) Characterization of AG–13, a newly reported anastomosis group of *Rhizoctonia solani*. Phytopathology 92:893–899

Carling DE, Kuninaga S, Brainard KA (2002b) Hyphal anastomosis reactions, rDNA–internal transcribed spacer sequences, and virulence level s among subsets of *Rhizoctonia solani* anastomosis group–2 (AG2) and AG–BI. Phytopathology 92:43–50

Ceresini PC, Shew HD, Vilgalys RJ, Cubeta MA (2002) Genetic diversity of *Rhizoctonia solani* AG–3 from potato and tobaco in North Carolina. Mycologia 94:437–449

Chakrabarty PK, Chavhan RL, Sable SV, Narwade AV, Monga D, Khadi BM (2007) Development of sensitive molecular diagnostic tools for detection of economically important fungal pathogens of cotton. In: World Cotton Research Conference–4, Lubbock, Texas, USA, 10–14 Sept 2007. International Cotton Advisory Committee (ICAC)

Chand T, Logan C (1983) Cultural and pathogenic variation in potato isolates of *Rhizoctonia solani* in Northern Ireland. Trans Br Mycol Soc 81:585–589

Chang DCN, Chou LC (2007) Growth responses, enzyme activities and component changes as influenced by *Rhizoctonia* orchid mycorrhizal on *Anoectochilus formosanus* Hayata. Bot Stud 48:445–451

Chu PWG, Waterhouse PM, Martin RR, Gerlach WL (1989) New approaches to the detection of microbial plant pathogens. Biotechnol Genet Eng Rev 7:45–111

Cook RJ, Schillinger WF, Christensen NW (2002) *Rhizoctonia* root rot and take all of wheat in diverse direct–seed spring cropping systems. Can J Plant Pathol 24:349–358

Cubeta MA, Vilgalys R, Gonzalez D (1996) Molecular analysis of ribosomal RNA genes in *Rhizoctonia* fungi. In: Sneh B, Jabaji–Hare S, Neate S, Dijst G (eds) *Rhizoctonia* species: taxonomy, molecular biology, ecology, pathology and disease control. Kluwer Academic Publishers, Dordrecht, pp 81–86

Date H, Yagi S, Okamoto Y, Oniki M (1984) On the leaf blight of tomatoes by *Thanatephorus cucumeris* (Frank) Donk (*Rhizoctonia solani*). Ann Phytopathol Soc Jpn 50:339

Demirci E, Döken MT (1998) Host penetration and infection by the anastomosis groups of *Rhizoctonia solani* Kühn isolated from potatoes. Trans J Agric For 22:609–613

Dijst G, Schneider HM (1996) Flower bulb diseases incited by *Rhizoctonia solani*. In: Sneh B, Jabaji–Hare S, Neate S, Dijst G (eds) *Rhizoctonia* species: taxonomy, molecular biology, ecology, pathology and disease control. Kluwer Academic Publishers, Dordrecht, pp 279–288

Edel–Hermann V, Jobard M, Gautheron N, Friberg H, Steinberg C (2009) Real–time PCR assay for identification and quantification of *Rhizoctonia solani* AG–2–2 in soil. Multitrophic Interact Soil IOBC/wprs Bull 42:41–46

Erper I, Karaca GH, Turkkan M, Ozkoc I (2005) Characterization and pathogenicity of *Rhizoctonia* spp. from onion in Amasya, Turkey. J Phytopathol 154:75–79

Errampalli S, Saunders J, Cullen DW (2001) A PCR–based method for detection of potato pathogen, *Helminthosporium solani* in silver scurf–infected tuber tissue and soils. J Microbiol Methods 44:59–68

Fox RT (1993) Principles of diagnostic techniques in plant pathology. Cab International, Wallingford, UK

Galindo JJ, Abawi GS, Thurston HD (1982) Variability among isolates of *Rhizoctonia solani* associated with snap bean hypocotyls and soils in New York. Plant Dis 66(5):390–394

Gardes M, Bruns TD (1993) ITS primers with enhanced specificity for basidiomycetes application to the identification of mycorrhizae and rusts. Mol Ecol 2:113–118

Gonzáles García V, Portal Onco MA, Rubio Susan V (2006) Review. Biology and systematics of the form genus *Rhizoctonia*. Span J Agric Res 4:55–79

Gonzalez D, Carling DE, Kuninaga S, Vilgalys R, Cubeta MA (2001) Ribosomal DNA systematics of *Ceratobasidium* and *Thanatephorus* with *Rhizoctonia* anamorphs. Mycologia 93(6):1138–1150

Grisham MP, Anderson NA (1983) Pathogenicity and host specificity of *Rhizoctonia solani* isolated from carrots. Phytopathology 73:1564–1569

Grosch R, Schneider JH, Peth A, Waschke A, Franken P, Kofoet A, Jabaji-Hare SH (2007) Development of a specific PCR assay for the detection of *Rhizoctonia solani* AG 1-IB using SCAR primers. J Appl Microbiol 102(3):806–819

Guo Y, Li W, Sun H, Wang N, Yu H, Chen H (2012) Detection and quantification of *Rhizoctonia cerealis* in soil using real–time PCR. J Gen Plant Pathol 78(4):247–254

Hane JK, Anderson JP, Williams AH, Sperschneider J, Singh KB (2014) Genome sequencing and comparative genomics of the broad host–range pathogen *Rhizoctonia solani* AG8. PLoS Genet 10(5):e1004281

Hashiba T, Kobayashi T (1996) Rice diseases incited by *Rhizoctonia* species. In: Sneh B, Jabaji–Hare S, Neate S, Dijst G (eds) *Rhizoctonia* species: taxonomy, molecular biology, ecology, pathology and disease control. Kluwer Academic Publishers, Dordrecht, pp 331–340

Heard S, West JS (2014) New developments in identification and quantification of airborne inoculum. In: Detection and diagnostics of plant pathogens. Springer, Netherlands, pp 3–19

Hebert PD, Gregory TR (2005) The promise of DNA barcoding for taxonomy. Syst Biol 54(5):852–859

Herr LJ (1993) Host sources, virulence and overwinter survival of *Rhizoctonia solani* anastomosis groups isolated from field lettuce with bottom rot symptoms. Crop Prot 12:521–526

Herr LJ (1996) Sugarbeet diseases incited by *Rhizoctonia* spp. In: Sneh B, Jabaji–Hare S, Neate S, Dijst G (eds) Rhizoctonia species: taxonomy, molecular biology, ecology, pathology and disease control. Kluwer Academic Publishers, Dordrecht, pp 341–350

Herr LJ, Fulton MM (1995) *Rhizoctonia solani* AG–1–1A and AG–2–2 IIIB cause brown patch on tall fescue and creeping bentgrass in Ohio. Plant Dis 79:1186

Hide GA, Read PJ, Hall SM (1992) Stem canker (*Rhizoctoniu solani*) on three early and three maincrop potato cultivars: effects of seed tuber size on growth and yield. Ann Appl Biol 120:391–403

Higa T, Parr JF (1994) Beneficial and effective microorganisms for a sustainable agriculture and environment, vol 1. International Nature Farming Research Center, Atami

Hwang SF, Howard RJ, Chang KF (1996) Forage and oilseed diseases incited by *Rhizoctonia* species. In: Sneh B, Jabaji–Hare S, Neate S, Dijst G (eds) *Rhizoctonia* species: taxonomy, molecular biology, ecology, pathology and disease control. Kluwer Academic Publishers, Dordrecht, pp 289–301

Hwang SF, Gossen PD, Conner RL, Chang KF, Turnbull GD, Lopetinsky K, Howard RJ (2007) Management strategies to reduce losses caused by *Rhizoctonia* seedling blight of filed pea. Can J Plant Sci 87:145–155

Hyakumachi M, Ui T (1988) Development of the teleomorph of non-self-anastomosing isolates of *Rhizoctonia solani* by a buried-slide method. Plant Pathol 37:438–440

Jayasinghe CK, Wijayaratne SCP, Fernando THPS (2004) Characterization of cell wall degrading enzymes of *Thanatephorus cucumeris*. Mycopathologia 157:73–79

Kang S, Mansfield MA, Park B, Geiser DM, Ivors KL, Coffey MD, Grünwald NJ, Martin FN, Lévesque CA, Blair JE (2010) The promise and pitfalls of sequence–based identification of plant–pathogenic fungi and oomycetes. Phytopathology 100:732–737

Kataria HR, Verma PR, Gisi U (1991) Variability in the sensitivity of *Rhizoctonia solani* anastomosis groups to fungicides. J Phytopathol 133:121–133

Keijer J (1996) The initial steps of the infection process in *Rhizoctonia solani*. In: Sneh B, Jabaji–Hare S, Neate S, Dijst G (eds) *Rhizoctonia* species: taxonomy, molecular biology, ecology, pathology and disease control. Kluwer Academic Publishers, Dordrecht, pp 149–162

Khangura RK, Barbetti MJ, Sweetingham MW (1999) Characterization and pathogenicity of *Rhizoctonia* species on canola. Plant Dis 83:714–721

Khiyami MA, Almoammar H, Awad YM, Alghuthaymi MA, Abd–Elsalam KA (2014) Plant pathogen nanodiagnostic techniques: forthcoming changes? Biotechnol Biotechnol Equip 28(5):775–785

Kiewnick S, Jacobsen BJ, Braun–Kiewnick A, Eckhoff JLA, Bergman JW (2001) Integrated control of *Rhizoctonia* crown and root rot of sugar beet with fungicides and antagonistic bacteria. Plant Dis 85:718–722

Kluth C, Varrelmann M (2010) Maize genotype susceptibility to *Rhizoctonia solani* and its effect on sugar beet crop rotations. Crop Prot 29:230–238

Kodama T, Horimoto K, Ogoshi A (1982) On the brown spot of eggplant caused by *Thanatephorus cucumeris* (Frank) Donk (*Rhizoctonia solani*) AG3. Ann Phytopathol Soc Jpn 48:356

Kristensen R, Gauthier G, Berdal KG, Hamels S, Remacle J, Host–Jenson A (2007) DNA microarray to detect and identify trichothecene– and moniliformin–producing *Fusarium* species. J Appl Microbiol 102:1060–1070

Kuhn J (1858) Die Krankenheiten der Kulturwachse, ihre Ursachen und ihre Verhutung. Gustav Bosselman, Berlin, p 312

Kuhn J, Rippel R, Schmidhalter U (2009) Abiotic soil properties and the occurrence of *Rhizoctonia* crown and root rot in sugar beet. J Plant Nutr Soil Sci Z Pflanzenernah Bodenkd 172:661–668

Kumar S, Sivasithamparam K, Sweetingham MW (2002) Prolific production of sclerotia in soil by *Rhizoctonia solani* anastomosis group (AG) 11 pathogenic on lupin. Ann Appl Biol 141:11–18

Kuninaga S (1996) DNA base sequence relatedness. In: Sneh B, Jabaji–Hare S, Neate S, Dijst G (eds) *Rhizoctonia* species: taxonomy, molecular biology, ecology, pathology and disease control. Kluwer Academic Publishers, Dordrecht, pp 73–80

Kuninaga S, Yokosawa R (1985a) DNA base sequence homology in *Rhizoctonia solani* Kühn. VI. Genetic relatedness among seven anastomosis groups. Ann Phytopathol Soc Jpn 51:127–132

Kuninaga S, Yokosawa R (1985b) DNA base sequence homology in *Rhizoctonia solani* Kühn. VII. Genetic relatedness between AG–BI and other anastomosis groups. Ann Phytopathol Soc Jpn 51:133–138

Kuninaga S, Natsuaki T, Takeuchi T, Yokosawa R (1997) Sequence variation of the rDNA ITS regions within and between anastomosis groups in *Rhizoctonia solani*. Curr Genet 32:237–243

Lakshman DK, Alkharouf N, Roberts DP, Natarajan SS, Mitra A (2012) Gene expression profiling of the plant pathogenic basidiomycetous fungus *Rhizoctonia solani* AG 4 reveals putative virulence factors. Mycologia 104:1020–1035

Lehtonen MJ (2009) *Rhizoctonia solani* as a potato pathogen – variation of isolates in Finland and host response. University of Helsinki Finland, (PhD thesis)

Liu ZL, Sinclair JB (1991) Isolates of *Rhizoctonia solani* anastomosis group 2–2 pathogenic to soybean. Plant Dis 75:682–687

Liu ZL, Domier LL, Sinclair JB (1995) Polymorphism of genescoding for nuclear 18s ribosomal-RNA indicates genetic distinctiveness of anastomosis group-10 from other groups in the *Rhizoctonia solani* species complex. Appl Environ Microbiol 61:2659–2664

Lopez N, Castano J (2012) Characterization of phytopathogenic fungi, bacteria, nematodes and virus in four commercial varieties of heliconia (*Heliconia* sp.). Rev Fac Agron 65(2):6697–6710

Lori M (2013) Producing fruits: the technology of farming. Heinemann Library, Chicago

Lübeck M (2004) Molecular characterization of *Rhizoctonia solani*. In: Arora DK, Khachatourians GG (eds) Applied mycology and biotechnology, vol 4, Fungal Genomics. Elsevier B.V, Amsterdam, pp 205–224

Lübeck M, Lübeck PS (2005) Universally primed PCR (UP–PCR) and its applications in mycology. In: Deshmukh SK, Rai MK (eds) Biodiversity of fungi: their role in human life. Science Publishers, Enfield, pp 409–438

Lübeck M, Poulsen H (2001) UP–PCR cross blot hybridization as a tool for identification of anastomosis groups in the *Rhizoctonia solani* complex. FEMS Microbiol Lett 201:83–89

MacNish GC, Neate SM (1996) *Rhizoctonia* bare patch of cereals – an Australian perspective. Plant Dis 80:965–971

MacNish GC, Carling DE, Sweetingham MW, Ogoshi A, Brainard KA (1995) Characterization of anastomosis group 10 (AG-10) of *Rhizoctonia solani*. Aust Plant Pathol 24:252–260

Majer D, Mithen R, Lewis BG, Vos P, Oliver RP (1996) The use of AFLP fingerprinting for the detection of genetic variation in fungi. Mycol Res 100:1107–1111

Matsumoto T, Yamamoto W, Hirane S (1932) Physiology and parasitology of the fungi generally referred to as Hypochnus Sasakii Shirai: differentiation of the strains by means of hyphal fusion and culture in differential media. J Soc Trop Agric 4:370–388

Mazzola M, Smiley RW, Rovira AD, Cook RJ (1996) Characterization of *Rhizoctonia* isolates, diseases occurrence and management in cereals. In: Sneh B, Jabaji–Hare S, Neate S, Dijst G (eds) *Rhizoctonia* species: taxonomy, molecular biology, ecology, pathology and disease control. Kluwer Academic Publishers, Dordrecht, pp 259–267

McCartney HA, Foster SJ, Fraaije BA, Ward E (2003) Molecular diagnostics for fungal plant pathogens. Pest Manag Sci 59:129–142

Montealegre J, Valderrama L, Sánchez S, Herrera R, Besoain X, Pérez LM (2010) Biological control of *Rhizoctonia solani* in tomatoes with *Trichoderma harzianum* mutants. Electron J Biotechnol 13:2, http://www.ejbiotechnology.info/content/vol13/issue2/full/6/

Moore RT (1987) The genera of *Rhizoctonia*–like fungi: *Asorhizoctonia*, *Ceratorhiza* gen. nov., *Epulorhiza* gen. nov., *Moniliopsis* and *Rhizoctonia*. Mycotaxon 29:91–99

Murray GM, Brennan JP (2009) Estimating disease losses to the Australian wheat industry. Aust Plant Pathol 38:558–570

Muyolo NG, Lipps PE, Schmitthenner AF (1993) Anastomosis grouping and variation of virulence associated with dry bean and soybean in Ohio and Zaire. Phytopathology 83:438–444

Narayanasamy P (2001) Plant pathogen detection and disease diagnosis, 2nd edn. Marcel Dekker, New York

Narayanasamy P (2011) Detection of fungal pathogens in plants. In: Microbial plant pathogens–detection and disease diagnosis. Springer, Netherlands, pp 5–199

Nelson B, Helms T, Christianson T, Kural I (1996) Characterization and pathogenicity of *Rhizoctonia* on soybean. Plant Dis 80:74–80

Ogoshi A (1975) Grouping of *Rhizoctonia solani* Kühn and their perfect stages. Rev Plant Prot Res 8:93–103

Ogoshi A (1987) Ecology and pathogenicity of anastomosis and intraspecific groups of *Rhizoctonia solani* Kühn. Ann Rev Phytopathol 25:125–143

Ogoshi A (1996) Introduction – the genus *Rhizoctonia*. In: Sneh B, Jabaji–Hare S, Neate S, Dijst G (eds) *Rhizoctonia* species: taxonomy, molecular biology, ecology, pathology and disease control. Kluwer Academic Publishers, Dordrecht, pp 1–9

Ogoshi A, Oniki M, Araki T, Ui T (1983) Anastomosis groups of binucleateRhizoctonia in Japan and North America and their perfect states. Mycol Soc Jpn 24:79–87

Okubara PA, Schroeder KL, Paulitz TC (2008) Identification and quantification of *Rhizoctonia solani* and *R. oryzae* using real–time polymerase chain reaction. Phytopathology 98:837–847

Olaya G, Abawi GS (1994) Characteristics of *Rhizoctonia solani* and binucleate *Rhizoctonia* species causing foliar blight and root rot on table beets in New York state. Plant Dis 78(8):800–804

Pannecoucque J, Höfte M (2009) Detection of rDNA ITS polymorphism in *Rhizoctonia solani* AG 2-1 isolates. Mycologia 101:26–33

Parmeter JR (1970) *Rhizoctonia solani*, biology and pathology. University of California Press, Berkeley

Parmeter JR Jr, Whitney HS (1970) Taxonomy and nomenclature of the imperfect state. Pages 7–19. In: Parmeter JR Jr (ed) Biology and pathology of *Rhizoctonia solani*. University of California Press, Berkeley, p 255

Parmeter JRJ, Sherwood RT, Platt WD (1969) Anastomosis grouping among isolates of Thanatephorus cucumeris. Phytopathology 59:1270–1278

Paulitz TC, Schroeder KL (2005) A new method for the quantification of *Rhizoctonia solani* and *R. oryzae* from soil. Plant Dis 89:767–772

Peltonen S (1995) Comparison of xylanases production by fungal pathogens of barley with special preference to *Bipolaris sorokiniana*. Mycol Res 99:717–723

Perkins AJ, Mcgee PA (1995) Distribution of the orchid mycorrhizal fungus, *Rhizoctonia solani*, in relation to its host, *Pterostylis acuminata*, in the field. Aust J Bot 43(6):565–575

Priyatmojo A, Escopalao VE, Tangonan NG, Pascual CB, Suga H, Kageyama K, Hyakumachi M (2001) Characterization of a new subgroup of *Rhizoctonia solani* anastomosis group 1 (AG–1–ID), causal agent of a necrotic leaf spot on coffee. Phytopathology 91:1054–1061

Read PJ, Hide GA, Firmager JP, Hall SM (1989) Growth and yield of potatoes affected by severity of stem canker (*Rhizoctonia solani*). Potato Res 32:9–15

Rothrock CS (1996) Cotton diseases incited by *Rhizoctonia solani*. In: Sneh B, Jabaji–Hare S, Neate S, Dijst G (eds) *Rhizoctonia* species: taxonomy, molecular biology, ecology, pathology and disease control. Kluwer Academic Publishers, Dordrecht, pp 269–277

Rush CM, Carling DE, Haverson RM, Mathieson JT (1994) Prevalence and pathogenicity of anastomosis groups of *Rhizoctonia solani* from wheat and sugar beet. Plant Dis 18:349–352

Salazar O, Julian MC, Rubio V (2000) Primers based on specific rDNA–ITS sequences for PCR detection of *Rhizoctonia solani*, *R. solani*AG–2 subgroups and ecological types, and binucleate *Rhizoctonia*. Mycol Res 104:281–285

Sayler RJ, Yang Y (2007) Detection and quantification of *Rhizoctonia solani* AG-1 IA, the rice sheath blight pathogen, in rice using real-time PCR. Plant Dis 91:1663–1668

Sharon M, Kuninaga S, Hyakumachi M, Sneh B (2006) The advancing identification and classification of *Rhizoctonia* spp. using molecular and biotechnological methods compared with the classical anastomosis grouping. Mycoscience 47:299–316

Singh DV (2008) Introductory plant pathology. NISCAIR, New Delhi

Sneh B, Burpee L, Ogoshi A (1998) Identification of *Rhizoctonia* species. The APS, St. Paul

Stodart BJ, Harvey PR, Neate SM, Melanson DL, Scott ES (2007) Genetic variation and pathogenicity of anastomosis group 2 isolates of *Rhizoctonia solani* in Australia. Mycol Res 111:891–900

Strashnov Y, Elad Y, Sivan A, Rudich Y, Chet I (1985) Control of *Rhizoctonia solani* fruit rot of tomatoes by *Thrichoderma harzianum* Rifai. Crop Prot 4:359–364

Sue MJ, Yeap SK, Omar AR, Tan SW (2014) Application of PCR–ELISA in molecular diagnosis. BioMed Res Int 2014:1–6. http://dx.doi.org/10.1155/2014/653014

Sumner DR, Phatak SC, Carling DE (2003) Characterization and pathogenicity of a new anastomosis subgroup AG-2-3 of Rhizoctonia solani Kuün isolated from leaves of soybean. Plant Dis 87:1264

Tilman D, Cassman KG, Matson PA, Naylor R, Polasky S (2002) Agricultural sustainability and intensive production practices. Nature 418(6898):671–677

Toda T, Mushika T, Hyakumachi M (2004) Development of specific PCR primers for the detection of *Rhizoctonia solani* AG 2-2 LP from the leaf sheaths exhibiting large-patch symptom on zoysia grass. FEMS Microbiol Lett 232(1):67–74

Tom S (2009) An edible history of humanity. Walker and Company, New York

Tomaso–Peterson M, Trevathan LE (2007) Characterization of *Rhizoctonia*–like fungi isolated from agronomic crops and turf grasses in Mississippi. Plant Dis 91:260–265

Trigiano RN, Ament MH, Habera LF, Caetano–Anolles G (2004) Molecular techniques used to study systematics, ecology, and evolution of plant pathogens. In: Trigiano RN, Windham MT, Windham AS (eds) Plant pathology concepts and laboratory exercises. CRC Press, Boca Raton, pp 217–218

Tu CC, Hsieh TF, Chang YC (1996) Vegetable diseases incited by *Rhizoctonia* spp. In: Sneh B, Jabaji–Hare S, Neate S, Dijst G (eds) *Rhizoctonia* species: taxonomy, molecular biology, ecology, pathology and disease control. Kluwer Academic Publishers, Dordrecht, pp 369–377

Valkonen JPT, von Heiroth W, Savela ML (1993) Fungi and gram–negative bacteria as soilborne minor pathogens of goat rue (Galega orientalis Lam.). Ann Appl Biol 123:257–269

Verma PR (1996a) Biology and control of *Rhizoctonia solani* on rapeseed: a review. Phytoprotection 77:99–111

Verma PR (1996b) Oilseed rape and canola disease incited by *Rhizoctonia* species. In: Sneh B, Jabaji–Hare S, Neate S, Dijst G (eds) *Rhizoctonia* species: taxonomy, molecular biology, ecology, pathology and disease control. Kluwer Academic Publishers, Dordrecht, pp 249–258

Vilgalys R (1988) Genetic relatedness among anastomosis groups in *Rhizoctonia* as measured by DNA/DNA hybridization. Phytopathology 78:698–702

Vilgalys R, Gonzalez D (1990) Organization of ribosomal DNA in the basidiomycete *Thanatephorus praticola*. Curr Genet 18:277–280

Vilgalys R, Hopple JS, Hibbett DS (1994) Phylogenetic implications of generic concepts in fungal taxonomy: the impact of molecular systematic studies. Mycol Helv 6:73–91

Vos P, Hogers R, Bleeker M, Reijans M, van de Lee T, Hornes M, Frijters A, Pot J, Peleman J, Kuiper M, Zabeau M (1995) AFLP: a new technique for DNA fingerprinting. Nucleic Acids Res 23:4407–4414

Weinhold AR, Sinclair JB (1996) *Rhizoctonia solani*: penetration, colonization and host response. In: Sneh B, Jabaji–Hare S, Neate S, Dijst G (eds) *Rhizoctonia* species: taxonomy, molecular biology, ecology, pathology and disease control. Kluwer Academic Publishers, Dordrecht, pp 163–174

West JS, Heard S, Canning GGM, Fraaije BA, Hammond Kosack K (2013) New developments in identification and quantification of airborne inoculums. Acta Phytopathol Sin 43(Suppl):16, International Congress of Plant Pathology 2013 abstract O01.002

Whisson DL, Herdina S, Francis L (1995) Detection of *Rhizoctonia solani* AG-8 in soil using a specific DNA-probe. Mycol Res 99:1299–1302

Willocquet L, Savary S (2011) Resistance to rice sheath blight (*Rhizoctonia solani* Kuhn) [(teleomorph: Thanatephorus cucumeris (A.B. Frank) Donk.] disease: current status and perspectives. Euphytica 178:1–22

Wittwer CT, Herrmann MG, Moss AA, Rasmussen RP (1997) Continuous fluorescence monitoring of rapid cycle DNA amplification. Biotechniques 22:130–138

Wong DH, Sivasithamparam K (1985) *Rhizoctonia* spp. associated with root rots of subterranean clover. Trans Br Mycol Soc 85:156–158

Yang G, Li C (2012) General description of Rhizoctonia species complex. In: Cumagun CJ (ed) Plant Pathology, pp 41–52

Yang XB, Berggren GT, Snow JP (1990) Seedling infection of soybean by isolates of *Rhizoctonia solani* AG–1, causal agent of aerial blight and web blight of soybean. Plant Dis 74:485–488

Yang GH, Chen JY, Pu WQ (2007) First report of head rot of cabbage and web blight of snap bean causedby*Rhizoctoniasolani*AG–4HGI.PlantPathol56:351.doi:10.1111/j.1365-3059.2007.01543.x

Zhao YQ, Wu YH, Zhao XX, An MN, Chen JG (2014) Study on the taqman real–time PCR to the detection and quantification of *Rhizoctonia solani* AG–3 of tobacco target spot. Adv Mater Res 1010–1012:80–83

Webliography

"Farming Pollution." WWF Global 2013. Accessed 30 Jun 2015

"Growing a Nation: The Story of American Agriculture." AG Classroom. Accessed: 30 Jun 2015

Fusarium moniliforme Associated with Sugarcane Leaf Binding Disease in India and Its Possible Management

Ajay Kumar Tiwari, Kavita Kumari, Surendra Pratap Singh, Nishtha Mishra, Atul Singh, and B.L. Sharma

Introduction

Plant pathogens like viruses, fungi, bacteria, phytoplasma and nematodes have become serious threat to commercial crops including sugarcane. Sugarcane is highly demanding crop in India and widely cultivated in many states of the India, because of it's cash generating capability to the farmers. Being an annual crop, sugarcane faces many biotic and abiotic stresses. More than 100 plant diseases have been reported on sugarcane crop across the world which significantly reduce yields (Tiwari et al. 2010; Vishwanathan and Rao 2011). Among the biotic stress, the most devastating pathogens of the sugarcane are fungi which cause enormous economical losses to sugarcane.

Leaf binding disease of sugarcane was first observed in 1918 in Argentina (Abbott 1964) caused by *Myrigenospora aciculispore* V. and it was called *Myrigenospora* leaf binding disease or Tangle top disease. In India so far no report is available about the leaf binding disease, In the present study we identified the causal organism s characteristics and also attempted the control prospective of this disease by utilizing fungicides.

F. moniliforme also known as *Fusarium verticillioides* belongs to the subdivision *Deuteromycotina* of class *Sardariomycetes* and has earlier been reported to be associated with the wilt, *Pokkah boeng* and Knife knut disease in sugarcane in India (Vishwanathan and Rao 2011; Teher Khani et al. 2013).

A.K. Tiwari (✉) • K. Kumari • N. Mishra • B.L. Sharma
Central Lab, UP Council of Sugarcane Research, Shahjahnapur, Uttar Pradesh 242001, India
e-mail: ajju1985@gmail.com

S.P. Singh
UP Council of Sugarcane Research, Shahjahnapur, Uttar Pradesh 242001, India

A. Singh
Division of Plant Pathology, UP Council of Sugarcane Research, Shahjahnapur, Uttar Pradesh 242001, India

P. Kumar et al. (eds.), *Current Trends in Plant Disease Diagnostics and Management Practices*, Fungal Biology, DOI 10.1007/978-3-319-27312-9_7

Materials and Methods

Surveys were conducted during June to September 2011 and 2012 in sugarcane fields at Sugarcane Research Institute (SRI), Shahjahnapur, U.P., India. Three sugarcane varieties CoS 07250, CoS 98259 and CoSe 01434 showing leaf binding disease symptoms were collected and brought to laboratory for proper identification of the causal pathogen(s).

The small cut leaf bits were washed under running tap water for 5 min, surface sterilized with 0.1 % mercuric chloride for 1 min and finally washed with distilled water. The bits were placed on Oat Meal Agar (OMA) in Petri-dishes under aseptic conditions and maintained at 25 ± 2 °C temperature until the mycelial growth initiated. These cultures were further purified by a single spore isolation and regularly transferred at an interval of 10–12 days to the fresh OMA medium and incubated at 25 ± 2 °C until sporulation. The cultures were submitted to Culture Type Collection Repository, Agarkar Research Institute (ARI), Pune, India for its identification. The associated pathogen was also identified with the help of its morphological and cultural characteristics.

Fungal DNA was extracted from 0.5 g of fresh mycelial mat using a cetyltrimethylammonium bromide (CTAB) protocol (Saghai-Maroof et al. 1984), with a slight modification i.e., in place of lyophilized fungal mat, fresh mycelial mat was ground in liquid nitrogen. The DNA concentration was determined through agarose gel electrophoresis using known concentration of k-uncut DNA as standard. Equal amount of three such isolations were pooled and used as templates in the polymerase chain reaction (PCR).

PCR was carried out in 50 μl reaction volume containing 1 μl DNA, 1 μl of each primers, 25 μl PCR Master Mix (Takara, Japan) and rest water. Amplification was performed using an universal primer pairs ITS-1 and ITS-4 in a DNA Engine thermal cycler (BIOCHEM) with the conditions: initial denaturation at 94 °C for 5 min, denaturation at 94 °C for 30 s, annealing at 60 °C for 30 s, and extension at 72 °C for 40 s. A final extension was given at 72 °C for 10 min.

The PCR amplified products were resolved on 2 % (w/v) agarose gel containing 0.5 μg/ml ethidium bromide using 1X TBE buffer. Resolved products were visualized and photographed under UV light source using an alpha-imager system. The amplified products were eluted (PCR clean-up, Germany) and directly sequenced and consensus sequences were submitted in the GenBank. These sequences were used as query sequences in BLAST search.

Foliar spray of Copper oxychloride (@ 0.25 %), Mancozeb (@0.25 %) and Bavistin (@0.1 %) was applied twice during the grand growth period (July–September) at fortnightly interval to check the incidence of the disease in affected plots.

Results and Discussion

During the field survey at SRI, Shahjahanpur, UP, India three varieties (CoS 07250, CoS 98259, CoSe 01434) were found with the symptoms of mechanical binding with necrosis of leaves (Fig. 1). Early death of affected shoots and the dwarfing and fascination of affected clumps were also noticed. No symptoms were observed on the stalks of the affected sugarcane cultivars. The growth of the affected plants was

Fig. 1 Sugarcane leaves showing typical symptoms of mechanical binding (**a**, **b**), symptoms on standing crop (**c**)

Fig. 2 Cottony growth and pink discoloration of culture

checked due to the disease as compared to the non symptomatic plants standing in the same field. The occurrence of the disease was only observed during monsoon period (June last week to September) which is the grand growth period of sugarcane.

The incidence of the disease varied from 8 to 15 % on the varieties CoS 07250, CoS 98259 and CoSe 01434. It was higher in variety CoS 07250 which was around 15 %, however other two varieties viz.CoS 98259 and CoSe 01434 were found with the incidence of 8 and 11 %, respectively (incidence was calculated on clump basis). Culture of infected leaves of such symptomatic plants showed Cottony growth and pinkish colour appeared in Petri dishes on placing the symptomatic leaf on PDA 10 days after inoculation at 25 ± 2 °C (Fig. 2). In the microscopic examination, hyaline, septate and branched hyphae having abundance microconidia with large setae were observed (Fig. 3).

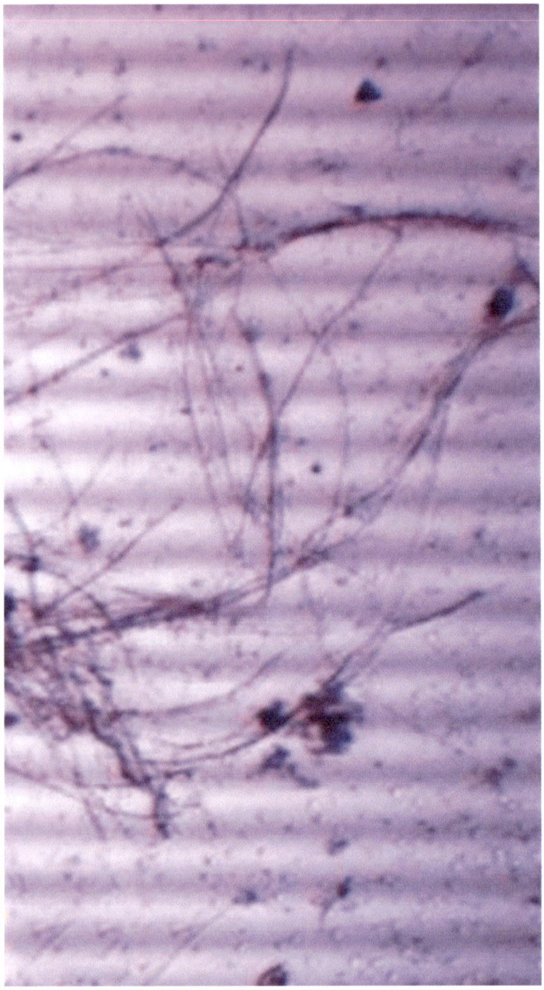

Fig. 3 Microscopic view of the isolated culture

The associated pathogen was confirmed as *F. moniliforme* in all the three collected symptomatic samples in all three test sugarcane varieties (CoS 07250, CoS 98259, CoSe 01434). One such culture was submitted at Type Culture Collection facility at ARI, Pune, India with the accession no NFCCI 3093.

In PCR analysis ~0.6 kbp amplicon was observed in all three symptomatic leaf samples (culture maintained in laboratory). All the three amplicons were directly sequenced and were found 99 % identical to each other. One of the sequence from variety CoS 07250 was submitted in the GenBank with the accession number KM382420. BLASTn analysis of the sequence showed highest 97–98 % identity with several isolates of *Fusarium* sp. (JQ388287, JN254793, HQ631057, HQ630965, JQ885453), *F. proliferatum* (EU821492, EU821467, EU821478, EU272509, FN868470, EU821469), *F. fujikuroi* (KJ000437, KJ000444, KJ000433),

F. oxysporum (FJ867936, FJ466709), and *F. temperatum* (KC179827, KC179826). These, biological, microscopic and molecular detection strongly confirmed the association of *F. moniliforme* with the leaf binding disease of three susceptible varieties of sugarcane plants in India. Literature surveys revealed that there have been three published records on the occurrence of *Myriogenospora* leaf binding disease of sugarcane from Brazil (Vizioli 1926; Freise 1930) and Louisiana (Abbott and Tippett 1941). Affected plants were characterized by extreme stunting and the adherence of tips of the unfolding leaf to adjacent older leaves. The fascination might be appeared due to mechanical binding by fungus tissues of the adjacent leaves and clumps during the growth of the shoots. In affected plants, growing points of few shoots were killed resulting in the death of the entire shoots. Diehl (1934) suggested that leaf binding disease of sugarcane is probably a stage of Pokkah boeng disease which is caused by different fungus and similar type of disease was abundantly noticed by him in *Andropogan scopairus* plants from Louisiana but the disease was found unable to infect sugarcane plants.

About management practices, twice foliar spray of Bavistin during the grand growth period (July–September) at fortnightly interval (@0.1 % w/v suspension) was found to be most effective (100 %) in controlling the disease followed by Copper Oxychloride (84 %) and Mancozeb (80 %) fungicide.

Acknowledgments First author thanks to Department of Science & Technology, New Delhi, India for financial assistance.

References

Abbott EV (1964) *Myriogenospora* leaf binding. In: Hughes CG, Abbott EV, Wismer CA (eds) Sugar-cane disease of the world, vol 2. Elsevier Publishing Company, Amsterdam, pp 40–42

Abbott EV, Tippett RL (1941) *Myriogenospora* on sugar cane in Louisina. Phytopathology 31:564–566

Diehl WW (1934) The *Myriogenospora* disease of grasses. Phytopathology 24:677–681

Freise FW (1930) Cane disease and plagues in Brazil. Facts About Sugar 25:613–614

Saghai-Maroof MA, Soliman KM, Jorensen AR, Allard RW (1984) Ribosomal DNA spacer length polymorphism: In: Barley: mendelian inheritance, chromosomal location and population dynamics. Proc Nat Acad Sci USA 81(4):8014–8018

Teher Khani K, Alizadeh A, Farokhi Nejad R, Sharifi Tehrani A (2013) Pathogenicity of *fusarium proliferatum*, a new causal agent of pokkah boeng in Sugarcane. Proc Int Soc Sugar Cane Technol 28:1–6

Tiwari AK, Bharti YP, Mishra N, Tripathi S, Lal M, Sharma PK, Rao GP, Sharma ML (2010) Biotechnological approaches for improving sugarcane crop with special reference to disease resistance. Acta Phytopathol Hun 45:235–249

Vishwanathan R, Rao GP (2011) Disease scenario and management of major sugarcane diseases in India. Sugar Tech 13:336–353

Vizioli R (1926) Estudio preliminary sorbe un novo pyrenomyceto parasite de canna. Bol Agr Sao Paulo 27:60–69

Macrophomina phaseolina: The Most Destructive Soybean Fungal Pathogen of Global Concern

Vibha

Charcoal rot caused by the fungus, *Macrophomina phaseolina,* have emerged as serious concern for cultivation of soybean under climate change scenario worldwide. *Macrophomina phaseolina* causes huge annual losses to the crop and can survives in the soil mainly as microsclerotia for 2 years or longer and; germinate repeatedly during the crop-growing season. The pathogen generally attacks the young plants when their growth is retarded due to unfavourable conditions. Moreover, charcoal rot is usually most severe in older plants which have been subjected to stressful environmental conditions such as high temperature, drought, or poor fertility. The disease severity is directly related to the humidity, temperature, tillage practices and soil nutrient conditions. This review deals with the details of pathogen and its management approaches. The management of disease through stress management is the most viable solution to overcome the menace of it. Although, the fungicide is the means of disease prevention but cultural practices, irrigation management during drought and resistant cultivars are the most practical means of control as the pathogen have more than 500 plant species to inhabit. The possibilities in substantial yield reduction under present changing climate underscore the need for further research.

Introduction

Charcoal rot, caused by the fungus *Macrophomina phaseolina* (Tassi) Goidanich, is a cosmopolitan soil saprophyte and is well known as a facultative, opportunistic plant pathogen that infects plants exposed to certain stress conditions (Tesso et al. 2005). It is ranked among the five top most important soybean diseases, causing

Vibha
Department of Plant Pathology, JNKVV, Jabalpur, MP, India
e-mail: vibhapandey93@gmail.com

P. Kumar et al. (eds.), *Current Trends in Plant Disease Diagnostics and Management Practices*, Fungal Biology, DOI 10.1007/978-3-319-27312-9_8

huge annual losses (Wrather et al. 1997, 2001). *M. phaseolina* (Tassi) Goidanich, is one of the most important soil borne pathogens, infecting over 500 plant species in more than 100 plant families around the world (Smith and Wyllie 1999). It can survive as microsclerotia (masses of fungal tissue) for 2 or more years in dry soil, but not more than 7–8 weeks in wet soils and mycelium not more than 7 days (Sinclair 1982). Being seed-borne (Kunwar et al. 1986) in nature, it is found both on the seed coat and cotyledons (Reuveni et al. 1983) and causes charcoal rot by infecting the roots due to the adherence of microsclerotia to the seed coat during germination and emergence (De Mooy and Burke 1990). Positive correlations have been reported between the inoculum level of *M. phaseolina* in the seedbed and disease severity (Khan 2007). Temperatures near 30 °C and dry conditions make this pathogen prevalent in regions with arid subtropical and tropical climates such as in Pakistan (Khan 2007), China (Xiaojian et al. 1988) and India (Suriandraselvan et al. 2006) where yield losses caused by this fungus can reach even 90 % of yield.

Owing to higher variability among the isolates of this pathogen, no commercial resistant soybean variety is yet available for effective management of this disease. Therefore, reducing drought stress during the reproductive stages of growth of soybean plants can help in minimizing the risk from charcoal rot. This can be done by following production systems like no-till that conserve soil moisture, maintaining proper plant populations, growing drought tolerant varieties and maintaining soil fertility. Fields with a history of severe charcoal rot should be rotated for 1–2 years with non-host crops (cereals). Fungal propagules exposed to energy stress, lose endogenous C by respiration and exudation resulting in energy (nutrient) stress, with demand for nutrients during germination, viability loss and decreased pathogenic aggressiveness (Mondal and Hyakumachi 1998). In addition, the beneficial bacterial live in rhizosphere (i.e., the region around the root) which is rich in nutrients due to the exudation of plant nutrients from the roots can influence the plant bi-directionally. One direct influence may be stimulation of plant growth and other plant health promotion (i.e. indirect influence). Hence, biological control of plant pathogens and deleterious microbes occurs through the production of antibiotics, lytic enzymes, hydrogen cyanide and siderophore or through mycoparasitism, competition for nutrients and space by bioagents that results in plant health promotion significantly. Soil application of biocontrol agents' viz., *Trichoderma viride, T. harzianum, Pseudomonas fluorescens* and *Bacillus subtilis* effectively reduced root rot caused by soil borne pathogens in several crops (Thilgavathi et al. 2007; Loganathan et al. 2010).

Pathogen

The binomial nomenclature of *M. phaseolina* is applied to both the microsclerotial and pycnidial anamorphs, however the microsclerotial phase is the one predominantly observed worldwide (Dhingra and Sinclair 1978). Different synonyms have been ascribed to the fungus *M. phaseolina* (Tassi) that includes *M. phaseoli* (Maubl.) Ashby, *Macrophoma conchoci* Swada, *Sclerotium bataticola* Taub. and *Rhizoctonia bataticola* (Taub.) (Mihail 1992). The lack of a known teleomorph has stalled its taxonomy over the years (Kulkarni and Patil 1966; Crous et al. 2006); however, a thorough phylogenetic study of 113 members of the family Botryosphaeriaceae using ribosomal DNA sequences was able to separate the genera *Macrophomina* and *Tiarosporella* (Crous et al. 2006). Although, only one species is recognized within the genus (Mihail and Taylor 1995), great variability in morphology and pathogenicity was recognized among isolates from different host species and between isolates from different parts of the same plant (Fernandez et al. 2006). Efforts were also been made to characterize the fungus population in different parts of the world based on its pathogenic variability (Karunanithi et al. 1999), morphological characteristics (Fernandez et al. 2006), as well as the molecular characteristics (Almeida et al. 2003; Jana et al. 2003; Purkayastha et al. 2006). The unstable B chromosome may be one of the mechanisms for generating variation in fungi (Miao et al. 1991). In addition, the mature hyaline and pigmented hyphal cells of *Macrophomina* are uninucleate, but young, growing hyphal cells and hyphal tip cells are usually multinucleate (Knox- Davies 1967). Hyphal fusion heterokaryosis after mitotic segregation and recombination may explain the occurrence of cultural types or physiological races (Punithalingam 1983). Double-stranded RNA (dsRNA) has also been reported in *M. phaseolina* with sizes ranging from 0.4 to 10 kbp and the number of dsRNA ranging from 1 to 10 (Pecina et al. 2000). Variations exist in pathogenicity among *M. phaseolina* from different geographical regions (Dhingra and Sinclair 1973). The phytotoxin produced in cultures of *M. phaseolina* is Botryodiplodin. Phaseolinone was not detected, which suggested that botryodiplodin may be the phytotoxin that facilitates infection (Ramezani et al. 2007). Moreover, the host species-specific conservation of a family of repeated DNA sequences in the genome of a fungal plant pathogen is a possible mechanism (Hamer et al. 1989) for isolates distinctness in this pathogen. Jana and coworkers (2005) reported the use of microsatellites markers as potential diagnostic markers for the study of the variability within closely related isolates of *M. phaseolina* population specific to soybean and cotton.

Symptoms

Charcoal rot can infect soybeans at any growth stage; however, the worst infection is typically seen during the reproductive phase. *M. phaseolina* overwinters as sclerotia in the soil and infected plant debris and can remain viable for several years (at least 2 years). Under favorable conditions (e.g., higher soil temperatures and low water potential), the sclerotia germinate and colonize the plants (Olaya et al. 1996). *M. phaseolina* can grow rapidly in infected plants and produce large amount of sclerotia that clog the vascular tissue, resulting in disease symptoms ranging from leaf yellowing, wilting to plant death (Wyllie 1989). Charcoal rot symptoms usually appear under high temperature conditions (28–35 °C) and low soil moisture, or when unfavourable environmental circumstances stress the plant (Wyllie 1988; Sinclair and Backman 1989). Although initial infection occurs at the seedling stage, it usually remains latent until the soybean plant approaches maturity (growth stages R5–R7) (Short et al. 1978). Diseased plants may wilt and prematurely die with senesced leaves remaining attached to petioles. Seed yield is frequently reduced under these conditions. The diagnostic symptoms of charcoal rot on prematurely dying or dead plants are the sloughing of cortical tissues from the lower stem and tap root and the speckled grey appearance of these infected tissues due to abundant formation of microsclerotia in vascular, cortical, and pith tissues (Smith and Carvil 1997). Other soybean diseases such as Sudden Death Syndrome (SDS), Brown Stem Rot (BSR) or stem canker may cause these symptoms as well. The distinguishing characteristic of charcoal rot is black speckling within the lower stem from microsclerotia. These black specks look like charcoal briquettes, thus the name charcoal rot. Additionally, reddish-brown to black streaks form in the vascular tissue as well. Charcoal rot is a root and stem disease that commonly occurs in hot, dry weather conditions. Therefore, symptoms of charcoal rot are also referred as dry-weather wilt or summer wilt, because it often occurs when plants are under heat and drought stresses (Smith and Wyllie 1999). These stresses can also occur in irrigated soybeans causing losses from 6 % to 33 % in experimental plots (Mengistu et al. 2011) and the combination of stress and the presence of *M. phaseolina* cause higher yield loss on soybeans than drought alone. This disease is most severe when plants are stressed from lack of moisture or nutrients, at excessive plant populations or where soil compaction, other diseases or nematodes or improperly applied pesticides impair root development. Charcoal rot symptoms typically appear when soybeans approach maturity. The earliest symptoms are smaller than normal sized leaves, which become chlorotic, then turn brown, but remain attached to the petiole giving the entire plant a dull greenish-yellow appearance. In many cases, these plants wilt and die. The pathogen attacks the plant throughout the season, often causing progressive debilitation of the host. After flowering, a light gray or silvery discoloration of the epidermal and sub-epidermal tissues develops in the taproot and the lower part of the stem. The best diagnostic symptom is found when the epidermis is peeled away from the stem exposing numerous small, black bodies of microsclerotia that are frequently produced in the xylem and pith of the stem and may block water flow.

Infection Process

The first reports of Mp (*M. phaseolina*) infection process in soybean were made by Ammon et al. (1974, 1975), which were based on scanning electron microscopy analyses. They suggested that penetration through soybean cell walls occurred as a result of mechanical pressure and/or chemical softening. Ilyas and Sinclair (1974) described the formation of intra- xylem sclerotia in wound-inoculated soybean plants, lacking the characterization of the initial penetration stages. Inside host tissues, Mp develops thin hyaline walls, which are presumed to be more permeable with increased potential for resource exchange with the host (Barrow and Aaltonen 2001; Barrow 2003). The development of Mp structures with swelled and pigmented walls, produced either inter- and intracellularly. These structures, similar to appressoria, were previously described as hyphopodia by Howard (1997). He described appressoria as structures that develop from swellings at the tips of conidial germ tubes and hyphopodia as structures that arise from mature vegetative hyphae. Hyphopodia have been defined as structures that allow the spreading of the fungus after infection of the plant, and they might enhance penetration or survival (Howard 1997; Solomon et al. 2006). Hyphopodia generally are melanized and deposition of melanin, for instance, in the fungal cell wall of appressoria is associated with the generation of intracellular turgor pressure that provides the necessary force for plant penetration (Money et al. 1998). The fungus has been shown to infect cotyledons, roots or stems either pre-emergence or post-emergence stages and microsclerotia form appressoria over host epidermal cells. The developing hyphae enter and grow between the epidermal cells inter- and intracellularly, and attack cells by mechanical or enzymatic action. However, the intracellular colonization occurs after lamella and cell wall disintegration (Ammon et al. 1974). Following epidermal and cortex invasion, *M. phaseolina* colonizes the vascular system developing microsclerotia on xylem vessels which may lead to their blocking causes wilt symptoms in soybean and other hosts (Ilyas and Sinclair 1974).

Epidemiology

The mycelium in the soil is not considered to be a primary source of inoculum (Meyer et al. 1974), however, the sclerotia serve as the prime sources of inocula (Papavizas and Klag 1975) for disease initiation. The occurrence of sclerotia in plant debris allows the fungus to live in soil, even in the absence of a host for 2 or more years, depending on soil conditions (Wantanabe et al. 1970). Seed, soil and plant remains are the sources of primary inoculum (Reuveni et al. 1983) and the severity of the disease is directly related to the number of live sclerotia in the soil. Under dry soil conditions, the fungus can remain viable as sclerotia for more than 10 months. Pathogenicity is optimal between 28 and 35 °C (Dhingra and Sinclair 1978) and host water stress is another principal factor favouring development of the disease (Pearson et al. 1984; Mayek-Perez et al. 2002). In addition, charcoal rot incidence is much higher when plants are exposed to prolonged drought and high temperature stress during grain development (Tesso et al. 2005). Mechanical injury, high plant density and insect attacks are considered to be predisposing factors for transmission of the disease (Ahmad et al. 1991). The severity of infection depends on relative humidity, temperature, the nature of the isolate, climatic region and host cultivar. In some agricultural systems in which soil is generally low in easily available nutrients and consequently poor in microbial biomass, activity and diversity, such as those systems under conventional tillage, the suppression of soil borne plant pathogens is more difficult to attain (Vargas Gil et al. 2008). Almeida and co-workers (2003) reported higher densities of microesclerotia in soybean roots, in plots under conventional tillage, and stated that tillage has also been considered an important factor in the spread of fungal propagules in soil. Moreover, according to those authors, high temperature has also been mentioned as a factor that predisposes plants to infection by *M. phaseolina*, low soil moisture being the most important factor for infection. It is well known that no-till systems are cooler than conventional ones, mainly due to the crop residue layer on the soil surface. Direct seeded systems do not provide suitable soil conditions for the spread of the pathogen and reduce the stress conditions of the plants. Accordingly, a significant negative correlation between charcoal rot and water holding capacity but a positive correlation between the disease and sand content, which is reasonably considered that sandy soils usually retains less water than silty or clayey ones (Perez-Brandán et al. 2012). Charcoal rot undergoes rapid development under strong water content depletion (Pedgaonkar and Mayee 1990), therefore, cultivars that show reduced water depletion rates and a stable cellular turgor are resistant to charcoal rot (Mayek-Perez et al. 2002). Besides, the pathogen specialization to the host also seems to be related to stem nitrogen composition, and is promoted at low water availability (Pearson et al. 1987). Infection by nematodes can provide a favourable substrate for the development of the fungus by disrupting and damaging the vascular tissues and bringing physiological changes, and therefore increasing the severity of charcoal rot. Ross (1965) documented the interaction of *Heterodera glycines* Inchinohe (soybean cyst nematode) and *M. phaseolina* separately in disease complexes and explained that disruption of vascular tissues resulting from infection by *H. glycines* increased the susceptibility of the host to water stress. Stress-related nitrogenous compounds such as asparagines and prolines are utilized efficiently by *M. phaseolina* and this could explain the positive correlation between *H. glycines* and population of fungus (Pearson et al. 1987) and as a result of interaction with fungi, the populations of sedentary nematodes are suppressed (Powell 1971). Two season soybeans crop or late planting may add greater severity to charcoal rot.

Management

Plant diseases are considered as an important biotic constraint, where an interaction between host, pathogen and the environment occur and leads to significant crop losses worldwide. Most plants are immune or completely resistant to almost all pathogens. However, owing to co-evolution of host and pathogen, pathogens overcome the natural resistance of particular hosts through mechanism of specialization under favourable environmental conditions. Therefore, the success of any disease management strategy should focus on the host, the pathogen and/or the environment. Integrated disease management (IDM), which combines crop improvement, biological, cultural, physical and chemical control strategies in a holistic way rather than using a single component strategy proved to be more effective and sustainable. Hence, an 'Integrated Disease Management' approach can be helpful in selection and application of a harmonious range of control strategies that would minimize losses and maximize returns. Fields with a history of severe charcoal rot should be rotated for 1–2 years with non-host cereal crops.

Efforts to manage charcoal rot in soybean through adjusting planting dates, crop rotation, planting densities, and irrigation have all been suggested as means of control (Mengistu et al. 2007) as no commercial resistant soybean variety is yet available for effective management of this disease. Managing the population of microsclerotia in the soil is the primary management strategy. Avoiding excessive seed rates and maintaining adequate soil fertility reduces loss from the diseases by maintaining healthy and vigorous plants ecosystem. The best way to avoid issues with charcoal rot is to limit drought stress during the reproductive stages of growth by managing production systems like no-till that conserve soil moisture may also reduce losses by charcoal rot. Planting corn for 3 or more years can decrease disease pressure followed by a yearly rotation to keep populations low. Charcoal rot exhibited a negative and significant relationship with soil organic matter, total N, K and Ca that suggests the soil systems with high levels of biological diversity and activity, and with high internal nutrient cycling, such as no-tillage systems, allow the development of plants with healthier root systems and can avoid the infection by a soil borne pathogen, because this system becomes more resilient to disturbance than conventional tillage systems (Perez-Brandán 2012).

Deshpande and Murumkar (2008) found a reduction in microbial growth and abundance, and at the same time an increase of the pathogen *M. phaseolina*, which resulted in an increase of root rot in sorghum. High microbial diversity agricultural soils have been associated with suppression of soil-borne plant diseases, and this kind of suppression may be due to general competition or antagonism, which may be non-specific and active against a wide range of soil-borne pathogens (van Bruggen et al. 2006). Patil and Kamble (2011) examined the effect of UV light on the hostile/antagonistic action of *Trichoderma koningii* against *M. phaseolina*, using five *T. koningii* mutants, and found that *T. koningii 2* showed maximum antagonistic activity against the charcoal rot pathogen when tested by dual culture method. Seed treatment with *P. flourescens* along with soil amendment like mustard cake, vermicompost and FYM provided a better protection against *Macrophomina* root rot of chickpea (Khan and Gangopadhyay 2008). Similarly, soil application of $ZnSO_4$ followed by combined application of *T. viride* + $ZnSO_4$ significantly reduced root rot incidence (Sundaravadana 2002). Almeida and co-workers (2003) stress that alternative control practices of charcoal rot could be the modification of the soil environment, which would favour antagonists interfering with the biology or survival of the pathogen. The analysis of the fungal and bacterial sequences detected in DS (Direct seeded) treatment showed that the most frequently found fungi are effective biological control agents of plant pathogens (Perez-Brandán et al. 2012). *Plectosphaerella cucumerina* and *Paecilomyces marquandii* are nematophagous fungi (Atkins et al. 2003), and *Bionectria ochroleuca* is a mycoparasite (Chaverri et al. 2011) were recorded under no-till soil. In addition, the most frequent bacterial clone detected is also related to plant protection, such as *Bradyrhizobium* sp., inducing effective systemic resistance and protecting the host plant against pathogen attacks (Cartieaux et al. 2008). Studies in various pathosystems indicate that auxin signalling is required for host resistance against some necrotrophs, whereas for pathogenic bacteria and biotrophic and hemibitrophic fungi, auxin signalling promotes susceptibility (Karzan and Manners 2009). Auxin is a plant hormone that is involved in many aspects of plant development, and the cross talk between auxin with other plant hormones such as JA and SA is important for balancing plant growth versus defense (Wang et al. 2007; Bari and Jones 2009). These hormones trigger the activation of induced systemic resistance and systemic acquired resistance (SAR) to necrotrophic pathogens (Fey and Parker 2000; Glazebrooki 2005). The SAR is an effective defense mechanism against a broad range of pathogens and insects incurred by host. Genes involved in SA response such as hydroxyl-2-methyl-2(E) butenyl 4-diphosphate. HopW1-1- Interacting protein 1 (WIN1) were identified (Lee et al. 2008). The SA pathway, which is considered one of the major pathways involved in defense against necrotrophic pathogens, regulates the expression of defense defector genes and systemic acquired resistance through the repression of the auxin signaling pathway (Gill et al. 2005). Another hormone that seems to play a role in the resistance of stem rot is abscissa acid (ABA). While ABA was described as a susceptibility factor, other studies (Wiese et al. 2004) showed that it activates plant defense by priming for callose deposition or by restricting the progression of the fungus *Cochliobolus miyabeanus* in the mesophyll of rice (De Vleesschauwer et al. 2010). Other signaling genes involved in SAR that induce numerous defense genes included apoplectic lipid transfer protein, basic chitinase etc. (Zander et al. 2010). The third category of genes with stem rot tissues includes genes involved in early response as part of the HR. Among these are transcripts encoding proteins such as ATPase transporter, kinases, carbonic anhydrase, AMMECR1, MIPS1, voltage-dependent anion channel, 2-deoxy-D-arabinoheptulosonate 7-phosonate (DAHP) synthase and glutathione peroxidase that were reported previously to be involved in the hypersensitivity resistance (HR) and cell death in plants under pathogenic attack (La Camera et al. 2009). Reactive oxygen species (ROS) seems to be induced following *M. phaseolina* infection as several genes involved in oxidative stress (alpha-dioxygenase, fumarase, cytosolic GADPH (C subunit), cytosolic ascorbate peroxidase APX1) had more abundant transcripts. Furthermore, several pathogenesis related (PR) genes such as elicitor activated gene 3-1 (EL13), aromatic alcohol: NADP+ oxidoreductase, thaumain, pathogenesis-related and antifungal chitin binding protein had differentially abundant transcripts in diseased versus healthy tissues (Biswas et al. 2014). PR proteins, of which some have antimicrobial functions (Sels et al. 2008) are mainly induced in localized pathogen attack around HR lesion.

Sinclair (1989) examined the effect of thermotherapy on the growth of seed-borne fungi in soybean by immersing infected seeds in heated palm, sunflower and soybean oil as a means of eliminating seed-borne fungi. Glyphosate (N-[phosphonomethyl]glycine) application on glyphosate-resistant crops has been shown to enhance and in a few cases reduce severity (Johal and Huber 2009) of selected soybean diseases. Shahda et al. (1991) studied the in-vitro effect of certain fungicides such as Benlate T, 2-(4-thiazolyl)-1H-benzimidazole (Thiabendazole), N trichtoromethylthio–cyclohexene-1,2-dicarboximide (Captan 75), 5,6-dihydro-2-methyl-N-phenyl-1, 4-oxathiin-3-carboxamide and tetramethylthiuram disulfide (Vitavax 200), 5,6-dihydro-2 methyl-Nphenyl-1,4-oxathiin-3-carboxamide and N-trichtoromethylthio– cyclohexene-1,2-dicarboximide (Vitavax 300), on mycelial growth of seed-borne fungi of sunflower and 5,6-dihydro-2-methyl-N-phenyl-1,4-oxathiin-3-carboxamide and tetramethylthiuram disulfide (Vitavax 200) was found to be most effective for *M. phaseolina.* Moreover, Brooker et al. (2007) screened six derivatives of coumarin, for their antifungal activity against *M. phaseolina* and *Pythium* species and observed that these derivatives have higher antifungal activities and stability as compared with either the original coumarin or sesamol compounds alone.

Conclusion

During the last five decades, extensive progress has been made by researchers in areas of etiology, epidemiology, biology and biocontrol of the ascomycete fungus *M. phaseolina*. Charcoal rot epidemics are common under stress conditions such as water scarcity and other biotic and abiotic stresses. The basic knowledge of the biology of *M. phaseolina* has provided the foundation for developing sustainable strategies to control the disease. Efforts are needed to develop a biocontrol technology for practical use in the management of charcoal rot diseases. In addition, it has become increasingly clear that among integrating several effective control methods, breeding and biological control methods, could be the best strategy for managing this important disease. Compared with major technological, environmental, and socioeconomic changes affecting agricultural production during next century, climate change may be more important; it will however, add another layer of complexity and uncertainty onto a system that is already exceeding difficulty to manage on a sustainable basis. Research on climate change and its interaction with pathogenically different isolates from different geographical regions of *M. phaseolina* could result in improved understanding and management of pathogen in face of current and future climate extremes.

References

Ahmad I, Burney K, Asad S (1991) Current status of sunflower diseases in Pakistan. In: Ghaffar A, Shahzad S (eds) Proceedings of the national symposium on status of plant pathology in Pakistan, 31 December 1991. Department of Botany, University of Karachi, Karachi, p 53

Almeida AMR, Amorim L, Bergamin A, Torres E, Farias JRB, Benato LC, Pinto MC, Valentim N (2003) Progress of soybean charcoal rot under tillage and no-tillage systems in Brazil. Fitopatol Bras 28:131–135

Ammon V, Wyllie TD, Brown MF Jr (1974) An ultra structural investigation of pathological alterations induced by *Macrophomina phaseolina* (Tassi)Goidin seedlings of soybean, *Glycine max (L.)*. Merrill Physiol Plant Pathol 4:1–4

Ammon V, Wyllie TD, Brown MF Jr (1975) Investigation of the infection process of *Macrophomina phaseolina* on the surface of soybean roots using scanning electron microscopy. Mycopathologia 55:77–81

Atkins SD, Clark IM, Sosnowska D, Hirsch PR, Kerry BR (2003) Detection and quantification of *Plectosphaerella cucumerina*, a potential biological control agent of potato cyst nematodes, by using conventional PCR, Real-Time PCR, selective media, and baiting. Appl Environ Microbiol 69(8):4788–4793

Bari R, Jones J (2009) Role of plant hormones in plant defense responses. Plant Mol Biol 69:473–488

Barrow JR (2003) A typical morphology of dark septate fungal root endophytes of Bouteloua in arid south western USA range lands. Mycorrhiza 13:239–247

Barrow JR, Aaltonen RE (2001) Evaluation of the internal colonization of *Atriplex canescens* (Pursh) Nutt. roots by dark septate fungi and the influence of host physiological activity. Mycorrhiza 11:199–205

Biswas C, Dey P, Karmakar PG, Satpathy S (2014) Next-generation sequencing and micro RNAs analysis reveal SA/ MeJA1/ ABA pathway genes mediated systemic acquired resistance (SAR) and its master regulation via production of phased, trans-acting siRNAs against stem rot pathogen *Macrophomina phaseolina* in RIL population of jute (*Corchorus capsularis*). Physiol Mol Plant Pathol 87:76–85

Brooker NL, Kuzimichev Y, Lass J, Pavlis R (2007) Evaluation of coumarin derivatives as anti fungal agents against soil borne fungal pathogens. Commun Agric Appl Biol Sci 72:785–793

Cartieaux F, Contesto C, Gallou A, Desbrosses G, Kopka J, Taconnat L, Renou JP, Touraine B (2008) Simultaneous interaction of *Arabidopsis thaliana* with *Bradyrhizobium* sp. strain ors278 and *Pseudomonas syringae* pv. Tomato dc3000 leads to complex transcriptome changes. Mol Plant Microbe Interact 21:244–259

Chaverri P, Salgado C, Hirooka Y, Rossman AY, Samuels GJ (2011) Delimitation of Neonectria and Cylindrocarpon (Nectriaceae, Hypocreales, Ascomycota) and related genera with Cylindrocarpon-like anamorphs. Stud Mycol 68:57–78

Crous PW, Slippers B, Wingfield MJ, Rheeder J, Marasas WFO, Philips AJL, Alves A, Burgess T, Barber P, Groenewald JL (2006) Phylogenetic lineages in the Botryosphaeriaceae. Stud Mycol 55:235–253

De Mooy CJ, Burke DW (1990) External infection mechanism of hypocotyls and cotyledons of cowpea seedling by *Macrophomina phaseolina*. Plant Dis 74:720

De Vleesschauwer D, Yang Y, Cruz CV, Hofte M (2010) Abscisic acid-induced resistance against the brown spot pathogen *Cochlibolus miyabeanus* in rice involves MAP kinase-mediated repression of ethylene signaling. Plant Physol 154:2036–2052

Deshpande AN, Murumkar DR (2008) Effect of tillage practice and nutrient source on microorganism counts in the rhizosphere of winter sorghum (*Sorgum bicolor*). Indian J Agron Sci 78:470–472

Dhingra OD, Sinclair JB (1973) Variation among isolates of *Macrophomina phaseolina* (*Rhizoctonia bataticola*) from different regions. J Phytopathol 76:200–204

Dhingra OD, Sinclair JB (1978) Biology and pathology of *Macrophomina phaseolina*. Imprensa Universitária, Universidade Federal de Viçosa, Viçosa

Fernandez RB, De Santiago A, Delgado SH, Perez NM (2006) Characterization of Mexican and non-Mexican isolates of *Macrophomina phaseolina* based on morphological characteristics, pathogenicity on bean seeds and endoglucanase gene. J Plant Pathol 88:1

Fey BJ, Parker JE (2000) interplay of signaling pathways in plant disease resistance. Trends Genet 16:449–455

Gill MJ, Mauch-Mani B, Jorda L, Vera P (2005) The Arabidopsis csb3 mutant reveals a regulatory link between salicylic acid- mediated disease resistance and the methyl-erythriol-4-phosphate pathway. Plant J 44:155–166

Glazebrooki J (2005) Contrasting mechanism of defense against biotropic and necrotrophic pathogens. Annu Rev Phytopathol 43:205–227

Hamer JE, Farral L, Orbach MJ, Valent B, Chumley FG (1989) Host species-specific conservation of a family of repeated DNA sequences in the genome of a fungal plant pathogen. Proc Natl Acad Sci U S A 86:9981–9985

Howard RJ (1997) Breaching the outer barriers: cuticle and cell wall penetration. In: Carroll GC, Tudzynski P (eds) Themycota. Plant relationships, vol 5, Part A. Springer, Berlin, pp 43–60

Ilyas MB, Sinclair J (1974) Effects of plant age upon development of necrosis and occurrence of intraxylem sclerotiain soybean infected with *Macrophomina phaseolina*. Phytopathology 64:156–157

Jana TK, Sharma TR, Prasad RD, Arora DK (2003) Molecular characterization of *Macrophomina phaseolina* and Fusarium species by using single primer RAPD technique. Microbiol Res 158:249–257

Jana T, Sharma TR, Singh NK (2005) SSR-based detection of genetic variability in the charcoal root rot pathogen *Macrophomina phaseolina*. Mycol Res 109(1):81–86

Johal GS, Huber DM (2009) Glyphosate effects on diseases of plants. Eur J Agron 31:144–152

Karunanithi K, Muthusamy M, Seetharaman K (1999) Cultural and pathogenic variability among the isolates of *Macrophomina phaseolina* causing root rot of sesame. Plant Dis 14:113–117

Kazan K, Manners JM (2009) Linking development to defense: auxin in plantpathogen interactions. Trends Plant Sci 14:373–382

Khan SN (2007) *Macrophomina phaseolina* as causal agent for charcoal rot of sunflower. Mycopathology 5:111–118

Khan MA, Gangopadhyay S (2008) Efficacy of *Pseudomonas fluorescens* in controlling root rot of chick pea caused by *Macrophomina phaseolina*. J Mycol Plant Pathol 38:580–587

Knox-Davies PS (1967) Mitosis and aneuploidy in the vegetative hyphae of *Macrophomina phaesoli*. Am J Bot 54:1290–1295

Kulkarni NB, Patil BC (1966) Taxonomy and discussion on the nomenclature of *Macrophomina phaseoli* (Maubl.) ashby and its isolates from India. Mycopathology 28:257–264

Kunwar IK, Singh T, Machado CC, Sinclair JB (1986) Histopathology of Soybean seed and seedling infection by Macrophomina phaseolina. Phytopathology 76:532–535

La Camera S, Balague C, Gobel C, Geoffroy P, Legrand M, Feussner I (2009) The Arabidopsis paratin-like protein 2(PLP2) plays an essential role in cell death execution and differentially affects biosynthesis of oxylipins and resistance to pathogens. Mol Plant Microbe Interact 22:469–481

Lee MW, Jelenska J, Greenberg JT (2008) Arabidopsis protein important for modulating defense response to *Pseudomonas syringe* HopW1-1. Plant J 54:452–465

Loganathan M, Sible GV, Maruthasalam S, Saravanakumar D, Raguchander T, Sivakumar R, Samiyappan R (2010) *Trichoderma* and chitin mixture based bioformulation for the management of head rot (*Sclerotinia sclerotiorum* (Lip.) deBary)-root knot (*Meloidogyne incognita* Kofoid and White) chitwood complex disease of cabbage. Arch Phytopathol Plant Proc 43:1011–1024

Mayek-Perez N, Gracia-Espiosa R, Lopez-Castaneda C, Acosta-Gallegis JA, Simpson J (2002) Water relations, histopathology and growth of common bean (*Phaseolus vulgaris* L.) during pathogenesis of *Macrophomina phaseolina* under drought stress. Physiol Mol Plant Pathol 60:185–195

Mengistu A, Ray JD, Smith JR, Paris RL (2007) Charcoal rot disease assessment of soybean genotypes using a colony forming unit index. Crop Sci 47:2453–2456

Mengistu A, Smith JR, Ray JD (2011) Seasonal progress of charcoal rot and its impact on soybean productivity. Plant Dis 95:1159–1166

Meyer WA, Sinclair JB, Khare MN (1974) Factors affecting charcoal rot of soybean seedlings. Phytopathology 64:845–849

Miao VP, Covertand SF, Vanetten HD (1991) Fungal gene for antibiotic resistance on a dispensable ('B') chromosome. Science 254:1773–1776

Mihail JD (1992) *Macrophomina*. In: Singleton LS, Mihail JD, Rush CM (eds) Methods for research on soil-borne phytopathogenic fungi. Am Phytopathol Soc Press, St. Paul, pp 134

Mihail JD, Taylor ST (1995) Interpreting variability among isolates of *Macrophomina phaseolina* in pathogenicity, pycnidium production and chlorate utilization. Can J Bot 73:1596–1603

Mondal SN, Hyakumachi M (1998) Carbon loss and germinability of chlamydospores of *Fusarium solani* f. sp. *phaseoli* after exposure to soil at different pH levels, temperatures and matric potentials. Phytopathology 88:148–155

Money NP, Frederick B, Henson JM (1998) Melanin synthesis is associated with changes in hyphopodial turgor, permeability, and wall rigidity in *Gaeumannomycesg raminis var. graminis*. Fungal Genet Biol 24:240–251

Olaya G, Abawi GS, Barnard J (1996) Influence of water potential on survival of sclerotia in soil and on colonization of bean stem segments by *Macrophomina phaseolina*. Plant Dis 80:1351–1354

Papavizas GC, Klag NG (1975) Isolation and quantitative determination of *Macrophomina phaseolina* from soil. Phytopathology 65:182–187

Patil VB, Kamble SS (2011) The influence of ultraviolet light on antagonistic activity of Trichoderma koningii against *Macrophomina phaseolina* causing charcoal rot of Charcoal rot on sunflower 5 sweet potato. Int J Acad Res 3:702–704

Pearson CAS, Schwenk FW, Crowe FJ, Kelly K (1984) Colonization of soybean roots by *Macrophomina phaseolina*. Plant Dis 68:1086–1088

Pearson CAS, Leslie JF, Schwenk FW (1987) Host preference correlated with chlorate resistance in *Macrophomina phaseolina*. Plant Dis 71:828–831

Pecina V, Alvarado MJ, Alanis HW, Almaraz RT, Vandemark GJ (2000) Detection of double-stranded RNA in *Macrophomina phaseolina*. Mycologia 92:900–907

Pedgaonkar SM, Mayee CD (1990) Stalk water potential in relation to charcoal rot of Sorghum. Indian Phytopathol 43:192–196

Perez-Brandán C, Arzeno JL, Huidobro J, Grümberg B, Conforto C, Hilton S, Bending GD, Meriles JM, Vargas-Gil S (2012) Long-term effect of tillage systems on soil microbiological, chemical and physical parameters and the incidence of charcoal rot by *Macrophomina phaseolina* (Tassi)Goid in soybean. Crop Prot 40:73–82

Powell NT (1971) Interactions between nematodes and fungi in disease complexes. Annu Rev Phytopathol 9:253–274

Punithalingam E (1983) The nuclei of *Macrophomina phaseolina* (Tassi) Goid. Nova Hedwig 38:339–367

Purkayastha S, Kaur B, Dilbaghi N, Chaudhury A (2006) Characterization of *Macrophomina phaseolina*, the charcoal rot pathogen of cluster bean, using conventional techniques and PCRbased molecular markers. Plant Pathol 55:106–116

Ramezani M, Shier WT, Abbas HK, Tonos JL, Baird RE, Sciumbato GL (2007) Soybean charcoal rot disease fungus *Macrophomina phaseolina* in Mississippi produces the phytotoxin (–)-botryodiplodin but no detectable phaseolinone. J Nat Prod 70:128–129

Reuveni R, Nachmias A, Krikun J (1983) The role of seedborne inoculum on the development of Macrophomina phaseolina on melon. Plant Dis 67:280–281

Ross JP (1965) Predispositions of soybeans to Fusarium wilt by *Heterodera glycines* and *Meloidogyne incognita*. Phytopathology 55:361–364

Sels J, Mathys J, De Coninck BM, Cammue BP, De Bolle MF (2008) Plant pathogenesis related (PR) proteins: a focus on PR peptide. Plant Physiology and Biochemistry 46(11):941–950

Shahda WT, Tarabeih AM, Michail SH, Hemeda AAH (1991) Fungi associated with sunflower seeds in Egypt with reference to chemical control measures. J King Saud Univ Agric Sci 3:287–293

Short GE, Wyllie TD, Ammon VD (1978) Quantitative enumeration of *Macrophomina phaseolina* in soybean tissues. Phytopathology 68:736–741

Sinclair JB (1982) Compendium of Soybean disease. 2nd ed. by American Phytopathology Society, St. Paul, Minnesota, USA.

Sinclair JB (1989) Vegetable oil thermotherapy for soybean seeds. In: Raychaudhri SP, Verma JP (eds) Review of tropical plant pathology: diseases of fiber and oilseed crops, vol 5. Today and Tomorrow's Printers and Publishers, New Delhi

Sinclair JB, Backman PA (1989) In: Sinclair JB, Backman PA (eds) Compendium of soybean diseases, 3rd edn. APS Press, St. Paul, p 106, Am Phytopathol Soc, EE. UU

Smith GS, Carvil ON (1997) Field screening of commercial and experimental soybean cultivars for their reaction to *Macrophomina phaseolina*. Plant Dis 81:363–368

Smith GS, Wyllie TD (1999) Charcoal rot. In: Hartman GL, Sinclair JB, Rupe JC (eds) Compendium of soybean disease, 4th edn. American Phytopathological Society, St. Paul, pp 29–31

Solomon PS, Wilson TJG, Rybak K, Parker K, Lowe RGT, Oliver RP (2006) Structural characterization of the interaction between *Triticum aestivum* and the dothideomycete pathogen *Stagono sporanodorum*. Eur J Plant Pathol 114:275–282

Suriandraselvan M, Salalrajan F, Aiyyanathan KEA (2006) Relationship between morphological variations and virulence in the isolates of *Macrophomina phaseolina* causing charcoal rot of sunflower. Madras Agric J 93(1–6):63–67.

Sundaravadana S (2002) Management of blackgram (*Vigna mungo* (L.) Hepper) root rot *Macrophomina phaseolina* (Tassi) Goid with bioagents and nutrients. M. Sc. (Ag.) Thesis, Tamil Nadu Agriculture University, Coimbatore

Tesso TT, Claflin LE, Tuinstra MR (2005) Analyses of stalk rot resistance and genetic diversity among drought tolerant sorghum genotypes. Crop Sci 45:645–652

Thilgavathi R, Saravanakumar D, Ragupathy N, Samiyappan R (2007) Integration of biocontrol agents for the management of dry root rot (*Macrophomina phaseolina*) disease in greengram. Phytopathol Mediterr 46:157–167

Van Bruggen AHC, Semenov AM, Van Diepeningen AD, de Vos OJ, Blok WJ (2006) Relation between soil health, wave-like fluctuations in microbial populations, and soil-borne plant disease management. Eur J Plant Pathol 115:105–122

Vargas Gil S, Pedelini R, Oddino C, Zuza M, Marinelli A, March GJ (2008) The role of potential biocontrol agents in the management of peanut root rot in Argentina. J Plant Pathol 90:35–41

Wang D, Pajerowska-Mukhtar K, Culler AH, Dong XN (2007) Salicylic acid inhibits pathogen growth in plants through repression of the auxin signaling pathway. Curr Biol 17:1784–1790

Wantanabe T, Smith RS, Snyder WC (1970) Population of *Macrophomina phaseoli* [sic] in soil as affected by fumigation and cropping. Phytopathology 60:1717–1719

Wiese J, Kranz T, Schubert S (2004) Induction of pathogen resistance in barley by abiotic stress. Plant Biol 6:529–536

Wrather JA, Anderson TR, Arsyad DM, Gai J, Ploper LD, Porta-Puglia A, Ram HH, Yorinori JT (1997) Soybean disease loss estimates for the top 10 soybean producing countries in 1994. Plant Dis 81:107–110

Wrather JA, Anderson TR, Arsyad DM, Tan Y, Ploper LD, Porta-Puglia A, Ram HH, Yorinori JT (2001) Soybean disease loss estimates for the top 10 soybean producing countries in 1998. Can J Plant Pathol 23:115–121

Wyllie TD (1988) Charcoal rot of soybeans-current status. In: Wyllie TD, Scott DH (eds) Soybean diseases of the North Central Region. APS Press, St. Paul, pp 106–113

Wyllie TD (1989) Charcoal rot. In: Sinclair JB, Backman PA (eds) Compendium of soybean diseases, 3rd edn. APS Press, St. Paul, p 30

Xiaojian L, Liu LI, Baidnun O, Derong Z (1988) Geographical distribution of sunflower diseases in China. In: Proc 12th Int Sunfl Conf Novi Sad, Serbia, pp 16–20

Zander M, La Camera S, Lamorte O, Metraux JP, Gatz C (2010) *Arabdiopsis thaliana* class II TGA transcription factors are essential activator of jasmonic acid/ethylene-induced defense response. Plant J 61:200–210

Colletotrichum gloeosporioides: Pathogen of Anthracnose Disease in Mango (*Mangifera indica* L.)

Madhu Kamle and Pradeep Kumar

Introduction

The mango (*Mangifera indica* L.) world-wide well known for its excellent exotic flavor and commonly referred as the "King of fruits." Mango is a dicotyledonous fruit tree belongs to order sapindales in the family *Anacardiaceae.* It grows throughout the tropical and subtropical countries and regarded as one of the world's most important edible fruit crops. Mango fruit is popular in the international market for its strong aroma, intense peel coloration, delicious taste, excellent flavor, attractive fragrance, beautiful colour and high nutritional values (Tharanathan et al. 2006). The consumption of mangoes can provide significant amount of bioactive compounds with antioxidant activity. The top five prominent mango producing countries include India, China, Thailand, Pakistan and Mexico. India is the leading country in Mango production. Unfortunately, production of mango rigorously experience inconsistent fruit maturity, ripening variability and causes huge pre and post-harvest economic losses. Anthracnose disease caused by *Colletotrichum gloeosporioides* is the most serious biological constraint to mango production. It delivers generous losses to young shoots, flowers and fruits under favorable conditions with high humidity, frequent rains and temperature ranges from 24 to 32 °C. Anthracnose causes about 30–60 % damage and sometimes increased up to 100 % in fruit produced under wet or very humid conditions. It contributes

M. Kamle (✉)
Department of Dryland Agriculture and Biotechnology, Ben Gurion University of the Negev, Beersheva 84105, Israel
e-mail: madhu.kamle18@gmail.com

P. Kumar
Department of Biotechnology Engineering, Ben Gurion University of the Negev, Beersheva 84105, Israel

P. Kumar et al. (eds.), *Current Trends in Plant Disease Diagnostics and Management Practices*, Fungal Biology, DOI 10.1007/978-3-319-27312-9_9

significantly to pre-harvest and post-harvest losses in mango and other fruit crops such as cashew, pomegranate, guava, acid lime and papaya. It is also revealed commonly as Bird's eye disease (leaf spot), blossom blight or fruit rot (Prakash et al. 1996).

Colletotrichum gloeosporioides Penz.: The Pathogen

The ubiquitous filamentous fungus *Colletotrichum gloeosporioides* Penz. belongs to order melanoconiales. The pathogen is widely distributed and the anamorph stage (asexual stage of the pathogenic fungus) causes anthracnose disease in mango (Prakash et al. 1996; Fitzell and Peak 1984; Jefferies et al. 1990) and other major tropical and sub-tropical fruit crops (Prusky 1996). *Colletotrichum* exists in two stages, mostly observed in the vegetative (asexual) stage and the perfect (sexual) stage *Glomerulla cingulata* rarely identified for most species (Alahakoon et al. 1994). In general, the sexual stage is liable for the presence and absence of genetic variability and asexual anamorph stage is for fungal spore dispersal. Sexual reproduction is complex phenomenon in *Glomerella* than other filamentous ascomycete fungi.

Morphology

The morphology of the *Colletotrichum* described by (Palo 1932; Vaillancourt et al. 2000) the spore of *C. gloeosporioides* to be 8.3–27.4 μm in length and 2.0–6.6 μm in width (mean 14.2×4.4 μm). The acervuli is highly variable in size and upon maturation exude pink masses of conidia under moist conditions (Palo 1932). The acervuli measure 115–467×95–22 μm (Bose et al. 1973), 80–250 μm. The conidia are borne on distinct, well-developed hyaline conidiophores. The conidia are straight, cylindrical or oval, 8–20×5–7 μm hyaline and size varies from 11 to 16×4–6 μm (Sattar and Malik 1939). Bose et al. 1973 reported conidia 11.9–17.0×3.6–5.8 μm. (mean 13.8×4.8 μm) broad, oblong with rounded ends, 11.1–17.7×3.1–5.0 μm. (mean 14.0×3.7 μm) for *C. gloeosporioides*. The hyphae of the *C. gloeosporioides* are hyaline, septate, full of oil globules and both inter and intra cellular (Fig. 1).

Symptomatology

The anthracnose disease symptoms visible on ripe-fruits, young leaves, twigs, petioles and inflorescence (panicles). The main source of infection are dead leaves entangled in the tree canopy, defoliated branch terminals, mummified panicles, fruits and

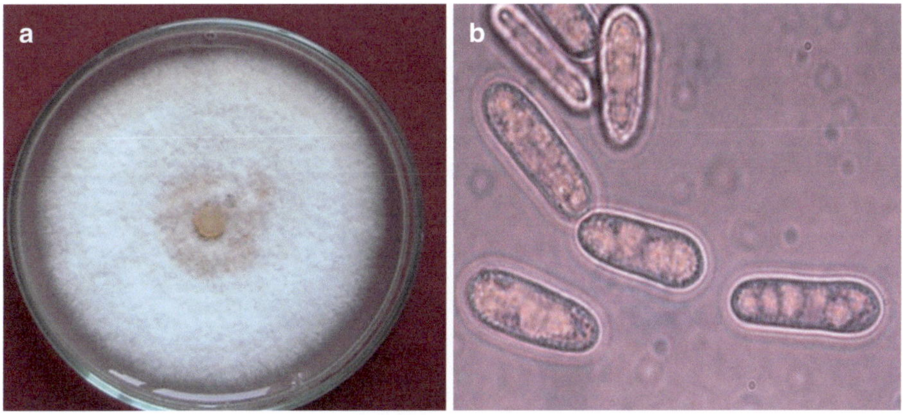

Fig. 1 (**a**) *Colletotrichum gloeosporioides* culture in medium (**b**) spore

bracts. In mango orchards, conidia spread by means of irrigation and light rain, heavy dew, rainy season being conducive to conidia production, dispersal and infection (Simmonds 1965). The blossom as well as peduncle blight is the most destructive phase of this disease, as it affects fruit set and ultimately the yield. Ripe fruits affect most significant and develop sunken, prominent, dark brown to black decay spots post-harvest. Fruits may drop from trees prematurely. The fruit spots typically do coalesce and ultimately penetrate deep into the fruit, resulting in extensive fruit rotting. Most green fruit infections remain dormant and largely invisible until ripening. The first symptoms on panicles are small black or dark-brown spots, which then enlarge, coalesce and destroy the flowers before fruit-set and results significant yield loss. Stems petioles and twigs are also susceptible to anthracnose and develop the typical black, expanding lesions found on leaves, flowers and fruits (Fig. 2).

Second symptoms type on fruits consists of a "tear stain" symptom that shows linear necrotic regions on the fruit that may or may not be associated with superficial cracking of the epidermal layer and gives an "alligator-skin" effect. These cracks sometimes develops into deep cracks extend towards pulp. Lessons on stems, fruits may produce conspicuous, pinkish-orange spore masses under wet, humid conditions and warm weather conditions favor post-harvest anthracnose development. On leaves, lesions start as small, angular, brown-black spots that can expand to form widespread dead areas. Necrosis manifests across or between leaf veins on leaf margins and at tips mainly and lesions may drop out of leaves during dry weather.

Colletotrichum gloeosporioides Biology of Infection

C. gloeosporioides causing anthracnose disease and grounds huge economic losses of mango fruits all over the world. The interaction between pathogen *Colletotrichum* and host mango fruit is a complex phenomenon. The pathogen is present at the

Fig. 2 Symptoms of anthracnose disease caused by *Colletotrichum gloeosporioides* in mango (**a**) Panicles (**b**) twig (**c**) leaves showing Bird's eye spots and (**d**) sunken *dark brown-black* spots on mango ripe fruit

surface in quiescent stage and with the onset of favorable climatic conditions the pathogen causes infection. The infection starts when the conidia that resides on the surface of mango fruit adheres and germinates to produce a germ-tube that forms terminal appresorium. Then, an infection peg develops which penetrate inside the outer layer and cuticle of the skin of fruit. During this stage of quiescence, due to presence of some antifungal compounds in the exocarp of unripe fruits called as dienes, the pathogen is unable to progress colonization (Prusky and Lichter 2007) In actual, this is the hemibiotrophic stage where the pathogen remains in quiescence stage and changes into necrotrophic (tissue disintegration) stage with the onset of fruit ripening. Fruit ripening, results into the loss of dienes with the resumption of pathogen growth and disease progression. Infection peg penetrates through the epidermal cell wall of the lumen of the cells that results in maceration and cell death. Production of cell wall degrading enzymes pectate lyase coupled with the transition of *Colletotrichum* from hemibiotrophic stage to necrotrophic attack (Prusky and Keen 1993) and causes anthracnose disease.

Epidemiology

The optimum temperature for conidial germination and infection is around 25–30 °C when free moisture is available. There is considerable variation in optimum temperature for germination and appressoria formation among the isolates of *C. gloeosporioides* from different locations. The injury caused by the anthracnose pathogen is dependent on humidity, rain, misty condition or heavy dews at the time of blossoming. Continuous wet weather during flowering causes serious blossom blight. Relative humidity above 95 % for 12 h. is essential for infection and development of C. *gloeosporioides* on mango fruit. Infection progresses faster in wounded tissues, ripe fruits and spread by rain splashing and irrigation.

Life Cycle of *Colletotrichum gloeosporioides* in Mango

Colletotrichum gloeosporioides display a range of nutritional strategies and lifestyles, including plant associations that occupy a continuum from necrotrophy to hemibiotrophy and endophytism. During the anamorph (asexual) stage the pathogen is typically haploid and becomes diploid during transition towards teleomorph (sexual) stage.

The Life Cycle of *Colletotrichum* on Mango Fruits Divide into Three Distinct Phases

Phase I: Dissemination: The *Colletotrichum gloeosporioides* spores in the form of conidia are colonizes to dead twigs and injured plant tissues. Conidia can disperse over relatively short distances passively by rain splashing and irrigation.

Phase II: Pathogen Inoculation: The ascospore are airborne and travel up to long distances and lands on infection sites like panicles, inflorescence, twigs, terminal branches, leaves and fruits.

Phase III: Pathogen Infection and reproduction: On the young unripe fruits, leaves and panicles, the conidia starts germinate to produce appressoria and penetrate through the cuticle and epidermis to ramify through the tissues cause quiescent infection. The quiescent infection lead to tissue necrosis and tissue is subsequently

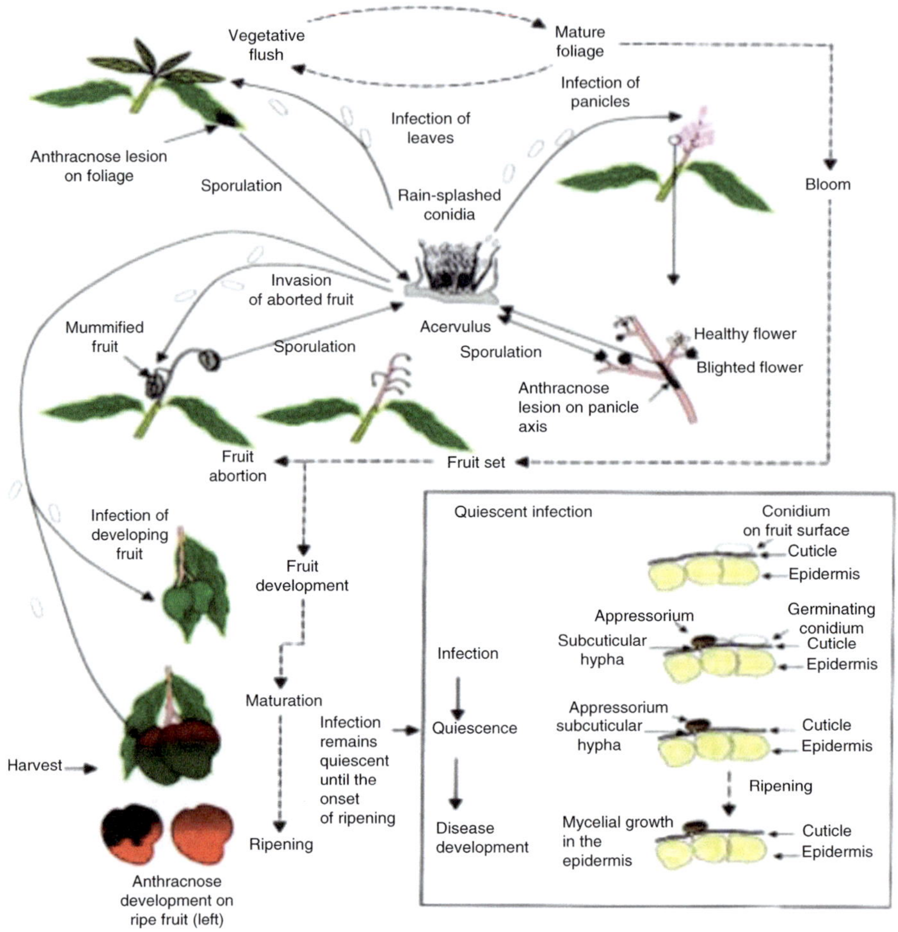

Fig. 3 Anthracnose disease cycle in mango (*Mangifera indica* L.) (Source: Arauz (2000))

colonized, acervuli are formed and thus pathogen completes its life cycle. The disease cycle of pathogen becomes polycyclic when conidia continues to multiply on many organs during the entire season and the pathogen survived on the dead leaves and defoliated branch terminals.

Disease occurrence: Fruits with quiescent infection remains asymptomatic before harvesting. Once the climacteric period of the fruit starts, lesions began to develop. The black-brown sunken spots that gradually darkens and enlarges to form a concentric ring pattern on the affected area. There is generally no fruit to fruit infection, hence the post-harvest anthracnose is monocyclic. However, in case of pre-harvest mango anthracnose disease, the developing fruits are infected in the field with some virulent and pathogenic isolates that causes pre-harvest losses (Coates et al. 1993) (Fig. 3).

Integrated Disease Management (IDM)

The management of tree fruit diseases depends on the level of disease itself, environmental conditions, effective cultural manipulations and knowledge of pathogens, sensitivity to fungicides and proper selection and timing of fungicidal sprays. The accurate identification and early detection of plant pathogens are the cornerstones of successful disease management. Control and effective management of mango anthracnose disease consists five major approaches:

1. **Resistant varieties**: Resistance is defined as an incompatible interaction between host and pathogen. The resistant has not been consistently used as a means of control of mango anthracnose. Because of variability in the presence and occurrence of virulent isolates and cultivars to the disease from one to another location. Although mostly all cultivars are susceptible to mango anthracnose but, some less susceptible and resistant cultivars include Tommy Atkins, Van Dyke, Alphonso, Baramasi, Carabao, Carrie early gold, Kent, Kishan Bhog, Rad and Saigon are resistant to infection caused by *Colletotrichum gloeosporioides* (Gantotti and Davis 1993).
2. **Cultural Practices in the field**

   (a) **Pruning**: Pruning of trees on yearly basis and always remove dead and fallen leaves, debris and burnt from the ground area near mango tree. Plant vigour plays an important role in keeping the plants free from twig and flower infection. Therefore, proper irrigation and fertilizer application are essential to maintain the tree vigour.
   (b) **Intercropping**: Inter-planting mango trees with other plants that are non-hosts of mango anthracnose will lead to avoid chances of epidemics.

3. **Fungicide Spray**: Fungicides spray involves chemical method and was not recommended in rainy season. Fungicides spray applied at the interval of 14–28 days in the orchards effective control measure. Spraying twice with Carbendazirn (Bavistin 0.1 %) at 15 days interval during flowering controls blossom infection. In general, copper fungicides are used to control pre-harvest anthracnose. Spraying of copper fungicides (0.3 %) is recommended for the control of foliar infection. Mancozeb, a dithiocarbamate fungicide is highly effective for the control of anthracnose (Dinh et al. 2003). Mancozeb is a contact fungicide and thus required to be applied at shorter intervals to achieve same results as systemic fungicides. Plocloraz can be used effectively when weather condition is favorable to *C. gloeosporioides*.
4. **Post-Harvest treatments**: The major strategy for effective management of post-harvest mango anthracnose disease is regular planned fungicide sprays and hot water treatment after harvest may reduce anthracnose. Anthracnose could be controlled by dip treatment of fruits in (0.1 %) Carbendazim in hot water at 52 °C for 15 min have been recommended for export quality mangoes. Cultivars also vary in their tolerance to the hot water and temperature treatments should never exceed 55 °C for 5 min. Hot water treatment (HWT) as a decay control treatment is applied commercially in few countries due to its efficacy. Bagging

of mango fruit before harvest and post-harvest treatment for 10 min in HWT (52–55 °C) was reported to reduce anthracnose infection successfully by 83 %. The temperature and duration of the treatment depend on the size or weight of the fruit, stage of maturity, cultivar type growing conditions and intensity of disease severity. It is recommended that fruit subjected to HWT within 24 h after harvest (Arauz 2000). HWT technology is widely acceptable due to increased profit and lower damage cause and high market value of fruit. This technology is consumer friendly, easy accessible and adopted frequently by farmers also due to its environment friendly approach (Arauz 2000).

5. **Irradiation and fruit quality**: Irradiation is recommended as quarantine or phytosanitary treatments. The purpose of irradiation is to kill or to sterilize microbes or insects by damaging their DNA. According to FDA the approved dosage for irradiation treatments on fresh produce is 1 KGy (100 krad). However, 1 KGy may not effective to kill the microbes and such high doses may affect negatively the fruit quality. Generally, gamma rays (from Co^{60}) are used for food irradiation because they can penetrate deeply into the pallet loads of the fruit. The effectiveness of irradiation on mango fruit quality depends upon the irradiation dose, cultivar and fruit maturity stage (Aveno and Orden 2004). Although incidence of anthracnose during storage was reduced with irradiation doses up to 600 Gy (Johnson et al. 1990). One kilogray failed to provide a complete control of anthracnose in mangoes. However, integrated treatments using a dose of 750 Gy with HWT at 40 °C for 20 min or 50 °C for 5 min' was effective in controlling anthracnose (Mitcham and Yahia 2009).
6. **Disease Predictive Model:** Two predictive models based on temperature and moisture requirement for infection on mango by *C. gloeosporioides* have been developed. These models are the basis of two forecasting systems for mango anthracnose that is employed in the field during fungicides application. In Australia, Fitzell (Prakash et al. 1996) studied the requirements of temperature and wetness duration for production of dark appressoria form conidia applied to detached young mango leaves under laboratory conditions. Similar system developed in Philippines based on the studies by Dodd and his coworkers (Fitzell 1979). It differs from the Australian system and tested under field conditions for the control of post-harvest anthracnose in the Philippines. Benomyl, Prochloraz (both at the rate of 0.47 mg/ml), or triforinge (0.375 mg/ml) was applied following a predicted infection period with a threshold of 40 % of conidia forming dark appressoria. One time application of benomyl or Prochloraz was as effective as six calendar based sprays of either fungicide (Dodd et al. 1991).

Molecular Identification and Characterizations

The morphological identification of *Colletotrichum* is often difficult and time-consuming and requires expertise. Molecular approaches based detection techniques are more efficient and accurate. ITS region considered as the primary fungal barcode marker for various reasons, including pragmatism – the number of existing fungal ITS sequences is far greater than that for any other gene and consider as Gold-standard for fungal identification and characterization. Many other genes/gene fragments have been used for diagnostic purposes in the Fungi, especially beta-tubulin (TUB2) and calmodulin for e.g. *Aspergillus* and *Penicillium* (Estrada et al. 1996; Samson et al. 2007; Peterson 1986), TEF1 for *Fusarium* (Houbraken et al. 2011; Geiser et al. 2004) and COX1 *Penicillium* (O'Donnell et al. 2009). Most of them currently used for phylogenetic analysis in *Colletotrichum* causing mango anthracnose disease. Mango anthracnose in India was reported to be caused by "*C. gloeosporioides*", based on either morphology or ITS gene-sequence data (Seifert et al. 2007; Kumar et al. 2007; Sangeetha and Rawal 2009; Gupta et al. 2010; Laxmi et al. 2011; Kamle et al. 2013a) and recently using multi-locus sequence-data (ITS and tub2) (Kamle et al. 2013b). Multigene-based molecular characterization and sequence phylogenetic studies may confirm the *Colletotrichum* species associated with mango anthracnose disease. However, until now the dominant causal agent responsible for anthracnose or leaf spot disease in India is *C. gloeosporioides.* The *C. gloeosporioides* species complex involved morphological similar species but, totally different on genetically as referred by Cai (Freeman et al. 1998; Cai et al. 2009).

Colletotrichum MAT Genes

Fungal species that reproduce sexually can often be classified as either self-fertile (homothallic), or self-sterile (heterothallic). Based on extensive studies on genetics of mating system in *Glomerella cingulata* concluded that heterothallism obtained from homothallism through mutations in genes controlling steps in the morphogenetic pathway necessary for self-fertility (Johnson et al. 1990). Sexual reproduction is rarely documented from genus *Colletotrichum.* The genetics underlying the mating system in *Colletotrichum* are perplexing, fungi in this genus do not employ the canonical bipolar mating system characteristic of other ascomycete fungi (Wheeler 1954; Vaillancourt and Hanau 1992). In general, the bipolar model described to regulate sexual compatibility, that mating occurs when both idiomorph of mating type gene Mat 1 (Mat 1-1 and Mat 1-2) are present. However, till now only Mat 1-2 idiomorph with characteristic conserved high mobility group (HMG) binding domain is found in *Colletotrichum* (Vaillancourt et al. 2000). *Colletotrichum* species are also extensively studied as model organisms for research into genetics. This work has a long history; the first investigation into mating types in *Glomerella* was published a century ago (Martínez-Culebras et al. 2003; Edgerton 1912) and genetic mechanisms in *G. cingulata* were extensively studied in the 1940s and 1950s (Johnson et al. 1990; Edgerton 1914; Andes 1941; Lucas et al. 1944). Later, the phylogenetic diversity of the *C. gloeosporioides* species complex associated with *mangifera indica* from India based on the six-gene markers (Olive 1951). However, few research groups claimed *ApMat* marker provides better resolution as compared to the gene-markers to resolve the species identification issues in the *C. gloeosporioides* species complex (Weir et al. 2012; Silva et al. 2012; Doyle et al. 2013).

Colletotrichum Genomics and Transcriptomics

C. gloeosporioides is a notorious and destructive plant pathogen that exists as endophytes with mango and during favorable conditions cause anthracnose disease symptoms. With the availability of whole genome sequences of *Colletotrichum* in other hosts that enabled an idea of comparative transcriptome analysis at different stages of hemibiotrophic infection. The whole genome sequencing of mango anthracnose pathogen were used to investigate the gene –expression during the three growth stages of *C. gloeosporioides.* The deep sequencing at the three stages of development in anthracnose disease in mango are likely to be performed at (1) prepenetration (appressoria), (2) biotrophic hyphae and (3) necrotrophic stage (tissue damage) (Sharma et al. 2013). By exploring the transcriptomics approach we can demonstrate the factors affecting and set of genes responsible for the transition lifestyle of *C. gloeosporioides* stage from biotrophic to necrotrophic and then again to biotrophic. The transcriptomic analysis demonstrate the massive set of gene-expression underlying the developmental transition that occur in planta, from spore germination to necrotrophic (O'Connell et al. 2012). Thus, the lifestyle switch to necrotrophic stage is characterized by a considerable shift in fungal gene expression, with activation of large number of genes encoding enzymes and membrane transporter. To understand the signals transduction mechanism involved in activation of transition of lifestyle from biotrophic to necrotrophic mode during the fruit ripening which are yet an enigma.

Future Perspectives

Mango anthracnose disease caused by *Colletotrichum gloeosporioides* is one the most severe post-harvest disease caused in mango. The disease does not reflect any prior symptoms as the pathogen exist in the quiescent stage and with the onset of fruit ripening the disease progresses and caused serious losses especially during storage. The successful disease management practices with implementation of advanced integrated disease management practices proved effective. With the advances in genomics and transcriptomics sequence analysis in future there would be more new vistas in exploring molecular mechanism of disease occurrence. How certain signal molecules activated and leads to change of lifestyle from biotrophic to necrotrophic stage during ripening. To elucidate the signaling pathways involved in mango – *Colletotrichum* interaction. To unravel the molecules involved in disease incidence and then implication of certain genes towards genetic engineering to make future mango fruit resistant to anthracnose. By bringing genetically, improved anthracnose resistant mango will explore the markets in tropical and subtropical countries where the disease prevalence is extreme.

References

Alahakoon PW, Brown AE, Sreenivasaprasad S (1994) Cross-infection potential of genetic groups of Colletotrichum gloeosporioides on tropical fruits. Physiol Mol Plant Pathol 44:93–103

Andes JO (1941) Experiments on the inheritance of the "plus" and "minus" characters in *Glomerella cingulata*. Bull Torrey Bot Club 68:609–614

Arauz LP (2000) Mango anthracnose: economic impact and current options for integrated management. Plant Dis 84:600–611

Aveno JL, Orden MEM (2004) Hot water treatment of mango: a study of four export corporations in the Philippines. KMITL Sci Technol J. 4:1685–2044

Bose SK, Sindhan GS, Pandey BN (1973) Studies on the die back disease of mango in the Tarai region of Kumaon. Progress Hortic 5:41–53

Cai L, Hyde KD, Taylor PWJ, Weir BS, Waller J, Abang MM, Zhang JZ, Yang YL, Phoulivong S, Liu ZY, Prihastuti H, Shivas RG, McKenzie EHC, Johnston PR (2009) A polyphasic approach for studying Colletotrichum. Fungal Divers 39:183–204

Coates LM, Muirhead IF, Irwin JAG, Gowanlock D (1993) Initial infection processes by Colletotrichum gloeosporioides on avocado fruit. Mycol Res 97:1363–1370

Dinh SQ, Chongwungse J, Pongam P, Sangchote S (2003) Fruit infection by C. gloeosporioides and anthracnose resistance of some mango cultivars in Thailand. Aust Plant Pathol 32: 533–538

Dodd JC, Bugante R, Koomen I, Jeffeies P, Jeger MJ (1991) Pre and post-harvest control of mango anthracnose in the Philippines. Plant Pathol 40:576–583

Doyle VP, Oudemans PV, Rehner SA, Litt A (2013) Habitat and host indicate lineage identity in Colletotrichum gloeosporioides s.l. from wild and agricultural landscapes in North America. PLoS ONE 8:623–694

Edgerton CW (1912) Plus and minus strains in an ascomycete. Science 35:151

Edgerton CW (1914) Plus and minus strains in the genus *Glomerella*. Am J Bot 1:244–254

Estrada AB, Jeffries P, Dodd JC (1996) Field evaluation of a predictive model to control anthracnose disease of mango in the Philippines. Plant Pathol 45:294–301

Fitzell RD (1979) Colletotrichum acutatum as a cause of anthracnose of mango in New South Wales. Plant Disease Reporter 63:1067–1070.

Fitzell RD, Peak CM (1984) The epidemiology of anthracnose disease of mango: inoculum sources, spore production and dispersal. Ann Appl Biol 104:53–59

Freeman S, Katan T, Shabi E (1998) Characterization of Colletotrichumspecies responsible for anthracnose diseases of various fruits. Plant Dis 82:596–605

Gantotti BV, Davis MJ (1993) Pectic zymogram analysis for characterizing genetic diversity of the mango anthracnose pathogen. Acta Hortic 341:353–359

Geiser DM, Jiménez-Gasco M, Kang S, Makalowska I, Veeraraghavan N (2004) FUSARIUM-ID v. 1.0: a DNA sequence database for identifying *Fusarium*. Eur J Plant Pathol 110:473–479

Gupta VK, Pandey A, Kumar P, Pandey BK, Gaur RK, Bajpai V, Sharma N, Sharma S (2010) Genetic characterization of mango anthracnose pathogen Colletotrichum gloeosporioides Penz. By random amplified polymorphic DNA analysis. Afr J Biotechnol 9:4009–4013

Houbraken J, Frisvad JC, Samson RA (2011) Fleming's penicillin producing strain is not *P. chrysogenum* but *P. rubens*. IMA Fungus 2:87–95

Jefferies P, Dodd JC, Jeger MJ, Plumbley RA (1990) The biology and control of *Colletotrichum* species on tropical fruit crops. Plant Pathol 39:343–366

Johnson GI, Boag TS, Cooke AW, Izard M, Panitz M, Sangchote S (1990) Interaction of postharvest diseases control treatments and gamma irradiation of mangoes. Ann Appl Biol 116:245–257

Kamle M, Kumar P, Gupta VK, Tiwari AK, Misra AK, Pandey BK (2013a) Identification and phylogenetic correlation among *Colletotrichum gloeosporioides* pathogen of anthracnose for mango. Biocat Agric Biotechnol 2(3):285–287

Kamle M, Pandey BK, Kumar P, Muthukumar M (2013b) A species-specific PCR based assay for rapid detection of mango anthracnose pathogen *Colletotrichum gloeosporioides* Penz. and Sacc. J Plant Pathol Microbiol 4:6

Kumar AS, Reddy NPE, Reddy KH, Devi MC (2007) Evaluation of fungicidal resistance among Colletotrichum gloeosporioides isolates causing mango anthracnose in Agri Export Zone of Andhra Pradesh, India. Plant Pathol Bull 16:157–160

Laxmi BKM, Reddy PN, Prasad RD (2011) Cross infection potential of Colletotrichum gloeosporioides Penz. isolates causing anthracnose in subtropical fruit crops. Trop Agric Res 22:183–193

Lucas GB, Chilton SJP, Edgerton CW (1944) Genetics of *Glomerella*. I. Studies on the behavior of certain strains. Am J Bot 31:233–239

Martínez-Culebras PV, Querol A, Suarez-Fernandez MB, García-Lopez MD, Barrio E (2003) Phylogenetic relationships among *Colletotrichum* pathogens of strawberry and Design of PCR primers for their identification. J Phytopathol 151:135–143

Mitcham E, Yahia E (2009) Alternative treatments to hot water immersion for mango fruit report to the national mango board. http://www.mango.org/media/73313/alternatives%20to%20 hot%20water%20treatment-final%20report.pdf. Accessed 8 Oct 2010

O'Connell RJ, Thon MR, Hacquard S, Amyotte SG, Kleemann J et al (2012) Life-style transitions in plant pathogenic *Colletotrichum* fungi deciphered bygenome and transcriptome analyses. Nat Genet. doi:10.1038/ng.2372

O'Donnell K, Gueidan C, Sink S, Johnston PR, Crous PW et al (2009) A twolocus DNA sequence database for typing plant and human pathogens within the *Fusarium oxysporum* species complex. Fungal Genet Biol 46:936–948

Olive LS (1951) Homothallism and heterothallism in *Glomerella*. Trans N Y Acad Sci 13:238–242

Palo MA (1932) Sclerotium seed rot and seedling stem rot of mango. Philippines J Sci 52(3):237–261

Ploetz R (1999) Anthracnose: The most important disease in much of the mango-producing world. News Lett. Plant Pathol 3:1–6

Peterson RA (1986) Mango diseases. In: Proceedings of the CSIRO 1st Australian Mango Research Workshop. CSIRO, Cairns, pp 233–247

Prakash O, Misra AK, Kishun R (1996) Some threatening diseases of mango and their management. In: Management of threatening plant diseases of national importance. Malhotra publishing House, New Delhi, pp 179–205

Prusky D (1996) Pathogen quiescence in post-harvest diseases. Annu Rev Phytopathol 34:413–434

Prusky D, Keen NT (1993) Involvement of preformed antifungal compounds in the resistance of subtropical fruits to fungal decay. Plant Dis 77:114–119

Prusky D, Lichter A (2007) Activation of quiescent infection by postharvest pathogens during transition from the biotrophic to necrotrophic stage. FEMS Microbiol Lett 268:1–8

Samson RA, Noonim P, Meijer M, Houbraken J, Frisvad JC et al (2007) Diagnostic tools to identify black aspergilli. Stud Mycol 59:129–145

Sangeetha CG, Rawal RD (2009) Temperature requirements of different isolates of Colletotrichum gloeosporioides isolated from mango. Am Eurasian J Sci Res 4:20–25

Sattar A, Malik SA (1939) Some studies on anthracnose of mango caused by Glomerella cingulata (Stonem.) Spauld. Sch. (Colletotrichum gloeosporioides Penz.). Indian J Agric Sci 1:511–521

Seifert KA, Samson RA, deWaard JR, Houbraken J, Levesque CA (2007) Prospects for fungus identification using CO1 DNA barcodes, with *Penicillium* as a test case. Proc Natl Acad Sci U S A 104:3901–3906

Sharma G, Kumar N, Weir BS, Hyde KD, Shenoy BD (2013) The ApMat marker can resolve Colletotrichum species: a case study with Mangifera indica. Fungal Divers 61:117–138

Silva DN, Talhinas P, Várzea V, Cai L, Paulo OS, Batista D (2012) Application of the Apn2/MAT locus to improve the systematics of the *Colletotrichum gloeosporioides* complex: an example from coffee (*Coffea* spp.) hosts. Mycologia 104:396–409

Simmonds JH (1965) A study of the species of *Colletotrichum* causing ripe fruit rots in Queensland. Queensland J Agric Anim Sci 22:437–459

Tharanathan RN, Yashoda HM, Prabha TN (2006) Mango (*Mangifera indica* L.) "The King of Fruits" — an overview. Food Rev Int 22:95–123

Vaillancourt LJ, Hanau RM (1992) Genetic and morphological comparisonsof *Glomerella* (*Colletotrichum*) isolates from maize and from sorghum. Exp Mycol 16:219–229

Vaillancourt L, Wang J, Hanau R (2000) Genetic regulation of sexual compatibility in Glomerella graminicola. In: Prusky D, Freeman S, Dickman M (eds) Colletotrichum host specificity, pathology, and host-pathogen interaction. APS press, St. Paul, pp 29–44

Weir B, Johnston PR, Damm U (2012) The *Colletotrichum gloeosporioides* species complex. Stud Mycol 73:115–180

Wheeler HE (1954) Genetics and evolution of heterothallism in *Glomerella*. Phytopathology 44:342–345

Current Scenario of Mango Malformation and Its Management Strategies: An Overview

Pradeep Kumar and Ashok Kumar Misra

Introduction

Mango is regarded as the king of fruits in India; it has been cultivated for at least 4,000 years and has a great cultural and religious significance (Popenoe 1932; Purseglove 1972). Mango is an important commercial crop (Purseglove 1972) that currently ranks fifth among the major fruits cultivated worldwide (FAOSTAT 2013). Mango is cultivated commercially throughout the tropics and in many subtropical regions (Mukerjee and Litz 2009). India's contribution to the world's mango production is highest i.e., 18,431,300 mt from 2,516,000 ha and about 41280.01 mt of mango is exported of approx. value of 50.26 million USD with the productivity 7.3 mt/ha (National Horticulture Database 2014). Several factors affecting the mango production in India in which mango malformation is one of the major constrain. Flowering responses of mango differ significantly in tropical as opposed to subtropical environments. In the tropics, flowering can be induced chemically, while in the subtropics stimulation is ineffective and is primarily governed by chilling temperatures (Iyer and Schnell 2009). Mango malformation disease is one of the most important and destructive diseases of this crop (Kumar et al. 2011a; Litz 2009; Ploetz 2001). It affects vegetative and floral part of the plant. Although trees are not killed, the vegetative phase of the disease impedes canopy development and the floral phase reduces fruit yield dramatically; substantial economic losses can

P. Kumar (✉)
Department of Biotechnology Engineering, Ben Gurion University of the Negev, Beersheva 84105, Israel
e-mail: pkbiotech@gmail.com

A.K. Misra
Division of Crop Protection, Central Institute for Subtropical Horticulture, Lucknow, Uttar Pradesh, India
e-mail: misra_a_k@yahoo.co.in

P. Kumar et al. (eds.), *Current Trends in Plant Disease Diagnostics and Management Practices*, Fungal Biology, DOI 10.1007/978-3-319-27312-9_10

occur since malformed inflorescences do not bear fruit. Mango malformation first reported in India in 1981 from Darbhanga district of Bihar state. Later it has been also observed in Australia, Brazil, China, Egypt, El Salvador, India, Israel, Malaysia, Mexico, Myanmar, Nicaragua, Oman, Pakistan, Senegal, South Africa, Spain, Sri Lanka, Sudan, Swaziland, Uganda, and the United States (Anonymous 2013; Bastawros 1996; Crookes and Rijkenberg 1985; Freeman et al. 1999; Goldman et al. 1976; Kumar and Beniwal 1991; Kvas et al. 2008; Lim and Khoo 1985; Lima et al. 2008; Marasas et al. 2006; Nor et al. 2013; Otero-Colina et al. 2010; Rodríguez-Alvarado et al. 2013; Senghor et al. 2012; Sinniah et al. 2013; Zhan et al. 2012). In India, this problem is more acute in north-west regions including Jammu than the north-eastern and southern India (Chib et al. 1984; Varma et al. 1974a). The incident of disease also reported from different states of India mainly in Maharastra, Gujrat, Uttar Pradesh, Punjab, Jammu and Kashmir, Delhi, Bihar, Madhya Pradesh, Himachal Pradesh, Haryana and Andhra Pradesh (Kumar et al. 2011b). Thus, mango malformation has now become a national problem affecting production in almost all of the mango growing areas of the country. Malformation causes heavy damage to the trees as the inflorescence fails to produce fruits. The extent of damage varies from 50 to 60 % in some cases and in severe cases the loss may be up to 100 % (Misra and Singh 2002).

Symptom of Malformation

Broadly three distinct types of symptoms were described by various workers. These are bunchy top of seedlings, vegetative malformation and floral malformation. Later, these were grouped under two broad categories i.e. vegetative and floral malformation (Varma 1983) (Fig. 1). This affects vegetative and floral meristematic tissues (Chakrabarti 2011; Ploetz 2001). Vegetative malformation is most serious on seedlings and young trees in nurseries, especially where seedlings are grown beneath the canopies of affected trees (Ploetz et al. 2002; Youssef et al. 2007), but it also occurs on mature trees. The seedlings produce small shootlets bearing small scaly leaves with a bunch like appearance on the shoot apex. The multi-branching of shoot apex with scaly leaves is known as "Bunchy Top", also referred to as 'Witche's Broom" (Bhatnagar and Beniwal 1977). Seedling affected in early stages are remain stunted and die while, those infected in later stage resume normal growth above the malformed area (Kumar and Beniwal 1992).

Floral malformation is most important economically since affected inflorescences usually do not set fruit (Kumar et al. 1991; Youssef et al. 2007; Ploetz and Freeman 2009). The primary, secondary and tertiary rachises become short, thickened and hypertrophied. Such panicles are greener and heavier with increased crowded branching. Malformed panicles produce up to three times the normal number of flowers, which range from one-half to two times the normal size, and have an increased proportion of male to perfect flowers that are either sterile or eventually abort (Haggag et al. 2011). Malformed panicles may also produce dwarfed and

Fig. 1 Schematic representation of mango malformation disease symptom and pathogen (**a**) Healthy Inflorescence (**b**) malformed affected inflorescence (**c**) Healthy vegetative tissue (**d**) malformed affected vegetative tissue

distorted leaves (phyllody). Both healthy and malformed flowers appear on the same panicle or on the same shoot. The severity of malformation may vary on the same shoot from light to medium or heavy malformation of panicles (Varma et al. 1969). The heavily malformed panicles are compact and overcrowded due to larger flowers. They continue to grow and remain as black masses of dry tissue during summer but some of them continue to grow till the next season.

On the basis of compactness of panicles, malformed panicles are classified into different groups viz., heavy, medium and light (Varma et al. 1969; Majumder and Sinha 1972a); compact malformed panicle, elongated malformed panicle and slight malformed panicle (Rajan 1986) and small compact type and loose type (Kumar et al. 1993). Sometimes, a shoot tip may bear both types of panicles i.e. healthy as well as malformed. Less frequently, a healthy panicle may contain one or more malformed branches of a few malformed flowers or vice-versa. These partially infected panicles may bear fruits up to maturity (Kumar et al. 1993).

Susceptibility of Cultivars for Malformation

The susceptibility to malformation in mango varieties is variable and is governed by the different factors viz. temperature, age of the tree, time, etc. Wide ranges in susceptibility have been reported for different mango cultivars (Kumar and Beniwal 1991). However, cv. Amrapalli was found highly susceptible to mango malformation particularly in Delhi regions and north India (Yadava and Singh 1995). In general, late blooming varieties are less susceptible to malformation than the early blooming ones (Khurana and Gupta 1973). The level of polyphenol oxidase (PPO) in the early years of plant growth or in the flush of vegetative growth may provide an estimate of synthesis of phenolic compounds in the plants, which may be correlated to susceptibility or resistance to floral malformation (Sharma et al. 1994). The disease is serious in the north-west region where temperature lie from 10 to 150 C during December–January (winter) before flowering. Age of the flowering shoot also influences the incidence of floral malformation, as reported by Varma (1983). Misra et al. (2000) reported cv. Ellaichi as free from malformation and is now been used in breeding programme. Ewais', a popular cultivar in Egypt, was reported to be moderately susceptible to malformation (Bastawros 1996), but of low and high susceptibility in two other references (Chakrabarti 2011; Ploetz and Freeman 2009). Bastawros (1996) reported 0 % disease to 'Kent' and 'Keitt', however Freeman et al. (1999) reported that both are prone to natural and artificial inoculation malformation According to Bastawros (1996), two newly introduced cultivars in Egypt, 'Kent' and 'Keitt', were immune (0 % disease), even though they are susceptible to natural and artificial inoculation in other places. Cultivars listed as "resistant" may have been established from pathogen-free nursery stock or may have escaped infection once planted in the field (Ploetz 2001).

Etiology and Epideniology

The etiology of floral malformation in mango has always been controversial. However, the evidence of association of fungal pathogen(s), virus and mite has been suggested by various workers.

Role of Mites in Malformation

In India, Narasimhan (1954) for the first time claimed eriophyid mite to be the disease causing organism and found it inter and intracellularly in the meristem and tender regions (Narasimhan 1959). Mites were found to induce the disease in test plants when transfered from diseased (Singh et al. 1961) or even from healthy plants. Nariani and Seth (1962) successfully induced the disease on young seedlings by introduction of eriophyid mites but it was not certain whether these mites were vectors of some pathogen or is a direct cause. Sternlicht and Goldenberg (1976) advocated that Aceria mangiferae has no direct effect on inflorescence malformation, but it's interaction with another biotic factor (a fungus) is not ruled out. In Egypt, Wabha et al. (1986) suggested that A. mangiferae does not cause the malady directly but may play a role in the overall malformation process. Labuschang et al. (1993) could not correlate that presence of mites and increase in the incidence of malformation. However, role of mite as a carrier of fungus (*Fusarium* monilifome var. sublgutinans) was advocated by Summanwar (1967), Summanwar and Raychoudhury (1968) and Pinkas and Gazit (1992). The pathogen was previously recovered from the mite on culture media (Crookes and Rijkenberg 1985), and was recently shown to adhere to its body (Gamliel-Atinsky et al. 2010). Gamliel-Atinsky et al. (2009) reported that the mite could not ingest the pathogen, due to its small mouth, but experimentally dispersed conidia of F. mangiferae to infection courts within mango buds, probably as a body adherent. Wounds caused by the mites' feeding could facilitate infection of buds by the pathogen (Crookes and Rijkenberg 1985; Gamliel-Atinsky et al. 2010). In Israel, A. mangiferae did not appear to play a significant role in disseminating the pathogen among trees. Mites were not found in traps that were designed to monitor their movement in malformed affected orchard, although high numbers of F. mangiferae conidia were trapped (Gamliel-Atinsky et al. 2007). Whether, and under what circumstances, the mite plays a role in spreading malformation among trees and orchards in other mango-production areas should be determined due to the potential impact these factors would have on malformation management strategies. Other arthropods that frequent infected panicles may serve as dispersal agents even though no conidia were detected on wind borne mango bud mites originating from infected panicles (Gamliel-Atinsky et al. 2009, 2010).

Fungus as Causal Agent of Mango Malformation

Summanwar et al. (1966) reported for the first time, a fungus *Fusarium moniliforme* Sheld, associated with malformation (floral and vegetative) and proved its pathogenicity and it has been isolated from the various part of malformed affected plants (Varma et al. 1974b). Growth of fungus was inhibited at higher temperature in summer or even at room temperature (Varma et al. 1971). Vegetative malformation (Prasad et al. 1972) and floral malformation (Varma et al. 1974a, b) can be initiated in the healthy test plants by artificial inoculation of aerial branches with the fungus as it is mostly intercellular and occasionally forms intercellular agglomerates in the cortex and phloem regions and the fungus form globose bodies similar to chlamydospores, particularly in the cortex when inoculated with spore suspension (Varma et al. 1972, 1974b). Typical bunchy top symptoms can be produced in seedling by inoculating the fungus through soil. The fungus is systemically present in parenchymatous cells of the pith region of malformed tissues (Bhatnagar and

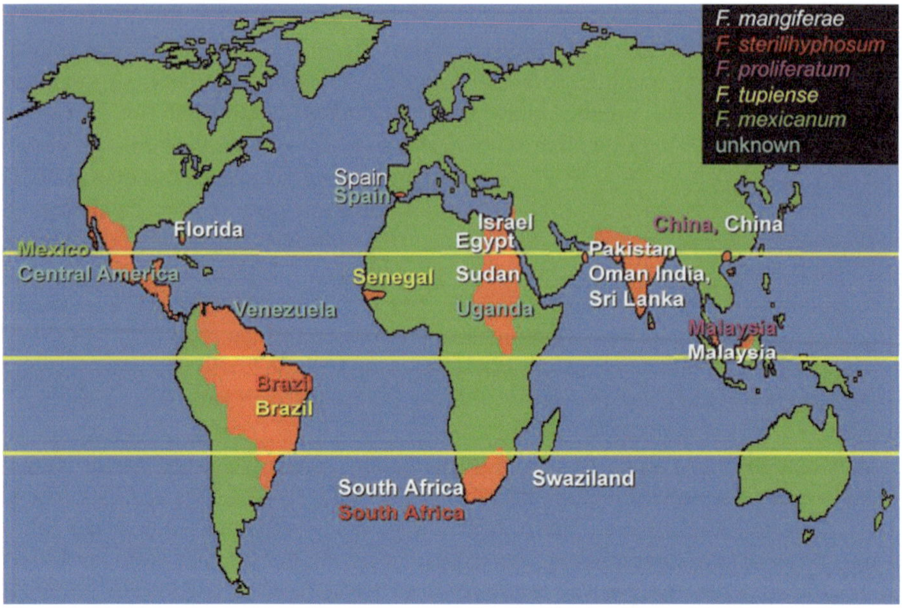

Fig. 2 Distribution of the different Fusarium species pathogens causing mango malformation disease worldwide (Freeman et al. 2014b)

Beniwal 1977). There are several classification and synonyms updated in the literature, including F. Subglutinans (Wollenweb and Reinking) Nelson, Toussoun and Marasas and it was redescribed as new species, F. mangiferae Britz, Winfield and Marasas, in 2002 with isolates from Egypt, Florida, Israel, Malaysia and south Africa (Steenkamp et al. 2000; Britz et al. 2002; Marasas et al. 2006). *Fusarium* mangiferae was also identified in India (O'Donnell et al. 1998; Zheng and Ploetz 2002), Oman (Kvas et al. 2008), Spain (Crespo et al. 2012), Sri Lanka (Sinniah et al. 2013), and China (Zhan et al. 2012), and appear to most common causal agent of mango malformation worldwide (Fig. 2). *F. sterilihyphosum* Britz, Winfield and Marasas is the second malformation causal agent in South Africa isolated from limited place (Britz et al. 2002) and it also isolated from Brazil (Lima et al. 2009) where it was shown to cause malformation after artificial inoculations (Lima et al. 2008). *F. mexicanum* sp. nov., was describe from Mexico as another pathogen for malformation and its multilocus sequencing shown to differ significantly from taxa in the *G. fujikuroi* species complex (Otero-Colina et al. 2010; Rodríguez-Alvarado et al. 2007). A fourth recently described species, *F. tupiense* sp. nov., and its cause malformation in Brazil (Lima et al. 2012) and Senegal (Senghor et al. 2012). *F. proliferatum* (Matsushima) Nirenberg (Zhan et al. 2010; Lv et al. 2013) and *F. pseudocircinatum* (Freeman et al. 2014a) have been also reported as a causal agent of mango malformation. Inoculation of a *Fusarium* spp. in mango seedlings showed that the disease is neither systemic, nor completely localized but behaved erratically. Internal spread is always acropetal and is supposed to be facilitated through active cell divisions of terminal growth under environmental conditions favorable to both host and pathogen. Effect of *F. moniliforme* var. *subglutinans* infection on mangiferin production in the twigs of *Mangifera indica* was studied by Chakrabarti and Ghosal (1985). Infected twigs contain less mangiferin than twigs of healthy plants. In both cases, mangiferin concentration was high during cooler months and low during hotter months. Mass isolation and recovery of *F. moniliforme* var. *subglutinans* above 82 % colonies from affected sample (Misra and Singh 1998, 2002; Kumar et al., 2011b). They also isolated pathogen from normal apical shoot and healthy panicles and advocated its systemic nature. Freeman et al. (1999) prove that the apical buds were the primary sites for the infection by artificial inoculation of F. mangiferae that were transformed with GUS reporter gene (B-glucorunidase) Youssef et al. (2007) reported that conidium survival rapidly declined in soil during summer and only 50 % population recovered after 10 day from the original and it survive longer in infected plant tissues under soil. Freeman et al. (2014a) reported that on the basis of available evidence that F. mangiferae is generally restricted to apical and meristematic tissue of the host and localized but not systemic infection of these bud take place.

Molecular Diversity and Detection of Mango Malformation

In recent years, DNA polymorphisms have increasingly been used to complement traditional markers for the analysis of genetic identity, variability and relatedness in fungi (Jamil et al. 2000). Molecular markers reveal information concerning the genetic structure of pathogen populations. DNA fingerprinting has evolved as a major tool in fungal characterization. Attempts to characterize fungal isolates through Random Amplified Polymorphic DNA (RAPD) have been reported (Williams et al. 1990; Grajal Martin et al. 1993). The development of RAPD has allowed the rapid generation of reliable, reproducible DNA fragments or fingerprints in wide variety of species, including several fungi (Crowhurst et al. 1991; Kush et al. 1992). It has proved successful in generating amplification patterns specific to *F. mangiferae* (Ploetz et al. 2002) causing mango malformation. Zheng and Ploetz (2002) examined genetic diversity among 74 *F. subglutinans* isolates collected from Brazil, Egypt, USA, India, Israel and South Africa. A pair of 20-mer primers developed from a RAPD fragment amplified a specific 608 bp fragment for 51 out of 54 mango isolates. Ploetz et al. (2002) amplified genomic DNA of *F. mangiferae* isolates with 33 arbitrary decamer primers. Most of the primers produced reproducible polymorphic banding patterns. Unique Major bands were generated with primers OPZ-5 and OPJ-l. RAPD analysis divided the isolates into two genetically distinct clusters. Saleem (2004) identified 20 isolates of *F. mangiferae* from malformed tissues of mango obtained from different areas of Pakistan. No site specific correlation was found. The potential of RAPD to identify DNA markers related to intraspecific diversification of the pathogens led to study the genetic diversity within *F. mangiferae* population. These investigations found an application in distinguishing isolates of *Fusarium* spp. within the section Liseola. Jana et al. (2003) studied genetic variation in 22 isolates of *Fusarium* species collected from geographically distinct regions over a range of hosts, using RAPD markers. UPGMA analysis classified these isolates into five major groups using primer OPA-13. Isolates of pathogenic and non-pathogenic *Fusarium* species of different formae-speciales and races were distinguished among each other. A robust and repeatable PCR-assay was developed for the detection and differentiation of *F. sporotrichiodes* from other *Fusarium* species based on sequence determined from differentially amplified RAPD-PCR products. These assays were able to detect both species in samples of grain taken from the field. Freeman and Maymon (2000) and Sabir (2006) finding showed that low level variability among *Fusarium* isolates collected from different geographical locations. Iqbal et al. (2006) has reported a low level of genetic variation in *Fusarium* isolates collected from malformed tissues of different geographical locations in Pakistan based on UPGMA clustering. Arif et al. (2011) also reported variability among the *F. moniliforme* var. *subglutinans* isolates of mango malformation by using RAPD but this study was limited to assessment of genetic diversity of *Fusarium moniliforme* isolates of Pantnagar, India. Kumar et al. (2014) also isolated and identify the *F. moniliforme* var. *subglutinans* collected from either the same or different regions and studies the variability among the all isolates showed high similarity values indicating that there existed narrow molecular variation and almost all the isolates are genetically related. There are different reports suggesting the existence of genotypic diversity among the isolates of *Fusarium* (Saharan et al. 2006). Moreover, because of this low level variability there was no possibility of different races among the isolates even though they are from different agro-climatic regions. Haggag et al. (2011) use different primers of RAPD and used them as for discriminating between *Fusarium* isolates and degree of relationship between *F. sterilihyphosum* and *F. proliferatum*; between *F. moniliforme* and *F. subglutinans*; between *F. oxysporum* and *F. chlamydospore;* the degree of relationship among *F. subglutinans*, *F. proliferatum* and *F. sterilihyphosum* and degree of relationship among *F. moniliforme*, *.sterilihyphosum*, *F. proliferatum* and *F. subglutinans*. Zhan et al. (2010) amplify the 500 bp band with primers pITS1 and pITS4 in F. proliferatum isolated from the mango malformation tissue from the South china. By using the BLAST sequence alignment algorithms from the NCBI website and sequence phylogeny of ITS-rDNA confirmed similarity with fungal species in *Fusarium* section Liseola, isolates of F. proliferatum.

Management Strategies' for Mango Malformation

There are several control measures have been reported to management of mango malformation but the measure problem is either they are not reproducible nor prove 100 % cure malformation. There are various approaches tried to control the malformation few of them describe below.

Disease Management Through Nutrients

Nutrient application has improved of nutritional status thus improving yield in some cases. A direct inhibitory effect of chemicals against pathogen is inferred together with secondary control through improved nutritional status of trees. The combined effect of potassium sulphate as soil application and monocrotophos as trunk injection cured the malady. Result of a 10 year trial of NPK fertilization on panicle malformation in mango cv. Dashehari indicated that increasing nitrogen doses reduced panicle malformation whereas the effect of phosphorus and potassium was just the reverse (Minessy et al. 1971). Partial control of the disease has been achieved in India by spraying the malformed parts with mangiferin-Zn^{2+} and mangiferin-Cu^{2+} chilates (Chakrabarti and Ghosal 1989).

Disease Management Through Deblossoming and Prunning

Deblossoming at the bud burst stage alone or in combination with spraying of 200 ppm NAA was reported to be very effective in controlling malformation. Deblossoming at bud burst stage gives substantial reduction in malformation (Singh et al. 1974; Singh and Dhillon 1986, 1988b). Deblossoming between January and February regenerated new panicles in the same season. Regenerated panicles bore fruits similar to healthy once. Deblossoming after February failed to regenerate panicles (Tripathi and Ram 1998).

Since, the process of deblossoming is cumbersome it is advisable to develop a chemical for deblossoming. Application of 200 and 500 ppm etherel completely control malformation (Chadha et al. 1979). 250 ppm of cycloheximide was also very effective in deblossoming the panicles (Pal and Chadha 1982). 750–6,000 ppm dikegulac and 500 ppm etherel at bud burst stage were ineffective (Singh and Dhillon 1986). Pruning reduced malformation (Narasimhan 1959; Desai et al. 1962; Singh et al. 1983). Pruning however had no effect on malformation (Bindra and Bakhetia 1971). Pruning followed by spraying with the mixture of fungicide (Captan 0.1 %), miticide (Akar 338–0.1 %) and Sticker (Tenae) helped considerably in controlling at least the spread of the disease (Summanwar 1967). Pruning of diseased parts and spraying with diazinon were reported to control the malady (Rai and Singh 1967; Yadav 1972). Regardless of the extent to which sanitation is imposed, it reduces MMD, pre- sumably by reducing inoculum in an orchard. Although it is difficult to impose this treatment on large trees with panicles that are difficult to access, growers may be unwilling to devote the effort that is required to ensure that this approach succeeds. Nonetheless, we regard sanitation as an important component of any integrated strategy to manage this disease.

Insecticides and Fungicides

Pruning followed by a spray of insecticides viz., follidol and/or metasystox as a control measure was recommended (Giani 1965). *Aceria mangiferae* was effectively controlled by a spray of 0.15 % phosphon or Formothion (Wafa and Osman 1972). Significant reduction in the malady was reported by applying various insecticides (Giani 1965; Diekmann et al. 1982).

Fungal theory necessitated the use of systemic fungicides (Varma et al. 1971). It was concluded that copper fungicides were superior to organic fungicides due to excellent tenacity under monsoon conditions. The inhibitory effect of different fungicides viz., Fytolan, hexaferb and captan on mango plants artificially inoculated with *F. moniliforme* var. *subglutinans* was studied (Chattopadhyay and Nandi 1977). Fytolan was maximally effective at all concentration. In view of experimental evidences and economy of fytolan, it may be recommended for controlling malformation of mango in areas where the disease is a serious problem (Chattopadhyay and Nandi 1977). Benomyl failed to control the problem in South Africa (Diekmann et al. 1982) and in India but some success in reducing disease severity by spray application of benomyl has been reported in India (Siddiqui et al. 1987) and in Israel (Pinkas and Gazit 1992). Carbendazim through trunk injection or soil application, either alone or in combination with cultural practices such as root pruning showed no improvement (Kumar and Beniwal 1992).

In in vitro evaluation Carbendazim was found to check the growth of *F. subglutinans* completely at 0.1 %. Hence, a trial was laid out for the control of malformation with spray of Carbendazim (Bavistin 50 W.P.) 0.1 % at 10, 15 and 30 days interval starting from October to February (consisting of 13, 9 and 5 sprays respectively) during the flower bud differentiation stage. It was revealed from the data, that malformation incidence was least in 10 and 15 days interval spray interval schedule and between the two spray schedules, there was no difference. Control of the malformation was up to 76.93 % in 10 and 15 days spray schedule, while it was 65.39 % in 30 days spray schedule over check. Thus, 15 days interval spray schedule is recommended for the control of the disease Misra et al. (2002).

Biopesticides

Usha et al. (2009) clearly showed strong antifungal activity of a concoction brewed from *Datura stramonium, Calotropis gigantea, Azadirachta indica (neem)* and cow manure (T_1) followed by methanol-water (70/30 v/v) extracts of *Datura stramonium, Calotropis gigantea* and *Azadirachta indica* (T_2) against *Fusarium* mangiferae. It was proved that the concoction-brewed compost (T1) is effective, inexpensive, easy to prepare and constitutes a sustainable and eco-friendly approach to control floral malformation in mango when it is sprayed at bud break stage and again at fruit set stage. Kumar et al. 2009 further evaluated leaf extract of 23 plants for their antifungal activity against *F. moniliforme* var. *subglutinans.* Although, all the leaf extracts checked the radial growth of test fungus, extracts of *Azadirachta indica* A. Juss., *Achyrenthes roseus* and *Calotropis gigantea* were found more effective against *F. subglutinans under in vitro conditions.* However, the leaf extract from *Aegle marmelos* (L.) Corr., *Ricinus communis* L. and *Ficus racemosa* L. were found less effective.

Three different species of *Trichoderma (T.viride, T. virens* and *T. harzianum)* were tested against the malformation pathogen *in vitro.* All the three bioagents were effective against all evaluated isolates of *Fusarium* and inhibit the growth. However, out of the three bioagents best result was obtained with *T. harzianum* followed by *T. virens* and *T. viride.* Results clearly showed that the per cent inhibition of *Fusarium* isolates by *T. harzianum* was significantly superior to *T. viride* for all the isolates (Kumar et al. 2012).

Conclusion

Fusarium complex is the dominant pathogen of mango malformation disease. Based on the background knowledge it is further utilize in genomics and transcriptome approaches towards the understanding of Fusarium-mango (host-pathogen) relationship. Identification of key pathogenicity responsive genes, pathways of signal transduction and what were the possible solutions that can be employed for eradication of mango malformation disease at genetic level and help orchardists for disease free saplings and effective quarantine measures.

Acknowledgments Author (Pradeep Kumar) would like to thanks Ben-Gurion University of the Negev, Israel and Director, CISH (ICAR), Lucknow, India.

References

Andotra PS, Chib HS, Gupta BR (1984) Studies on the fungal etiology and behaviour of mango malformation under Jammu conditions. Indian J Mycol Plant Pathol 14:90–92

Anonymous (2013) Accessed 2 Oct 2013 at: http://www.daff.qld.gov.au/plants/health-pests diseases/a-z-significant/mango-malformation-disease

Arif M, Pani DR, Zaidi NW, Singh US (2011) PCR-Based Identification and Characterization of Fusarium sp. Associated with Mango Malformation. Biotechnol Res Int 2011:Article ID 141649. doi:10.4061/2011/141649

Bastawros MK (1996) Mango malformation in Egypt. Acta Hortic 455:566–574

Bhatnagar SS, Beniwal SPS (1977) Involvement of *Fusarium oxysporum* in causation of mango malformation. Plant Dis Report 61:894–898

Bindra OS, Bakhetia DRC (1971) Investigations on the etiology and control of mango malformation. Indian J Hort 28:80–85

Britz H, Steenkamp ET, Coutinho TA, Wingfield BD, Marasas WFO, Wingfield MJ (2002) Two new species of *Fusarium* section Liseola associated with mango malformation. Mycologia 94:722–730

Chadha KL, Pal RN, Sahay RK (1979) Effect of abscission promoting chemicals on deblossoming of mango. Indian J Hortic 36:238–244

Chakrabarti DK (2011) Mango malformation. Springer Press, New York

Chakrabarti DK, Ghosal S (1985) Effect of *Fusarium moniliforme* var. *subglutinans* infection on mangiferin production in the twigs of *Mangifera indica*. Phytopathol Z 113:47–50
Chakrabarti DK, Ghosal S (1989) The disease cycle of mango malformation induced by *Fusarium moniliforme* var. *subglutinans* and the curative effects of mangiferin metal chilates. J Phytopathol 125:238–246
Chattopadhyay NC, Nandi B (1977) Nucleic acid content in malformed mango inflorescence caused by Fusarium moniliforme var. subglutinans. Phytopath Z 89:256–260
Chattopadhyay NC, Nandi B (1981) Nutrition *in Fusarium moniliforme* var. *subglutinans* causing mango malformation. Mycologia 73:407–414
Chib HS, Aandotra PS, Gupta BR (1984) Survey report on the incidence and extent of mango malformation on mango growing areas of Jammu and Kashmir state. Indian J Mycol Plant Pathol 14:86–88
Crespo M, Cazorla FM, Hermoso JM, Guirado E, Maymon M, Torés JA, Freeman S, de Vicent A (2012) First report of mango malformation disease caused by *Fusarium* mangiferae in Spain. Plant Dis 96:286
Crookes CA, Rijkenberg FHJ (1985) A literature review of the distribution, symptomatology, cause and control of mango blossom malformation. S Afr Mango Grower's Assoc Res Rep 5:15–24
Crowhurst RN, Hawthron BT, Rikkerink EHA, Templeton MD (1991) Differentiation of *Fusarium solani* f. sp. *Cucurbitae* races1 and 2by Random amplification of polymorphic DNA. Curr Genet 20:391–396
Desai MV, Patel KP, Patel MK, Thirumalacher MJ (1962) Control of mango malformation in Gujarat. Curr Sci 31:392–393
Diekmann F, Manicom BO, Coetzee K (1982) An attempt to control blossom malformation of mangoes with chemical sprays. Citrus and subtropical Fruit Institute. Informatio Bulletin No. 117:15–16
FAOSTAT (2013) http://faostat.fao.org/site/291/default.aspx FAO, Rome. Accessed Sept 2013
Freeman S, Maimon M, Pinkas Y (1999) Use of GUS transfomants of *Fusarium subglutinans* for determining etiology of mango malformation. Phytopathology 89:456–460
Freeman S, Maimon M, Pinkas Y (2000) Etiology of mango malformation disease using GUS transformants of Fusarium subglutinans. Acta Hort 509:731–758
Freeman S, Maymon M (2000) Reliable detection of the fungal pathogen Fusarium oxisporum f. sp. albedinis, causal agent of bayound disease of date palm using molecular technique. Phytoparasitica 28:341–384
Freeman S, Otero-Colina G, Rodríguez-Alvarado G, Fernández-Pavía S, Maymon M, Ploetz RC, Aoki T, O'Donnell K (2014a) First report of mango malformation disease caused by *Fusarium* pseudocircinatum in Mexico. Plant Dis 98(11):1583
Freeman S, Shtienberg D, Maymon M, Levin AG, Ploetz RC (2014b) New insights into mango malformation disease epidemiology lead to a new integrated management strategy for subtropical environments. Plant Dis 98(11):1456–1466
Gamliel-Atinsky E, Freeman S, Sztejnberg A, Maymon M, Belausov E, Palevsky E (2007) Interactions of the mango bud mite, *Aceria mangiferae*, with *Fusarium mangiferae*, the causal agent of mango malformation disease. IOBC WPRS Bull 30:23–28
Gamliel-Atinsky E, Freeman S, Sztejnberg A, Maymon M, Ochoa R, Belausov E, Palevsky E (2009) Interaction of *Aceria mangiferae* with *Fusarium mangiferae*, the causal agent of mango malformation disease. Phytopathology 99:152–159
Gamliel-Atinsky E, Freeman S, Maymon M, Belausov E, Ochoa R, Skoracka A, Peña J, Palevsky E (2010) The role of eriophyoids in fungal pathogen epidemiology, mere association or true interaction? Exp Appl Acarol 51:191–204
Giani MA (1965) Malformation of mango inflorescence. Pak J Agric Res 3:248–251
Goldman T, Horin M, Pinkas Y (1976) Mango malformation disease in Israel. Alon Hanotea 9:583–589
Grajal-Martin MJ, Simon CJ, Muehlbauer FJ (1993) Use of random amplified polymorphic DNA (RAPD) to characterize race 2 of Fusarium oxysporum f. sp. pisi. Phytopathology 83:612–614

Haggag MW, Zakaria A, Hazza M, Sehab FS, El-Wahab MA (2011) Histopathological evaluation of the *Fusarium* species causing malformation disease of Mangifera indica. J Plant Pathol Microbiol 2:103

Iqbal Z, Rahman UM, Dasti AA, Saleem A, Zafar Y (2006) RAPD of *Fusarium* isolates causing mango malformation disease in Pakistan. World J Microb Biotechnol 22:1161–1167

Iyer CPA, Schnell RJ (2009) Breeding and genetics. In: Litz RE (ed) The mango: botany production and uses, 2nd edn. CAB International, Wallingford, pp 67–96

Jamil FF, Sarwar N, Sarwar M, Khan JA, Geistlinger J, Kahl G (2000) Genetic and pathogenic diversity within *Ascochyta rabiei* (Pass.) Lab. population in Pakistan causing blight of chikpae (*Cicer arientium* L.). Physiol Mol Plant Pathol 57:243–254

Jana T, Sharma TR, Prasad RD, Arora DK (2003) Molecular characterization of Macrophomina phaseolina and Fusarium species by a single primer RAPD technique. Micrbiol Res 158:249–257

Khurana AD, Gupta OP (1973) Mango malformation in India. Pesticides 7:12–13

Kumar J, Beniwal SPS (1991) Mango malformation. In: Kumar J, Chaube HS, Singh US, Mukhopadhyay AN (eds) Plant diseases of international importance, vol III, Diseases of Fruit Crops. Prentice Hall, Englewood Cliffs, pp 357–393

Kumar R, Beniwal SPS (1992) Role of Fusarium species in the etiology of mango malformation. In: 4th International mango symposium, Miami, 5–10 July, 133 pp (Abst.) p 17

Kumar J, Singh US, Beniwal SPS (1993) Mango malformation: one hundred years of research. Annu Rev Phytopathol 31:217–232

Kumar P, Misra AK, Pandey BK, Misra SP, Modi DR (2009) Eco- friendly management of mango malformation *Fusarium moniliforme* var. *subglutinans* from leaf extracts. J Eco-Friendly Agric 4(1):61–64

Kumar P, Misra AK, Modi DR (2011a) Current status of mango malformation in India. Asian J Plant Sci 10:1–23

Kumar P, Misra AK, Srivastava AK, Modi DR (2011b) Mapping of *Fusarium moniliforme var. subglutinans* from normal and malformed panicles and seedlings of mango by recovery method. Plant Arch 11:567–569

Kumar P, Misra AK, Modi DR, Gupta VK (2012) Biocontrol potential of Trichoderma species against mango malformation pathogens. Arch Phytopathol Plant Prot 45:1237–1245

Kumar P, Gupta VK, Misra AK, Modi DR (2014) Molecular characterization of *Fusarium* moniliforme var. subglutinans isolates. J Environ Biol 35(1):211–216

Kush RS, Beeker E, Wach M (1992) DNA amplification polymorphisms of the cultivated mushroom Agaricus bisporus. Appl Environ Microbiol 58:2971–2977

Kvas M, Steenkamp ET, Al Adawi AO, Deadman ML, Al Jahwari AA, Marasas WFO, Wingfield BD, Ploetz RC, Wingfield MJ (2008) Fusarium mangifarae associated with mango malformation in Sultanate of Oman. Eur J Plant Pathol 121:195–199

Labuschang TI, Joubert MH, Steyn A (1993) Role of the mango bud mite, *Aceria mangefera* (Sayed) in mango malformation. Inlightings Bull Instituut-Trop-Subtrop Gewaise 246:19–24

Lim TK, Khoo KC (1985) Diseases and disorders of mango in Malaysia. Tropical Press, Kuala Lumpur

Lima CS, Pfenning LH, Costa SS, Campos MA, Leslie JF (2008) A new lineage within the Gibberella fujikuroi species complex is the main causal agent of mango malformation disease in Brazil. Plant Pathol 58:33–42

Lima CS, Monteiro JHA, Crespo NC, Costa SS, Leslie JF, Pfenning LP (2009) VCG and AFLP analyses identify the same groups in the casual agents of mango malformation in Brazil. Eur J Plant Pathol 123:17–26

Lima CS, Pfenning LH, Costa SS, Abreu LM, Leslie JF (2012) *Fusarium* tupiense sp. nov., a member of the Gibberella fujikuroi complex that causes mango malformation in Brazil. Mycologia 104:1408–1419

Litz RE (ed) (2009) The mango: botany, production and uses, 2nd edn. CAB International, Wallingford

Lv YC, Pu JJ, Qia YX, Xie YX, Lu Y, Zhang X, Zhang H, Zhang HQ (2013) *Fusarium* proliferatum caused mango malformation disease in Panzhihua and Huaping Provinces of China. Acta Hortic 992:423–428

Majumder PK, Sinha GC (1972a) Studies on the effect of malformation on growth, sex ratio, fruit set and yield of mango. Acta Hort 24:230–234

Marasas WFO, Ploetz RC, Wingfield MJ, Wingfield BD, Steenkamp ET (2006) Mango malformation disease and the associated *Fusarium* species. Phytopathology 96:667–672

Minessy FA, Biely MP, El-Fahl A (1971) Effect of iron chelates in correcting malformation of terminal bud growth in mango. Sudan Agric J 6:71–74

Misra AK, Singh VK (1998) Fusarium moniliforme var. subglutinans in relation to mango malformation. In: National symposium on Mango Production and Export, June 25–27, Central Institute for Subtropical Horticulcure, Lucknow, pp 71–72

Misra AK, Singh VK (2002) *Fusarium subglutinans* (*F. moniliforme var. subglutinans*) in relation to mango malformation. Indian J Plant Pathol 20:81–83

Misra AK, Negi SS, Rajan S, Ram K (2000) Cultivar Elaichi-a new source of resistance to mango malformation. Proceedings Indian Phytopathological Society-Golden Jubilee, International Conference on Integrated Disease Management for Sustainable Agriculture Vol. II Pub. Indian Phytopathological Society, New Delhi, pp 750–751

Mukerjee SK, Litz RE (2009) Introduction, botany and importance. In: Litz RE (ed) The mango: botany production and uses, 2nd edn. CAB International, Wallingford, pp 1–18

Narasimhan MJ (1954) Malformation of panicles in mango incited by spp. of *Eriophyes*. Curr Sci 23:297–298

Narasimhan MJ (1959) Control of mango malformation disease. Curr Sci 28:254–255

Nariani TK, Seth ML (1962) Role of eriophyid mites in causing malformation disease of mango. Ind Phytopathol 15:231–234

National Horticulture Board (2014) Handbook on horticulture statistics 2014, government of India, ministry of agriculture. Department of agriculture and cooperation, New Delhi. http://agricoop.nic.in/imagedefault/whatsnew/handbook2014.pdf

Nor NMIM, Salleh B, Leslie JF (2013) *Fusarium* species associated with mango malformation in peninsular Malaysia. J Phytopathol 161:617–624

O'Donnell K, Cigelnik E, Nirenberg HI (1998) Molecular systematics and phylogeography of the Gibberella fujikuroi species complex. Mycologia 90:465–493

Otero-Colina G, Rodríguez-Alvarado G, Fernández-Pavía S, Maymon M, Ploetz RC, Aoki T, O'Donnell K, Freeman S (2010) Identification and characterization of a novel etiological agent of mango malformation disease in México, *Fusarium* mexicanum sp. nov. Phytopathology 100:1176–1184

Pal RN, Chadha KL (1982) Deblossoming mangoes with cycloheximide. J Hortic Sci 57: 331–332

Pinkas Y, Gazit S (1992) Mango malformation control strategies. In: 4th Int. mango symposium, Miami, 5–10 Jul (Abstr) P 22, p 133

Ploetz RC (2001) Malformation: a unique and important disease of mango, *Mangifera indica* L. In: Summerell BA, Leslie JF, Backhouse D, Bryden WL (eds) *Fusarium*: Paul E. Nelson memorial symposium. American Phytopathological Society, St Paul, pp 233–247

Ploetz RC, Freeman S (2009) Foliar, floral and soilborne diseases. In: Litz RE (ed) The mango: botany production and uses, 2nd edn. CAB International, Wallingford, pp 231–302

Ploetz R, Zheng QI, Vazquez A, Abdel Sattar MA (2002) Current status and impact of mango malformation in Egypt. Int J Pest Manag 48:279–285

Popenoe W (1932) Manual of tropical and subtropical fruits. Macmillan Co., New York

Prasad A, Nirvan RS, Singh S (1972) Mango malformation: a review of work done at Horticultural Research Institutes, Saharanpur (U.P.). Acta Hortic 24:227–229

Purseglove JW (1972) Mangoes west of India. Acta Hortic 24:107–174

Rai BK, Singh N (1967) An observation on recovery from malformation in mango sapling. Curr Sci 36:525–526

Rajan S (1986) Biochemical basis of mango (Mangifera indica L.) malformation. Ph.D. Thesis, I.A.R.I., New Delhi, pp 67

Rodríguez-Alvarado G, Fernández-Pavía SP, Ploetz RC, Valenzuela-Vázquez M (2007) A *Fusarium* sp., different from *Fusarium* oxysporum and F. mangiferae, is associated with mango malformation in Michoacán, Mexico. Plant Pathol 57:781

Rodríguez-Alvarado G, Fernández-Pavía S, Otero-Colina G, Ploetz RC, Aoki T, O'Donnell K, Maymon M, Freeman S (2013) Identification and characterization of *Fusarium mexicanum* causing mango malformation disease in México. Acta Hortic 992:377–384

Sabir JSM (2006) Genotype identification for some *Fusarium sambusinum* strain isolated from wheat in upper Egypt. World J Agric Sci 2:6–10

Saharan MS, Naef A, Kumar J, Tiwar R (2006) Characterization of variability among isolates of *Fusarium graminearum* associated with head scab of wheat using DNA markers. Curr Sci 92:230–235

Saleem A (2004) Studies on mango malformation pathogen. 2nd ALP international workshop. NARC, Islamabad, p 35

Senghor AL, Sharma K, Kumar PL, Bandyopadhyay R (2012) First report of mango malformation disease caused by *Fusarium tupiense* in Senegal. Plant Dis 96:1582

Sharma RR, Sharma HC, Goswami AM (1994) Polyphenol oxidase activity and phenolic content pattern during shoot development of grape (*Vitis vinifera* L.) in different growing seasons. J Plant Biochem Biol 3:145–147

Siddiqui S, Sandooja JK, Mehta N, Yamadagni R (1987) Biochemical stages during malformation in mango cultivars as influenced by various chemicals. Pesticides 21:17–19

Singh Z, Dhillon BS (1986) Effect of naphthalene acetic acid, ethrel, dikegulac and hand deblossoming on floral malformation, flowering, yield and fruit quality of mango (*Mangifera indica* L.). Acta Hortic 175:307–313

Singh Z, Dhillon BS (1988b) Influence of deblossoming on floral malformation of mango (*Mangifera indica* L.). Tropical Agric 65:295–296

Singh RN, Majumder PK, Sharma DK, Sinha GC, Bose PC (1974) Effect of de-blossoming on the productivity of mango. Sci Hortic 2:399–403

Singh LS, Singh SM, Nirvan RS (1961) Studies on mango malformation review, symptoms, extent, intensity and cause. Hort Adv 5:197–207

Sinniah GD, Adikaram NKB, Vithanage ISK, Abayasekara CL, Maymon M, Freeman S (2013) First report of mango malformation disease caused by *Fusarium* mangiferae in Sri Lanka. Plant Dis 97:427

Steenkamp ET, Britz H, Coutinho TA, Wingfield BD, Marasas WFO, Wingfield MJ (2000) Molecular characterization of *Fusarium* subglutinans associated with mango malformation. Mol Plant Pathol 1:187–193

Sternlicht M, Goldenberg S (1976) Mango eriophyid mites in relation to inflorescence. Phytoparasitica 4:45–50

Summanwar AS (1967) Mango malformation: serious economic consequences. Indian Hortic 11:12–16

Summanwar AS, Raychaudhuri SP (1968) The role of eriophyid mite (*Aceria mangiferae*) in the causation of mango malformation. Indian Phytopathol 21:463–464

Summanwar AS, Raychaudhuri SP, Pathak SC (1966) Association of fungus *Fusarium moniliforme* Sheld. with the malformation in mango. Indian Phytopathol 19:227–228

Tripathi PS, Ram S (1998) Effect of deblossoming on panicle regeneration and control of malformation in mango. Prog Hortic 30:40–43

Usha K, Singh B, Praseetha P, Deepa N, Agarwal DK, Agarwal A, Nagaraja A (2009) Antifungal activity of *Dhatura stramonium, Calotropis gigantean* and *Azadirachta indica* against *Fusarium mangiferae* and floral malformation in mango. Eur J Plant Pathol 124:637–657

Varma A (1983) Mango malformation. In: Singh KG (ed) Exotic plant quarantine pests and procedures for introduction of plant materials. Asean Plant Quarantine Centre and Training Institutes, Serdang, pp 173–188

Varma A, Lele VC, Majumder PK, Ram A, Sachchidananda J, Shukla US, Sinha GC, Yadava TD, Raychaudhuri SP (1969) Mango malformation. In: Proceedings of ICAR workshop on fruit research, Ludhiana, 28th–30th Apr, pp 1

Varma A, Lele VC, Raychaudhuri SP, Ram A (1971) Preliminary investigations on epidemiology and control of mango malformation. Proc Indian Natl Sci Acad 37:291–300

Varma A, Raychaudhuri SP, Lele VC, Ram A (1972) Towards the understanding of the problem of mango malformation. Acta Hortic 24:237

Varma A, Lele VC, Goswami BK (1974) Mango malformation. In: Raychaudhury SP, Varma JP (eds) Current trends in plant pathology, Lucknow S.N. Dasgupta Birthday Celebration Committee. pp 196–208

Varma A, Lele VC, Raychaudhuri SP, Ram A, Sang A (1974b) Mango malformation-a fungal disease. Phytopathol Z 79:254–257

Wabha ML, El-Enany MAM, Farrag AMI (1986) Five mango varieties as affected by malformation phenomenon and bud mite infestation in Egypt. Agric Res Rev 61:193–201

Wafa AK, Osman RS (1972) Control of bud mite *Aceria mangiferae* Sayed on Mango tree in Egypt (Acarina). Bull Ent Soc Egypt Econ Ser 6:231–238

Williams JGK, Kubelik AR, Livak KJ, Rafalski JA, Tingey SV (1990) DNA polymorphisms amplified by arbitrary primers are useful as genetic markers. Nucleic Acids Res 8:6531–6535

Yadav TD (1972) Role of mango bud mite, *Aceria mangiferae*, in mango malformation. Acta Hortic 24:238

Yadava RBR, Singh VK (1995) Extent of floral malformation and its relationship with physiochemical components bin mango cultivars. Indian J Plant Physiol 38(4):328–330

Youssef SA, Maymon M, Zveibil A, Klein-Gueta D, Sztejnberg A, Shalaby AA, Freeman S (2007) Epidemiological aspects of mango malformation disease caused by *Fusarium mangiferae* and source of infection in seedlings cultivated in orchards in Egypt. Plant Pathol 56:257–263

Zhan RL, Yang SJ, Ho HH, Liu F, Zhao YL, Chang JM, He YB (2010) Mango malformation disease in South China caused by *Fusarium proliferatum*. J Phytopathol 158:721–725

Zhan RL, Yang SJ, Liu F, Zhao YL, Chang JM, He YB (2012) First report of *Fusarium* mangiferae causing mango malformation in China. Plant Dis 96:762

Zheng Q, Ploetz R (2002) Genetic diversity in the mango malformation pathogen and development of a PCR assay. Plant Pathol 51:208–216

Paradigm Shift in Plant Disease Diagnostics: A Journey from Conventional Diagnostics to Nano-diagnostics

Prachi Sharma and Susheel Sharma

Introduction

Disease diagnosis can be defined as either the identification of a disease or the identification of the agent that causes the disease. The early and rapid identification of a plant pathogen is the key for appropriate management practices to be applied prior for avoiding further spread of the disease or its introduction due to transboundary movement of plant material. The demand for rapid, accurate, sensitive, standard, high throughput and simultaneous detection assays of plant pathogens has risen in the last few decades due to intensive monocropping. Conventional methods relied on study of symptoms and morphological studies. However, it is well known fact that symptoms are not always unique and can be confused with other diseases. Conventional methods are often time-consuming, laborious, and require expert taxonomist. The limitations posed by conventional diagnostics have led to the development of techniques with improved accuracy and reliability. Present era demands fast and sensitive methods for detection and identification of specific fungal pathogens. Accordingly, Plant disease diagnosticians have an array of methodologies that allow much faster, more specific, more sensitive, more accurate and multiple detection of plant fungal pathogens, leaving the need of skilled taxonomist. The present chapter highlights several techniques developed that have revolutionized the field of plant diagnostics.

P. Sharma (✉)
Division of Plant Pathology, Sher-e-Kashmir University of Agricultural Sciences and Technology of Jammu (SKUAST-J), Jammu, Jammu and Kashmir, India
e-mail: prachisharma15@gmail.com

S. Sharma
School of Biotechnology, Sher-e-Kashmir University of Agricultural Sciences and Technology of Jammu (SKUAST-J), Jammu, Jammu and Kashmir, India

P. Kumar et al. (eds.), *Current Trends in Plant Disease Diagnostics and Management Practices*, Fungal Biology, DOI 10.1007/978-3-319-27312-9_11

Conventional Diagnostics

The conventional way of identification of plant pathogenic fungi involves interpreting visual symptoms of disease or signs like presence of structures: mycelia, sporophore, spores and fructifications of the pathogen (Nezhad 2014). It may be followed by pathogen confirmation using microscopic and cultural techniques. Most of the traditional fungi were identified by taxonomists based on spore morphology and conidiogenesis (Kendrick 1971; Barnett and Hunter 1972; Agrios 2005). The methods related to fungal morphology require extensive knowledge of classical taxonomy. Other limitations include the difficulty of some species to be cultured in vitro, and the inability to accurately quantify the pathogen (Goud and Termorshuizen 2003). Based on symptomatology, cultivar susceptibility and epidemiology *Fusarium oxysporum* f. sp. *radicis-cucumerinum* (FORC) was identified as a different *forma specialis* and was distinguished from *F. oxysporum* f.sp. *cucumerinum* (FOC) (Vakalounakis 1996). The other conventional methods namely: direct inspection of dry seeds, washing test, soaking test, incubation tests, blotter tests, embryo count test and; filter and centrifuge extraction technique were often used for detection of seed borne fungal pathogens (Castro et al. 1994). Visual inspection is relatively easy when symptoms clearly are characteristic of a specific disease. The direct inspection method has been used for detection of seed borne pathogens e.g. Presence of sclerotia of *Sclerotinia sclerotiorum* and *Claviceps purpurea*. Incubation methods such as blotter and agar plate were most popular and frequently used for the detection of a great number of seed-transmitted pathogens (De Tempe and Binnerts 1979; Majumder et al. 2013). Culturing is another traditional method requiring a few days or weeks to detect the presence of a pathogen in a plant. Direct agar plating technique (DAPT) in which acidified potato dextrose agar (APDA having pH=3.5 amended with 25 % strength lactic acid per litre of medium) has been used to detect latent infections in asymptomatic nuts and fruits (Michailides et al. 2005). These methods are cheap and simple but are time consuming, laborious and require skilled labour and cannot diagnose pathogens before the symptoms are observable.

Physical Diagnostics

Electron Microscopy (EM)

EM is one of the most important tool since its first use in Germany in 1939. In electron microscope, a beam of short wave electrons are used instead of visible light passing through conventional light microscopes. It works under high vacuum and focusing is done by electromagnetic/electrostatic lenses (Bos 1983). The specimen to be studied is mounted on a copper grid containing apertures covered with a thin film of plastic (formvar) (Noordam 1973).

Scanning Electron Microscopy (SEM)

Introduction of scanning electron microscopy (SEM) has revolutionized the study of the microscopic world owing to high quality three dimensional images, large magnitude of increase from 10 to 1,000,000 times, rapid processes of image digitalization and acquisition, easiness to prepare and operate samples, as well as affordable costs (Bozzola and Russell 1999). SEM has been potentially used for identification and detection of seed-borne fungi e.g. *Colletotrichum lindemuthianum* in seeds of common bean (*Phaseolus vulgaris* L.), *Colletotrichum truncatum* in maize (*Zea mays* L.) and *Colletotrichum gossypii* var. *cephalosporioides* in cotton (*Gossypium hirsutum* L.) (De Carvalho Alves and Pozza 2012). SEM was also employed effectively for the detection of destructive pathogen *Sphaeropsis sapinea* (anamorph-*Diplodia pinea*) infecting conifers. The pathogen differes in conidial morphology which cannot be recognized by light microscope. However, SEM was able to group 30 isolates of *S. sapinea* into Type-A having smooth conidial surface and Type-B having pits distributed over the conidial surface (Wang et al. 1985).

Cryo-Scanning Electron Microscopy (Cryo -SEM)

Cryo-SEM is used for imaging of samples containing moisture without causing drying artifacts such as extraction, solubilization and shrinkage. This is a rapid method which enables three-dimensional in situ visualization of fungal invasion within roots and is broadly applicable for identification of plant diseases caused by necrotrophic fungi. This method was effectively used to visualize internal infection of wheat (*Triticum aestivum*) roots by the pathogenic fungus *Rhizoctonia solani* AG-8. Cryo-SEM helped in retaining fungal hyphae and root cells in situ in disinteged root tissues, avoiding the distortions that are usually introduced during conventional preparation by chemical fixation, dehydration and embedding (Refshauge et al. 2006). Cryo-SEM has also been used to show that hyphae of *Bipolaris sorokiniana* remain adhered to the wax surfaces of barley leaves by means of an extensive extracellular matrix (Jansson and Akesson 2003).

Serological Diagnostics

Serology is the use of specific antibodies to detect their respective antigens in test samples. Antibodies are composed of immunoglobulin (Ig) proteins produced in the body of the vertebrate in response to the antigens which are usually foreign proteins, complex carbohydrates, polynucleotides or lipopolysaccharides. Each antibody is specific to a particular antigen and binds to it. Antibodies are produced by B lymphocytes and include five classes- IgG, IgM, IgA, IgE and IgD (Hull 2002). The major soluble antibody IgG (Gamma immunoglobulins) is the most commonly used and is Y shaped molecule with two antigen binding sites (Dickinson 2005). Serum containing antibodies is known as antiserum which is of two types: polyclonal, containing antibodies to all the available epitopes on the antigen, and monoclonal, containing antibodies to single epitope (Hull 2002) Polyclonal antisera was first produced in rabbits as they are most convenient animal, easy to keep, easy to inject with antigens and relatively a straight-forward process to extract blood containing antibodies by slight cut to the ear vein. Later, other animals such as cows, rats, mice and chickens were also used (Singh 2005). Monoclonal antiserum is raised in mice (Harlow and Lane 1988) and is highly specific against a single type of antigen. Both polyclonal and monoclonal antibodies are now available commercially against a wide range of fungal pathogens from companies such as Adgen®, Neogen® Agdia®, Loewe® and Bioreba®. Numerous serological techniques have been developed and are being used for specific detection and identification of plant fungal pathogens as:

Immuno-sorbent Electron Microscopy (ISEM)

This technique was introduced by Derrick (1973) as serologically specific electron microscopy (SSEM) and has been widely used in plant virology (Milne 1972; Milne and Luisoni 1977). Because of its similarity with solid phase immunoassays, the method was known as immunosorbent electron microscopy by Roberts and Harrison (1979). ISEM has been extensively used for detection and in situ characterization of phytoplasma (Musetti and Favali 2004). ISEM combines the specificity of serological assays with the visualization capabilities of the EM. It is an ideal confirmatory test requiring small amount of samples, if the EM facility and specific antisera are available. ISEM method involves the production of antibodies against the fungal pathogen/antigen and linkage to the antibodies to protein A-gold complexes to locate the antigen (Narayanasamy 2011) e.g. Monoclonal antibodies have been raised against the species-specific eipitopes on the surface of zoospores and cysts *of Phytophthora cinnamomi* to detect six isolates of *P. cinnamomi* and six species of *Pythium* using immunoflorescence (Hardham et al. 1986).

Enzyme Linked Immuno-sorbent Assay (ELISA)

ELISA, a solid phase heterogeneous immunoassay has been proved to be a valuable serological tool in detection of plant fungal pathogens (Casper and Mendgen 1979; Johnson et al. 1982). ELISA is based on the specific recognition capabilities of antibodies. These antibodies are usually derived from the immunization of animals (usually rabbits, mice, chicken or goat) with certain immunogens such as culture filtrates or mycelial compounds. After repeated injections of the immunogen, blood samples are taken and the serum is used either as a whole or it is applied after certain clean-up steps for the ELISA tests. Polyclonal antibodies are mostly used, but often lead to problem of high background due to reactions with host proteins. However, monoclonal antibodies increasingly available from commercial companies have overcome the problem of background reactions with host proteins. Different monoclonal antibodies are being used to detect specific strains of a pathogen or to detect a group of strains or species. Many variations of ELISA have been developed and include: direct and indirect ELISA procedures. They differ in the way the antigen–antibody complex is detected, but the underlying theory and the final results are the same. Mostly, double antibody sandwich form of ELISA (DAS-ELISA), which is a direct form of ELISA, as described by Clark and Adams (1977) is used for the detection of plant pathogens. In direct ELISA procedures, the antibodies (usually as an IgG fraction of the antiserum) bound to the well surface of the microtitre plate (polystyrene inflexible rigid plates or polyvinyl chloride, flexible plates) capture the fungus. The captured fungus sample is detected by incubation with an antibody-enzyme conjugate followed by addition of color development reagents (substrate or substrate/dye combination). The capturing and detecting antibodies can be from the same or different sources. Since the pathogen is sandwiched between two antibody molecules, this method is called the double antibody sandwich (DAS-ELISA). DAS-ELISA procedure is known to be highly specific and often detects closely related strains (Koenig 1978). There are also several alternative indirect forms of ELISA available for fungus detection. One among them includes: direct antigen coating procedure (DAC-ELISA), the method developed by Mowat (1985) with minor modifications (Hobbs et al. 1987). In this method, plant extracts prepared in a carbonate buffer are applied directly to the wells and antibodies raised in two different animal species are used. This method is by far the simplest of all the forms of ELISA test (Reddy et al. 1988). There are number of examples in which ELISA has been employed for specific detection of plant pathogenic fungi. E.g. ELISA method was used as a tool to detect fungal pathogens *Rhizoctonia solani*, *Pythium* spp. and *Sclerotinia homoeocarpa* in turfgrass (Fidanza and Dernoeden 1995; Shane 1991) and for early detection of karnal bunt pathogen in wheat when the infection levels are very low (Varshney 1999). DAC-ELISA was standardized and used for detection of *Collectotrichum falcatum* causing red rot of sugarcane (Hiremath and Naik 2003). Polyclonal antibodies (IgG K91) were raised to detect a quarantine pathogen of strawberry – *Colletotrichum acutatum* using ELISA (Kratka et al. 2002) and against *Aspergillus oryzae*, the common plant pathogenic fungi found associated with wheat, sorghum and other crops (Kamraj et al. 2012) for use in ELISA.

Dot Immuno-binding Assay (DIBA)

DIBA technique is similar to ELISA in principle except that the plant extracts are spotted on to a nitrocellulose or nylon membrane rather than using a polystyrene plate as the solid support matrix. Unlike in ELISA, where a soluble substrate is used for color development, a precipitating (chromogenic) substrate is used for probing pathogen. Hydrolysis of chromogenic substrates results in a visible coloured precipitate at the reaction site on the membrane. DIBA procedure was applied for the detection of the resting spores of *Plasmodiophora brassicae*, causing club root disease of crucifers (Orihara and Yamamoto 1998).

Tissue Immuno Blot Assay (TIBA)

TIBA is a variation of DIBA in which a freshly-cut edge of a leaf blade, stem, leaf, tuber, root or an insect is blotted on the membrane, followed by detection with labelled antibodies. This method is simple, does not require elaborate sample preparation or extraction, and even can provide information on the distribution of fungal pathogen in plant tissues. The disadvantages of DIBA and TIBA are possible interference of sap components with the subsequent diagnostic reactions. However, their sensitivity, the relatively short time required to assay large numbers of samples, the need for minimum laboratory facilities for the assay, the ability to store blotted membranes for extended periods, and low costs favor TIBA and DIBA as useful diagnostic techniques. The other advantage is that the samples can be blotted onto the membranes right in the field and such membranes can be carried or shipped by mail for further processing at a central location either within the country or in a different country. A direct tissue immuno blot assay (DTIBA) procedure has been developed to detect *Fusarium* spp. in the transverse sections from stems and crown of tomato and cucumber plants by employing a combination of the MAb (AP19-2) and Fluorescein 5-isothiocyanate (FITC)-conjugated antimouse IgM-sheep IgG (Arie et al. 1995).

Lateral Flow Assay (LFA)

LFA is a one step, fast, simple, versatile based on the serological specificity of polyclonal or monoclonal antibodies. Lateral flow assay also referred to as Dipstick method is equivalent to medical detection systems such as pregnancy kits (Dickinson 2005). The advantages of these devices are that they are simple to use and results are quick, usually in less than 10 min. Lateral Flow Devices (LFDs) typically consist of a porous nitrocellulose membrane bound to a narrow plastic strip on which pathogen-specific antibodies are immobilized in a band partway up the strip. Species-specific antibodies bound to microparticles of latex, colloidal gold, or silica are placed between the band of immobilized antibodies and a sample application pad. The lateral-flow assay kit usually has an inlet for receiving the pathogen infected sap. After the sample fluid is placed in the sample inlet, the sample flows from the sample pad through embedded reagents, in which specific chemical reactions occur by capillary forces. The reaction product continues to flow through the membrane arriving at the capture reagents. The capture reagents are immobilized on the membrane as a band shape. The captured reaction product generates visually distinguishable color on the bands. Typically two bands are formed on the membrane, one of which is a test band for detecting the sample by its concentration, and the other is the control band for confirming the success of the assay. Sample fluids may continue to flow and can be collected in an absorbent pad. The test kit does not require a permanent dedicated space, high-priced instruments and skillful operators. An LFA test was developed using an IgM monoclonal antibody to detect *Rhizoctonia solani* (sensitivity 3 ng ml^{-1} of antigen) (Thornton et al. 2004).

Polymerase Chain Reaction (PCR) Based Diagnostics

PCR is a popular molecular biology in vitro technique developed in 1983 by Kary Mullis for enzymatically replicating DNA. The technique allows a small amount of DNA molecule to be amplified many times, in an exponential manner by repeated cycles of denaturation, polymerisation and elongation at different temperatures using specific oligonucleotides (primers), deoxyribonucleotide triphosphates (dNTPs) and a thermostable *Taq* DNA polymerase in the adequate buffer (Mullis and Faloona 1987). Oligonucleotides, flanking part of the genome of the pathogen, are extended by a thermostable DNA polymerase to increase the copies of target DNA (Webster et al. 2004). PCR technique is extremely sensitive, fairly inexpensive and requires minimal skill to perform. The presence of a specific DNA band of the expected size indicates the presence of the target pathogen in the sample. In 1993, Mullis was awarded the Nobel prize in Chemistry for his work on PCR (Bartlett and Stirling 2003). PCR is commonly carried out in small reaction tubes (0.2–0.5 ml volumes) in a machine called as thermal cycler. The thermal cycler alternatively heats and cools the reaction tubes to achieve the temperatures required at each step of the reaction. Many modern thermal cyclers make use of the Peltier's effect, which permits both heating and cooling of the block holding the PCR tubes simply by reversing the electric current (Rahman et al. 2013). PCR allows the amplification of millions of copies of specific DNA sequences. The amplified DNA fragments are visualized by electrophoresis in agarose gel stained with ethidium bromide (Capote et al. 2012). Several attempts have been made to develop species-specific PCR primers for fungal plant pathogens (Henson and French 1993). Fungal mitochondrial DNA has been widely used as a source of molecular markers for evolution (Bruns et al. 1991), taxonomy (Martin and Kistler 1990) and genetic diversity studies (Forster and Coffey 1993). DNA region mostly targeted for PCR based diagnostic include ribosomal DNA (rDNAs) as it is present in all organisms at high copy number, inter transcribed spacers (ITS) region for developing DNA barcodes to identify the fungal species and β-tubulin genes used extensively for phylogenetics (Sanchez-Ballesteros et al. 2000; Hirsch et al. 2000; Fraaije et al. 2001; White et al. 1990; El-Sheikha and Ray 2014). The intergenic spacer sequence (IGS) primers based PCR have been used to detect and identify *Verticillium dahliae* and *V. alboatrum* (Schena et al. 2004) and to distinguish pathogenic and non-pathogenic

Fusarium oxysporum in tomato (Validov et al. 2011). The ITS region has also been widely used in fungal taxonomy and is known to show variation between species e.g., between *Pythium ultimum* and *P. helicoides* (Kageyama et al. 2007); *Peronospora arborescens* and *P. cristata* (Landa et al. 2007); *Colletotrichum gloeosporioides* and *C. acutatum* (Kim et al. 2008). PCR technology has many applications in plant pathology and several variants of PCR are being used for detection (El-Sheikha and Ray 2014):

Nested PCR

Nested PCR is a modification of standard PCR involving two consecutive PCR runs, in which the first round PCR products are subjected to a second round PCR amplification with more specific primers. Initial primer pair is used to generate PCR products, which may contain products amplified from non-target areas. The products from the first PCR are used as template in a second PCR, using one (hemi-nesting) or two different primers whose binding sites are located (nested) within the first set, thus increasing specificity. In nested PCR mode, two primer pairs are used; one for the outer fragment and other for the inner fragment. These include outer forward primer, outer reverse primer, forward inner primer and reverse inner primer. Usually, the products of the first amplification are transferred to another tube before the nested PCR is carried out using one or two internal primers. Nested PCR requires more detailed knowledge of the sequence of the target and aims to reduce the product contamination due to the amplification of unintended primer binding sites (mispriming). The nested-PCR is an ultrasensitive technique for detection of several plant pathogenic bacteria, fungi and phytoplasma. E.g. Nested PCR primers based on microsatellite regions were designed for *Monilinia fructicola*, the causal agent of brown rot of stone fruits, and *Botryosphaeria dothidea*, the causal agent of panicle and shoot blight of pistachio (Ma et al. 2003). Intra and inter specific variations in *Ustilainoidea virens*, the causal agent of false smut/green smut of rice were utilized for its detection using nested PCR (Young-Li 2004). Similarly, nested PCR primers have been designed for detection of *Gremmeniella abietina*, the causal agent of stem canker and shoot dieback of conifers namely Abies, Picea, Pinus, Larix, Pseudotsuga, and Tsuga based on 18S rDNA sequence variation pattern in (Zeng et al. 2005) and for detection of *Colletotrichum gloeosporioides*, the causal agent of anthracnose in *Camellia oleifera* based on ITS region (Liu et al. 2009). A rapid nested PCR based diagnostic was developed for detection of *Ramularia collo-cygni*, the causal agent of leaf spot of *Hordeum vulgare* based on species specific primers developed from entire nuclear ribosomal internal transcribed spacer and 5.8S rRNA gene (Havis et al. 2006). Nested PCR was successfully developed for detection of *Verticillium dahliae*, the causal agent of verticillium wilt of strawberry (Kuchta et al. 2008). Single nucleotide polymorphism (SNP) in the FOW1 gene in *Fusarium oxysporum* f. sp. *chrysanthemi*, an economically important pathogen of ornamentals namely *Gerbera jamesonii*, *Osteospermum sp.*, and *Argyranthemum frutescens* was exploited for nested PCR (Li et al. 2010).

Multiplex PCR

Multiplex PCR allows the simultaneous and sensitive detection of different DNA targets in a single reaction reducing time and cost (Webster et al. 2004; Lopez et al. 2009). Multiplex PCR is useful in plant pathology because different fungi frequently infect a single host and consequently sensitive detection is needed and helps in reducing the number of tests required (James et al. 2006). Different fragments specific to the target fungi are simultaneously amplified and identified based on molecular sizes on agarose gels but care is needed to optimize the conditions so that respective amplicons can be generated efficiently. E.g. To detect and quantify four foliar fungal pathogens in wheat namely *Septoria tritici* (leaf blotch) and *S. nodorum* (leaf and glume blotch), the β-tubulin gene was used as the target region (Fraaije et al. 2001). Multiplex PCR has been effectively utilized for detecting *Phytophthora lateralis* in cedar trees (Winton and Hansen 2001), for determining the mating type of the pathogens *Tapesia yallundae* and *T. acuformis* (Dyer et al. 2001), to differentiate two pathotypes of *Verticilliun albo-atrum* infecting hop (Radisek et al. 2006) and for distinguishing 11 taxons of wood decay fungi infecting hardwood trees (Guglielmo et al. 2007). Multiplex PCR technique was also used for the simultaneous detection and differentiation of powdery mildew fungi: *Podosphaera xanthii* and *Golovinomyces cichoracearum* infecting sunflower (Chen et al. 2008).

Cooperational PCR (Co-PCR)

Co-PCR is a highly sensitive method of amplification, originally developed for detection of plant viruses. This technique involves the use of four primers and reaction process consists of the simultaneous reverse transcription of two different fragments from the same target, one containing the other. Four amplicons are produced by the combination of the two pair of primers, one pair external to other and largest fragment is produced due to the co-operational action of amplicons (Olmos et al. 2002) it can be coupled with dot blot hybridization, making it possible to characterize the nucleotide sequence (Bertolini et al. 2007). E.g. Co-operational PCR coupled with dot blot hybridization was developed for the detection of *Phaeomoniella chlamydospora*, causing petri disease of grapevine. Co-PCR was able to amplify the partial region of the fungal rDNA including the internal transcribed spacer (ITS) region for detection of *P. chlamydospora* and 17 additional grapevine-associated fungi belonging to the genera *Botryosphaeria*, *Cryptovalsa*, *Cylindrocarpon*, *Dematophora*, *Diplodia*, *Dothiorella*, *Eutypa*, *Fomitiporia*, *Lasiodiplodia*, *Neofusicoccum*, *Phaeoacremonium*, *Phomopsis* and *Stereum*, based on the use of primer pairs NSA3/NLC2 (external pair) and NSI1/NLB4 (inner pair). A specific probe (Pch2D) targeting the ITS2 region in the rDNA was further developed for carrying out dot blot for specific detection of *P. chlamydospora* (Martos et al. 2011).

Real-Time PCR

Real-time PCR, also called quantitative PCR (qPCR) is a laboratory technique based on the PCR, used to amplify and quantify a targeted DNA molecule (Paplomatas 2006). Real-Time PCR is based on the principle of PCR except that the progress of the reaction can be monitored on-line by a camera or detector while they accumulate at each reaction cycle, without the need of post-reaction processing such as gel electrophoresis. Target DNA is quantified using a calibration curve that relates threshold cycles to a specific amount of template DNA. Amplicons are detected using several chemistries based on the emission of fluorescence signal produced proportionally during the amplification (Heid et al. 1996; Mackay et al. 2002; Makkouk and Kumari 2006). Four main chemistries are currently used in real time PCR: SYBR Green I (amplicon sequence non-specific) and; TaqMan, Molecular beacons, and Scorpion-PCR method (sequence specific) (Mackay et al. 2002; Thelwell et al. 2000; Schaad and Frederick 2002; Schaad et al. 2003). Amplicon sequence nonspecific methods are based on the use of a dye that emits fluorescent light when intercalated into double stranded DNA while amplicon sequence specific methods are based on the use of oligonucleotide probes labeled with a donor fluorophore and an acceptor dye (quencher) (Schena et al. 2004). All of these methods are based upon the hybridization of fluorescently labelled oligonucleotide probe sequences to a specific region within the target amplicon that is amplified using traditional forward and reverse PCR primers. TaqMan® probes, developed by Applied Biosystems (Foster City, California, USA), consist of single-stranded oligonucleotides that are complementary to one of the target strands (Lopez et al. 2003). Molecular beacons are the simplest hairpin probes and have complementarity nucleotide sequences that are complementary to the target amplicon (Alemu 2014). Scorpion probes covalently couple the stem-loop structure to a PCR primer due to intramolecular hybridization of probe sequence with PCR amplicon (Lopez et al. 2003). A number of plant pathogenic fungi such as *Helminthosporium solani* were detected in soil and in tubers using TaqMan probe based real time PCR (Cullen et al. 2001). Real time PCR was also used for detection of *Rhizoctonia solani* in soil samples (Lees et al. 2002) and to quantify different species of *Fusarium* in wheat kernels using TaqMan chemistry (Waalwijk et al. 2004). Even oomycete plant pathogen and *Phytophthora ramorum*, the cause of sudden oak death disease were detected by Cepheid SmartCycler real time PCR (Nezhad 2014).

Isothermal Nucleic Acid Amplification Based Diagnostics

Isothermal nucleic acid amplification facilitates rapid target amplification through single-temperature incubation, reducing system complexity compared to PCR-based methods. The method differs in terms of complexity (multiple enzymes or primers), sensitivity, and specificity. Isothermal DNA amplification produces longer

DNA fragments with higher yields than the conventional PCR technique and has greater amplification efficiency owing to undisrupted and sustained enzyme activity. Isothermal nucleic acid amplification includes several methods as:

Nucleic Acid Sequence-Based Amplification (NASBA)

NASBA is a novel transcription based isothermal amplification method developed by Compton (1991). NASBA also known as self sustained sequence replication (3SR) and Transcription Mediated Amplification (TMA) is a sensitive transcription-based amplification system (TAS) for the specific replication of nucleic acids in vitro (Guatelli et al. 1990; Gill and Ghaemi 2008). The assay targets rRNA, which is more stable than mRNA (Zhang 2013). The reaction involves two-stage protocol: the initial phase of denaturation and primer annealing at 65 °C, and the cycle phase for target amplification at the predefined temperature of 41 °C (Chang et al. 2012). NASBA requires three enzymes namely *Avian myeloblastosis virus reverse transcriptase* (AMV-RT), *RNase* H and T7 *DNA dependent RNA polymerase* (DdRp) and two primers. The first primer (P1) carrying the binding/promotor sequence is used to initiate the RNA reverse-transcription (RT) reaction, catalyzed by a reverse-transcriptase after which RNA–cDNA hybrid molecules are degraded by *RNase*H. The remaining cDNA is accessible to the second primer (P2) which initiates the synthesis of the complementary strand. A third enzyme, T7 RNA Polymerase, docks the double strand DNA on the sequence at the 5′ end of P1, transcribing many RNA copies of the gene. This process, i.e. the cycle of first strand synthesis, RNA hydrolysis, second strand synthesis and RNA transcription, is repeated indeterminately starting from the newly transcribed RNA. RNA and double stranded cDNA accumulates exponentially and can be detected by EtBr/agarose gel electrophoresis (Fakruddin et al. 2012). This technology was initially applied for detection of a number of plant viruses such as *Apple stem pitting virus* (Klerks et al. 2001), *Plum pox virus* (Olmos et al. 2007), *Potato virus Y*, *Arabis mosaic virus* (ArMV) and the bacteria *Clavibacter michiganensis* subsp. *Sepedonicus* and *R. solanacearum* (Szemes and Schoen 2003). Recently, NASBA combined with real time has also been used for detection of fungi such as *Candida* sp. and *Aspergillus* sp. (Zhao and Perlin 2013).

Loop-Mediated Isothermal Amplification (LAMP)

LAMP assay first described by Notomi et al. (2000) is a novel DNA amplification technique that amplifies DNA with high specificity, efficiency and rapidity under isothermal conditions. LAMP is based on the principle of autocycling strand displacement DNA synthesis performed by the *Bst* polymerase derived from *Bacillus stearothermophilus* (isolated from hot springs having temperature 70 °C, with polymerization and 5′–3′ exonuclease activity) for the detection of a specific DNA sequence (Chang et al. 2012). The technique makes use of four specially designed primers, a pair of outer primers and a pair of inner primers, which together recognize six distinct sites flanking the amplified DNA sequence. As inner primers possess hybrid design, amplified DNA structures take on a loop configuration at one or both ends of the elongated strands, which in turn serve as stem-loop structured templates for further displacement DNA synthesis. The final amplified product consists of a mixture of stem loop DNA strands with various stem lengths and structures with multiple loops (De Boer and Lopez 2012). Amplification can be carried out in a simple and inexpensive device like a water bath at temperatures between 60 and 65 °C (Rigano et al. 2014). LAMP products can be directly observed by the naked eye or using a UV transilluminator in the reaction tube by addition of SYBR Green I stain to the reaction tube separately (Tsui et al. 2011). LAMP assay has been employed for the specific detection of *Fusarium graminearum*, the major causative agent of *Fusarium* head blight of small cereals based on the *gaoA* gene (galactose oxidase) of the fungus (Niessen and Vogel 2010) and for detection of *Ganoderma lucidum* associated with the basal stem rot disease of coconut based on primers targeting small subunit ribosomal RNA gene (Sharadraj et al. 2015).

Molecular Inversion Probe (MIP) Assay Based Diagnostics

MIPs were initially used for high-throughput analysis of single nucleotide polymorphisms, DNA methylation, detection of genomic copy number changes and other genotyping applications (Diep et al. 2012; Hardenbol et al. 2003). Now, the methodology is being utilized for the detection of plant pathogens and can detect as little as 2.5 ng of pathogen DNA due to high specificity (Lau et al. 2014). MIPs originally called as Padlock probes (PLPs), are single-stranded DNA molecules containing two regions complementary to the target DNA that flank SNP in question. Each probe contains universal primers' sequences separated by endoribonuclease recognition site and a 20-nt tag sequence. During the assay the probes undergo a unimolecular rearrangement as: circularization due to filling of gaps with nucleotides corresponding to the SNP in four separate allele-specific polymerization (A, C, G and T) and ligation reactions followed by linearization due to mode of enzymatic reaction. As a result they become inverted followed by PCR amplification step. Further processing of the probes depends on specific assay (Absalan and Ronaghi 2007). MIPs have high accuracy due to fidelity of both polymerase and ligase in the gap-fill step, high specificity due to hybridization, polymerization and ligation (Thiyagarajan et al. 2006). A specific assay has been developed based on padlock probes along with microarray having detection limit of 5 pg of pathogen DNA for the detection of economically important plant pathogens including oomycetes (*Phytophthora* spp. and *Pythium* spp.), fungi (*Rhizoctonia* spp., *Fusarium* spp. and *Verticillium* spp.) and a nematode (*Meloidogyne* spp.) (Szemes et al. 2005). Two padlock probes have been designed to target species-specific single nucleotide polymorphisms (SNPs) located at the inter-generic spacer two region and large subunit of the rRNA respectively, to discriminate the two fungal species, *Grosmannia clavigera* and *Leptographium longiclavatum*, intimately associated with the mountain pine beetles (*Dendroctonus ponderosae*) in western Canada (Tsui et al. 2010). MIP technology was used as a diagnostic tool to screen the plant pathogens, *Fusarium oxysporum* f.sp. *conglutinans*, *Fusarium oxysporum* f.sp. *lycopersici* and *Botrytis cinerea* (Lau et al. 2014).

Hybridization Based Diagnostics

Fluorescent In Situ Hybridization (FISH)

FISH is a powerful method for in situ detection of pathogens which combines microscopic observation of pathogen along with the specificity of hybridization and is dependent on the hybridization of DNA probes to species-specific regions (Wullings et al. 1998; Volkhard et al. 2000). FISH probes often target sequences of ribosomal RNA or mitochondrial genes as they are abundant in sequence databases and exist in multiple copies in each cell (Tsui et al. 2011). The major step of FISH involves the preparation of biological samples and labeling (incorporation of a fluorescent label/marker e.g. carboindocyanine dye) of a nucleic acid sequence to form a probe. The probe is hybridized to the DNA or RNA in biological materials to form a double-stranded molecule under controlled experimental conditions followed by detection of hybridization (Amann et al. 1995). The first FISH probe targeting a living fungus was designed and used for detection of *Aureobasidium pullulans* on the phylloplane of apple seedlings (Li et al. 1996).

Array Based Diagnostics

Arrays both, microarrays and macroarrays, hold promise for quick and accurate detection and identification of plant pathogens due to multiplex capabilities of the system (Saikia and Kadoo 2010). Array refers to reverse dot blot assays in which assorted DNA probes are bound to a fixed matrix (e.g. nylon membrane or microscope slides for microarrays) in a highly regular pattern (De Boer and Lopez 2012). Macro arrays are generally membrane-based with spot sizes greater than 300 μm while microarrays are high-density arrays with spot sizes smaller than 150 μm. The macroarray technology is now commercially available in four European countries under the name DNA Multiscan (http://www.dnamultiscan.com) for the test of plant pathogens (Tsui et al. 2011).

A typical microarray slide can contain up to 30,000 spots (Webster et al. 2004). ssDNA probes are irreversibly fixed as an array of discrete spots to a surface of glass, membrane or polymer. Each probe is complementary to a specific DNA sequence (genes, ITS, ribosomal DNA) and hybridization with the labeled complementary sequence provides a signal that can be detected and analyzed. Arrays printed with probes corresponding to a large number of fungal pathogens can be utilized to simultaneously detect all the pathogens within the tissue of an infected host. The steps include extraction of nucleic acid, RT-PCR and labeling with a fluorescent probe such as fluorescin, Cy3 or Cy5. The labeled target molecule is denatured and allowed to hybridize with the arrayed probes. Specific patterns of fluorescence in the form of spots are detected by a microarray reader which allows the identification of specific gene sequences found only in the pathogen of interest (Schmitt and Henderson 2005). DNA array technology was developed for rapid and efficient detection of tomato vascular wilt pathogens *Fusarium oxysporum* f.sp. *lycopersici*, *Verticillium albo-atrum*, and *V. dahliae*. The array successfully detected the tomato wilt pathogens from complex substrates like soil, plant tissues, and irrigation water as well as samples collected from tomato growers (Lievens et al. 2003). Tambong et al. (2006) developed macroarray for simultaneous detection of most of the known *Pythium* species. Recently, *Magnaporthe grisea* array was developed and is commercially available from Agilent Technologies having genome-wide coverage of *Magnaporthe grisea* and inclusion of relevant rice genes in a single microarray with 60-mer oligo probe length (http://www.agilent.com/).

Sequencing Based Diagnostics

Routine sequencing is likely to play an increasingly important role in species identification. PCR amplicons can often be sequenced relatively inexpensively and rapidly. Genetic databases available on the internet such as GenBank allow rapid comparison of one's sample sequence to extensive and growing libraries of sequences (Vincelli and Tisserat 2008). With the advancement in the field of sequencing, full genome sequencing of plant pathogens is possible at lower rate and offers a means for pathogen detection.

Next-Generation Sequencing (NGS)

NGS techniques also referred to as second-generation sequencing (SGS) emerged in 2005 using commercial Solexa sequencing technology. In this technique, sequencing reaction is detected on amplified clonal DNA templates by emulsion or solid phase PCR methods (Nezhad 2014). It involves isolation of total DNA or RNA from diseased plant, elimination of host nucleic acid, enrichment of pathogen DNA, and exploitation of different NGS technologies (Adams et al. 2009; Studholme et al. 2011). Three platforms: Roche/454 FLX, the Illumina/Solexa genome analyzer and the applied biosystems SOLID™ system were widely used and recently, two more parallel platforms came into existence: Helicos Heliscope™ and Pacific Biosciences SMRT instruments (Mardis 2008). Nunes et al. (2011) applied 454 sequencing technology to elucidate and characterize the small RNA transcriptome (15–40 nt) of mycelia and appressoria of *Magnaporthe oryzae*. A number of both known and unknown plant pathogenic fungi have been detected using NGS e.g. *Pyrenophora teres* f. sp. *teres* and *Phytophthora infestans* in sweet potato (Zhou and Holliday 2012; Neves et al. 2013). The draft genome of the soil borne *Pyrenochaeta lycopersici* causing corky root rot (CRR) disease in tomato and affecting other solanaceous species including pepper, eggplant and tobacco, as well as other cultivated crops such as melon, cucumber, spinach and safflower was characterized based on paired-end Illumina reads is highly effective in reconstructing contigs containing almost full length genes (Aragona et al. 2014).

Third Generation Sequencing (TGS)

More recent single molecule sequencing technologies are known as third-generation sequencing (TGS). TGS also referred as single molecule sequencing (SMS) uses single DNA molecules for sequence reactions without the need for DNA template amplification. TGS has been used in plant genomics and pathogen detection (Pan et al. 2008; Rounsle et al. 2009). TGS is superior to SGS as it simplifies the sample preparation, increases the detection accuracy by eliminating PCR-caused errors, and generates longer sequence reads by better throughput platforms. Oxford Nanopore technology and recently IBM's plan of silicon-based nanopores are the recent devices developed for third generation DNA sequencing (Kircher and Kelso 2010).

Biosensor Based Diagnostics

Immunosensors are those biosensors in which the recognition element is antibody and offers direct label free pathogen detection. It is a device comprising of an antigen or antibody species coupled to a single transducer which detects the binding of the complementary species (Priyanka et al. 2013). Different types such as surface plasmon resonance (SPR), quartz crystal microbalance (QCM) and cantilever-based sensors are currently the most promising (Skottrup et al. 2008).

Quartz Crystal Microbalance (QCM) Immune-Sensors

In this novel technique of plant pathogen detection, a quartz crystal disk is coated with pathogen-specific antibodies. Voltage is applied across the disk, making the disk warp slightly via a piezoelectric effect (Webster et al. 2004). Adsorption of pathogen to the crystal surface changes its resonance oscillation frequency in a concentration dependent manner. It is both qualitative and quantitative method. The QCM term broadly includes bulk acoustic wave (BAW), quartz crystal resonance sensors (QCRS) and thickness shear mode (TSM). The difference between BAW, QCRS and TSM acoustic sensors, is their mode of wave propagation (Cooper and Singleton 2007).

Surface Plasmon Resonance (SPR) Sensors

SPR uses surface plasmons, which are electromagnetic waves that can be excited by light at gold sensor interfaces. As incoming light interacts with the gold interface at angles larger than the critical angle, the reflected light displays a characteristic decrease, the so-called SPR minimum, due to resonant energy transfer from the incoming photons to surface plasmons. SPR sensors have been shown to be rapid, label-free, and selective tools for the detection (Skottrup et al. 2008). Surface plasmon resonance (SPR) was first used for detection of fungal spores (urediniospores) of *Puccinia striiformis* f.sp. *tritici* (Pst). The approach involved the use of a mouse monoclonal antibody (Pst mAb8) and a SPR sensor for label-free detection of spores (Skottrup et al. 2007). SPR sensor based on DNA hybridization was also used for the detection of *Fusarium culmorum*, a fungal pathogen of wheat (Zezza et al. 2006; Pascale et al. 2013).

Cantilever-Based Sensors

The use of a cantilever as a sensor dates back to 1943 when Norton proposed a hydrogen gas sensor based on a cantilever and was initially used in atomic force microscopy (AFM) for surface characterization (Datar et al. 2009). An AFM measures the forces between the tip of a cantilever and the sample surface using the tip deflection (contact mode AFM) or changes in the resonance frequency of a vibrating cantilever (dynamic mode AFM). Cantilever technology has been used for biosensing applications using antibodies (Waggoner and Craighead 2007). Cantilevers can be operated in either (a) static mode, which measures cantilever bending upon analyte binding or (b) dynamic mode, which measures resonance frequency changes when analytes binds the surface (similarly to QCM sensors). Cantilever sensors have been applied to the detection of relatively small analytes such as nucleic acids and disease proteins (Waggoner and Craighead 2007). Cantilever sensors use have been demonstrated for detection of fungus *Aspergillus niger* at 10^3 cfu/ml using resonance changes (Nugaeva et al. 2007). A specific micromechanical cantilever array system has been used for detection of *Saccharomyces cerevisiae* (Banik and Sharma 2011).

Spectroscopic and Imaging Technique Based Diagnostics

A number of spectroscopic and imaging techniques are being used for forecasting the occurrence of disease and detection of pathogen such as fluorescence spectroscopy where the fluorescence from the object of interest is measured after excitation with a beam of light (usually ultraviolet spectra), visible and infrared spectroscopy whereby visible and infrared rays are used for a rapid, non-destructive, and cost-effective method for the detection of plant diseases (Sankaran et al. 2010). Hyperspectral imaging has also found application in precision agriculture whereby the spectral reflectance is acquired for a range of wavelengths in the electromagnetic spectra profiling (Okamoto et al. 2009). However, these techniques require trained person having the knowledge of softwares related to image data analysis and requirement of high efficiency computers. Imaging spectroscopy has been used to scan wheat kernels for head blight disease through machine vision techniques (Delwiche and Kim 2000) while spectral and fluorescence data has been employed to monitor winter wheat yellow rust (Moshou et al. 2005).

Volatile Organic Compounds (VOCs) Based Diagnostics

Plants emit many low molecular weight biomolecules in gaseous phase called as volatile organic compounds (VOCs) from their surfaces into their immediate surroundings that serve essential functions (Baldwin et al. 2006). VOC profiling is an emerging innovative avenue and has potential applications in disease diagnosis. The emitted VOC profiles of healthy plants are significantly different than those infected ones (Martinelli et al. 2014). The electronic nose (e-nose) is a platform for VOCs profiling.

Electronic Nose

In recent years, the development of innovative devices such as electronic nose (e-nose) based on different electronic aroma detection (EAD) principles and mechanisms has been investigated and implemented for diverse disciplines within the plant sciences by many researchers (Wilson et al. 2004). The electronic nose is often referred to as an intelligent device, able to mimic the human olfaction functions and may be used for detection, recognition and classification of volatile compounds and odours. This type of electronic olfactory system was introduced in 1982 by Dodd and Persaud from the Warwick Olfaction Research Group, UK (Troy Nagle et al. 1998). A complete electronic-nose system typically consists of several integrated and/or interfaced components including a multisensor array (composed of several gas sensors with broad sensitivity and cross-reactivity or partially-overlapping selectivity), a data-processing and analysis unit such as an artificial neural network (ANN), software having digital pattern-recognition algorithms, and often aroma reference-library databases containing stored files with digital fingerprints of specific aroma reference (signature) patterns (Wilson 2013). An electronic nose incorporating artificial intelligence to detect basal stem rot (BSR) disease caused by *Ganoderma boninense* fungus affecting oil palm plantations in South East Asia was developed based on three types of odour samples for both healthy and infected oil palm trees, namely odour of the air surrounding the tree, odour of bored tree trunk and odour of soil surrounding the base of the tree trunks (Markom et al. 2009). A Cyranose® 320 was developed to detect postharvest fungal diseases namely gray mold caused by *Botrytis cinerea*, anthracnose caused by *Colletotrichum gloeosporioides* and fruit rot caused by *Alternaria* spp. in blueberries (Li et al. 2009). Electronic Noses (ENs) have also been used to analyse the Volatile Organic Compounds (VOCs) of both healthy and infected powdery mildew infected tomato (*Solanum lycopersicum*) crops (Ghaffari et al. 2010). In order to detect contamination of wheat by *Fusarium* species, an electronic nose based on an array of metalloporphyrin coated quartz microbalances was developed for detection of *Fusarium* species, *F. cerealis*, *F. graminearum*, *F. culmorum* and *F. redolens* based on release of toxic metabolites, especially the mycotoxin deoxynivalenol (DON, Vomitoxin) and zearalenone (ZEA) (Eifler et al. 2011).

Nano-diagnostics

Nano diagnostics is the use of nano-biotechnology to diagnose plant diseases.

Quantum Dots (QDs)

QDs are nanometer scale semiconductor nanoparticles that fluoresce when stimulated by an excitation light source and are defined as particles with physical dimensions smaller than the exciton Bohr radius (Jamieson et al. 2007). QDs are ultrasensitive nanosensor based on fluorescence resonance energy transfer (FRET) can detect very low concentration of DNA and do not require separation of unhybridized DNA (Khiyami et al. 2014). QDs are linked to specific DNA probes to capture target DNA. The target DNA strand binds to a fluorescent-dye (fluorophore) labeled reporter strand and thus forming FRET donor-acceptor assembly. Unbound DNA strand produce no fluorescence but on binding of even small amount of target DNA (50 copies) may produce very strong FRET signal (Chun-Yang Zhang et al. 2005). QD specific antibody sensor was developed for rapid detection of *Polymyxa betae*, an obligate parasite of sugarbeet roots and vector of *Beet necrotic yellow vein virus* (BNYVV), the causal agent of rhizomania (Safarpour et al. 2012).

Nanoprobes

Fluorescent silica nanoprobes have potential for rapid diagnosis of plant diseases. Fluorescent rubpy doped silica nanoparticles (FSNP) at 50 ± 4.2 nm conjugated with the secondary antibody of goat anti-rabbit IgG (using microemulsion method) has been used for successful detection of a bacterial plant pathogen *Xanthomonas axonopodis* pv. *vesicatoria*, the causal agent of bacterial spot disease in solanaceous plants (Yao et al. 2009). In future, nanoprobes can be utilized for detection of other plant pathogens also.

Portable Devices and Kits

On site diagnosis of plant pathogens require portable devices and such a portable system in the form of PCR termed as Palm PCR was developed by Ahram Biosystems Company in Korea in which DNA can be amplified in less than 25 min. The portable system presents a highly functional and user-friendly way to perform different types of PCR tests for both beginners and experienced researchers.

Lab on a Chip

A Lab on a chip is a new micro technique which possess several advantages such as portability, low reagent consumption, short reaction times and on site diagnosis. A large number of samples can be processed directly in the field itself (Figeys and Pinto 2000; Kricka 2001). The first lab-on-a-chip system in the field of plant pathology was developed for rapid diagnosis of *Phytophthora* species (Julich et al. 2011). A portable real-time microchip PCR system was developed for detection of *Fusarium oxysporum* f. sp. *lycopersici* (Fol) strains. The system included fluorescence detector and a battery-powered microcontroller unit for PCR. The entire system was 2,561,668 cm^3 in size and weighs under 850 g (Koo et al. 2013).

Lab in a Box

Lab in a box also termed as nanodiagnostic kits refers to a briefcase sized kit that can be carried to the field to search for pathogens. Nanodiagnostic kit equipment can easily and quickly detect potential serious plant pathogens, allowing experts to help farmers in prevention of disease epidemics from breaking out.

Phytophthora Test Kits

The Alert test kit for *Phytophthora* has been used to detect all common *Phytophthora* species. Pathogen detection can be accomplished when as little as 0.5 % of a plant's roots are infected (http://danrcs.ucdavis.edu.)

Conclusion

Plant pathogenic fungi are becoming more widespread globally due to modern high input monocropping based agriculture and easy transboundary movement. In order to manage plant pathogens and restrict their movement as well as secondary spread in new geographical areas, early and timely detection is pre-requisite. Thus, plant disease diagnosis and detection of plant pathogen are critical and integral part of successful disease management and serve as the first and crucial line of defense. Once the pathogen is identified, appropriate control measures can be employed. In the past, detection of fungal pathogens involved time consuming biological indexing for days/weeks at a certain temperature on the appropriate medium or grow out tests for seed borne fungi. These processes are extremely cumbersome and cannot be adopted for routine diagnosis of large number of samples. The constraints posed by these traditional biological indexing methods led to profound advancement in the development of affordable and simple new improved methods which served as powerful tools for detection and identification of phytopathogenic fungi. New innovative detection technologies have been formulated and demonstrated that are accurate, cost effective, portable, rapid, robust, sensitive, and high throughput for routine plant disease diagnosis. Several techniques have been developed which have an edge over the traditional methods of plant pathogen diagnosis; these include physical diagnostic tools (EM; SEM etc.), serological techniques (DIBA; ELISA etc.), molecular techniques (PCR), lateral flow assays, hybridization based assays, nano-based kits, electronic nose etc. are gaining momentum and have potential applications. The era of Next/Third Generation Sequencing, in which the entire DNA or RNA sequences of organisms can be traced, has provided an ocean full of diagnostic techniques involving the complementation of bioinformatics approaches for authentic identification *vis-a-vis* characterization of plant pathogenic fungi. On site molecular diagnostics is in its infancy but is surely evolving faster and will become a boon due to user friendliness. Despite availability of array of frontier tools and techniques for plant pathogen detection in the era of biotechnology, conventional methods can't be completely ignored in some instances. The best approach for both disease diagnosis and detection of plant pathogenic fungi demands blend of diverse range of conventional and advanced unconventional methods. With this continuous evolving spectrum of advanced techniques, the major challenge in the future for phytopathologists will be to choose a particular modus operandi among array of these techniques for specific detection of pathogen.

References

Absalan F, Ronaghi M (2007) Molecular inversion probe assay. Methods Mol Biol 396:315–330

Adams IP, Glover RH, Monger WA, Mumford R, Jackeviciene E, Navalinskiene M, Boonham N (2009) Next-generation sequencing and metagenomic analysis: a universal diagnostic tool in plant virology. Mol Plant Pathol 10:537–545

Agrios G (2005) Plant pathology. Elsevier Academic Press, London

Alemu K (2014) Real-time PCR and its application in plant disease diagnostics. Adv Life Sci Technol 27, www.iiste.org ISSN 2224-7181 (Paper) ISSN 2225-062X (Online)

Amann RI, Ludwig W, Schleifer KH (1995) Phylogenetic identification and in situ detection of individual microbial cells without cultivation. Microbiol Mol Biol Rev 59:143–169

Aragona M, Minio A, Ferrarini A, Valente MT, Bagnaresi P, Orru L, Tononi P, Zamperin G, Infantino A, Vale G, Cattivelli L, Delledonne M (2014) *De novo* genome assembly of the soil-borne fungus and tomato pathogen *Pyrenochaeta lycopersici*. BMC Genomics 15:313

Arie T, Hayashi Y, Yoneyama K, Nagatani A, Furuya M, Yamaguchi I (1995) Detection of *Fusarium* spp. in plants with monoclonal antibody. Ann Phytopathol Soc Jpn 61:311–317

Baldwin IT, Halitschke R, Paschold A, Von-Dahl CC, Preston CA (2006) Volatile signaling in plant-plant interactions: "talking trees" in the genomics era. Science 311(5762):812–815

Banik S, Sharma P (2011) Plant pathology in the era of nanotechnology. Indian Phytopathol 64:120–127

Barnett HL, Hunter BB (1972) Illustrated genera of imperfect fungi. Burgess, Minneapolis, 218 p

Bartlett JM, Stirling D (2003) A short history of the polymerase chain reaction. In: PCR protocols. Humana Press Inc.Totowa, NJ. doi:10.1385/1-59259-384-4:3

Bertolini E, Torres E, Olmos A, Martin MP, Bertaccini A, Cambra M (2007) Co-operational PCR coupled with dot blot hybridization for detection and 16SrX grouping of phytoplasmas. Plant Pathol 56:677–682

Bos L (1983) Introduction to Plant Virology. Pudoc, Wageningen, in conjunction with Longman, London/New York.

Bozzola JJ, Russell LD (1999) Electron microscopy: principles and techniques for biologists. Jones and Bartlett, Boston

Bruns TD, White TJ, Taylor JW (1991) Fungal molecular systematics. Annu Rev Ecol Syst 22:525–564

Capote N, Aguado A, Pastrana AM, Sanchez-Torres P (2012) Molecular tools for detection of plant pathogenic fungi and fungicide resistance. Plant Pathology, Dr. Christian Joseph Cumagun (Ed.), ISBN: 978-953-51-0489-6, InTech, doi:10.5772/38011

Casper R, Mendgen K (1979) Quantitative serological estimation of hyperparasite: detection of Verticillium lecanii in yellow rust infected wheat leaves by ELISA. Phytpathol Z 94:89–91

Castro C, Schaad NW, Bonde MR (1994) A technique for extracting Tilletia indica teliospores from contaminated wheat seeds. Seed Sci Technol 22:91–98

Chang CC, Chen CC, Wei SC, Lu HH, Lian YH, Lin CW (2012) Diagnostic devices for isothermal nucleic acid amplification. Sensors 12:8319–8337

Chen RS, Chu C, Cheng CW, Chen WY, Tsay JG (2008) Differentiation of two powdery mildews of sunflower (Helianthus annuus) by PCR mediated method based on ITS sequences. Eur J Plant Pathol 121:1–8

Clark MF, Adams AN (1977) Characteristics of the microplate method of enzyme-linked immunosorbent assay for the detection of plant viruses. J Gen Virol 34:475–483

Compton J (1991) Nucleic acid sequence-based amplification. Nature 350:91–92

Cooper MA, Singleton VT (2007) A survey of the 2001 to 2005 quartz crystal microbalance biosensor literature: applications of acoustic physics to the analysis of biomolecular interactions. J Mol Recognit 20:154–184

Cullen DW, Lees AK, Toth IK, Duncan JM (2001) Conventional PCR and real-time quantitative PCR detection of Helminthosporium solani in soil and on potato tubers. Eur J Plant Pathol 107:387–398

Datar R, Kim S, Jeon S, Hesketh P, Manalis S, Boisen A, Thundat T (2009) Cantilever sensors: nanomechanical tools for diagnostics. MRS Bull 34:449–454

De Boer SH, Lopez MM (2012) New grower-friendly methods for plant pathogen monitoring. Annu Rev Phytopathol 50:197–218

De Carvalho Alves M, Pozza EA (2012) Scanning electron microscopy detection of seed-borne fungi in blotter test. In: Méndez-Vilas A (ed) Current microscopy contributions to advances in science and technology. Formatex, Badajoz, Spain, pp 230–238

De Tempe J, Binnerts J (1979) Introduction to methods of seed health testing. Seed Sci Technol 7:601–636

Delwiche SR, Kim MS (2000) Hyperspectral imaging for detection of scab in wheat. Proc SPIE 4203:13–20

Derrick KS (1973) Quantitative assay for plant viruses using serologically specific electron microscopy. Virology 56:652–653

Dickinson M (2005) Molecular Plant Pathology. BIOS Scientific Publishers, Taylor and Francis Group, 11 New Fetter Lane, London.

Diep D, Plongthongkum N, Gore A, Fung HL, Shoemaker R (2012) Library-free methylation sequencing with bisulfite padlock probes. Nat Methods 9:270–272

Dyer PS, Furneaux PA, Douhan G, Murray TD (2001) A multiplex PCR test for determination of mating type applied to the plant pathogens *Tapesia yallundae* and *Tapesia acuformis*. Fungal Genet Biol 33:173–180

Eifler J, Martinelli E, Santonico M, Capuano RS, Natale CD (2011) Differential detection of potentially hazardous fusarium species in wheat grains by an electronic nose. PLoS ONE 6, e21026. doi:10.1371/journal.pone.0021026

El-Sheikha, Ray RC (2014) Is PCR-DGGE an innovative molecular tool for the detection of microbial plant pathogens? In: Neeta Sharma (ed) Biological controls for preventing food deterioration: strategies for pre- and post harvest management. John Wiley & Sons Ltd, doi:10.1002/9781118533024, pp 409–433

Fakruddin M, Mazumdar RM, Chowdhury A, Mannan KSB (2012) Nucleic acid sequence based amplification prospects and applications. Int J Life Sci Pharma Res 2:L116–L121

Fidanza MA, Dernoeden PH (1995) Evaluation of an enzyme-linked immunosorbent assay method for predicting brown patch infection in turf. HortSci 30:1263–1265

Figeys D, Pinto D (2000) Lab-on-a-chip: a revolution in biological and medical sciences. Anal Chem 72:330A–335A

Forster H, Coffey MD (1993) Molecular taxonomy of *Phytophthora megasperma* based on mitochondrial and nuclear DNA polymorphisma. Mycol Res 97:1101–1112

Fraaije B, Lovell D, Coelho JM (2001) PCR-based assays to assess wheat varietal resistance to blotch (Septoria tritici and Stagonospora nodorum) and rust (Puccinia striiformis and Puccinia recondita) diseases. Eur J Plant Pathol 107:905–917

Ghaffari R, Zhang F, Iliescu D, Hines E, Leeson M, Napier R, Clarkson J (2010) Early detection of diseases in tomato crops: an electronic nose and intelligent systems approach. In Proceedings of the International Joint Conference on Neural Networks (IJCNN), Barcelona, 18–23 July 2010, pp 1–6

Gill P, Ghaemi A (2008) Nucleic acid isothermal amplification technologies—a review. Nucleosides Nucleotides Nucleic Acids 27:224–243

Goud JC, Termorshuizen AJ (2003) Quality of methods to quantify microsclerotia of Verticillium dahliae in soil. Eur J Plant Pathol 109:523–534

Guatelli J, Whitfield K, Kwoh D, Barringer K, Richman D, Gingeras T (1990) Isothermal, in vitro amplification of nucleic acids by a multi enzyme reaction modeled after retrovirus replication. Proc Natl Acad Sci U S A 87:1874–1878

Guglielmo F, Bergemann SE, Gonthier P, Nicolotti G, Garbelotto M (2007) A multiplex PCR-based method for the detection and early identification of wood rotting fungi in standing trees. J Appl Microbiol 103:1490–1507, ISSN 1364-5072

Hardenbol P, Baner J, Jain M, Nilsson M, Namsaraev EA (2003) Multiplexed genotyping with sequence-tagged molecular inversion probes. Nat Biotechnol 21:673–678

Hardham AR, Suzaki E, Perkin JL (1986) Monoclonal antibodies to isolate-, species- and genus-specific components on the surface of zoospores and cysts of the fungus Phytophthora cinnamomi. Can J Bot 64:311–321
Harlow E, Lane E (1988) Antibodies: a laboratory manual. Cold Spring Harbor Press, Cold Spring Harbor, Chapter 12
Havis ND, Oxley SP, Piper SR, Langrell SRH (2006) Rapid nested PCR-based detection of Ramularia collo-cygni direct from barley. FEMS Microbiol Lett 256(2):217–223. doi:10.1111/j.1574-6968.2006.00121.x
Heid CA, Stevens J, Livak KJ, Williams PM (1996) Real time quantitative PCR. Genome Res 6:986–994
Henson J, French R (1993) The polymerase chain reaction and plant disease diagnosis. Annu Rev Phytopathol 31:81–109
Hiremath L, Naik GR (2003) DAC-ELISA technique for detection of red rot pathogen in sugarcane var. CoC671. Indian J Biotechnol 1:363–366
Hirsch PR, Mauchline TH, Mendum TA, Kerry BR (2000) Detection of the nematophagous fungus Verticillium chlamydosporium in nematode-infested plant roots using PCR. Mycol Res 104:435–439
Hobbs HA, Reddy DVR, Rajeshwari R, Reddy AS (1987) Use of direct antigen coating and protein A coating ELISA procedures for detection of three peanut viruses. Plant Dis 71: 747–749
Hull R (2002) Matthews Plant Virology. Academic Press, San Diego
James D, Varga A, Pallas V, Candresse T (2006) Strategies for simultaneous detection of multiple plant viruses. Can J Plant Pathol 28:16–29
Jamieson T, Bakhshi R, Petrova D, Pocock R, Imani M, Seifalian AM (2007) Biological applications of quantum dots. Biomaterials 28:4717–4728
Jansson HB, Akesson H (2003) Extracellular matrix, esterase and the phytotoxin prehelminthosporol in infection of barley leaves by Bipolaris sorokiniana. Eur J Plant Pathol 109: 599–605
Johnson MC, Pirone TP, Siegel MR, Varney DR (1982) Detection of Epichloe typhina in tall fescue by means of enzyme linked immunoassay. Phytopathology 72:647–650
Julich S, Riedel M, Kielpinskia M, Urbana M, Kretschmerc R, Wagnerb S, Fritzschea W, Henkela T, Mollerc R, Werres S (2011) Development of a lab-on-a-chip device for diagnosis of plant pathogens. Biosens Bioelectron 26:4070–4075
Kageyama K, Senda M, Asano T, Suga H, Ishiguro K (2007) Intra-isolate heterogeneity of the ITS region of rDNA in *Pythium helicoides*. Mycol Res 111:416–423
Kamraj M, Rajeshwari S, Govarthanan M (2012) Raising of polyclonal antibody against the Aspergillus oryzae proteins in albino rat. Int J Pharma Bio Sci 3:343–348
Kendrick B (ed) (1971) Taxonomy of fungi imperfecti. Toronto Univ. Press, Toronto, 309 p
Khiyami MA, Almoammar H, Awad YM, Alghuthaymi MA, Abd-Elsalam KA (2014) Plant pathogen nanodiagnostic techniques: forthcoming changes? Biotechnol Biotechnol Equip 28:775–785. doi:10.1080/13102818.2014.960739
Kim JT, Park SY, Choi W, Lee YH, Kim HT (2008) Characterization of *Colletotrichum* isolates causing anthracnose of pepper in Korea. Plant Pathol J 24:17–23
Kircher M, Kelso J (2010) High-throughput DNA sequencing–concepts and limitations. Bioessays 32:524–536
Klerks MM, Leone G, Linder JL, Schoen CD, Heuvel FJM (2001) Rapid and sensitive detection of apple stem pitting virus in apple trees through RNA amplification and probing with fluorescent molecular beacons. Phytopathology 91:1085–1091
Koenig R (1978) ELISA in the study of homologous and heterologous reactions of plant viruses. J Gen Virol 40:309–318
Koo C, Malapi-Wight M, Kim HS, Cifci OS, Vaughn-Diaz VL (2013) Development of a real-time microchip PCR system for portable plant disease diagnosis. PLoS ONE 8, e82704. doi:10.1371/journal.pone.0082704

Kratka J, Pekarova-Kynerova B, Kudlikova I, Slovacek J, Zemankova M (2002) Utilisation of immunochemical methods for detection. Plant Prot Sci 38:55–63

Kricka LJ (2001) Microchips, microarrays, biochips and nanochips: personal laboratories for the 21st century. Clin Chim Acta 307:219–223

Kuchta P, Jecz T, Korbin M (2008) The Suitability of PCR-based techniques for detecting *Verticillium dahliae* in strawberry plants and soil. J Fruit Ornamental Plant Res 16:295–304

Landa BB, Montes-Borrego M, Munoz-Ledesma FJ, Jimenez-Diaz RM (2007) Phylogenetic analysis of downy mildew pathogens of opium poppy and PCR based *in planta* and seed detection of *Peronospora arborescens*. Phytopathology 97:1380–1390

Lau HY, Palanisamy R, Trau M, Botella JR (2014) Molecular inversion probe: a new tool for highly specific detection of plant pathogens. PLoS ONE 9, e111182. doi:10.1371/journal.pone.0111182

Lees AK, Cullen DW, Sullivan L, Nicolson MJ (2002) Development of conventional and quantitative real-time PCR assays for the detection and identification of *Rhizoctonia solani* AG-3 in potato and soil. Plant Pathol 51:293–302

Li S, Cullen D, Hjort M, Spear R, Andrews JH (1996) Development of an oligonucleotide probe for Aureobasidium pullulans based on the small subunit rRNA gene. Appl Environ Microbiol 62:1514–1518

Li C, Krewer G, Kays SJ (2009) Blueberry postharvest disease detection using an electronic nose. ASABE Paper No. 096783, ASABE Annual International Meeting, Reno, June 21–June 24, 2009

Li Y, Garibaldi A, Gullino ML (2010) Molecular detection of *Fusarium oxysporum* f. sp. *chrysanthemum* on three host plants: *Gerbera jasmonii*, *Osteospermum* sp. and *Argyranthemum frutescens*. J Plant Pathol 92:525–530

Lievens B, Brouwer M, Vanachter ACRC, Levesque CA, Cammue BPA, Thomma BPHJ (2003) Design and development of a DNA array for rapid detection and identification of multiple tomato vascular wilt pathogens. FEMS Microbiol Lett 223:113–122

Liu J, He L, Zhou G (2009) Specific and rapid detection of *Camellia oleifera* anthracnose pathogen by nested-PCR. Afr J Biotechnol 8:1056–1061

Lopez MM, Bertolini E, Olmos A, Caruso P, Gorris MT, Llop PLR, Cambra M (2003) Innovative tools for detection of plant pathogenic viruses and bacteria. Int Microbiol 6:233–243

Lopez MM, Llop P, Olmos A, Marco-Noales E, Cambra M, Bertolini E (2009) Are molecular tools solving the challenges posed by detection of plant pathogenic bacteria and viruses? Curr Issues Mol Biol 11:13–46

Ma Z, Luo Y, Michailides TJ (2003) Nested PCR assays for detection of *Monilinia fructicola* in stone fruit orchards and *Botryosphaeria dothidea* from pistachios in California. J Phytopathol 151:312–322

Mackay IM, Arden KE, Nitsche A (2002) Real-time PCR in virology. Nucleic Acids Res F30:1292–1305

Majumder D, Rajesh T, Suting EG, Debbarma A (2013) Detection of seed borne pathogens in wheat: recent trends. Aust J Crop Sci 7:500–507

Makkouk K, Kumari S (2006) Molecular diagnosis of plant viruses. Arab J Plant Protect 24(2): 135–138

Mardis ER (2008) Next-generation DNA sequencing methods. Annu Rev Genomics Hum Genet 9:387–402

Markom MA, Shakaff AY, Adomb AH, Ahmadc MN, Hidayat W, Abdullahb AH, Fikri NA (2009) Intelligent electronic nose system for basal stem rot disease detection. Comput Electron Agric 66:140–146

Martin FN, Kistler HC (1990) Species-specific banding pattern of restriction endonuclease-digested mitochondrial DNA from the genus *Pythium*. Exp Mycol 14:32–46

Martinelli F, Scalenghe R, Davino S, Panno S, Scuderi G, Ruisi P, Villa P, Stroppiana D, Boschetti M, Goulart LR, Davis CE, Dandekar AM (2014) Advanced methods of plant disease detection. Rev Agron Sustain Dev. doi:10.1007/s13593-014-0246-1

Martos S, Torres E, El Bakali MA, Raposo R, Gramaje D, Armengol J, Luque J (2011) Co-operational PCR coupled with dot blot hybridization for the detection of *Phaeomoniella chlamydospora* on infected grapevine wood. J Phytopathol 159:247–254

Michailides TJ, Morgan DP, Ma Z, Luo Y, Felts D, Doster MA, Reyes H (2005) Conventional and molecular assays aid diagnosis of crop diseases and fungicide resistance. http://Californiaagriculture.ucop.edu

Milne RG (1972) Electron microscopy of viruses. In: Kado CI, Agarwal HO (eds) Principles and techniques in plant virology. Van Nostrand Reinhold Company, New York, pp 76–128

Milne RG, Luisoni L (1977) Rapid immune electron microscopy. In: Maramorosch K, Koprowski H (eds) Methods in virology, vol 6. Academic, New York, pp 265–281

Moshou D, Bravo C, Oberti R, West J, Bodria L, McCartney A, Ramon H (2005) Plant disease detection based on data fusion of hyper-spectral and multi-spectral fluorescence imaging using Kohonen maps. Real-Time Imag 11:75–83

Mowat WP (1985) Simplified enzyme immunoassay for plant virus detection and identification. Report of the Scottish Crop Research Institute for 1984, Scottish Crop Research Institute, Dundee, p 188

Mullis KB, Faloona FA (1987) Specific synthesis of DNA in vitro via a polymerase-catalyzed chain reaction. Methods Enzymol 155:335–350

Musetti R, Favali MA (2004) Microscopy techniques applied to the study of phytoplasma diseases: traditional and innovative methods. Curr Issues Multidiscip Microsc Res Educ 2:72–80

Narayanasamy P (2011) Microbial plant pathogens: detection and disease diagnosis: bacterial and phytoplasmal pathogens, vol 2. Springer, Dordrecht

Neves LG, Davis JM, Barbazuk WB, Kirst M (2013) Whole genome targeted sequencing of the uncharacterized pine genome. Plant J 75:146–156

Nezhad AS (2014) Future of portable devices for plant pathogen diagnosis. Lab Chip R Soc Chem. doi:10.1039/c4lc00487f

Niessen L, Vogel RF (2010) Detection of *Fusarium graminearum* DNA using a loop-mediated isothermal amplification (LAMP) assay. Int J Food Microbiol 140:183–191

Noordam D (1973) Identification of plant viruses methods and experiments. Centre for Agricultural Publishing and Documentation (Pudoc), Wageningen

Notomi T, Okayama H, Masubuchi H (2000) Loop-mediated isothermal amplification of DNA. Nucleic Acids Res 28, E63

Nugaeva N, Gfeller KY, Backmann N, Düggelin M, Lang HP, Güntherodt HJ, Hegner M (2007) An antibody-sensitized microfabricated cantilever for the growth detection of *Aspergillus niger* spores. Microsc Microanal 13:13–17

Nunes CC, Gowda M, Salisbery J, Xue M, Chen F, et al. (2011) Diverse and tissue-enriched small RNAs in plant pathogenic fungus, Magnaporthe oryzae. BMC Genomics 12:288. doi: 10.1186/1471-2164-12-288

Okamoto H, Suzuki Y, Kataoka T, Sakai K (2009) Unified hyperspectral imaging methodology for agricultural sensing using software framework. Acta Horticult 824:49–56

Olmos A, Bertolini E, Cambr M (2002) Simultaneous and co-operational amplification (Co-PCR): a new concept for detection of plant viruses. J Virol Methods 106:51–59

Olmos A, Bertolini E, Capote N, Cambra M (2007) Molecular diagnostic methods for plant virus. In: Punja ZK, De Boer SH, Sanfac¸on H (eds) Biotechnology and plant disease management. CABI, Oxfordshire, pp 227–249

Orihara S, Yamamoto T (1998) Detection of resting spores of Plasmodiophora brassicae from soil and plant tissues by enzyme immunoassay. Nippon Shokubutsu Byori Gakkaiho 64:569–573

Pan Q, Shai O, Lee LJ, Frey BJ, Blencowe BJ (2008) Deep surveying of alternative splicing complexity in the human transcriptome by high-throughput sequencing. Nat Genet 40:1413–1415

Paplomatas EJ (2006) Molecular diagnostics of fungal pathogens. Arab J Plant Protect 24: 158–147

Pascale M, Zezza F, Perrone G (2013) Surface plasmon resonance genosensor for the detection of *Fusarium culmorum*. Methods Mol Biol 968:155–165

Priyanka S, Shashank P, Muhammad Aslam MK, Prashant S, Krishan P, S (2013) Nanobiosensors: diagnostic tool for pathogen detection. Int Res J Biol Sci 2:76–84

Radisek S, Jakse J, Javornik B (2006) Genetic variability and virulence among verticillium albo-atrum isolates from hop. Eur J Plant Pathol. doi:10.1007/s10658-006-9061-0

Rahman MT, Uddin MS, Sultana R, Moue A, Setu M (2013) Polymerase chain reaction (PCR): a short review. AKMMC J 4:30–36

Reddy DVR, Nambiar PTC, Rajeswari R, Mehan VK, Anjaiah V, McDonald D (1988) Potential of enzyme-linked immunosorbent assay for detecting viruses, fungi, bacteria, mycoplasma-like organisms, mycotoxins, and hormones. ICRISAT (International Crops Research Institute for the Semi-Arid Tropics). 1988. Biotechnology in tropical crop improvement: proceedings of the International Biotechnology Workshop, 12–15 Jan 1987, ICRISAT Center, India. Patancheru, A.P. 502 324, India: ICRISAT

Refshauge S, Watt M, McCully ME, Huang CX (2006) Frozen in time: a new method using cryo-scanning electron microscopy to visualize root–fungal interactions. New Phytol 172:369–374

Rigano LA, Malamud F, Orce IG, Filippone MP, Marano MR, Do Amaral AM, Castagnaro AP, Vojnov AA (2014) Rapid and sensitive detection of Candidatus Liberibacter asiaticus by loop mediated isothermal amplification combined with a lateral flow dipstick. BMC Microbiol 14:86

Roberts IM, Harrison BD (1979) Detection of potato leafroll and potato moptop viruses by immunosorbent electron microscopy. Ann Appl Biol 93:289–297

Rounsle S, Marri PR, Yu Y, He R, Sisneros N, Goicoechea JL, Wing RA (2009) De novo next generation sequencing of plant genomes. Rice 2:35–43

Safarpour H, Safarnejad MR, Tabatabaei M, Mohsenifar A, Rad F, Basirat M, Shahryari F, Hasanzadeh F (2012) Development of a quantum dots FRET based biosensor for efficient detection of Polymyxa betae. Can J Plant Pathol 34(4):507–515. doi:10.1080/07060661.2012.709885

Saikia R, Kadoo N (2010) Molecular detection and identification of *Fusarium oxysporum*. In: Molecular identification of fungi. Springer, Berlin, pp 131–157

Sanchez-Ballesteros J, Gonzalez V, Salazar O (2000) Phylogenetic study of Hypoxylon and related genera based on ribosomal ITS sequences. Mycologia 92:964–977

Sankaran S, Mishra A, Ehsani R, Davis C (2010) A review of advanced techniques for detecting plant diseases. Comput Electron Agric 72:1–13

Schaad NW, Frederick RD (2002) Real-time PCR and its application for rapid plant disease diagnostics. Can J Plant Pathol 24:50–258

Schaad NW, Frederick RD, Shaw J, Schneider WL, Hickson R, Petrillo MD, Luster DG (2003) Advances inmolecular-based diagnostics in meeting crop biosecurity and phytosanitary issues. Annu Rev Phytopathol 41:305–324. doi:10.1146/annurev.phyto.41.052002.095435

Schena L, Nigro F, Ippolito A, Gallitelli D (2004) Real-time quantitative PCR: a new technology to detect and study phytopathogenic and antagonistic fungi. Eur J Plant Pathol 110:893–908

Schmitt B, Henderson L (2005) Diagnostic tools for animal diseases. Rev Sci Tech Off Int Epiz 24:243–250

Shane WW (1991) Prospects for early detection of Pythium blight epidemics on turfgrass by antibody-aided monitoring. Plant Dis 75:921–925

Sharadraj KM, Nidhina K, Nair S, Hegde V (2015) Loop-mediated isothermal amplification (LAMP) for detection of Ganoderma lucidum causing basal stem rot of coconut. In national symposium on Understanding host-pathogen interaction through science of omics, ICAR-Indian Institute of Spices Research, Kozhikode, Kerala. 16–17 Mar

Singh RP (2005) Plant Pathology. Kalyani Publishers, New Delhi, 9788127223168. ISBN 8127223166

Skottrup P, Hearty S, Frokiaer H, Leonard P, Hejgaard J, O'Kennedy R, Nicolaisen M, Justesen AF (2007) Detection of fungal spores using a generic surface plasmon resonance immunoassay. Biosens Bioelectron 22:2724–2729

Skottrup PD, Nicolaisen M, Justesen AF (2008) Towards on-site pathogen detection using antibody-based sensors. Biosens Bioelectron 24:339–348

Studholme DJ, Glover RH, Boonham N (2011) Application of high-throughput DNA sequencing in phytopathology. Annu Rev Phytopathol 49:87–105

Szemes M, Schoen CD (2003) Design of molecular beacons for AmpliDet RNA assay-Characterization of binding stability and probe specificity. Analytical Biochemistry 315: 189–201. doi:10.1016/S0003-2697(02)00692-9

Szemes M, Bonants P, de Weerdt M, Baner J, Landegren U, Schoen CD (2005) Diagnostic application of padlock probes — multiplex detection of plant pathogens using universal microarrays. Nucleic Acids Res 33, e70

Tambong JT, De Cock A, Tinker NA, Levesque CA (2006) Oligonucleotide array for identification and detection of Pythium species. Appl Environ Microbiol 72:2691–2706

Thelwell N, Millington S, Solinas A, Booth J, Brown T (2000) Mode of action and application of Scorpion primers to mutation detection. Nucleic Acids Res 28:3752–3761

Thiyagarajan S, Karhanek M, Akhras M, Davis RW, Pourmand N (2006) PathogenMIPer: a tool for the design of molecular inversion probes to detect multiple pathogens. BMC Bioinforma 7:500

Thornton CR, Groenhof AC, Forrest R, Lamotte R (2004) A one-step, immunochromatographic lateral flow device specific to Rhizoctonia solani and certain related species, and its use to detect and quantify R. solani in soil. Phytopathology 94(3):280–288

Troy Nagle H, Schiffman SS, Gutierrez-Osuma R (1998) The how and why of electronic nose. IEEE Spectr 35:22–34

Tsui CKM, Wang B, Khadempour L, Alamouti SM, Bohlmann J, Murray BW, Hamelin RC (2010) Rapid identification and detection of pine pathogenic fungi associated with mountain pine beetles by padlock probes. J Microbiol Methods 83:26–33

Tsui CKM, Woodhall J, Chen W, Levesque CA, Lau A, Schoen CD, Baschien C, Najafzadeh MJ, Hoog GS (2011) Molecular techniques for pathogen identification and fungus detection in the environment. IMA Fungus 2:177–189

Vakalounakis DJ (1996) Root and stem rot of cucumber caused by Fusarium oxysporum f. sp. radicis-cucumerinum. Plant Dis 80:313–316

Validov SZ, Kamilova FD, Lugtenberg BJJ (2011) Monitoring of pathogenic and nonpathogenic Fusarium oxysporum strains during tomato plant infection. Microb Biotechnol 4:82–88

Varshney GK (1999) Characterization of genetic races of Karnal bunt (Tilletia indica) of wheat by immunobinding assay procedure. M.Sc Thesis, G. B. Pant Univ. of Agric. & Tech, Pantnagar

Vincelli P, Tisserat N (2008) Nucleic acid–based pathogen detection in applied plant pathology. Plant Dis 92:660–669

Volkhard A, Kempf J, Trebesius K, Autenrieth IB (2000) Fluorescent *in situ* hybridization allows rapid identification of microorganisms in blood cultures. J Clin Microbiol 38:830–838

Waalwijk C, Heide R, van der deVries I, Lee T, van der Schoen C, Coainville GC, Hauser-Hahn I, Kastelein P, Kohl J, Demarquet T, Kema GH (2004) Quantitative detection of Fusarium species in wheat using TaqMan. Eur J Plant Pathol 110:481–494

Waggoner PS, Craighead HG (2007) Micro-and nanomechanical sensors for environmental, chemical, and biological detection. Lab Chip 7:1238–1255

Wang CG, Blanchette RA, Jackson WA, Palmer MA (1985) Differences in conidial morphology among isolates of Sphaeropsis sapinea. Plant Dis 69:838–841

Webster CG, Wylie SJ, Jones MGK (2004) Diagnosis of plant viral pathogens. Curr Sci 86:1604–1607

White TJ, Bruns T, Lee S, Taylor J (1990) Amplification and direct sequencing of fungal ribosomal RNA genes for phylogenetics. In: Innis MA, Gelfand DH, Sniosky JJ, White TJ (eds) PCR protocols: a guide to methods and applications. Academic, San Diego, pp 315–322

Wilson AD (2013) Diverse applications of electronic-nose technologies in agriculture and forestry. Sensors 13:2295–2348. doi:10.3390/s130202295

Wilson AD, Lester DG, Oberle CS (2004) Development of conductive polymer analysis for the rapid detection and identification of phytopathogenic microbes. Phytopathology 94:419–431

Winton LM, Hansen EM (2001) Molecular diagnosis of *Phytophthora lateralis* in trees, water, and foliage baits using multiplex polymerase chain reaction. For Pathol 31:275–283

Wullings BA, Beuningen AR, Van-Janse JD (1998) Detection of *Ralstonia solanacearum*, which causes brown rot of potato, by fluorescent in situ hybridization with 23S rRNA-targeted probes. Appl Environ Microbiol 64:4546–4554

Yao KS, Li SJ, Tzeng KC, Cheng TC, Chang CY, Chiu CY, Liao CY, Hsu JJ, Lin ZP (2009) Fluorescence silica nanoprobe as a biomarker for rapid detection of plant pathogens. Adv Mater Res 79:513–516, Volume-Multi-Functional Materials and Structures II

Young Li Z (2004) Specific and sensitive detection of the fungal pathogen Ustilaginoidea virens by nested PCR. Mycosystema 23:102–108

Zeng QY, Hansson P, Wang XR (2005) Specific and sensitive detection of the conifer pathogen *Gremmeniella abietina* by nested PCR. BMC Microbiol 5:65. doi:10.1186/1471-2180-5-65

Zezza F, Pascale M, Mulè G, Visconti A (2006) Detection of Fusarium culmorum in wheat by a surface plasmon resonance-based DNA sensor. J Microbiol Methods 66:529–537

Zhang SX (2013) Enhancing molecular approaches for diagnosis of fungal infections. Future Microbiol 8:1599–1611

Zhang H, Balaban M, Portier K, Sims CA (2005) Quantification of spice mixture compositions by electronic nose: Part II. Comparison with GC and sensory methods. J Food Sci 70:E259–E264

Zhao Y, Perlin DS (2013) Quantitative detection of *Aspergillus* spp. by real-time nucleic acid sequence-based amplification. Methods Mol Biol 968:83–92

Zhou L, Holliday JA (2012) Targeted enrichment of the black cottonwood (*Populus trichocarpa*) gene space using sequence capture. BMC Genomics 13:703

Advance Detection Techniques of Phytopathogenic Fungi: Current Trends and Future Perspectives

Pardeep Kumar, Jameel Akhtar, A. Kandan, Sandeep Kumar, Ritu Batra, and S.C. Dubey

Introduction

Plant disease is the result of a complex interaction between the host plant, pathogen and the environmental conditions which not only cause yield loss but also affect the social and cultural structure of human being. It is illustrated by the Irish famine of potato 1846–1850 which took as many as one million lives from hunger and disease. The combined forces of famine, disease and emigration caused drop in Ireland's population from eight million to five million and changed the social and cultural structure of Ireland in profound ways. Plant diseases are caused by fungi, bacteria, viruses, nematodes and parasite plants. But most of plant diseases (around 85 %) are caused by pathogenic fungi. Out of 100,000 known fungal species, more than 10,000 species can cause diseases in plants whereas, approximately 50 species causes diseases in human beings as well as animals. The enormous diversity of pathogenic interaction between plant and fungus was reflected by the establishment of parasitic interaction in lower Devonian i.e. approximately 400 million years ago (Taylor et al. 1992). Within a phytopathogenic fungal species, different formae specialis are found. For example in case of *Fusarium oxysporum*, 120 different formae specialis

P. Kumar (✉) • J. Akhtar • A. Kandan • S.C. Dubey
Plant Quarantine Division, ICAR-National Bureau of Plant Genetic Resources,
New Delhi 110012, India
e-mail: pardeepkumar@nbpgr.ernet.in

S. Kumar
Germplasm Evaluation Division, ICAR-National Bureau of Plant Genetic Resources,
New Delhi 110012, India

R. Batra
Molecular Biology, Biotechnology and Bioinformatics Department,
CCS Haryana Agricultural University, Hisar 125004, India

P. Kumar et al. (eds.), *Current Trends in Plant Disease Diagnostics and Management Practices*, Fungal Biology, DOI 10.1007/978-3-319-27312-9_12

have been reported (Agrios 2005). The different forms of strains and the formae specialis of a fungal pathogen make their detection and identification more complicated and require specific methods or techniques. Rapid and accurate detection of phytopathogenic fungal pathogen up to species level is essential for executing proper disease management strategies.

The conventional methods available for detection and identification of phytopathogenic fungi are time consuming and not always very specific. An advanced plant disease diagnostics can provide rapid, accurate, and reliable detection of plant diseases in early stages for avoiding economic yield losses and healthy crop production. Early detection and identification of these pathogens is crucial for their timely management. The conventional methods of detection and identification of phytopathogenic fungi mostly relied on symptoms, isolation and culturing followed by morphological observations and biochemical tests (Pearson et al. 1987; Tan et al. 2008). The study of fungal biology and their interaction with host plant has experienced great advances in recent years because of the development of modern holistic and high throughput techniques. Some of the important advanced techniques of fungal disease detection and diagnosis available are polymerase chain reaction (PCR), real time-PCR, nested PCR, magnetic capture-hybridisation (MCH)-PCR, reverse transcriptase PCR etc. High levels of sensitivity, specificity and simplicity, have made the PCR-based assay as the technique of choice for routine and large scale application in pathogen detection. As plants are generally infected by more than one pathogens, so multiplexing by multiplex PCR and DNA array will reduce the time and cost of detection. Although the identification procedures that involve molecular tools, are shortened by avoiding culture cultivation, yet the whole process of molecular analysis of target genes remains time consuming and costly. The latest technologies that hold the key for fungal disease detection and development of sensor are based on spectroscopy and imaging, mass metabolite and volatile profiling. This chapter emphasizes the importance of rapid, sensitive, specific and reliable techniques for detection, differentiation and quantification of phytopathogenic fungal pathogen in early stages of infection.

Molecular Techniques for Detection of Phytopathogenic Fungi

Now a days, molecular techniques are the most commonly used advanced techniques for detection of fungi like conventional PCR, real-time PCR, nested PCR, reverse transcriptase and LAMP. Other molecular techniques include MCH PCR, PCR-RFLP, in situ PCR, PCR DGGE, co-operational PCR, multiplex PCR, DNA arrays etc.. The molecular techniques are highly sensitive, specific and rapid for the detection of fungal pathogen as PCR can detect the concentration that is as low as 10 pg of DNA of fungus Lin et al. (2009). The commonly used molecular techniques for fungal detection are conventional PCR, real-time PCR, nested PCR, reverse transcriptase and LAMP. Other molecular techniques include MCH PCR, PCR-RFLP, in situ PCR, PCR DGGE, co-operational PCR, multiplex PCR, DNA arrays etc.

Polymerase Chain Reaction

Polymerase chain reaction (PCR) is a simple technique capable of amplifying specific DNA sequences exponentially by in vitro DNA synthesis. PCR was discovered by Karry Mullis in 1980s which is being applied widely in field of biological sciences including plant pathology. The usefulness of PCR depends upon its ability to amplify a specific DNA in vitro from trace amounts of complex templates. High levels of sensitivity, specificity and simplicity have made the PCR-based assays as the techniques of choice for routine and large scale application in detection of phytopathogenic fungal pathogen.

Designing suitable primers is a critical step in PCR based detection assay. The primers, which are unique for the target sequence to be amplified, should fulfil certain criteria such as primer length, GC content, annealing and melting temperature, 5′ end stability, 3′ end specificity, etc. (Dieffenbach et al. 1993). Specific primers are derived from sequences of either amplified or cloned DNA or RNA from target pathogen species (Table 1). Ribosomal genes and the spacers between them provide targets of choice for molecular detection, since they are present in high copy numbers, contain conserved as well as variable sequences and can be amplified and sequenced with universal primers based on their conserved sequences (Stackebrandt et al. 1992). Similarly, greater sequence differences have been noted in the non-transcribed spacer (NTS) regions between the rDNA repeat units and also in the intergenic spacer (IGS) region.

Internal transcribed spacer (ITS) primers are the most commonly used primers for detection of phytopathogenic fungi. Depending on the primer pair, either ITS1or ITS2, or both ITS1 and ITS2 regions can be amplified. But ITS primers have to be selected very carefully because some primers, e.g. ITS1-F, ITS1 and ITS5 are biased towards amplification of basidiomycetes whereas others, e.g. ITS2, ITS3 and ITS4 are biased towards ascomycetes. Further, ascomycetes are more easily amplified than basidiomycetes using these regions as targets due to systematic length differences in the ITS2 region as well as the entire ITS. Larena et al. (1999) developed ITS 4A (A is for ascomycetes) primer which can amplify only ascomycetes with ITS1F as forward primer. Similarly ITS 4B (B is for basidiomycete) amplifies only basidiomycetes (Gardes and Bruns 1993). Some of the ITS primers, such as ITS1-F, were hampered with a high proportion of mismatches relative to the target sequences, and most of them appeared to introduce taxonomic biases during PCR (Bellemain et al. 2010). So, different primer combinations or different parts of the ITS region should be analyzed in parallel or alternative ITS primers should be used.

Although the ITS region is the main target, other housekeeping genes with higher variability like LSU, SSU, and RPB1 are being more extensively used to develop diagnostics for fungi. Six DNA regions including the above four markers along with two protein markers (translation elongation factor 1-α and β-tubulin) were evaluated as potential DNA barcodes for fungi using newly generated sequences from 742 strains or specimens to analyze a smaller subset of about 200 fungi by a multinational and multilaboratory consortium. ITS will be formally proposed as the primary fungal barcode marker to the Consortium for the Barcode of Life, with the possibility that supplementary barcodes may be developed for particular narrowly

Table 1 List of ITS primers with sequence used in detection and identification of phytopathogenic fungi

Primer name	Primer sequence (5′–3′)	Reference
ITS1	TCCGTAGGTGAACCTGCGG	White et al. (1990)
ITS2	GCTGCGTTCTTCATCGATGC	White et al. (1990)
ITS3	GCATCGATGAAGAACGCAGC	White et al. (1990)
ITS4	TCCTCCGCTTATTGATATGC	White et al. (1990)
ITS5	GGAAGTAAAAGTCGTAACAAGG	White et al. (1990)
NS7	GAGGCAATAACAGGTCTGTGATGC	White et al. (1990)
LR3	CCGTGTTTCAAGACGGG	Vilgalys and Gonzalez (1990)
ITS-F	CTTGGTCATTTAGAGGAAGTAA	Gardes and Bruns (1993)
ITS-B	CAGGAGACTTGTACACGGTCCAG	Gardes and Bruns (1993)
ITS8mun	CTTCACTCGCCGTTACTA	Egger (1995)
ITS9mun	TGTACACACCGCCCGTCG	Egger (1995)
ITS10mun	GCTGCGTTCTTCATCGAT	Egger (1995)
NL6Amun	CAAGTGCTTCCCTTTCAACA	Egger (1995)
NL6Bmun	CAAGCGTTTCCCTTTCAACA	Egger (1995)
ITS-4A	CGCCGTTACTGGGGCAATCCCTG	Larena et al. (1999)
NSA3	AAACTCTGTCGTGCTGGGGATA	Martin and Rygiewicz (2005)
NSI1	GATTGAATGGCTTAGTGAGG	Martin and Rygiewicz (2005)
58A1F	GCATCGATGAAGAACGC	Martin and Rygiewicz (2005)
58A2F	ATCGATGAAGAACGCAG	Martin and Rygiewicz (2005)
58A2R	CTGCGTTCTTCATCGAT	Martin and Rygiewicz (2005)
NLB3	GGATTCTCACCCTCTATGA	Martin and Rygiewicz (2005)
NLB4	GGATTCTCACCCTCTATGAC	Martin and Rygiewicz (2005)
NLC2	GAGCTGCATTCCCAAACAACTC	Martin and Rygiewicz (2005)
ITS1-F_KYO1	CTHGGTCATTTAGAGGAASTAA	Toju et al. (2012)
ITS1-F_KYO2	TAGAGGAAGTAAAAGTCGTAA	Toju et al. (2012)
ITS2_KYO1	CTRYGTTCTTCATCGDT	Toju et al. (2012)
ITS2_KYO2	TTYRCTRCGTTCTTCATC	Toju et al. (2012)
ITS3_KYO1	AHCGATGAAGAACRYAG	Toju et al. (2012)
ITS3_KYO2	GATGAAGAACGYAGYRAA	Toju et al. (2012)
ITS4_KYO1	TCCTCCGCTTWTTGWTWTGC	Toju et al. (2012)
ITS4_KYO2	RBTTTCTTTTCCTCCGCT	Toju et al. (2012)
ITS4_KYO3	CTBTTVCCKCTTCACTCG	Toju et al. (2012)

circumscribed taxonomic groups. The most commonly used animal barcode i.e. mitochondrial cox I gene is not included in fungal barcode because it is difficult to amplify in fungi, often includes large introns, and can be insufficiently variable (Schoch et al. 2012).

Pathogen specific primers have been developed for many fungal pathogens from conserved sequences (Table 2). To design species-specific PCR primers, sequences of target fungal rDNA can be obtained from some databases (e.g. NCBI), or by amplifying the target rDNA fragment with conserved primers (White et al. 1990). After comparing the sequences of the target rDNA with those of related fungal spe-

Table 2 Species specific PCR primers for detection of some important phytopathogenic fungi

Fungal species	Primer name: sequence (5′–3′)	Target DNA	Product size (bp)	References
Botrytis cinerea	C729+: AGCTCGAGAGAGATCTCTGA C729-: CTGCAATGTTCTGCGTGGAA	SCAR	700	Rigotti et al. (2002)
Fusarium oxysporum f.sp. *ciceris*	SC FOC1: F-CCTCGCCAGCCTTGTACTTGCG R- CGGTACCGGATGGCCCTGCAA SC FOC2: F-ATGGCTCAGTGAGGCGTCCGGA R-GTGTGGGGGATAGAGCATTG	SCAR	1,400 1,300	Durai et al. (2012)
Fusarium oxysporum f.sp. *niveum*	Fn1:TACCACTTGTTGCCTCGGC Fn2: TTGAGGAACGCGAATTAAC	ITS region	327	Zhang et al. (2005)
Magnaporthe oryzae	pfh2a: CGTCACACGTTCTTCAACC pfh2b: CGTTTCACGCTTCTCCG	Pot2 transposon	687	Harmon et al. (2003)
Phomopsis longicolla	Phom.1: GAGCTCGCCACTAGATTTCAGGG Phom.2: GGCGGCCAACCAAACTCTTGT	ITS region	337	Zhang et al. (1997)
Puccinia reconditia	BR3: TCCCAAAGCAAGCCCAAATACACG 300 BR2: GAATGTTTCACAGCAGCTGCTGGT	β-tubullin gene	300	Fraaije et al. (2001)
Puccinia striiformis	YRNT1: CTTCAAGATCGGTGGCCTGACCGA YRNT2: GTGAGCTGTGAAGGGATCGCGGGA	β-tubullin gene	351	Fraaije et al. (2001)
Puccinia thlaspeos	F63: GCATATCAATAAGCGGAGGAAAAG RUST1:GCTTACTGCCTTCCTCAATC	28S rDNA	560	Kropp et al. (1995)
Rhizoctonia solani	SCAR-GS forward: GTGGA ACCAA GCATA ACACT GA SCAR-GS reverse: AGTTT CAACA ACGGA TCTTT GG	SCAR	285	Ganeshamorthi and Dubey (2013)
Tilletia caries	Tcar2A: ACAACAGCAAATACGCCAAT Tcar2B: TCCCGTACTTGACATGGACC	26S rRNA	296	Eibel et al. (2005)
Tilletia indica	TI17M1: TCCCCTTGGATCAGAACGTA TI17M2: AGAAGTCTAACTCCCCCCTCT	mtDNA region	825	Smith et al. (1996)
Ustilago maydis	UM11F: GAACCTTTCTGGCCTCCTTT UM11R: CCTTGGTTTCCGTTCCGTAC	Repetitive sequence of genomic DNA	900	Xu et al. (1999)

cies by alignment with a computer program, species-specific PCR primers can be developed using different computer software programmes (e.g. Primer3, PrimerQuest etc.).

Some anonymous unique DNA regions are also used for designing species specific PCR primers. Randomly amplified polymorphic DNA (RAPD) markers are used for comparing target pathogen with those of non-target organisms, unique bands specific to the target pathogen could be observed and cloned. Once unique bands have been detected, they are used as probes to check the presence of similar DNA sequences in related species. If the DNA fragment does not match, it can be sequenced, and species specific SCAR (Species-specific sequence characterized amplified region) marker can be developed (Ma and Michialides 2005). PCR primers designed based on conserved sequences have been used to detect more than 205 phytopathogenic fungal species of 101 genera (http://www.sppadbase.ipp.cnr.it/).

Real Time PCR

The development of real-time PCR in the early 1990s has revolutionized basic and applied research in all biological fields, including the detection of phytopathogenic fungi. Real time PCR enables the detection of amplicons through a specific fluorescent signal, thus eliminating the post-amplification processing steps needed in conventional PCR. This significantly reduces the time and cost of analyses and eliminates the use of harmful substances like ethidium bromide, which is still utilized to stain DNA in electrophoretic gels. Also, real time PCR is a versatile technique for the accurate, sensitive, and high throughput quantification of target DNA.

Real time PCR methods for the detection of phytopathogenic fungi can be grouped into sequence non-specific and specific methods based on chemistry used in the PCR (Schena et al. 2004). The sequence non-specific method is based on dyes that emit fluorescent light when interposed into double-stranded DNA (dsDNA). SYBR Green I is the most commonly used dye but several valid alternatives are also available (Gudnason et al. 2007). Since dyes do not discriminate between the different dsDNA molecules, the formation of non-specific amplicons, as well as primer dimmers, could lead to false positive results. As a consequence, amplification reactions need to be accurately optimized and examined at the end of the reaction by melting curve analysis. Sequence-specific methods primarily used are TaqMan probe, molecular beacons and scorpion probe. These methods are based on the use of oligonucleotide probes labeled with a fluorophore and quencher. The advantage of fluorogenic probes over DNA binding dyes is that specific hybridization between probe and target DNA sequence is required to generate a fluorescent signal. As a consequence, these methods guarantee higher specificity that enables the discrimination of single base pair mismatches.

A crucial step in real time PCR assay is the identification of appropriate target DNA regions for primers. A good target gene should be sufficiently variable to enable the differentiation of closely related species but, at the same time, should not contain intraspecific variation that would decrease sensitivity of the detection. A good target gene should readily be amplified and sequenced and, multicopied for sensitive detection. However, the use of multicopy genes is not always preferable since the variable number of copies can potentially complicate the development of quantitative assays (Schena et al. 2013).

The ITS regions provide attractive targets for primers because they are usually conserved within a species but variable enough to differentiate related taxa, easily amplified and sequenced using universal primers and occur in multiple copies (White et al. 1990). Different universal primers are available to amplify ITS regions and the selection should be made according to the taxonomic group (Cooke et al. 2007; Bellemain et al. 2010). The huge number of fungal ITS sequences currently deposited in the international nucleotide sequence databases provides a wide range of reference material for the identification of taxa and for the development of specific detection methods. IGS (Intergenic spacer) region can be a valid alternative to the ITS region when closely related taxa or even different species need to be differentiated or detected since it evolves faster than the ITS region and, as such, more sequence polymorphisms are present (Diguta et al. 2010; Bilodeau et al. 2012). Similar to ITS, the IGS region is also a multicopy but its length in fungi (approximately 2–4 kbp) provides considerable scope for primer and probe development. The wide utilization of the IGS region as target for developing specific molecular markers is primarily limited by difficulties in amplifying long fragments and the lack of effective universal primers. Apart from ITS and IGS region, many other nuclear and mitochondrial regions like *β-tubulin*, ras related proteins (*Ypt*1) genes and *mitochondrial small subunit* (*mtSSU*) *rRNA* genes have been used as targets in qRT PCR methods for fungal detection. For instance, the *β-tubulin* gene is one of the most frequently utilized targets for fungi and the use of this target is favoured by the availability of universal primers designed in conserved coding regions and amplified fragments also contain variable regions (mainly introns) that proved to effectively differentiate closely related taxa (Aroca et al. 2008). The detection limits of real time PCR reported in literature is mainly between 10 pg/mL and 10 fg/mL of target DNA. Real time PCR is more sensitive as compared to conventional PCR as, real time PCR commonly amplify very short DNA fragments (70–100 bp) which favour a higher level of PCR efficiency and sensitivity compared to conventional PCR. The production of very short amplicons is best avoided in conventional PCR due to their confusion with primer dimers and difficulty in visualization in agarose gels. Some examples of real time based detection of phytopathogenic fungi are detailed in Table 3.

Magnetic Capture-Hybridisation (MCH)-PCR

PCR sensitivity is limited by inhibitors like humic acids, phenolic compounds etc. In order to reduce the effects of inhibitors and non-target DNA on PCR amplification, a novel magnetic capture-hybridization (MCH)-PCR technology was introduced by

Table 3 Real-time PCR based assay for detection of phytopathogenic fungi in crop plants

Fungal pathogen	Real time chemistry	Host plant	Reference
Cladosporium fulvum	SYBR Green	Tomato	Yan et al. (2008)
Fusarium avenaceum	SYBR Green	Wheat	Moradi et al. (2010)
Fusarium avenaceum	TaqMan MGB	Cereals	Kulik et al. (2011)
Fusarium oxysporum	SYBR Green	Chickpea	Dubey et al. (2014)
Fusarium equiseti	Molecular Beacons	Barley	Macía-Vicente et al. (2009)
Mycosphaerella graminicola	TaqMan MGB	Wheat	Bearchell et al. (2005)
Macrophomina phaseolina	TaqMan MGB	Chickpea, soybean, pigeonpea	Babu et al. (2011)
Phialophora gregata	TaqMan	Soybean	Malvick and Impullitti (2007)
Phoma sclerotioides	SYBR Green	Wheat	Larsen et al. (2007)
Phytophthora erythroseptica	TaqMan	Potato	Nanayakkara et al. (2009)
Puccinia graminis and *P. striiformis* sensu stricto	TaqMan	Wheat	Liu et al. (2015)
Pythium irregular	SYBR Green	Wheat, barley	Schroeder et al. (2006)
Rhizoctonia oryzae	SYBR Green	Cereals	Okubara et al. (2008)
Sclerotinia sclerotiorum	SYBR Green	Oilseed rape	Yin et al. (2009)
Verticillium dahliae	SYBR Green	Potato	Attallah et al. (2007)

Jacobsen in 1995. MCH combines an initial DNA extraction and purification step, including hybridization with a single-stranded DNA probe on magnetic beads, and a subsequent PCR amplification step of the extracted gene. In this novel process, paramagnetic streptavidin-coated beads conjugated with biotin labeled oligonucleotide probe are used to capture single-stranded target DNA from crude DNA preparations (Jacobsen 1995). After the magnetic capture-hybridisation, PCR amplification was carried out using species-specific primers. This increased the efficiency of isolation of target DNA. MCH-PCR has the advantages over direct PCR in that the MCH process purifies and concentrates the DNA of interest while removing non-target DNA and other substances that can inhibit the amplification reaction of PCR. Langrell and Barbara (2001) used this method to detect *Nectria galligena* in apple and pear trees. Walcott et al. (2004) detected *Botrytis aclada* fungus causing Botrytis neck rot disease by using this technique in onion seed. MCH-PCR reduces the time required to test onion seeds from 10 to 14 days to less than 24 h. Additionally, MCH-PCR detected fungal DNA from aqueous solutions containing 100 fg DNA/mL.

PCR-ELISA

PCR-ELISA combines both PCR and ELISA into a single analytical technique and its application is very much similar to ELISA except that this method allows the detection of nucleic acid instead of protein (Shamloul and Hadidi 1999).

PCR-ELISA is an immunological method to quantify the PCR product directly after immobilization of biotinylated DNA on a microplate. PCR-ELISA method uses forward primer coated with biotin and reverse primer coated with an antigenic group (e.g. fluorescein) at their 5′ end (Landgraf et al. 1991). PCR amplified DNA can be immobilized on avidin or streptavidin-coated microtiter plates via the biotin moiety of the forward primer and then can be quantified by an ELISA specific for the antigenic group of the reverse primer. The majority of steps occur in 96-well microtiter plates for mass screening of PCR products making them very suitable for routine diagnostic purposes. Also, PCR-ELISA does not require electrophoretic separation and dot hybridisation, and can be easily automated. This procedure has been used for detection of several species of *Phytophthora* and *Pythium* (Bonants et al. 1997; Bailey et al. 2002).

In Situ PCR

In situ PCR combines two techniques of molecular biology, PCR and in situ hybridisation (ISH) for amplification of specific gene sequences within intact cells or tissues (Long 1998). The in situ PCR technique links PCR amplification to the light microscope image. The amplified tissue is stained, thus confirming which morphotype has been amplified. The improved sensitivity of this technique allows the localization of one target copy per cell (Haase et al. 1990) but background signal is very high. It is a time-consuming due to hybridisation step and technically demanding such as light microscopy. Bindslev et al. (2002) used in situ PCR technique to identify *Blumeria graminis* spores and mycelia causing the powdery mildew disease on barley leaves.

Polymerase Chain Reaction- Denaturing Gradient Gel Electrophoresis (PCR-DGGE)

In PCR-DGGE target DNA from fungal pathogen are firstly amplified by PCR and then subjected to denaturing electrophoresis. DGGE use chemical gradient such as urea to denature and separate DNA samples when they are moving across an acrylamide gel. Sequence variants of particular fragments migrate at different rate in the denaturing gradient gel, allowing a very sensitive detection. In addition, PCR-DGGE primers contain a GC rich tail at their 5′ end to improve the detection of small variations (Myers et al. 1985). This method is however time-consuming, poorly reproducible and fragments with different sequences but similar melting behaviour are not always correctly separated. This technique is most commonly used for diversity study but it is also used for detection of pathogenic fungi. Elsas et al. (2000) detected *Trichoderma harzianum* spores and *Arthrobotrys oligospora* hyphal fungus from soil using PCR-DGGE for about 14 days and

2 months, respectively. Recently, PCR-DGGE is also applied to detect multiple species of *Phytophthora* from plant material and environmental samples (Rytkönen et al. 2011).

Nested PCR

Nested PCR approach is used when an improvement of the sensitivity and specificity of detection is necessary. This method consists of two consecutive rounds of amplification in which two external primers amplify a large amplicon that is then used as a target for a second round of amplification using two internal primers (Porter-Jordan et al. 1990). The two reactions can be performed in single tubes supporting high throughput. The annealing temperature of inner primers should be more than outer primers so that inner primers should not bind at initial stage of amplification. Nested PCR is reported to successfully detect *Phytophthora fragariae* in naturally infected strawberry tissues and the sensitivity is 1,000–10,000 times more as compared to conventional PCR (Bonants et al. 2004). Zeng et al. (2005) developed two nested PCR systems. The first system employed universal fungal primers to enrich the fungal DNA targets in the first round, followed by a second round selective amplification of the pathogen while the other system employed *G. abietina*-specific primers in both PCR steps. Both approaches can detect the presence of *G. abietina* in composite samples with high sensitivity, as little as 7.5 fg *G. abietina* DNA in the host genomic background.

Co-operational PCR

Co-operational PCR (Co-PCR) is developed based on the simultaneous action of four primers and uses ten times fewer reagents than conventional PCR. The reaction process consists of the simultaneous reverse transcription of two different fragments from the same target, one containing the other. The production of four amplicons by the combination of the two pair of primers, one pair external to other and producing maximum possible largest fragment by co-operational action of amplicons (Olmos et al. 2002). Co-operational PCR is also used to increase sensitivity and specificity of PCR like nested PCR but fragment produced in Co-PCR is larger than nested PCR. In both Co-PCR and nested PCR methods, the use of external primers can be used for generic amplification and the internal primers for specific amplification of product. The technique was first developed and used successfully for the detection of plant RNA viruses, such as citrus tristeza virus (CTV), cucumber mosaic virus (CMV), cherry leaf roll virus (CLRV) and strawberry latent ring spot virus (SLRSV) (Olmos et al. 2002). Co-PCR is usually coupled with dot blot hybridisation by using a specific probe to enhance the specificity of the detection and provide a sensitivity level similar to nested PCR method. Martos et al. (2011) used this method for sensitive and specific detection of *Phaeomoniella chalmydospora* from infected grapevine wood.

Reverse Transcription (RT)-PCR

Reverse transcription (RT)-PCR is a diagnostic technique which detects target mRNA. An important limitation of DNA based PCR is the inability to distinguish living or dead fungi. So, results from detection and identification of fungal plant pathogens generally validated by pathogenicity tests. RT–PCR based detection is considered an accurate indicator of cell viability because mRNA is degraded rapidly in dead cells (Sheridan et al. 1998). In RT-PCR, the RNA is reverse transcribed using the enzyme reverse transcriptase to produce stable cDNA. The resulting cDNA is then amplified using conventional or any other PCR-based method. Although, the most frequent application of this technique in phytopathology is the analysis of fungal gene expression during disease development (Yang et al. 2010) but RT-PCR is also used to detect viable populations of fungal pathogen like *Mycosphaerella graminicola* in wheat (Guo et al. 2005) and *Oidium neolycopersici* in tomato (Matsuda et al. 2005).

Polymerase Chain Reaction- Restriction Fragment Length Polymorphism (PCR-RFLP)

PCR-RFLP combines the amplification of a target region with the further restriction enzyme digestion of the PCR products. This is followed by separation of the fragments by electrophoresis in agarose or polyacrilamide gels to detect differences in the size of DNA fragments. PCR primers specific to the genus *Phytophthota* were used to amplify and further digest the resulting amplicons yielding a specific restriction pattern of 27 different *Phytophthora* species (Drenth et al. 2006). It also allowed the differentiation of pathogenic and non-pathogenic strains of *Pythium myriotolum* (Gómez-Alpizar et al. 2011).

Fingerprinting Based Detection

Fingerprinting analyses are generally used to study the phylogenetic structure of fungal populations. However, these techniques also allow the screening of random regions of the fungal genome for identifying species-specific sequences when conserved genes have not enough variation to successfully identify species

(McCartney et al. 2003). Any band that appears to be unique to particular fungal species may be labelled and tested for use as specific probes for fungal detection. There are various markers like RFLP, RAPD, AFLP and microsatellites which generally used for phylogentic study are also used for diagonsis of fungal pathogens. RAPD analysis can provide markers to identify and differentiate fungal pathogens. Specific DNA bands were selected as probes from the RAPD profiles of 13 *formae speciales* of *Fusarium oxysporum*. The *formae specialis*-specific probe OPC18300c and OPC18520f were used to identify *Fusarium oxysporum* f.sp. *cucumerinum* (FOC) and *F. oxysporum* f.sp. *luffae* (FOL) infecting cucumber and *Luffa cylindrica* respectively by RAPD-PCR followed by dot blot hybridization (Wang et al. 2001). *Pyrenophora teres*, causing net blotch disease in barley leaves was identified by AFLP fingerprinting technique. Specific primers were designed based on the sequences of AFLP fragments. The primers amplified the DNA from *P. teres* f. *teres* (net form), but not from the closely related *P. teres* f. *maculata* (spot form), indicating the specificity of detection for distinguishing closely related fungal pathogen species/strains (Leisova et al. 2005). Microsatellite markers specific for *Phytophthora ramorum* were employed to distinguish between A1 and A2 mating types isolates of this pathogen from two different geographic origins (Prospero et al. 2004). Sequence tagged microsatellites (STMs) technique was developed to reduce the cost of developing microsatellites. STMs is amplified by PCR using a single primer specific to the conserved DNA sequence flanking the microsatellite repeat in combination with a universal primer that anchors to the 5′-ends of the microsatellites (Hayden et al. 2002). STMs have been developed for the detection of *Pyrenophora teres* causing barley net blotch disease (Keiper et al. 2007).

Multiplex PCR

Most of the molecular diagnostic assays used in fungal pathology target one specific pathogen. However, because crops can be infected by numerous fungal pathogens which are often present in plants as complexes. So, it is desirable to develop assays that can detect multiple pathogens simultaneously. Multiplex PCR method combines multiple species-specific primers in a single PCR tube and then amplified product is resolved by gel electrophoresis. It helps in reducing the number of tests required (James et al. 2006), but care is needed to optimize the conditions so that all amplicons can be generated efficiently. Primer designing and selection of primer is very critical in multiplex PCR. As we are using multiple primers, so there is chance of formation of primer dimmers. Also, it is very important that all primer should have nearly same annealing temperature and amplified product should be of different size.

Fraaije et al. (2001) developed a multiplex PCR assay to detect and quantify four foliar fungal pathogens; *Septoria tritici*, *Stagonospora nodorum*, *Puccinia striiformis* and *P. reconditain* of wheat. Luo and Mitchell (2002) developed multiplex PCR to identify simultenously multiple fungal pathogen in a single reaction. Five sets of species specific primers were desined from internal transcribed regions, ITS1 and ITS2, of the rRNA gene to identify *Candida albicans*, *C. glabrata*, *C. parapsilosis*, *C. tropicalis*, and *Aspergillus fumigatus*. Another set of previously published ITS primers, CN4 and CN5 were used to identify *Cryptococcus neoformans*. Six different species of pathogenic fungi were identified by using two multiplex PCR. *Puccinia graminis* f. sp. *tritici*, *Puccinia triticina*, and *Blumeria graminis* f. sp. *tritici*, are three of the most economically damaging fungal diseases causing wheat stem rust, wheat leaf rust, and wheat powdery mildew, respectively. Single-step multiplex PCR was also developed by Chen et al. (2014) for the simultaneous detection of the three pathogens in a single reaction.

DNA Arrays

DNA arrays offer a far greater capacity for multiplexing than multiplex PCR. In conventional PCR multiplexing is limited by the detection of different size product in an agarose gel while in real time PCR limited by the availability of dyes emitting fluorescence at different wavelengths on one hand, and the monochromatic character of the energizing light source in real-time PCR instruments on the other hand (Varga and James 2006). As a result, detection of more than a few pathogens per assay is currently not possible using conventional and real time PCR.

DNA arrays, originally designed to study gene expression or to generate SNP profiles, can be used to detect large no. of different organisms in parallel (Lievens et al. 2005). A pathogen detection array typically consists of many discretely located pathogen-specific detector sequences that are immobilized on a solid support, such as a nylon filter or a glass slide, to create a macroarray or a microarray, respectively. For signal amplification, in general the target DNA to be tested is amplified using consensus primers that target a genomic region containing the pathogen-specific sequences, and is labelled simultaneously or subsequently. Hence, it is possible to differentiate a large number of organisms using a single PCR, provided that sufficient discriminatory potential exists within the region that is used.

There are only a few reports for DNA array study in phytopathogenic fungi as this technology is in developmental stage. The high cost associated with microarray production and result reading limits its application. Lievens et al. (2003) developed DNA array technology for rapid and efficient detection of these vascular wilt pathogens i.e. *Fusarium oxysporum* f. sp. *lycopersici*, and *Verticillium albo-atrum or Verticillium dahlia*). They show the utility of this array for the sensitive detection of these pathogens from complex substrates like soil, plant tissues and irrigation water, and samples that are collected by tomato growers in their greenhouses conditions. Zhang et al. (2008) used a membrane-based macroarray technology to detect 25 pathogens of solanaceous crops. Based on the internal transcribed spacer sequences of the rRNA genes, 105 oligonucleotides (17–27 bases long) specific for 25 pathogens of solanaceous crops were designed and spotted on a nylon membrane. Their results indicate that the DNA-based macroarray detection system is a reliable and effective method for pathogen detection and diagnosis even when multiple pathogens are present in a field samples.

Loop-Mediated Isothermal Amplification (LAMP)

Loop-mediated isothermal amplification is a simple, rapid, specific and cost effective nucleic acid amplification method that amplifies a few copies of target DNA under isothermal conditions, using a set of four specially designed primers and a *Bst* DNA polymerase from *Geobacillus stearothermophilus* with strand displacement activity (Notomi et al. 2000; Tomita et al. 2008). The cycling reactions can result in the accumulation of 10^9 to 10^{10}-folds copies of target in less than an hour. LAMP is an alternative amplification technology and tenfolds more sensitive than conventional PCR (Duan et al. 2014a)

Four primers are used in LAMP that recognizes six distinct regions in the target DNA, instead of two primers in conventional PCR. Annealing of the four primers to the target DNA is a very crucial step for the efficiency of LAMP. The design of these four primers is therefore critical for a successful LAMP assay. PrimerExplorer is most commonly used programme for primer designing for LAMP assay. Two inner and two outer primers are required for LAMP. In the initial steps of the LAMP reaction, all four primers are employed, but in the later cycling steps only the inner primers are used for strand displacement DNA synthesis. The outer primers are referred to as F3 and B3, while the inner primers are forward inner primer (FIB) and backward inner primer (BIP). FIP contains F1C and F2 distinct sequences while BIP contains B1C and B2 distinct sequences corresponding to the sense and antisense sequences of the target DNA, one for priming in the first stage and the other for self-priming in later stages (Notomi et al. 2000). The Tm values of the outer primers F3 and B3 have to be lower than those of F2 and B2 to assure that the inner primers start synthesis earlier than the outer primers. Additionally, the concentrations of the inner primers are higher than the concentrations of the outer primers (Notomi et al. 2000; Tomita et al. 2008).

LAMP amplicons can be visualized directly and seen with the naked eye or under ultraviolet trans-illumination. The simplest way of detecting LAMP products is to inspect the white turbidity that results from magnesium pyrophosphate accumulation, as a by-product of the reaction, by naked eye (Mori et al. 2001). Some dyes like SYBR Green I, hydroxynaphthol blue (HNB), ethidium bromide are also used for detecting LAMP product (Table 4). SYBR Green I dye will be expected to show bright green fluorescence positive reactions, while a negative reaction would remain light orange. As LAMP is conducted at one temperature and no electrophoresis is required so, LAMP has the potential to implement early detection of fungal pathogens at field level, instead of well equipped laboratory.

Biochemical Techniques for Detection of Phytopathogenic Fungi

Recent technological developments in agricultural sector have lead to a demand for a new era of automated, sensitive and non-destructive methods of plant disease detection. It is desirable that the plant disease detection tool should be rapid, specific

Table 4 Some recent examples of detection of phytopathogenic fungi using LAMP techniques

Fungal pathogen	Primers	Target gene	Dye used	References
Fusarium oxysporum f. sp. *ciceris*	F3: ACAACCTCAATGAGTGCG B3: CATGAGCGACAACATACCA FIP:CCAGGCGTACTTGAAGGAACCGTCAAGCAGTCACTAACCAT BIP:AGCGTGAGCGTGGTATCACACGGTGACATAGTAGCGA LoopF: GCTCAGCGGCTTCCTATT 18 LoopB: CTCTGGAAGTTCGAGCATCC	*EF-1 alpha*	HNB	Ghosh et al. (2015)
Sclerotinia sclerotiorum	F3 :GACTTCGCTACCAAAGATAGCC B3: AGATCTAGACCCCGGTATCG FIP:GGCACGGAGAGATGGTAGAGAAGG CCAATCGAATGAAGCTCCCAC BIP: TCGACCATGACAGCATCTTCTACCC ACCAGGTCTTGGTCGTCTG	*Ssos5*	HNB	Duan et al. (2014a)
Botrytis cinerea	F3: CTACACAACGACCACAGT B3: CCACCAGGTAGTTTCAATCC FIP: GCCCTCCAGATAATCCATCTATGG-CCCGCGACAATATCATCA BIP:CATGCGACCTCCACCACAAA TTCTTGCGCTTAGTCCAC	Bcos5	HNB	Duan et al. (2014b)
Fusarium oxysporum f. sp. lycopersici	F3: GCCTGGATTCTGGTGTGG B3: CCGTCCCAGCCAAGATCA FIP:TACCAGACGTCTGGCTACCCTGT GTAGGGTAGGGTAGGCGTA BIP:ATGGGGGTGTAGGGTAGGTC TCCTCCCTCGACTAAAGCGAAG	28S rRNA	HNB, SYBR Green 1, Ethdium bromide	Almasi et al. (2013)
Botrytis cinerea	F3: TCGGAGTGTCCTAGGAATGC B3: TGAGATGGCCAACTCTCAGA FIP: GCCTGCTCACCGGTAGTAGTGTGTGAGCCCTTGGTCTAAAGC BIP: GCAGAATCTGTCCCCGGTGAGCGGGAGCAACAATTAATCGC F-Loop: TGGGGTTAACTAGTCACCTATACG B-Loop: AGGTCACCTTGCAATGAGTGGA	IGS	Betaine	Tomlinson et al. (2010)

FIP F1C + F2, *BIP* B1C + B2, *HNB* Hydroxynaphthol blue

and sensitive to a particular disease for detection at the early stage of the symptoms. Current focus is on the development of such technologies that can create a practical tool for a large-scale real-time disease monitoring under field conditions. The commonly used biochemical techniques for disease detection are spectroscopic techniques, imaging techniques, metabolite as a marker. Other advanced biochemical techniques include electronic nose, PLEX-ID system etc.

Spectroscopic Techniques

Spectroscopy and imaging techniques provide a platform along with other techniques which can be used for the development of non-destructive methods. Spectroscopy is the study of the interaction between matter and electromagnetic radiation (Crouch et al. 2007; Herrmann and Onkelinx 1986). At the end of the nineteenth century, spectroscopy was limited to the absorption, emission, and scattering of visible, ultraviolet and infrared electromagnetic radiations. During the twentieth century, the definition of spectroscopy has been extended to include other forms of electromagnetic radiations, including X-rays, microwaves and radio waves as well as energetic particles, such as electrons and ions (Harvey 2000). In case of plant disease detection tools, it is desirable that these should be rapid, specific to a particular disease, and sensitive for detection at the early stage of the symptoms developed (Lopez et al. 2003). Non-destructive methods fulfil these criteria with rapid analysis as minimal or no sample preparation is required. Current agricultural engineering research activities are working on the development of such technologies to create a practical tool for a large-scale real-time disease monitoring under field and laboratory conditions. There are many different types of spectroscopy techniques which provide unique tools for fungal pathogen detection (Table 5).

Visible, Infrared and Near Infrared Spectroscopy

Non-destructive methods based on visible, infrared and near infrared spectroscopy are becoming more popular as a tool for pathogen detection as these are rapid and cost-effective. In general, visible spectroscopy is used for disease detection in plants in combination with infrared/near infrared spectroscopy (Bravo et al. 2003; Larsolle and Muhammed 2007). Dowell et al. (1999) predicted scab, vomitoxin and ergosterol in single wheat kernels using NIR spectroscopy. Pettersson and Aberg (2003) described the application of NIR spectroscopy for measuring mycotoxins in cereals. Erukhimovitch et al. (2005) examined the potential of FTIR microscopy for an easy and rapid discrimination and identification of various fungi, which are responsible for a serious damage to agriculture. The results provided a unique and consistent spectral markers for each of the examined fungi. They showed that the spectral area ranged between 1,000 and 1,800 cm^{-1} can be considered as an important area for an

Table 5 Detection of phytopathogenic fungi using spectroscopic techniques

Technique	Crop	Disease/fungi	Optimum spectral range	Reference
Visible and Infrared spectroscopy	Wheat	Powdery mildew and take-all disease	490_{780},510_{780}, 516_{1300} and 540_{1300} nm	Graeff et al. (2006)
	Kiwifruit	Botrytis, sclerotinia rot	–	Costa et al. (2007)
	Trifoliate orange (*Poncirus trifoliata*), Sour orange (*Citrus aurantium*), Valencia orange (*C. sinensis*), and Bo tree (*Ficus religiosa*)	Sooty mold (*Capnodium spp.*)	blue (450 nm), green (550 nm), red (650 nm) and near-IR (850 nm), Near-IR/red image ratios	Summy and Little (2008)
	Grapefruit	Greasy spot (*Mycosphaerella citri*)		
	Muskmelon	Powdery mildew (*Sphaerotheca fuliginea*)		
Fluorescence Spectroscopy	Wheat	Powdery mildew and leaf rust	blue-to-green (F451/F522); blue-to-red (F451/F687); blue-to-far-red ratio (F451/F736)	Bürling et al. (2011a, b, Bürling et al. 2012)
	Barley	Powdery mildew (*Blumeria graminis*) or leaf rust (*Puccinia hordei*)	410–560, 470, 410, 440, 560, 530, 440–530 nm,	Leufen et al. (2014)
NMR Spectroscopy		Mycorrhiza (ectomycorrhizas and arbuscular mycorrhizas)	–	Pfeffer et al. (2001)

easy and reliable discrimination between the various examined fungi. Huang and Apan (2006) collected the hyperspectral data under field conditions using portable spectrometer to detect Sclerotinia rot disease in celery and the reflectance in the visible and infrared range from 400 to 1,300 nm was found sufficient in acquiring similar results as that of entire spectra (400–2,500 nm). Apart from compositional analysis, more complex applications of NIR are now being investigated, including analysis in breeding development and genetics, detection of adulteration and presence of weeds and insects in wheat and flour (Woodcock et al. 2008). In case of disease detection in greenhouse plants, Summy and Little (2008) analysed spectroradiometric measurements and color infrared (CIR) images of control, honeydew-coated, and sooty mold (*Capnodium* spp.) infested saplings and individual leaves from trifoliate orange (Poncirus trifoliata), sour orange (*Citrus aurantium*), Valencia orange (*C. sinensis*), and Bo tree (*Ficus religiosa*); grapefruit saplings and individual leaves infected with *Mycosphaerella citri* (greasy spot); and muskmelon foliage showing infection of powdery mildew (*Sphaerotheca fuliginea*) disease. All fungal biotic stressors generally resulted in variable spectral reflectance data in individual leaves, especially in the blue (450 nm) and green (550 nm) wavelengths. The values in the red (650 nm) tended to increase and values in the near-IR (850 nm) tended to decrease with stress. Near-IR/red image ratios were significantly reduced in stressed whole plant foliage and individual leaves relative to healthy controls. Wu et al. (2008) used NIR techniques to detect *Botrytis cinerea*-affected eggplant leaves prior to the visibility of symptoms under laboratory conditions. Salman et al. (2012) successfully used FTIR-ATR spectroscopy to differentiate among ten isolates of *Fusarium oxysporum*.

Fluorescence Spectroscopy

Fluorescence spectroscopy is a type of electromagnetic spectroscopy which analyzes fluorescence from the sample of interest. The sample is excited by using a beam of light which results in emission of light of a lower energy resulting in an emission spectrum which is used to interpret results (Ramanujam et al. 1994). Two types of fluorescence namely blue-green fluorescence (about 400–600 nm range) and chlorophyll fluorescence (about 650–800 nm range) are produced by green leaves. Fluorescence spectroscopy seems to be promising diagnostic technique with high sensitivity and specificity rate which makes it an ideal diagnostic tool. The fluorescence spectroscopy can be utilized to monitor nutrient deficiencies; environmental conditions based stress levels, and diseases in plants (Cerovic et al. 1999; Belasque et al. 2008). Leufen et al. (2014) investigated the potential of three optical devices namely fluorescence lifetime, image-resolved multispectral fluorescence and selected indices of a portable multiparametric fluorescence device for the proximal sensing of plant-pathogen interactions in four genotypes of spring barley in healthy leaves as well as leaves inoculated with powdery mildew (*Blumeria graminis*) or leaf rust (*Puccinia hordei*). They observed significant differences between healthy and diseased leaves.

Nuclear Magnetic Resonance (NMR) Spectroscopy

Greater sensitivity and spectral resolution have come with the advent of higher magnetic field. Advancements in technology have made it possible to combine different NMR techniques enabling metabolic, anatomical, and physiological information. The potential to spectroscopically differentiate host from fungal metabolites *in vivo* or in crude extracts without the need for separation or chemical derivatization is another useful aspect of NMR measurements in studying the biochemistry of mycorrhizas (fungi). Intact tissues analysis became possible through the implementation of high resolution solid state magic angle spinning nuclear magnetic resonance (HR-MAS NMR). Pfeffer et al. (2001) reviewed the Nuclear Magnetic Resonance (NMR) applied to the two main types of mycorrhiza (ectomycorrhizas and arbuscular mycorrhizas) to address the physiological question that these two mutualistic symbioses are sufficiently different. They reported that isotopic labelling NMR is able to examine the transfer of substrates between the symbionts both *in vivo* and *in vitro*, as well as the production of secondary metabolites in response to colonization and is also capable of determining the locations of the biosynthesis and translocations of storage compounds, such as polyphosphates, lipids and carbohydrates, in mycorrhizal fungi both in the free-living and in the symbiotic stages of their life cycle.

Imaging Techniques

On the other hand, imaging is the representation or reproduction of an object's form; especially a visual representation. The imaging techniques are an improvement over spectroscopic techniques as these methods acquire spectral information over a larger area and provide three-dimensional spectral information in the form of images. As improved form of spectroscopy, imaging techniques also provide unique tools for fungal pathogen detection are discussed in this chapter.

Fluorescence Imaging

Fluorescence imaging is an advancement of fluorescence spectroscopy, where fluorescence images (rather than single spectra) are obtained using a camera. A xenon or halogen lamp is used as a UV light source for fluorescence excitation and the fluorescence at specific wavelengths is recorded using the charge coupled device (CCD)-based camera system (Bravo et al. 2004; Chaerle et al. 2007). The regions of electromagnetic spectra that are commonly used for fluorescence imaging are blue (440 nm), green (520–550 nm), red (690 nm), far red (740 nm), and near infrared (800 nm) (Lenk and Buschmann 2006; Chaerle et al. 2007). Lenk et al. (2007) described the multispectral fluorescence and its possible application in monitoring disease symptoms in plants along with other traits. Bravo et al. (2004) used fluorescence imaging for detection of yellow rust in winter wheat and found that the difference between the fluorescence at 550 and 690 nm were higher in the diseased portion of the leaves, while it was very low for healthy regions. Hyperspectral reflectance imaging in combination with multispectral fluorescence imaging through sensor fusion was used to detect yellow rust (*Puccinia striiformis*) disease of winter wheat by Moshou et al. (2005). Summy and Little (2008) analysed muskmelon foliage showing low and high levels of powdery mildew (*Sphaerotheca fuliginea*) disease severity. When individual leaves were examined, all fungal biotic stressors generally resulted in variable spectral reflectance data, especially in the blue (450 nm) and green (550 nm) wavelengths; however, values in the red (650 nm) tended to increase and values in the near-IR (850 nm) tended to decrease with stress. Significantly reduced Near-IR/red image ratios were found in stressed whole plant foliage and individual leaves relative to healthy controls. Image acquisition and enhancement techniques may prove useful in large-scale production greenhouses where existing infrastructure and high plant populations require high throughput data analysis and identification of biotic stressors.

Hyperspectral Imaging

Hyperspectral imaging has gained considerable importance for its application in precision agriculture (Okamoto et al. 2009). In the hyperspectral imaging, the spectral reflectance of each pixel is acquired for a range of wavelengths in the electromagnetic spectra. The wavelengths may include the visible and infrared regions of the electromagnetic spectra. It is similar to multispectral imaging, with the difference that a broader range of wavelengths being scanned for each pixel. The resulting information in the form of an image is a set of pixel values in the form of intensity of the reflectance at each wavelength of the spectra. The major challenges in hyperspectral imaging-based plant disease detection are the selection of disease-specific spectral bands and selection of statistical classification algorithm for a particular application. Delwiche and Kim (2000) described the application of hyperspectral imaging for detecting Fusarium head blight in wheat using sensor with a wavelength range of 425–860 nm. To detect yellow rust disease (*Puccinia striiformis*) in wheat, the application of visible-NIR hyperspectral imaging was investigated by Bravo et al. (2003) wheras Moshou et al. (2004) used a spectrograph to acquire spectral images from 460 to 900 nm. Spectral imaging sensors combine image analysis with spectroscopic techniques. Based on reflection measurements of normal and head blight damaged kernels of three different varieties, wavelength bands were selected to differentiate the normal and damaged kernels (Sankaran et al. 2010). Blasco et al. (2007) applied multi-spectral computer vision using non-visible (ultraviolet, infrared and fluorescence) and visible multiple spectra for citrus sorting. The anthracnose was classified better with NIR images (86 %), whereas green mold was more accurately classified with fluorescence imaging (94 %). The stem-end injury was classified up to 100 % using the ultraviolet spectra in the same study. This study showed the utilization of hyperspectral

Table 6 Detection of phytopathogenic fungi of wheat using imaging techniques

Technique	Crop	Fungal disease	Optimum spectral range	Reference
Flourescence imaging	Wheat	Leaf rust	–	Buerling et al. (2010)
		Leaf rust and powdery mildew	–	Kuckenberg et al. (2007)
Hyperspectral imaging	Wheat	Fusarium head blight	568, 715 nm (550, 605, 623, 660, 697 and 733 nm)	Delwiche and Kim (2000)
	Wheat	Yellow rust	680, 725 and 750 nm	Moshou et al. (2005, 2006)

bands for detecting different aspects of a single problem. Similarly, the hyperspectral imaging could be used for detecting different features within a plant to identify diseases. Shafri and Hamdan (2009) used air-borne hyperspectral imaging for the detection of ganoderma basal stem rot disease in oil palm plantations. Mahlein (2010) examined the potential of hyperspectral imaging and non-imaging sensor systems for the detection, differentiation, and quantification of plant diseases. They repeatedly recorded the reflectance spectra of sugar beet leaves infected with the fungal pathogens *Cercospora beticola*, *Erysiphe betae*, and *Uromyces betae* causing cercospora leaf spot, powdery mildew, and sugar beet rust, respectively, during pathogenesis. Spectral reflectance of sugar beet was affected by each disease in a characteristic way, resulting in disease specific signatures. Hyperspectral imaging sensor provided extra information related to spatial resolution. Detection, identification and quantification of diseases was possible with high accuracy by spectral vegetation indices (SVIs) and spectral angle mapper classification, calculated from hyperspectral images which facilitated early detection and monitoring of Cercospora leaf spot and powdery mildew. Bauriegel et al. (2011) used hyperspectral imaging analysis for early detection of *Fusarium* head blight pathogen. Occurrence of head blight can be detected by spectral analysis (400–1,000 nm) before harvest. With this information, farmers could recognize *Fusarium* contaminations. Wheat plants were analyzed using a hyper-spectral imaging system under laboratory conditions (Table 6). Principal component analysis (PCA) was applied to differentiate spectra of diseased and healthy ear tissues in the wavelength ranges of 500–533 nm, 560–675 nm, 682–733 nm and 927–931 nm, respectively. They concluded that head blight could be successfully recognized during the development stages and derived disease index, which uses spectral differences in the ranges of 665–675 nm and 550–560 nm, can be a suitable outdoor classification method for the recognition of head blight. Bauriegel et al. (2014) compared the two methods viz. chlorophyll fluorescence and hyperspectral imaging in view of their usability for the detection of *Fusarium* on wheat, both in the field and in the laboratory. They highlighted that the modification of spectral signatures due to fungal infection allowed its detection by hyperspectral imaging, the decreased physiological activity of tissues resulting from *Fusarium* impacts provided the base for CFI analyses.

Metabolites as Biomarkers

The unique changes at metabolite levels can be used as biomarkers to detect fungal pathogens at an early stage of disease progression and to manage diseases in storage as well as outbreak of seed borne disease (Table 7). De Lacy Costello et al. (2001) identified many disease-specific compounds in potatoes infected with *Phytophthora infestans* namely, butanal, 3-methyl butanal, undecane and verbenone, while those infected with *Fusarium coeruleum* produced 2-pentyl furan and capaene. Prithiviraj et al. (2004) assessed the variability in the volatiles profile released from onion bulbs infected with bacterial (*Erwinia carotovora*) causing soft rot and fungal species (*Fusarium oxysporum* and *Botrytis allii*) causing basal and neck rots, respectively and found that 25 volatile compounds among the 59 consistently detected

Table 7 Detection of phytopathogenic fungi using mass spectroscopic techniques

Crop	Disease	Equipment	Profiling	Reference
Apple	*Botrytis cinerea* *Mucor piriformis* *Penicillium expansum* *Monilinia* sp.	GC-MS	methyl acetate; 4-methyl-1-Hexane; 2-methyltetrazole and butyl butanoate, and 3,4-dimethyl-1-hexene and fluorethene	Vikram et al. (2004)
Mango	*Lasiodiplodia theobromae* and *Colletotrichum gloeosporioides*	GC-MS	Volatile organic compounds (35)	Moalemiyan et al. (2007)
Potato	*Phytophthora infestans* *Fusarium coeruleum*	GC-MS	Butanal, 3-methyl butanal, undecane and verbenone 2-pentyl furan and capaene	De Lacy Costello et al. (2001)
Tomato	*A. niger* yielded 11; *A. flavus* yielded 15 different volatile metabolites while that inoculated with *F. oxysporum*	GC-MS	11, 15, 8	Ibrahim et al. (2011)
Onion bulbs	*Fusarium oxysporum* and *Botrytis cinerea*	GC-MS	Volatile organic compounds (25)	Prithiviraj et al. (2004)
Vegetable, oil yielding and seed spice crops	*Alternaria* species such as *A. solani*, *A. porri*, *A. brassicicola*, *A. brassicae*, *A. sesame*, *A. alternata*, *A. macrospora*, *A. ricini*, *A. carthami* and *A. brunsii*	HPLC-MALDI-TOF-MS	Species-specific markers	Lakshmi et al. (2014)

compounds can be used to identify the disease. In apples, methyl acetate was found to be unique to fruits inoculated with *Botrytis cinerea*, 4-methyl-1-hexene to fruits inoculated with *Mucor piriformis*, 2-methyltetrazole and butyl butanoate to fruits inoculated with *Penicillium expansum*, and 3,4-dimethyl-1-hexene and fluorethene to fruits inoculated with *Monilinia* sp. (Vikram et al. 2004). Moalemiyan et al. (2006, 2007) employed volatile organic compound (VOC) profiling to detect fungal diseases namely *Lasiodiplodia theobromae* causing stemend rot and *Colletotrichum gloeosporioides* causing anthracnose in mangoes. Ibrahim et al. (2011) studied volatile metabolites of tomato fruits inoculated with three toxigenic fungi isolated from spoilt tomatoes using GC-MS and observed differences in the number and amount of volatile metabolites. They obtained a total of 52 different volatile metabolites. Healthy ripe tomato fruits yielded 28 metabolites viz. oleic acid amide (10.89 %), 9-octadecenoic acid (9.83 %), methyl cis-9-octadecenoate (7.73 %), and 2, 3-Heptanedione (0.32 %). Tomato fruits inoculated with *A. niger* yielded 11 %; *A. flavus* yielded 15 different volatile metabolites while that inoculated with *F. Oxysporum* yielded eight volatile metabolites. Among them only five volatile metabolite occurred relatively consistent in fruits inoculated with *A. niger* and *A. flavus* while adogen 73 and 9-Octadecenoic acid (Z) occurred relatively consistently in fruits inoculated with the three fungi. Hexadecanoic acid and 6-Methyl-2,4-di – tert – butyl – phenol was common in fruits inoculated with *F. oxysporum* and *A. niger* with that of *A. niger* having the highest value (9.67 %) for hexadecanoic acid while fruits inoculated with *F. oxysporum* had the highest (2.66 %) for 6-Methyl-2,4-di – tert – butyl – phenol. Ten metabolites were unique to *A. flavus* while *A. niger* and *F. oxysporum* had four metabolites unique to each of them. Lakshmi et al. (2014) examined the secondary metabolite profiling of 50 fungal isolates belonging to 10 plant pathogenic *Alternaria* species such as *A. solani*, *A. porri*, *A. brassicicola*, *A. brassicae*, *A. sesame*, *A. alternata*, *A. macrospora*, *A. ricini*, *A. carthami* and *A. brunsii* isolated from vegetable, oil yielding and seed spice crops for classification and identification purposes and obtained characteristic 'species-specific metabolite fingerprints' that could be adopted as chemotaxonomic markers in species identification. Further, application of MALDI-MS offers relatively high tolerance against sample impurities (salts and detergents), as well as fast and accurate molecular mass determination and the possibility of automation, which makes it a powerful alternative to classical biological methods. Chalupova et al. (2013) reviewed the identification of fungal microorganisms by matris assisted laser desorption/ionization- time of flight (MALDI-TOF) mass spectrometry.

Electronic Nose

Fungi synthesize a very large range of organic (carbon-based) compounds that are categorized into many different chemical classes. These diverse organic chemicals are produced as a result of biochemical or metabolic processes that take place within specialized cells of fungi. Some chemical monomeric compounds are linked together to form various types of structural or functional biopolymers such as carbohydrates, lipids, proteins, and nucleic acids. These polymeric compounds generally have low volatility as a result of their high molecular weight. Other smaller intermediates of biochemical processes are modified to form a variety of primary and secondary metabolites performing many cellular or biochemical functions (Wilson 2013). Small molecular weight organic compounds, generally <350 Da (Cagni and Ghizzoni 2000), may contain various polar and nonpolar functional groups that contribute to volatility. Compounds having high vapour pressure (low boiling point), called volatile organic compounds (VOCs), are particularly conducive to e-nose detection because they are easily vaporized (made airborne as gases), greatly increasing their accessibility for detection within sampled air. The electronic nose instrumentation was developed in the early 1980s (Persaud and Dodd 1982) to mimic mammalian olfactory systems. Electronic noses are comprised of sensor arrays that are capable of detecting a selection of compounds (e.g. ketones, aldehydes, aromatic and aliphatic compounds) produced during the growth stages of fungi on a certain substrate. The first developed sensor array was a metal oxide semiconductor, which detected 20 odours (Persaud and Dodd 1982). The electronic nose system comprises a set of active materials that detects the odour and transduces the chemical vapours into electrical signals. These are capable of detecting, identifying, and discriminating many types and sources of a wide diversity of chemical species and mixtures of compounds, including VOCs most commonly produced and released from such organic sources as living microbes and multicellular organisms. The odour profile obtained can then be analysed using various statistical methods like principal components analysis, cluster analysis, neural network algorithms etc. Electronic noses offer great potential for the detection of different microbial species (Table 8). Some chemical products are specific to fungal and bacterial species and are commonly used as a useful diagnosis tool. Electronic noses originally were not designed for the purpose of identifying individual chemical species within the sample mixture, but were engineered to recognize the sample as a whole, or as a collective simple or complex air mixture released by any source that is identifiable by its unique electronic signature. Technological advances made it possible to create sensor arrays (from 6 up to 32 sensors) with different materials, processing thousands of smells.

Recent advances in electronic nose technologies based on many different electronic aroma detection principles and mechanisms have made possible the development of a wide variety of electronic nose (Zhu et al. 2006). The various applications that have proven useful in a range of diverse field including the agricultural, food industries, cosmetics industries, environmental, pharmaceutical, and in many fields of applied sciences (Ge et al. 2007; Wilson and Baietto 2009). Advances in the use of electronic-nose instruments in biomedical applications are no exceptions and now also used in agriculture.

Electronic noses have a number of advantages over traditional analytical instruments. Electronic nose sensors do not require chemical reagents, have good sensitivity and specificity, provide rapid repeatable (precise) results, and allow non-destructive sampling of gas odorants or analytes (García-González and Aparicio 2002). Furthermore, e-noses generally are far less expensive than analytical systems,

Table 8 Detection of phytopathogenic fungi using electronic nose techniques

Fungal pathogen	Host	Electronic nose	Sensor/type	Data processing method	References
Aspergillus carbonarius	Tomato (processed)	EOS835	MOS	PCA	Concina et al. (2009)
Fusarium cerealis, F. graminearum, F. culmorum and F. redolens	Wheat grains	Enose	QMB	PLS-DA	Eifler et al. (2011)
Botrytis cinerea, Colletotrichum gloeosporioides, Alternaria spp	Blueberries fruit	Cyranose® 320	CP	PCA	Li et al. (2009)
Botrytis sp., Penicillium sp., Rhizopus sp.	Strawberry fruit	PEN3	MOS	PCA	Pan et al. (2014)
Penicillium chrysogenum and *Fusarium verticillioides*	Wheat grain	Libra nose	QMB	PLS-DA, PCA	Paolesse et al. (2006)
Fusarium verticillioides	Maize grain	EOS835	MOS	PCA	Falasconi et al. (2005)

CP Conducting polymer, *MOS* Metal oxide semiconductor, *QMB* Quartz crystal microbalance, *PLS-DA* partial least squares discriminant analysis, *PCA* Principal Components Analysis

easier and cheaper to operate, and have greater potential for portability and field use compared with complex analytical laboratory instruments (Wilson 2011). However, some disadvantages of e-nose sensing include problems with reproducibility, recovery, negative effects of humidity and temperature on sensor responses, and inability to identify individual chemical species within gas samples.

Thus, electronic noses probably will never completely replace complex analytical instruments, but offer quick real-time detection and discrimination solutions for applications requiring accurate, rapid and repeated determinations. Nevertheless, much more research is needed to develop and take full advantage of electronic nose instruments to bring them to the full potential of capabilities for fungal pathogen detection applications. Once these difficulties and logistics are resolved, electronic-nose devices should be capable of solving many fungal pathogen diagnostic problems.

PLEX-ID System

Recently, a novel technology, the PLEX-ID system has been developed which used broad-range PCR amplification coupled with electrospray ionization-mass spectrometry (ESI-MS) for the direct detection of pathogens without the need to wait for growth in culture. This system measures the mass-to-charge ratio of PCR amplicons generated from several different target loci, focusing on both conserved and species-specific regions, to identify base compositions that can then be compared to a database of fungal signatures. By using base composition as a unique molecular signature, the PLEX-ID system is able to identify single or multiple organisms at trace levels in a variety of specimen sources. It has been used to identify fungal culture isolated directly in clinical biochemistry (Simner et al. 2013). It can also be used in agriculture for rapid identification of fungal infections.

Conclusion

Several detection technologies for fungal pathogen are now available, but regardless of the approach, important questions need to be answered prior to their inclusion in experiments. These include sensitivity, specificity, robustness, user friendly and cost-effective. Despite many novel technologies being available, challenges remain same to identify unculturable fungi. There is always a need to detect and identify fungi quickly and accurately. The latest technologies including molecular methods, spectroscopic and imaging, and mass metabolite profiling volatile profiling-based fungal disease detection methods holds the key for development of sensors. Further, PLEX-ID technique which used broad-range PCR amplification coupled with electrospray ionization-mass spectrometry (ESI-MS) for the direct detection of pathogens and has been used in clinical biochemistry for identification of fungal strains can also holds potential for agriculture sector.

References

Agrios GN (2005) Plant pathology, 5th edn. Elsevier, New York

Almasi MA, Dehabadi SMH, Moradi A, Eftekhari Z, Ojaghkandi MA, Aghaei S (2013) Development and application of loop-mediated isothermal amplification assay for rapid detection of *Fusarium Oxysporum* f. Sp. lycopersici. J Plant Pathol Microbiol 4:177

Aroca A, Raposo R, Lunello P (2008) A biomarker for the identification of four *Phaeoacremonium* species using the beta-tubulin gene as the target sequence. Appl Microbiol Biotechnol 80: 1131–1140

Attallah ZK, Bae J, Jansky SH, Rouse DI, Stevenson WR (2007) Multiplex real-time quantitative PCR to detect and quantify *Verticillium dahliae* colonization in potato lines that differ in response to *Verticillium* wilt. Phytopathology 97:865–872

Babu BK, Mesapogu S, Sharma A, Somasani SR, Arora DK (2011) Quantitative real-time PCR assay for rapid detection of plant and human pathogenic *Macrophomina phaseolina* from field and environmental samples. Mycologia 103:466–473

Bailey AM, Mitchell DJ, Manjunath KL, Nolasco G, Niblett CL (2002) Identification to the species level of the plant pathogens *Phytophthora* and *Pythium* by using unique sequences of the ITS1 region of ribosomal DNA as capture probes for PCR Elisa. FEMS Microbiol Lett 207:153–158

Bauriegel E, Herppich WB (2014) Hyperspectral and chlorophyll fluorescence imaging for early detection of plant diseases, with special reference to *Fusarium* spec. infections on wheat. Agriculture 4:32–57

Bauriegel E, Giebel A, Geyer M, Schmidt U, Herppich WB (2011) Early detection of *Fusarium* infection in wheat using hyper-spectral imaging. J Comput Electron Agric 75:304–312
Bearchell SJ, Fraaije BA, Shaw MW, Fitt BDL (2005) Wheat archive links long-term fungal pathogen population dynamics to air pollution. PNAS USA 102:5438–5442
Belasque L, Gasparoto MCG, Marcassa LG (2008) Detection of mechanical and disease stresses in citrus plants by fluorescence spectroscopy. Appl Opt 47(11):1922–1926
Bellemain E, Carlsen T, Brochmann C, Coissac E, Taberlet P, Kauserud H (2010) ITS as an environmental DNA barcode for fungi: an *in silico* approach reveals potential PCR biases. BMC Microbiol 10:189
Bilodeau GJ, Koike ST, Uribe P, Martin FN (2012) Development of an assay for rapid detection and quantification of *Verticillium dahliae* in soil. Phytopathology 102:331–343
Bindslev L, Oliver RP, Johansen B (2002) *In situ* PCR for detection and identification of fungal species. Mycol Res 106:277–279
Blasco J, Alexios N, Gomez J, Molto E (2007) Citrus sorting by identification of the most common defects using multispectral computer vision. J Food Eng 83(3):384–393
Bonants P, Weerdt MH, van Gent-Pelzer M, Lacourt I, Cooke D, Duncan J (1997) Detection and identification of *Phytophthora fragariae* Hickman by the polymerase chain reaction. Eur J Plant Pathol 103:345–355
Bonants P, van Gent-Pelzer MPE, Hooftman R, Cooke D, Guy DC, Duncan JM (2004) A combination of baiting and different PCR formats, including measurement of real-time quantitative fluorescence, for the detection of *Phytophthora fragariae* in strawberry plants. Eur J Plant Pathol 110:689–702
Bravo C, Moshou D, West J, McCartney A, Ramon H (2003) Early disease detection in wheat fields using spectral reflectance. Biosyst Eng 84:137–145
Bravo C, Moshou D, Oberti R, West J, McCartney A, Bodria L, Ramon H (2004) Foliar disease detection in the field using optical sensor fusion. CIGR J Sci Res Dev 6:1–14
Buerling K, Hunsche M, Noga G (2010) Quantum yield of non-regulated energy dissipation in PSII (Y(NO)) for early detection of leaf rust (*Puccinia triticina*) infection in susceptible and resistant wheat (*Triticum aestivum* L.) cultivars. Precis Agric 11:703–716
Bürling K, Hunsche M, Noga G (2011a) Use of blue–green and chlorophyll fluorescence measurements for differentiation between nitrogen deficiency and pathogen infection in winter wheat. J Plant Physiol 168:1641–1648
Bürling K, Hunsche M, Noga G (2011b) UV-induced fluorescence spectra and lifetime determination for detection of leaf rust (*Puccinia triticina*) in susceptible and resistant wheat (*Triticum aestivum*) cultivars. Funct Plant Biol 38:337–345
Bürling K, Hunsche M, Noga G (2012) Presymptomatic detection of powdery mildew infection in winter wheat cultivars by laser-induced fluorescence. Appl Spectrosc 66: 1411–1419
Cagni D, Ghizzoni C (2000) The role of non-volatile compounds in flavor science: applications of HPLC-mass spectrometry technique. Flavour and fragrance chemistry. In: Lanzotti V, Taglialatela-Scafati O (eds) Flavour and fragrance chemistry. Kluwer Academic Publishers, Dordrecht, pp 221–226
Cerovic ZG, Samson G, Morales F, Tremblay N, Moya I (1999) Ultraviolet induced fluorescence for plant monitoring: present state and prospects. Agronomie 19:543–578
Chaerle L, Lenk S, Hagenbeek D, Buschmann C, Van Der Straeten D (2007) Multicolor fluorescence imaging for early detection of the hypersensitive reaction to tobacco mosaic virus. J Plant Physiol 164:253–262
Chalupova J, Raus M, Sedlarova M, Sebela M (2013) Identification of fungal microorganisms by MALDI-TOF mass spectrometry. Biotechnol Adv. doi:10.1016/j.biotechadv.2013.11.002
Chen S, Cao Y, Li T, Wu X (2014) Simultaneous detection of three wheat pathogenic fungal species by multiplex PCR. Phytoparasitica. doi:10.1007/s12600-014-0442-1
Concina I, Falasconi M, Gobbi E, Bianchi F, Musci M, Mattarozzi M, Pardo M, Mangia A, Careri M, Sbeveglieri G (2009) Early detection of microbial contamination in processed tomato by electronic nose. Food Control 20:837–880

Cooke DEL, Schena L, Cacciola SO (2007) Tools to detect, identify and monitor *Phytophthora* species in natural ecosystems. J Plant Pathol 89:13–28

Costa G, Noferini M, Fiori G, Spinelli F (2007) Innovative application of nondestructive techniques for fruit quality and disease diagnosis. Acta Horticulturae 753(1):275–282

Crouch S, Holler FA, Skoog DA (2007) Principles of instrumental analysis. Thomson Brooks/Cole, Belmont. ISBN 0-495-01201-7

De Lacy Costello BPJ, Evans P, Ewen RJ, Gunson HE, Jones PRH, Ratcliffe NM, Spencer-Phillips PTN (2001) Gas chromatography mass spectrometry analyses of volatile organic compounds from potato tubers inoculated with *Phytophthora infestans* or *Fusarium coeruleum*. Plant Pathol 50:489–496

Delwiche SR, Kim MS (2000) Hyperspectral imaging for detection of scab in wheat. Proc SPIE 4203:13–20

Dieffenbach CW, Lowe TMJ, Dveksler GS (1993) General concepts for PCR primer design. PCR Methods Appl Cold Spring Harb Lab 3:S30–S37

Diguta CF, Rousseaux S, Weidmann S, Bretin N, Vincent B, Guilloux-Benatier M, Alexandre H (2010) Development of a qPCR assay for specific quantification of *Botrytis cinerea* on grapes. FEMS Microbiol Lett 313:81–87

Dowell FE, Ram MS, Seitz LM (1999) Predicting scab, vomitoxin and ergosterol in single wheat kernels using near-infrared spectroscopy. Cereal Chem 76:573–576

Drenth A, Wagals G, Smith B, Sendall B, O'Dwyer C, Irvine G, Irwin JAG (2006) Development of a DNA-based method for the detection and identification of *Phytophthora* species. Australas Plant Pathol 35(2):147–159

Duan Y, Ge C, Zhang X, Wang J, Zhou M (2014a) A rapid detection method for the plant pathogen *Sclerotinia sclerotiorum* based on loop-mediated isothermal amplification (LAMP). Australas Plant Pathol 43:61–66

Duan YB, Ge CY, Zhang XK, Wang JX, Zhou MG (2014b) Development and evaluation of a novel and rapid detection assay for *Botrytis cinerea* based on loop-mediated isothermal amplification. PLoS One 9:e111094

Dubey SC, Priyanka K, Upadhyay BK (2014) Development of molecular markers and probes based on TEF-1α, β-tubulin and ITS gene sequences for quantitative detection of *Fusarium oxysporum* f. sp. ciceris by using real-time PCR. Phytoparasitica 42:355–366

Durai M, Dubey SC, Tripathi A (2012) Genetic diversity analysis and development of SCAR marker for detection of Indian populations of *Fusarium oxysporum* f. sp. ciceris causing chickpea wilt. Folia Microbiol 57:229–235

Egger KN (1995) Molecular analysis of ectomycorrhizal fungal communities. Canadian J Bot 73:1415–1415

Eibel P, Wolf GA, Koch AE (2005) Detection of *Tilletia caries*, causal agent of common bunt of wheat by ELISA and PCR. J Phytopathol 153:297–306

Eifler J, Martinelli E, Santonico M, Capuano R, Schild D et al (2011) Differential Detection of potentially hazardous *Fusarium* species in wheat grains by an electronic nose. PLoS One 6(6):e21026

Elsas DJ, Duarte GF, Keijzer-Wolters A, Smit E (2000) Analysis of the dynamics of fungal communities in soil via fungal-specific PCR of soil DNA followed by denaturing gradient gel electrophoresis. J Microbiol Methods 43:133–151

Erukhimovitch V, Tsror L, Hazanovsky M, Talyshinsky M, Mukmanov I, Souprun Y, Huleihel M (2005) Identification of fungal phyto-pathogens by Fourier-transform infrared (FTIR) microscopy. J Agric Technol 1:145–152

Falasconi M, Gobbi E, Pardo M, Torre MD, Bresciani A, Sberveglieri G (2005) Detection of toxigenic strains of *Fusarium verticillioides* in corn by electronic olfactory system. Sensors Actuators B Chem 108:250–257

Fraaije BA, Lovell DJ, Coelho JM, Baldwin S, Hollomon DW (2001) PCR-based assays to assess wheat varietal resistance to blotch (*Septoria tritici* and *Stagonospora nodorum*) and rust (*Puccinia striiformis* and *Puccinia recondite*) diseases. Eur J Plant Pathol 107:905–917

Ganeshamoorthi P, Dubey SC (2013) Phylogeny analysis of Indian strains of *Rhizoctonia solani* isolated from chickpea and development of sequence characterized amplified region (SCAR) marker for detection of the pathogen. Afr J Microbiol Res 7:5516–5525

Garcia-Gonzalez DL, Aparicio R (2002) Sensors: From biosensors to the electronic nose. Grasasy Aceites 53:96–114

Gardes M, Bruns TD (1993) ITS primers with enhanced specificity for Basidiomycetes: application to the identification of mycorrhizae and rusts. Mol Ecol 2:113–118

Ge CQ, Xie CS, Cai SZ (2007) Preparation and gas-sensing properties of Ce-doped ZnO thin film sensors by dip-coating. Mater Sci Eng B 137:53–58

Ghosh R, Nagavardhini A, Sengupta A, Sharma M (2015) Development of loop-mediated isothermal amplification (LAMP) assay for rapid detection of *Fusarium oxysporum* f. sp. ciceris - wilt pathogen of chickpea. BMC Res Notes 8:40

Gómez-Alpizar L, Saalau E, Picado I, Tambong JT, Saborio F (2011) A PCR-RFLP assay for identification and detection of *Pythium myriotylum*, causal agent of the cocoyam root rot disease. Lett Appl Microbiol 52(3):185–192

Graeff S, Link J, Claupein W (2006) Identification of powdery mildew (*Erysiphe graminis* sp. tritici) and take-all disease (*Gaeumannomyces graminis* sp. tritici) in wheat (*Triticum aestivum* L.) by means of leaf reflectance measurements. Cent Eur J Biol 1:275–288

Gudnason H, Dufva M, Bang DD, Wolff A (2007) Comparison of multiple DNA dyes for real-time PCR: effects of dye concentration and sequence composition on DNA amplification and melting temperature. Nucleic Acids Res 35:e127

Guo JR, Schnieder F, Beyer M, Verreet JA (2005) Rapid detection of *Mycosphaerella graminicola* in wheat using reverse transcription-PCR assay. J Phytopathol 153:674–679

Haase AT, Retzel EF, Staskus KA (1990) Amplification and detection of lentiviral DNA inside cells. PNAS USA 87:4971–4975

Harmon PF, Dunkle LD, Latin R (2003) A rapid PCR-base method for the detection of *Magnaporthe oryzae* from infected perennial ryegrass. Plant Dis 87:1072–1076

Harvey D (2000) Spectroscopic methods of analysis. In: Modern analytical chemistry Chapter 10. Mc Graw Hill Publishers, USA, pp 368–460

Hayden MJ, Good G, Sharp PJ (2002) Sequence tagged microsatellite profiling (STMP): improved isolation of DNA sequence flanking target SSRs. Nucleic Acids Res 30:e129

Herrmann R, Onkelinx C (1986) Quantities and units in clinical chemistry: nebulizer and flame properties in flame emission and absorption spectrometry (recommendations 1986). Pure Appl Chem 58:1737–1742

Huang JF, Apan A (2006) Detection of Sclerotinia rot disease on celery using hyperspectral data and partial least squares regression. J Spat Sci 51:129–142

Ibrahim AD, Hussaini H, Sani A, Aliero AA, Yakubu SE (2011) Volatile metabolites profiling to discriminate diseases of tomato fruits inoculated with three toxigenic fungal pathogens. Res Biotechnol 2:14–22

Jacobsen CS (1995) Microscale detection of specific bacterial-DNA in soil with a magnetic capture-hybridization and PCR amplification assay. Appl Environ Microbiol 61:3347–3352

James D, Varga A, Pallas V, Candresse T (2006) Strategies for simultaneous detection of multiple plant viruses. Can J Plant Pathol 28:16–29

Keiper FJ, Capio E, Grcic M, Wallwork H (2007) Development of sequence tagged microsatellites for the barley net blotch pathogen *Pyrenophora teres*. Mol Ecol Notes 7:664–666

Kropp B, Albee S, Flint M, Zambino P, Szabo L, Thomson SV (1995) Early detection of systemic rust infections of dyers woad (*Isatis tinctoria*) using the polymerase chain reaction. Weed Sci 43:6–12

Kuckenberg J, Tartachnyk I, Schmitz-Eiberger M, Noga G (2007) Early detection of leaf rust and powdery mildew infections on wheat (*Triticum aestivum* L.) leaves by PAM fluorescence imaging. Precis Agric 8:515–521

Kulik T, Jestoi M, Okorski A (2011) Development of TaqMan assays for the quantitative detection of *Fusarium avenaceum*/*Fusarium tricinctum* and *Fusarium poae* esyn1 genotypes from cereal grain. FEMS Microbiol Lett 314:49–56

Lakshmi MJ, Chowdappa P, Mahmood R (2014) Secondary metabolite profiling of plant pathogenic *Alternaria* species by matrix assisted laser desorption ionization-time of flight (MALDI-TOF) mass spectrometry. Indian Phytopathol 67:374–382

Landgraf A, Reckmann B, Pingoud A (1991) Direct analysis of polymerase chain reaction products using enzyme-linked-immunosorbent-assay techniques. Anal Biochem 198:86–91

Langrell SRH, Barbara DJ (2001) Magnetic capture hybridization for improved PCR detection of *Nectria galligena* from lignified apple extracts. Plant Mol Biol Report 19:5–11

Larena I, Salazar O, Gonza'lez V, Julia'n MC, Rubio V (1999) Design of a primer for ribosomal DNA internal transcribed spacer with enhanced specificity for ascomycetes. J Biotechnol 75: 187–194

Larsen JE, Hollingsworth CR, Flor J, Dornbusch MR, Simpson NL, Samac DA (2007) Distribution of *Phoma sclerotioides* on alfalfa and winter wheat crops in the north central United States. Plant Dis 91:551–558

Larsolle A, Muhammed HH (2007) Measuring crop status using multivariate analysis of hyperspectral field reflectance with application to disease severity and plant density. Precis Agric 8:37–47

Leisova L, Kucera L, Minarikova V, Ovesna J (2005) AFLP-based PCR markers that differentiate spot and net forms of *Pyrenophora teres*. Plant Pathol 54:66–73

Lenk S, Buschmann C (2006) Distribution of UV-shielding of the epidermis of sun and shade leaves of the beech (*Fagus sylvatica* L.) as monitored by multi-colour fluorescence imaging. J Plant Physiol 163(12):1273–1283

Lenk S, Chaerle L, Pfundel EE, Langsdorf G, Hagenbeek D, Lichtenthaler HK, Van Der Straeten D, Buschmann C (2007) Multispectral fluorescence and reflectance imaging at the leaf level and its possible applications. J Exp Bot 58(4):807–814

Leufen G, Noga G, Hunsche M (2014) Proximal sensing of plant-pathogen interactions in spring barley with three fluorescence techniques. Sensors 14:11135–11152

Li C, Krewer G, Kays SJ (2009) Blueberry postharvest disease detection using an electronic nose. ASABE Paper No. 096783, ASABE Annual International Meeting, Reno, NV, June 21–June 24

Lievens B, Brouwer M, Vanachter ACRC, Levesque CA, Thomman BPHJ (2003) Design and development of a DNA array for rapid detection and identification of multiple tomato vascular wilt pathogens. FEMS Microbiol Lett 223:113–122

Lievens B, Brouwer M, Vanachter ACRC, Levesque CA, Thomma BPHJ (2005) Quantitative assessment of phytopathogenic fungi in various substrates using a DNA macroarray. Environ Microbiol 7:1698–1710

Lin YH, Chang JY, Liu ET, Chao CP, Huang JW, Chang PL (2009) Development of a molecular marker for specific detection of *Fusarium oxysporum* f.sp. *cubense* race 4. Eur J Plant Pathol 123:353–365

Liu M, McCabe E, Chapados JT, Carey J, Wilson SK, Tropiano R, Redhead SA, Lévesque CA, Hambleton S (2015) Detection and identification of selected cereal rust pathogens by TaqMan® real-time PCR. Can J Plant Pathol 137:92–105

Long AA (1998) *In-situ* polymerase chain reaction: foundation of the technology and today's options. Eur J Histochem 42:101–109

López MM, Bertolini E, Olmos A, Caruso P, Gorris MT, Llop P, Penyalver R, Cambra M (2003) Innovative tools for detection of plant pathogenic viruses and bacteria. Int Microbiol 6: 233–243

Luo G, Mitchell TG (2002) Rapid identification of pathogenic fungi directly from cultures by using multiplex PCR. J Clin Microbiol 40:2860–2865

Ma Z, Michialides TJ (2005) Advances in undertanding molecular mechanisms of fungicide resistance and molecular detection of resistant genotypes in phytopathogenic fungi. Crop Prot 24:853–863

Macia-Vicente JG, Jansson HB, Talbot NJ, Lopez-Llorca LV (2009) Real-time PCR quantification and live-cell imaging of endophytic colonization of barley (*Hordeum vulgare*) roots by *Fusarium equiseti* and *Pochonia chlamydosporia*. New Phytol 182:213–228

Mahlein AK (2010) Detection, identification, and quantification of fungal diseases of sugar beet leaves using imaging and non-imaging hyperspectral techniques. Inaugural-Dissertation. Institute of Crop Science and Rescource Conservation – Phytomedicine. Rheinischen Friedrich-Wilhelms-Universität

Malvick DK, Impullitti AE (2007) Detection and quantification of *Phialophora gregata* in soybean and soil samples with a quantitative, real-time PCR assay. Plant Dis 91:736–742

Martin KJ, Rygiewicz PT (2005) Fungal-specific PCR primers developed for analysis of the ITS region of environmental DNA extracts. BMC Microbiol 5:28

Martos S, Torres E, El Bakali MA, Raposo R, Gramaje D, Armengol J, Luque J (2011) Co-operational PCR coupled with dot blot hybridization for the detection of *Phaeomoniella chlamydospora* on infected grapevine wood. J Phytopathol 159:247–254

Matsuda Y, Sameshima T, Moriura N, Inoue K, Nonomura T, Kakutani K, Nishimura H, Kusakari S, Takamatsu S, Toyoda H (2005) Identification of individual powdery mildew fungi infecting leaves and direct detection of gene expression by single conidium polymerase chain reaction. Phytopathol 95:1137–1143

McCartney HA, Foster SJ, Fraaije BA, Ward E (2003) Molecular diagnostics for fungal plant pathogens. Pest Manag Sci 59:129–142

Moalemiyan M, Vikram A, Kushalappa AC (2007) Detection and discrimination of two fungal diseases of mango (cv. Keitt) fruits based on volatile metabolite profiles using GC/MS. Postharvest Biol Technol 45:117–125

Moalemiyan M, Vikram A, Kushalappa AC, Yaylayan V (2006) Volatile metabolite profiling to detect and discriminate stem-end rot and anthracnose diseases of mango fruits. Plant Pathol 55:792–802

Moradi M, Oerke EC, Steiner U, Tesfaye D, Schellander K, Dehne HW (2010) Microbiological and Sybr®Green Real-Time PCR detection of major *Fusarium* head blight pathogens on wheat ears. Microbiology 79:646–654

Mori Y, Nagamine K, Tomita N, Notomi T (2001) Detection of loop-mediated isothermal amplification reaction by turbidity derived from magnesium pyrophosphate formation. Biochem Biophys Res Commun 289:150–154

Moshou D, Bravo C, Oberti R, West J, Bodria L, McCartney A, Ramon H (2005) Plant disease detection based on data fusion of hyper-spectral and multi-spectral fluorescence imaging using Kohonen maps. Real-Time Imaging 11:75–83

Moshou D, Bravo C, Wahlen S, West J, McCartney A, De Baerdemaeker J, Ramon H (2006) Simultaneous identification of plant stresses and diseases in arable crops using proximal optical sensing and self-organising maps. Precis Agric 7:149–164

Moshou D, Bravo C, West J, Wahlen S, McCartney A, Ramon H (2004) Automatic detection of 'yellow rust' in wheat using reflectance measurements and neural networks. Comput Electron Agr 44(3):173–188

Myers RM, Fischer SG, Lerman LS, Manialis T (1985) Nearby all base substitution in DNA fragments joined to a GC clamp can be detected by denaturing gradient gel electrophoresis. Nucleic Acids Res 13:3131–3145

Nanayakkara UN, Singh M, Al-Mugharabi KI, Peters RD (2009) Detection of *Phytophthora erythroseptica* in above-ground potato tissues, progeny tubers, stolons and crop debris using PCR techniques. Am J Potato Res 86:239–245

Notomi T, Okayama H, Masubuchi H, Yonekawa T, Watanabe K, Amino N, Hase T (2000) Loop-mediated isothermal amplification of DNA. Nucleic Acids Res 28:E63

Okamoto H, Suzuki Y, Kataoka T, Sakai K (2009) Unified hyperspectral imaging methodology for agricultural sensing using software framework. Acta Horticulturae 824:49–56

Okubara PA, Schroeder KL, Paulitz TC (2008) Identification and quantification of *Rhizoctonia solani* and *R. oryzae* using real-time polymerase chain reaction. Phytopathology 98:837–847

Olmos A, Bertolini E, Cambra M (2002) Simultaneous and co-operational amplification (CoPCR): a new concept for detection of plant viruses. J Virol Methods 106:51–59

Pan L, Zhang W, Zhu N, Mao S, Tu K (2014) Early detection and classification of pathogenic fungal disease in post-harvest strawberry fruit by electronic nose and gas chromatography-mass spectrometry. Food Res Int 62:162–168

Paolesse R, Alimelli A, Martinelli E, Natale CD, D'Amico A, D'Egidio MG, Aureli G, Ricelli A, Fanelli C (2006) Detection of fungal contamination of cereal grain samples by an electronic nose. Sensors Actuators B Chem 119:425–430

Pearson CAS, Leslie JE, Schweak EW (1987) Host preference co-related with chlorate resistance in *Macrophomina phaseolina*. Plant Dis 71:828–831

Persaud KC, Dodd G (1982) Analysis of discrimination mechanisms in the mammalian olfactory system using a model nose. Nature 299:352–355

Pettersson H, Åberg L (2003) Near infrared spectroscopy for determination of mycotoxins in cereals. Food Control 14:229–232

Pfeffer PE, Bago B, Shachar-Hill Y (2001) Exploring mycorrhizal function with NMR spectroscopy. New Phytol 150:543–553

Porter-jordan K, Rosenberg EI, Keiser JF, Gross JD, Ross AM, Nasim S, Garrett CT (1990) Nested polymerase chain reaction assay for the detection of cytomegalovirus overcomes false positives caused by contamination with fragmented DNA. J Med Virol 30:85–91

Prithiviraj B, Vikram A, Kushalappa AC, Yaylayan V (2004) Volatile metabolite profiling for the discrimination of onion bulbs infected by *Erwinia carotovora* sp. *carotovora*, *Fusarium oxysporum* and *Botrytis allii*. Eur J Plant Pathol 110:371–377

Prospero S, Black JA, Winton LM (2004) Isolation and characterization of microsatellite markers in *Phytophthora ramorum*, the causal agent of sudden oak death. Mol Ecol Resour 4:672–624

Ramanujam N, Mitchell MF, Mahadevan A, Thomsen S, Silva E, Richards-Kortum R (1994) Luorescence spectroscopy: a diagnostic tool for cervical intraepithelial neoplasia. Gynecol Oncol 52:31–38

Rigotti S, Gindro K, Richter H, Viret O (2002) Characterization of molecular markers for specific and sensitive detection of *Botrytis cinerea* Pers:Fr in strawberry (*Fragaria_ananassa* Duch) using PCR. FEMS Microbiol Lett 209:169–174

Rytkönen KT, Williams TA, Renshaw GM, Primmer CR, Nikinmaa M (2011) Molecular evolution of the metazoan PHD-HIF oxygen-sensing system. Mol Biol Evol 28:1913–1926

Salman A, Lapidot I, Pomerantz A, Tsror L, Hammody Z, Moreh R, Huleihel M, Mordechai S (2012) Detection of *Fusarium oxysporum* fungal isolates using ATR spectroscopy. J Spectrosc 27:551–556

Sankaran S, Mishra A, Ehsani R, Davis C (2010) A review of advanced techniques for detecting plant diseases. Comput Electron Agric 72:1–13

Schena L, Nigro F, Ippolito A, Gallitelli D (2004) Real-time quantitative PCR: a new technology to detect and study phytopathogenic and antagonistic fungi. Eur J Plant Pathol 110: 893–908

Schena L, Li Destri Nicosia MG, Sanzani SM, Faedda R, Ippolito A, Cacciola SO (2013) Development of quantitative pcr detection methods for phytopathogenic fungi and oomycetes. J Plant Pathol 95:7–24

Schoch CL, Seifert KA, Huhndorf S, Robert V, Spouge JL, Levesque CA, Chen W (2012) Nuclear ribosomal internal transcribed spacer (ITS) region as a universal DNA barcode marker for fungi. Proc Natl Acad Sci U S A 109:6241–6246

Schroeder KL, Okubara PA, Tambong JT, Levesque CA, Paulitz TC (2006) Identification and quantification of pathogenic *Pythium* spp. from soils in eastern Washington using real-time polymerase chain reaction. Phytopathology 96:637–647

Shamloul AM, Hadidi A (1999) Sensitive detection of potato spindle tuber and temperate fruit tree viroids by reverse transcription-polymerase chain reaction-probe capture hybridization. J Virol Methods 80:145–155

Sheridan GEC, Masters CI, Shallcross JA, Mackey BM (1998) Detection of mRNA by reverse transcription-PCR as an indicator of viability in *Escherichia coli* cells. Appl Environ Microbiol 64:1313–1318

Simner PJ, Uhl JR, Hall L, Weber MW, Walchak RC, Buckwalter S, Wengenack NL (2013) Broad-range direct detection and identification of fungi by use of the PLEX-ID PCR-Electrospray Ionization Mass Spectrometry (ESI-MS) system. J Clin Microbiol 51:1699–1706

Smith OP, Peterson GL, Beck RJ, Schaad NW, Bonde MR (1996) Development of a PCR-based method for identification of *Tilletia indica*, causal agent of karnal bunt of wheat. Phytopathology 86:115–122

Stackebrandt E, Liesack W, Witt D (1992) Ribosomal RNA and rDNA sequences analyses. Gene 115:255–260

Summy KR, Little CR (2008) Using color infrared imagery to detect Sooty Mold and fungal pathogens of glasshouse-propagated plants. Hortscience 43(5):1485–1491

Tan DHS, Sigler L, Gibas CFC, Fong IW (2008) Disseminated fungal infection in a renal transplant recipient involving *Macrophomina phaseolina* and *Scytalidium dimidiatum*: case report and review of taxonomic changes among medically important members of the Botryosphaeriaceae. Med Mycol 46:285–292

Taylor TN, Remy W, Hass H (1992) Parasitism in a 400-million-year-old green alga. Nature 357:493–494

Toju H, Tanabe AS, Yamamoto S, Sato H (2012) High-coverage ITS primers for the DNA-based identification of ascomycetes and basidiomycetes in environmental samples. PLoS One 7:e40863

Tomita N, Mori Y, Kanda H, Notomi T (2008) Loop-mediated isothermal amplification (LAMP) of gene sequences and simple visual detection of products. Nat Protoc 3:877–882

Tomlinson JA, Dickinson MJ, Boonham N (2010) Detection of *Botrytis cinerea* by loop-mediated isothermal amplification. Lett Appl Microbiol 51:650–657

Varga A, James D (2006) Real-time RT-PCR and SYBR Green I melting curve analysis for the identification of Plum pox virus strains C, EA, and W: effect of amplicon size, melt rate, and dye translocation. J Virol Meth 132:146–153

Vikram A, Prithiviraj B, Hamzehzarghani H, Kushalappa AC (2004) Volatile metabolite profiling to discriminate diseases of McIntosh apples inoculated with fungal pathogens. J Sci Food Agric 84:1333–1340

Vilgalys R, Gonzalez D (1990) Organization of ribosomal DNA in the basidiomycete *Thanatephorus praticola*. Curr Genet 18:277–280

Walcott RR, Gitaitis RD, Langston DB (2004) Detection of *Botrytis aclada* in onion seed using magnetic capture hybridization and the polymerase chain reaction. Seed Sci Technol 32:425–438

Wang PH, Lo HS, Yeh Y (2001) Identification of *F. oxysporum cucumerinum* and *F. oxysporum luffae* by RAPD-generated DNA probes. Lett Appl Microbiol 33:397–401

White TJ, Bruns T, Lee S, Taylor J (1990) Amplification and direct sequencing of fungal ribosomal RNA genes for phylogenetics. In: Innis MA, Gelfand DH, Sninsky JJ, White TJ (eds) PCR protocols a guide to methods and applications. Academic, San Diego, pp 315–322

Wilson AD (2013) Diverse applications of electronic-nose technologies in agriculture and forestry. Sensors 13:2295–2348

Wilson AD, Baietto M (2009) Applications and advances in electronic-nose technologies. Sensors 9:5099–5148

Woodcock T, Downey G, Donnell CPO, Summy KR, Little CR (2008) Near Infrared Spectrosc. Using color infrared imagery to detect sooty mold and fungal pathogens of glasshouse-propagated plants. Hortscience 43:1485–1491

Wu D, Feng L, Zhang C, He Y (2008) Early detection of Botrytis cinerea on eggplant leaves based on visible and near-infrared spectroscopy. Trans ASABE 51:1133–1139

Xu ML, Melchinger AE, Lubberstedt T (1999) Species-specific detection of the maize pathogens *Sporisorium reiliana* and *Ustilago maydis* by dot blot hybridization and PCR-based assays. Plant Dis 83:390–395

Yan L, Zhang C, Ding L, Ma Z (2008) Development of a real-time PCR assay for the detection of *Cladosporium fulvum* in tomato leaves. J Appl Microbiol 104:1417–1424

Yang F, Jensen JD, Svensson B, Jorgensen HJL, Collinge DB, Finnie C (2010) Analysis of early events in the interaction between *Fusarium graminearum* and the susceptible barley (*Hordeum vulgare*) cultivar Scarlett. Proteomics 10:3748–3755

Yin Y, Ding L, Liu X, Yang J, Ma Z (2009) Detection of *Sclerotinia sclerotiorum* in planta by a real-time PCR assay. J Phytopathol 157:465–469

Zeng Q, Hansson P, Wang X (2005) Specific and sensitive detection of the conifer pathogen *Gremmeniella abietina* by nested PCR. BMC Microbiol 5:65

Zhang AW, Hartman GL, Riccinoi L, Chen WD, Ma RZ, Pedersen WL (1997) Using PCR to distinguish *Diaporthe phaseolorum* and *Phomopsis longicolla* from other soybean fungal pathogens and to detect them in soybean tissues. Plant Dis 81:1143–1149

Zhang Z, Zhang J, Wang Y, Zheng X (2005) Molecular detection of *Fusarium oxysporum* f.sp. *niveum* and *Mycosphaerella melonis* in infected plant tissues and soil. FEMS Microbiol Lett 249:39–47

Zhang N, McCarthy ML, Smart CD (2008) A macroarray system for the detection of fungal and oomycete pathogens of solanaceous crops. Plant Dis 92:953–960

Zhu BL, Xie CS, Wu J, Zeng DW, Wang AH, Zhao XZ (2006) Influence of Sb, In and Bi dopants on the response of ZnO thick films to VOCs. Mater Chem Phys 96:459–465

Molecular Diagnostics and Application of DNA Markers in the Management of Major Diseases of Sugarcane

Sangeeta Srivastava

Introduction

Sugarcane (*Saccharum* spp. hybrid) is a major industrial crop grown in ~91 countries in tropical and subtropical regions across the globe. India is the second largest producer of sugarcane in the world and sugarcane occupies a powerful position as an agro-industrial crop of country, covering around 5.01 million hectare area with production 3,521.42 million tonne. Sugarcane is susceptible to a myriad of bacterial, fungal, viral and phytoplasmal diseases. There are about 160 fungal and 8 bacterial pathogens reported along with at least 7 recognized sugarcane diseases of unknown aetiology world-wide (Rott et al. 2000). In India, more than 50 diseases of sugarcane caused by bacteria, viruses, fungi, phytoplasmas and nematodes are known to occur that affect cane and foliar parts of the plant causing substantial losses in cane yield and quality (Rott et al. 2000; Viswanathan and Rao 2011). Major fungal diseases *viz.* red rot, smut and wilt pose a real challenge to successful cultivation of sugarcane. Besides fungal diseases, bacterial and viral diseases cause considerable damage to the crop. Ratoon stunting and leaf scald are two important bacterial diseases of sugarcane in India. The two viral diseases of sugarcane *viz.* mosaic and yellow leaf syndrome occur throughout the crop cycle in almost all parts of the country and are accountable for the progressive deterioration of crop performance leading to reduced life-span of many promising sugarcane varieties. Lack of precise diagnostic techniques and clear symptoms besides non-availability of information on genome of disease causing virus are the main reasons for delay in management of different viral diseases in sugarcane. Off late, sugarcane grassy shoot disease and leaf yellows caused by phytoplasma and leaf fleck have emerged

---

S. Srivastava
Division of Crop Improvement, ICAR-Indian Institute of Sugarcane Research,
Lucknow 226 002, India
e-mail: sangeeta_iisr@yahoo.co.in

P. Kumar et al. (eds.), *Current Trends in Plant Disease Diagnostics and Management Practices*, Fungal Biology, DOI 10.1007/978-3-319-27312-9_13

as major constraints for sugarcane growers. It is very difficult to introduce resistance against all pathogens through conventional breeding approaches. Sometimes an elite cultivar may be susceptible to more than one pathogen. Generally an integrated approach involving disease free seed material, resistant varieties, appropriate disease control measures and strict quarantine procedures is used to manage sugarcane diseases in most sugarcane growing countries.

Diagnosis and Detection of Pathogen

Early detection, proper identification and accurate diagnosis of the causal organism and reasons for the prevalence of a disease are prerequisites for appropriate and well-timed control and management of any plant disease. Conventional methods to identify pathogen have often relied on disease symptoms, isolation, culturing and morphology of pathogen and biochemical tests. Using healthy planting materials is a key management strategy for containing diseases in any crop. Because of vegetative propagation mode, sugarcane is amenable to transmission of pathogens through planting materials in field thus facilitating the entry of pathogen in fields. Slow accumulation of different viral/phytoplasmal pathogens in sugarcane causes deterioration in varietal performance. It is crucial to examine the planting material for presence of any probable pathogen before the planting of the crop. Proper diagnosis of the pathogen is also essential for epidemiological studies. It is relatively simple to detect pathogens in symptomatic plants if one has extensive experience with disease diagnosis and isolation of plant pathogens. Detection of disease causing organism in seeds or asymptomatic vegetative propagative materials, such as in sugarcane setts, can be extremely difficult if the titer of pathogen is very low. This necessitates sensitive techniques capable of detection of very low numbers of pathogen propagules (Srivastava and Sinha 2009). Traditional methods of disease diagnosis and pathogen identification are relatively slow and often require trained and experienced persons for reliably taxonomic classification of the pathogens at genus or species level. Delays are damaging when quick diagnosis is needed because timely detection allows one to judge the problem, identify the need of the hour and choose right commercial variety as well as appropriate disease control measures that may be taken to prevent the risk of its culmination into a potential epidemic (Ward et al. 2004), especially when high value cash crops such as sugarcane are at stake.

Progress in Development of Molecular Diagnostic Tools for Sugarcane Diseases

Rapid advancements in biotechnology have strengthened efforts in recent years to develop innovative methods for detection, diagnosis and identification of disease causing agents. New and improved molecular assays based on the DNA, RNA or proteins of the causal agents have increasingly improved the efficiency, precision, and speediness of detection, identification and characterization of the pathogens worldwide (Ward et al. 2004). These molecular diagnostic tools are highly sensitive and fairly easy to use compared with the old-style detection methods and can detect the pathogens even in asymptomatic plants with an extremely low titer of pathogen. Molecular approaches such as PCR based amplification of rDNA sequence especially the ITS (internal transcribed spacer) region have been widely used for detection of several pathogens (Henson and French 1993; Schaad and Frederick 2002). Diagnostic tools based on nucleic-acid (DNA/RNA) and protein have been developed against target pathogens in sugarcane (viruses, fungi, bacteria, phytoplasma, insect pests etc.) e.g. ratoon stunting disease (Fegan et al. 1998; Pan et al. 1998), grassy shoot disease (Srivastava et al. 2006), Fiji disease and mosaic (Smith and Van De Velde 1994), striate mosaic (Thompson et al. 1998), yellow leaf syndrome (Irey et al. 1997; Chatenet et al. 2001; Viswanathan et al. 2009), smut (Albert and Schenck 1996), and SCBV (sugarcane bacilliform viruses) (Braithwaite et al. 1995). Detection and identification of pathogen has been very useful and significantly improved with techniques of Dot-blot immunoassay (DBIA) to detect *Liefsonia. xyli* subsp. *xyli* (Schenck et al. 1997), PCR, RT-PCR, DAS-ELISA and tissue-blot immunoassay (TBIA) for SCYP and SCYLV (Lehrer et al. 2001), SCMV (Comstock et al. 2000) and *Xanthomonas albilineans* (Schenck et al. 1997), immunocapture-RT-PCR based assay to detect the presence of *Sugarcane streak mosaic virus* (SCSMV) and RT-PCR (reverse transcription-polymerase chain reaction) for SCYLV (Moonan et al. 1999).

In Columbia, research activities have been concentrated on development of molecular diagnostics for sugarcane mosaic virus, sugarcane yellows, Fiji disease, sugarcane streak mosaic, leaf scald and ratoon stunting disease. Work on molecular diagnostics for yellow leaf and smut as well as characterisation of strains of *Xanthomonas albilineans* causing leaf scald disease has been done in France. Several genetic transformation projects for resistance to sugarcane yellow leaf virus and ratoon stunting disease have been carried out at Texas A & M University, USA. Immunoassays and molecular diagnosis confirmed the presence of leaf fleck, leaf scald, RSD and yellow leaf diseases in commercial fields of sugarcane in Ecuador and in some imported varieties in quarantine (Garces et al. 2005). In order to identify the white grub species attacking sugarcane, the base pair sequence of the mitochondrial cytochrome c oxidase sub unit I (cox 1) gene of scarabaeid larvae collected from sugarcane fields was compared with scarabaeid adults of known species in KwaZulu-Natal, South Africa. NGS (next generation sequencing) techniques and specific real-time PCR assays have been used to identify and confirm the presence of a combination of sap transmissible viruses *viz.* Maize chlorotic mottle virus and Sugarcane mosaic virus, from six different maize fields in two different regions of Kenya (Adams et al. 2013). Recently, Wongkaew and Poosittisak (2015) used immobilized ssDNA probe as a specific sensor for DNA based voltammetric electrochemical verification of sugarcane white leaf infection within plants collected from field having infections up to the limit of an epidemic. These tools have proved immensely useful to several crop improvement and crop protection programmes such as evaluation of germplasm to find resistant genotypes, breeding for disease and pest resistance, quarantine, disease monitoring and surveillance programmes. Identification of unique markers known as "DNA barcodes" have been extremely useful as genetic markers for differential identification of pathogens and insect pests especially for rapid detection of new and emerging diseases/ pathogens.

Molecular Diagnostics and Pathogen Identification for Major Fungal Diseases of Sugarcane

Diseases such as red rot, smut, wilt and sett rot or pineapple disease are some important fungal diseases of sugarcane causing substantial yield loss in different states of India. Besides, Pokkah boeng, rust, stalk rot and seedling rot are some other major fungal diseases affecting sugarcane in different parts of world.

Red Rot of Sugarcane

Red rot disease of sugarcane is one of the major limitations in its cultivation in many states of India. Usually considered as stalk and seed-piece transmissible disease, it equally affects the cane growers and millers by impeding the cane growth and yield as well as deteriorating the quantity and quality of juice. Several promising varieties became susceptible to red rot and could not be cultivated anymore. The disease is caused by the fungus *Colletotrichum falcatum* Went (imperfect state); perfect/ascigerous state = *Glomerella tucumanensis* (Speg.) von Arx and Muller, which remains dormant in cane tissues unless its expression and makes it difficult to diagnose under field conditions. Moreover, incipient infection in stalks most often does not lead to symptom expression and the pathogen is transmitted through such canes which cannot be distinguished from healthy stalks. ELISA techniques and polyclonal antisera against *C. falcatum* have been used to identify the pathogen in cane tissues. PCR method of detecting incipient infection is a highly sensitive and specific diagnostic tool that can detect infection in the nodal tissue of seed cane even before its planting. A PCR based detection kit for red rot of sugarcane has been developed at IISR, Lucknow for the first time in the world which is highly specific to *C. falcatum* and can amplify even 1.0 ng of DNA of the pathogen (Srivastava and Sinha 2009). *C. falcatum* specific primer sets based on conserved gene sequences have been used to detect *C. falcatum* at SBI, Coimbatore (Malathi and Viswanathan 2012). PCR results confirmed that the primers were able to detect *C. falcatum* in sugarcane in mixed state of infection which may help to detect the fungal infections more accurately (Malathi et al. 2012).

Sugarcane Smut Disease

The disease affects both plant crop and ratoon but impact is more in ratoon. Nearly 60 % loss in cane weight has been reported in 22 varieties in India. Yield losses of 48 % in plant cane and up to 90 % in first ratoon of NCo 376 have been reported in South Africa. The smut affected plants showed reduction in cane quality parameters *viz.* Brix, sucrose, purity and juice extraction. The pathogen *Sporisorium scitamineum* (Syn *Ustilago scitaminea*), produces abundant tiny brownish black wind-borne teliospores which are spread in the standing cane fields and can infect newly planted setts in the soil. Trypan blue staining technique was used to detect the pathogen in bud or apical meristematic tissues of suspected canes well before symptom expression. PCR assays have been developed at IISR, Lucknow and South Africa to detect inactive infection of *Sporisorium scitamineum* in nodal bud scales and apical meristem of sugarcane. An extremely sensitive and specific PCR based detection kit for smut disease of sugarcane using rDNA has been developed at IISR, Lucknow to detect the presence of the pathogen in the nodal bud of seed cane before its planting even if the inoculum concentration is quite low. An amplicon of 459 bp of *bE* mating type gene was used at SBI, Coimbatore for rapid and early detection of latent infection of smut pathogen (Ramesh et al. 2012).

Other Fungal Diseases of Sugarcane

Wilt is a serious disease of sugarcane in 34 countries including India, where it was first reported from North India in 1913. Currently, it occurs in all the sugarcane growing states of India but its severity is high in tropical states of Gujarat, eastern coastal Andhra Pradesh, Tamil Nadu and Orissa, and subtropical states of Bihar, Haryana, Punjab and Uttar Pradesh. Many wilt susceptible elite cultivars have been eliminated from cultivation. The disease causing pathogen has not been clearly established. RAPD, ISSR and rDNA markers have been used at IISR, Lucknow and SBI, Coimbatore to identify and characterize the wilt pathogen. Studies have indicated that the pathogen belongs to *F. sacchari* (Duttamajumder and Srivastava 2007; Viswanathan et al. 2011). Pokkah boeng (PB) is another foliar fungal diseases caused by *Fusarium verticillioides* (*F. moniliforme* Sheldon) which infects young leaves and the sugarcane tops become twisted due to malformation. First reported in 1886 by Walker and Went in Java, the disease now occurs in almost all the sugarcane growing regions in the world including India. The pathogen is primarily air-borne and secondary transmission takes place through diseased setts, irrigation or rain water and soil. Wilt disease has also been reported as an after effect of pokkah boeng. Sett rot or pineapple disease caused by *Ceratocystis paradoxa* (de Seynes) Moreau is a minor fungal disease which is both sett and soil borne, affecting the sett germination at the early stages of planting. Sugarcane rust is another fungal disease caused by *Puccinia melanocephela*, an obligate parasitic fungus. The disease has caused considerable damage in USA. The search for molecular markers for rust (*Puccinia melanocephala*) and yellow spot (*Mycovellosiella koepkei*) is in progress at Mauritius.

Molecular and Biochemical Detection and Identification of Sugarcane Viruses

Mosaic disease, streak mosaic, leaf fleck disease and sugarcane yellow leaf syndrome are the major virus diseases of economic concern in India and other countries (Bhargava 1975; Rao et al. 2002; Viswanathan et al. 2007), and several diagnostic tools have been developed for detection of disease causing virus.

Sugarcane Mosaic Disease

Mosaic disease prevalent in almost all sugarcane cultivating regions of India was the only reported virus disease of sugarcane for several decades that affected yield and quality in sugarcane resulting in significant yield losses (Viswanathan et al. 2007; Jain et al. 1998). It is a seed piece transmissible disease that develops symptoms as interveinal chlorotic specks, streaks or stripes in young, rapidly growing leaves resulting in contrasting shades of green patterns on leaf blades, particularly near the basal portion of the leaf, but the older leaves tend to recover and appear as healthy. However, the symptoms may vary in intensity depending upon the host, cane growing conditions, temperature and strain of the virus. With more virulent strains, stunting, yellowing and chlorosis, and sometimes necrosis are also noticed. Strains of several *Sugarcane mosaic virus* (SCMV) and *Sorghum mosaic virus* (SrMV) in combination or separately have been reported as causal agents (Viswanathan et al. 2007). A new virus *Sugarcane streak mosaic virus* (SCSMV) was reported (Hema et al. 2001) which was characterized as a new genus "Susmovirus" in the family *Potyviridae* based on its distinct coat protein genome (Gaur et al. 2003) and based on host range of the species in the genus, was renamed as *Poacevirus* by ICTV (International Committee on Taxonomy of Viruses). Occurrence of nine strains of SCMV in India has also been established (Viswanathan et al. 2009). Molecular characterization of several SCMV and SCSMV isolates has established the variation in coat protein genome of the respective viruses for the first time in India.

Diagnosis of SCMV and SCSMV is well established by serological techniques like, DIBA, ELISA, ISEM (Immunosorbent electron microscopy) and western blots) and molecular techniques of RT-PCR, NASH and Southern hybridization (Viswanathan et al. 2007, 2008 Hema et al. 2001; Gaur et al. 2003). Recently, complete nucleotide sequence of an isolate of *Sugarcane streak virus* from India

(SCSMV-IND) was determined which revealed a 9,786 nucleotides long single stranded positive sense RNA genome of SCSMV- IND (excluding the poly (A) tail) comprising of a large open reading frame encoding polyprotein of 3,131 amino acid residues (Rao 2014). At South African Sugarcane Research Institute (SASRI), a rapid and cost-effective diagnostic tool *viz.* quantitative reverse transcriptase polymerase chain reaction (qRT-PCR) has been optimised for detection of SCMV-infected sugarcane leaves in the quarantine facility.

Sugarcane Yellow Leaf Disease

Yellow leaf disease (YLD) of sugarcane was first reported from East Africa by Rogers in 1969, but later reports came from Cuba (Peralta et al. 1999), Hawaii (Schenck 1990), South Africa (Cronjé et al. 1998) and India (Rao et al. 2014). In India, it is prevalent in most of the sugarcane growing area and its intensity has been recorded up to hundred per cent in certain susceptible sugarcane varieties (Rao et al. 2014). Referred to as yellow leaf syndrome (YLS), the disease is caused by *Sugarcane yellow leaf virus* (SCYLV) and sugarcane yellow leaf phytoplasmas (SCYLP) and is characterized by yellowish midrib on the lower surface and yellow discolouration in adjoining laminar region parallel to midrib. In severe cases, necrosis of discoloured laminar region and subsequent drying of entire leaf is also noticed (Viswanathan 2012). Association of the virus with this disease was established after sequencing the virus genome. Molecular diagnostic assays using diagnostic primers *viz.*, SCYLV-615F and SCYLV-615R were developed and their efficiency for detection has been validated (Viswanathan et al. 2009). The RT-PCR assays established that the diagnostic primers efficiently detected all the SCYLV population even in asymptomatic plants. Phylogenetic analyses later established that all the isolates belong to genotype SCYLV-IND which shared a very close relationship with CHNI, the genotype reported from China, and they showed a separate lineage, probably of Asian genotypes.

Sugarcane Leaf Fleck Disease

Badnavirus infecting sugarcane are collectively referred to as sugarcane bacilliform viruses (SCBV) and cause the leaf fleck disease of sugarcane which was reported first in Cuba (Rodriguez-Lema et al. 1985) but later in many sugarcane growing countries of the world including India (Comstock and Lockhart 1990; Viswanathan et al. 1996). The disease affects species level sugarcane germplasm and commercial *Saccharum* sp. hybrids. Two types of symptoms are observed after SCBV infection, (1) chlorotic specks that change to stripes after sometime and (2) mild mottling. More severe infection leads to stunted growth, reduced number of tillers, shorter internodes with deep cracks and bunchy top (Rishi 2009). The insect vectors *viz.* sugarcane mealy bug

(*Saccharicoccus sacchari*) and the grey sugarcane mealy bug (*Dysmicoccus boninsis*) (Lockhart and Autrey 1991) are responsible for the transmission of SCBV. The ICTV earlier recognized two species of SCBV; *Sugarcane bacilliform IM virus* (SCBIMV) and *Sugarcane bacilliform MO virus* (SCBMOV) and later added two more species *viz. Sugarcane bacilliform Guadeloupe A virus* (SCBGAV) and *Sugarcane bacilliform Guadeloupe D virus* (SCBGAV) from Guadeloupe (Muller et al. 2011). Recently five more complete genomes of SCBV comprising of three newly proposed species (SCBV-BB, SCBV-BO and SCBV-BR) have been reported from India (Karuppaiah et al. 2013). Various techniques have been tried to detect this disease and immunosorbent electron microscopy (ISEM) has been found to be the most reliable method of detecting SCBV. Though PCR techniques have proved to be quite suitable for detection of pathogen in general, it has been found to be not so suitable in this case because of the presence of much variability in the pathogen even if isolated from the same clone of sugarcane. Besides, the problem of virus integration into the host genome is also feared (Rishi 2009).

Molecular Detection and Discrimination of Phytoplasma Diseases in Sugarcane

Phytoplasmas are wall-less plant pathogenic prokaryotes of 530–1,350 kb genome size of the class Mollicutes (Marcone 2002) and were earlier known as mycoplasma-like organisms (MLOs). They are associated with grassy shoot disease (SCGS), white leaf disease (SCWL), yellow leaf syndrome/leaf yellows (YLS) and Ramu Stunt diseases in sugarcane. They are mostly transmitted by insect vectors (leafhoppers, planthoppers, and psyllids) and several disease transmitting insect vectors have been identified. The grassy shoot and white leaf diseases in Southeast Asia (Rao 2014; Wongkaew et al. 1997) are transmitted by the leafhopper *Matsumuratettix hiroglyphicus* (Sdoodee 2000), grassy shoot disease (SCGS) in India is transmitted by *Deltocephalus vulgaris* (Srivastava et al. 2006), Ramu stunt disease in Papua New Guinea (Suma and Jones 2000) by the delphacid *Eumetopina flavipes* (Kuniata et al. 1994) and yellow leaf phytoplasma in western and central Cuba by the delphacid planthopper *Saccharosydne saccharivora* (Arocha et al. 2005).

Sugarcane Grassy Shoot Disease

Sugarcane grassy shoot (SCGS) disease caused by phytoplasma has been reported from several countries, including Bangladesh, Malaysia, Nepal, Pakistan, Sri Lanka and India (Rishi and Chen 1989; Singh et al. 2002; Srivastava et al. 2003). It has become a very important disease in India, next only to the fungal diseases like red rot, wilt, and smut (Rao 2014; Rao and Dhumal 2002) as it has resulted in hundred percent loss of susceptible varieties in India in the states of U.P., Bihar, Chhattisgarh, Haryana, Maharashtra, Karnataka, Kerala, Andhra Pradesh and Tamil Nadu causing great concern to the farmers as well as mill owners. The disease is characterized by bunch of thin, slender, adventitious tillers bearing thin, narrow pale yellow or chlorotic leaves which give the plant the look of a perennial grass. It affects both, the plant as well as ratoon crop and fails to produce any millable canes. Vegetative mode of propagation and phloem feeding leafhopper vectors facilitate transmission of SCGS phytoplasma in the field (Srivastava et al. 2006). The meristem culture technique may be quite successful for elimination of SCYLV in infected seed nurseries of sugarcane (Mishra et al. 2010).

Many diagnostic measures such as ELISA, immunofluorescent techniques and PCR assays have been developed to detect SCGS-phytoplasma. ELISA was not successful due to lack of sensitivity of the technique and problems of purification and specificity of phytoplasma. Substantial cross-reactions of polyclonal antisera produced with partially purified SCWL and SCGS phytoplasma with antigens from healthy plants was observed (Viswanathan 2002). Due to variability in SCGS phytoplasma, the expression pattern of SCGS disease in fields is variable which causes mistaken disease diagnosis. PCR techniques have been more successful in disease detection as they are sensitive and rapid. At present PCR amplification using oligonucleotide primers based on rRNA genes (rDNAs), especially the 16S and 23S rRNA genes, and 16S-23S rRNA intergenic spacer region sequences derived from the phytoplasma (Bertaccini et al. 1997; Gundersen et al. 1996) provides one of the most reliable and acceptable method to detect and identify SCGS phytoplasma in infected plants. The sensitivity of detection has been increased by nesting of primers; using two universal phytoplasma-specific primer pairs P1 and P7 in first round and P4 and P7 primers in the second round for nested PCR assays of SCGS phytoplasma in host as well as disease transmitting insect-vector (*Deltocephalus vulgaris*) for the first time in the world at IISR, Lucknow (Srivastava et al. 2006). This molecular technique can be successfully employed for diagnosis of SCGS, which is usually confused with white leaf disease and mineral deficiency symptoms in sugarcane and hence is useful for diagnosis, forecast and prevention of sugarcane grassy shoot disease.

The sequences of SCGS-causing phytoplasma in India were found to be more than 99 % identical, and their homology with SCWL (Sugarcane white leaf) and BGWL (Bermuda grass white leaf) is between 98 % and 99 % (Nasare et al. 2007). Moreover, the sequence similarity between SCGS and SCWL phytoplasmas was 97.5–98.8 % indicating that these two belonged to the same group, whereas, the most closely related phytoplasma outside the genus *Saccharum* were BGWL ('*Candidatus* Phytoplasma cynodontis') and *Brachiaria* grass white leaf (BraWL) sharing more than 97 % sequence similarity (Rao et al. 2008). Such high level of sequence similarity suggests that the phytoplasma causing white leaf diseases in sugarcane, bermuda grass and brachiaria grass (SCWL, BGWL and BraWL), and sugarcane grassy shoot disease (SCGS) may belong to the same species or are taxonomically akin (Ariyarathna et al. 2007).

Sugarcane White Leaf Disease

Sugarcane white leaf (SCWL) disease caused by phytoplasma is highly destructive disease of sugarcane and is predominant in Thailand, Taiwan and Sri Lanka (Chen and Kusalwong 2000; Matsumoto et al. 1968; Nakashima et al. 1994, 1996). The symptoms of this disease occur as single cream or light green or white coloured stripes developing parallel to the midrib which later turn into several straight stripes extending along the entire leaf length, but rarely onto the upper portion of the leaf sheath and coalesce with each other resulting in mottled appearance and severe chlorosis of entire leaf and the plant vigour is reduced drastically. The disease causing phytoplasma is transmitted to the plant by the leafhopper (*Matsumuratettix hiroglyphicus* Matsumura) and it has been detected in eggs, nymphs and adults of vector in the first and second generations and the host. The genomes of the SCGS and SCWL phytoplasmas are phylogenetically close to each other (99 %) as well as to other phytoplasmas associated with white leaf diseases in gramineous weeds (Wongkaew et al. 1997) and generally induce similar symptoms in infected sugarcane plants thus the two diseases are often not distinguishable (Marcone 2002), but PCR technique has been found to be most suitable for diagnosis of SCWL disease. Rapid detection of the phytoplasmas from the symptomatic tissues using PCR is well established by nested-PCR assay of 16S-23S intergenic spacer region specific to SCWL phytoplasma using P1/P7 and R16F2n/R16R2 followed by sequencing (Hoat et al. 2012; Wongwarat et al. 2011).

Sugarcane Leaf Yellows Phytoplasma

Sugarcane leaf yellows (YLS) is another widely distributed disease in most sugarcane-growing countries including Australia, Egypt, Cuba, Hawaii, Brazil, South Africa, USA, India and Mauritius (Lehrer et al. 2001; Cronjé et al. 1998; Arocha et al. 2005, 1999; Aljanabi et al. 2001; Matsuoka and Meneghin 2000; Scagliusi and Lockhart 2000; Vega et al. 1997) caused by phytoplasma. The disease is characterized by yellowing of midrib portion of leaf and adjoining lamina along with development of irregular yellow patches while rest of the lamina still remains green. The presence of phytoplasma in this disease has been confirmed through PCR amplification of rDNA region from DNA extracted from symptomatic yellow leaves or leaf lamina with yellow patches from leaf yellows diseased plants using phytoplasma specific primer pairs P1 and P6 followed by nesting with primers R16F2n and R16R2. The presence of phytoplasma in association with sugarcane yellow leaf was established in Egypt also through nested PCR assay of 16S rRNA gene using phytoplasma-specific primer pairs P1/P7, and R16F2n/R16R2.

Biochemical and Molecular Diagnostics for Bacterial Diseases in Sugarcane

The bacterial diseases of sugar cane which are known to occur world-wide are leaf scald, ratoon stunting, gumming disease, red-stripe and mottled-stripe (Duttamajumder 2004). A new bacterial disease called spindle rot has been added of late. Leaf scald disease in sugarcane is caused by *Xanthomonas albilineans* (Ashby) Dowson which is a systemic bacterial disease occurring in many sugarcane producing countries of the world (Rott and Davis 2000a; Rott et al. 1997; Wang et al. 1999). The symptoms are chronic and acute phenotypes differing in severity from single affected leaf to death of entire plant resulting in heavy losses in cane yield and reduced juice quality (Ricaud and Ryan 1989). Visual diagnosis of disease is challenging at times due to latent infections. Conventional methods, ELISA based assay, DIBA, polyclonal antiserum against the bacterium and PCR protocols etc. have been used for the diagnosis of sugarcane leaf scald (Wang et al. 1999; Viswanathan et al. 1998) of which, PCR was found to be the best and time saving as compared to the rest. A fast and reliable PCR protocol was developed using the primers designed from the ITS (intergenic transcribed spacer) region between the 16S and 23S rRNA genes for detection and identification of leaf scald pathogen (Pan et al. 1997). Another PCR protocol using the conserved sequences of two adjacent tRNA genes along with variable spacer region between them was found to be very sensitive. The specificity of this PCR-based detection system was further enhanced by *X. albilineans*-specific amplification of the region between the 16S rRNA-tRNA^{ala}-tRNA^{ile}-23S rRNA gene by a nested PCR reaction (Honeycutt et al. 1995).

Ratoon stunting disease (RSD) in sugarcane caused by xylem-limited coryneform bacteria *Leifsonia xyli* subsp. *xyli* (other name *Clavibacter xyli* subsp. *xyli*) is considered as the most important disease affecting sugarcane ratoon crop worldwide as it can cause yield losses of 5–15 % which may go up to 50 % under stress conditions (Davis and Bailey 2000). The disease has no easily recognized external or internal symptoms except stunting of growth that may not always be visible in the field. The detection of *Leifsonia xyli* subsp. *xyli* (*Lxx*) is normally done by immunofluorescence microscopy, serology (evaporative-binding enzyme-linked immunoassay or EB-EIA) or PCR. DAC-ELISA, Dot-blot and tissue blot immunoassay techniques were used for detection of RSD in infected sugarcane samples in India (Viswanathan 2012). A quick and reliable PCR protocol employing primers designed from the ITS (intergenic transcribed spacer) region between the 16S-23S rRNA genes was developed for detection and identification of RSD pathogen (Pan et al. 1998). The consensus sequence derived from the ITS region between the 16S and 23S rRNA genes of *Clavibacter xyli* subsp. *xyli* and *C. xyli* subsp. *cynodontis* (causing RSD of sugarcane and Bermudagrass stunting disease) was used to design the primers Cxx1 and Cxx2 which were further used to amplify DNA of *C. xyli* subsp. *xyli* either directly from cultured cells or from vascular sap of RSD infected cane with no need of genomic DNA extraction (Pan et al. 1998). Another PCR-based assay developed for detection of *Clavibacter xyli* subsp. *xyli*. was highly specific and did not produce any amplification product with the template of either the closely related species *C. xyli* subsp. *cynodontis*, or of any other bacterial species. Further, a multiplex PCR was also developed to identify and detect *C. xyli* subsp. *xyli* and *C. xyli* subsp. *cynodontis* in one PCR reaction (Fegan et al. 1998). Recently, Loop-mediated isothermal amplification (LAMP) techniques has been used for detection of RSD (Ghai et al. 2013; Liu et al. 2013) and has proved to be as sensitive as ELISA but much rapid and specific, and can be judged visually by colour without sophisticated laboratories thus decreasing the cost and time taken in diagnosis. (Ghai et al. 2014) used transposase gene to design six specific primer sets for eight genomic sequences of pathogen and used xylem sap as template instead of genomic DNA, thus bypassing the requirement of genomic DNA isolation.

Red stripe (top rot) of sugarcane caused by *Pseudomonas avenae* Manns = *P. ubrilineans* Stapp and mottled stripe caused by *Herbaspirillum rubrisubalbicans* syn *Pseudomonas rubrisubalbicans* (Christopher and Edgerton) Krassilnikov are two other bacterial diseases of sugarcane of not much significance (Rott and Davis 2000b; Saumtally 2000). Another bacterial disease of sugarcane is spindle rot caused by *Acidovorax avenae* subsp. *spindulifoliens*, (syn. *Pseudomonas rubrilineans* pv. *spindulifoliens* pv. Nov) which is confined mainly to the leaf spindle and affects both, the plant crop and ratoon and has been reported from tropical and subtropical states of India (Patil 2004). The pathogen is transmitted aerially and infection takes place through stomata, causing a mild rotting. A synergistic association of spindle rot disease with pokkah boeng was also observed. Gumming disease of sugarcane caused by the bacterium *Xanthomonas campestris* pv *vasculorum* (Cobb) Dye was a major epidemic of sugarcane when noble canes were grown in Australia, Mauritius and Reunion causing great losses in yield and sugar recovery but the disease was swiped away with the introduction of inter-specific hybrids in the 1930s (Ricaud and Autrey 1989). For detection and variability study of its pathogen, monoclonal antibodies and nucleic acid technologies have been applied.

It can be said conclusively that PCR technology for pathogen detection is highly specific, sensitive and rapid. Progress in the development of molecular diagnostic tools for various diseases of sugarcane has been praiseworthy so far. Since elimination of diseased seed cane is a critical step in controlling the diseases, there is further need for development and application of highly sensitive, specific and cost effective molecular diagnostic protocols for key pathogens, which can be used for large scale sugarcane seed certification, germplasm quarantine programmes and diagnostic laboratories in order to detect these pathogens in seed cane.

References

Adams IP, Miano DW, Kinyua ZM, Wangai A, Kimani E, Phiri N, Reeder R, Harju V, Glover R, Hany U, Souza-Richards R, Deb Nath P, Nixon T, Fox A, Barnes A, Smith J, Skelton A, Thwaites R, Mumford R, Boonham N (2013) Use of next-generation sequencing for the identification and characterization of Maize chlorotic mottle virus and sugarcane mosaic virus causing maize lethal necrosis in Kenya. Plant Pathol 62:741–749

Albert H, Schenck S (1996) Development of a sensitive assay for detection of sugarcane smut disease caused by *Ustilago scitaminea*. Plant Dis 80:1175–1178

Aljanabi SM, Parmessur Y, Moutia Y, Saumtally S, Dookun A (2001) Further evidence of the association of a phytoplasma and a virus with yellow leaf syndrome in sugarcane. Plant Pathol 50:628–636

Ariyarathna HACK, Everard JMDT, Karunanayake EH (2007) Diseased sugarcane in Sri Lanka is infected with sugarcane grassy shoot and/or sugarcane white leaf phytoplasma. Aust Plant Dis Notes 2:123–125

Arocha Y, Gonzalez L, Peralta EL (1999) First report of a virus and phytoplasma pathogens associated with yellow leaf syndrome of sugarcane in Cuba. Plant Dis 83:1177

Arocha Y, Lopez FMM, Pinol B, Horta D, Peralta EL, Almeida R, Carvajal O, Picornell S, Wilson MR, Jones P (2005) Transmission of a sugarcane yellow leaf phytoplasma by the delphacid planthopper *Saccharosydne saccharivora*, a new vector of sugarcane yellow leaf syndrome. Plant Pathol 54:634–642

Bertaccini A, Vibio M, Pastore M, Recupero S, Guerrini S, Grimaldi D (1997) Nested-PCR assays for detection of phytoplasmas in strawberry. Acta Horticult 439:787–790

Bhargava KS (1975) Sugarcane mosaic -retrospect and prospects. Ind Phytopathol 28:1–9

Braithwaite KS, Egeskov NM, Smith GR (1995) Detection of sugarcane bacilliform virus using the polymerase chain reaction. Plant Dis 79:792–769

Chatenet M, Delage C, Ripolles M, Irey MS, Lockhart BEL, Rott P (2001) Detection of Sugarcane yellow leaf virus in quarantine and production of virus-free sugarcane by apical meristem culture. Plant Dis 85:1177–1180

Chen CT, Kusalwong A (2000) White leaf. In: Rott P, Bailey RA, Comstock JC, Croft BJ, Saumtally AS (eds) A guide to sugarcane diseases. CIRAD, ISSCT, France, pp 231–236

Comstock JC, Lockhart BEL (1990) Wide-spread occurrence of sugarcane bacilliform virus in U.S. sugarcane germplasm collections. Plant Dis 74:530

Comstock JC, Follis J, Miller JD (2000) Evaluation of sugarcane mosaic incidence in Florida. J Am Soc Sugar Cane Technol 2:15–21

Cronjé CPR, Tymon AM, Jones P, Bailey RA (1998) Association of a phytoplasma with a yellow leaf syndrome in sugarcane in Africa. Ann Appl Biol 133:177–186

Davis MJ, Bailey RA (2000) Ratoon stunting. In: Rott P, Bailey RA, Comstock JC, Croft BJ, Saumtally AS (eds) A guide to sugarcane diseases. CIRAD, ISSCT, France, p 340

Duttamajumder SK (2004) Bacterial diseases of sugarcane in India: a bird's eye view. In: Rao GP, Saumtally AS, Rott P (eds) Sugarcane pathology. Volume 3: bacterial and nematode diseases. Science Publishers, Enfield, pp 15–50

Duttamajumder SK, Srivastava S (2007) Final report, ICAR network project on 'Wilt of crops with special reference to cultural, morphological, molecular characterization and pathogenic variability of isolates in India: Sugarcane'

Fegan M, Croft BJ, Teakle DS, Hayward AC, Smith GR (1998) Sensitive and specific detection of *Clavibacter xyli* subsp. *xyli*, causative agent of ratoon stunting disease of sugarcane, with a polymerase chain reaction-based assay. Plant Pathol 47:495–504

Garces FF, Balladarez C, Quiridumbay G, Munoz C (2005) Diagnosis of leaf fleck, leaf scald, mosaic, ratoon stunting disease and yellow leaf of sugarcane in commercial fields and quarantine in Ecuador. Proc ISSCT 25:695–700

Gaur RK, Singh AK, Singh M, Singh AK, Upadhyaya PP, Rao GP (2003) Reliability of serological identification of sugarcane mosaic potyvirus and sugarcane yellow leaf luteovirus from cane stalk juice. Sugar Cane Int 5:18–21

Ghai M, Martin LA, McFarlane SA, van Antwerpen T, Rutherford RS (2013) Rapid diagnosis of ratoon stunting disease by loop-mediated isothermal amplification. Proc S Afr Sug Technol Ass 86:255–260

Ghai M, Singh V, Martin LA, McFarlane SA, van Antwerpen T, Rutherford RS (2014) A rapid and visual loop-mediated isothermal amplification assay to detect *Leifsonia xyli* subsp. *xyli* targeting a transposase gene. Lett Appl Microbiol 59:648–657

Gundersen DE, Lee IM, Schaff DA, Harrison NA, Chang CJ, Davis RE, Kingsbury DT (1996) Genomic diversity and differentiation among phytoplasma strains in 16S rRNA groups I (aster yellows and related phytoplasmas) and III (X-disease and related phytoplasmas). Int J Syst Bacteriol 46:64–75

Hema M, Savithri HS, Sreenivasalu P (2001) Antibody and nucleic acid probe-based techniques for detection of sugarcane streak mosaic virus causing mosaic disease of sugarcane in India. Curr Sci 81:1105–1108

Henson JM, French R (1993) The polymerase chain reaction and plant disease diagnosis. Annu Rev Phytopathol 31:81–109

Hoat TX, Bon NG, Quan MV, Hien VD, Thanh ND, Dickinson M (2012) Detection and molecular characterization of sugarcane grassy shoot phytoplasma in Vietnam. Phytoparasitica 40: 351–359

Honeycutt RJ, Sobral BWS, McClelland M (1995) tRNA intergenic spacers reveal polymorphisms diagnostic for *Xanthomonas albilineans*. Microbiology 141:3229–3239

Irey MS, Baucum LE, Derrick KS, Manjunath KL, Lockhart BE (1997) Detection of the luteovirus associated with yellow leaf syndrome (YLS), by a reverse transcriptase polymerase chain reaction and incidence of YLS in commercial varieties in Florida. Proc. ISSCT 5th Pathology and 2nd Molecular Biology Workshop

Jain RK, Rao GP, Varma A (1998) Present status of management of sugarcane mosaic virus. In: Hadidi A, Khetrapal RK, Koganezawa H (eds) Plant virus disease control. APS Press, St Paul, pp 495–506

Karuppaiah R, Viswanathan R, Ganesh Kumar V (2013) Genetic diversity of sugarcane bacilliform virus isolates infecting *Saccharum* spp. in India. Virus Genes 46:505–516

Kuniata LS, Young GR, Pais E, Jones P, Nagaraja H (1994) Preliminary observations of *Eumetopina* sp. (Hemiptera: Delphacidae) as a vector of Ramu stunt disease of sugarcane in Papua New Guinea. J Aust Entomol Soc 33:185–186

Lehrer AT, Schenck S, Pitch MMM, Moore PH, Komor E (2001) Distribution and transmission of sugarcane yellow leaf virus (SCYLV) in Hawaii and its elimination from seed cane. Proc Int Soc Sugar Cane Technol 14:439–443

Liu J, Xu L, Guo J, Chen R, Grisham MP, Que Y (2013) Development of loop-mediated isothermal amplification for detection of *Leifsonia xyli* subsp. *xyli* in sugarcane. Biomed Res Int 2013:357692. doi:10.1155/2013/357692

Lockhart BEL, Autrey LJC (1991) Mealy bug transmission of sugar cane bacilliform and sugar cane clostero-like viruses. 3rd ISSCT Sugar Cane Pathological Workshop. p 17

Malathi P, Viswanathan R (2012) Functional genomics based approaches in analysis of pathogenicity genes in *Colletotrichum*. In: Singh HP, Chowdappa P, Chakraborty BN, Podile AR (eds) Molecular approaches for plant fungal diseases management. Westville Publishing House, Delhi, pp 96–107

Malathi P, Arunkumaran A, Kaverinathan K, Viswanathan R (2012) Molecular based diagnostics of Colletotrichum falcatum causing sugarcane red rot. In: Proceedings of the International Conference on "Phyllosphere Biology", Amity University, Uttar Pradesh, 15–17 Feb 2012, p 08.

Marcone C (2002) Phytoplasma diseases of sugarcane. Sugar Technol 4:79–85

Matsumoto T, Lee CH, Teng WS (1968) Studies on white leaf disease of sugarcane-transmission by *Epitettix hiroglyphicus* Matsumura. Plant Prot Bull (Taiwan) 10:3–9

Matsuoka S, Meneghin S (2000) Yellow leaf syndrome and alleged pathogens: a casual but not a causal relationship. Proc Int Soc Sugar Cane Technol 24:382–388

Mishra S, Singh D, Tiwari AK, Lal M, Rao GP (2010) Elimination of sugarcane mosaic virus and sugarcane streak mosaic virus by tissue culture. Sugar Cane Int 28:119–122

Moonan F, Molina J, Mirkov TE (1999) Sugarcane yellow leaf virus: an emerging virus that has evolved by recombination between luteoviral and poleroviral ancestors. Virology 269:156–171

Muller E, Dupuy V, Blondin L, Bauffe F, Daugrois J-H, Nathalie L, Iskra-Caruana ML (2011) High molecular variability of sugarcane bacilliform viruses in Guadeloupe implying the existence of at least three new species. Virus Res 160:414–419

Nakashima K, Chaleeprom W, Wongkaew P, Sirithorn P (1994) Detection of mycoplasma-like organisms associated with white leaf disease of sugarcane in Thailand using DNA probes. Jpn Int Res Cent Agric Sci 1:57–67

Nakashima K, Hayashi T, Chaleeprom W, Wongkeaw P, Sirithorn P (1996) Complex phytoplasma flora in Northeast Thailand as revealed by 16S rDNA analysis. Ann Phytopathol Soc Jpn 62:57–60
Nasare K, Yadav A, Singh AK, Shivasharanappa KB, Nerkar YS, Reddy VS (2007) Molecular and symptom analysis reveal the presence of new phytoplasmas associated with sugarcane grassy shoot disease in India. Plant Dis 91:1413–1418
Pan YB, Grisham MP, Burner DM (1997) A polymerase chain reaction protocol for the detection of *Xanthomonas albilineans*, the causal agent of sugarcane leaf scald disease. Plant Dis 81:189–194
Pan YB, Grisham MP, Burner DM (1998) A polymerase chain reaction protocol for the detection of *Clavibacter xyli* subsp. *xyli*, the causal bacterium of sugarcane ratoon stunting disease. Plant Dis 82:285–290
Patil AS (2004) Bacterial spindle rot disease of sugarcane. In: Rao GP, Saumtally AS, Rott P (eds) Sugarcane pathology. Volume 3: bacterial and nematode diseases. Science Publishers, Enfield, pp 143–151
Peralta EL, Chinea A, Ortega E, Arocha Y, Piñón D (1999) Informe Final Proyecto 003-00-111. 'Contribución al Conocimiento del Síndrome del Amarilleamiento Foliar de la Caña de Azúcar (YLS) en Cuba'. National Cuba Science Academy, Cuba
Ramesh SA, Leonard BE, Malathi P, Viswanathan R (2012) A mini-review on smut disease of sugarcane caused by Sporisorium scitamineum. In: John KM (ed) Botany. InTech Publishers, Croatia, pp 107–128. ISBN 978-953-51-0355-4
Rao GP (2014) Characterization of viruses and phytoplasmas associated with sugarcane crops in India. In: Srivastava S, Swapna M, Kumar S (eds) Compendium of lectures: ICAR sponsored winter school on 'novel genomic tools and breeding approaches for sugar crops improvement. IISR, Lucknow, pp 209–228
Rao GP, Dhumal KN (2002) Grassy shoot disease of sugarcane. In: Singh SB, Rao GP, Easwaramoorthy S (eds) Sugarcane crop management. SCI Tech Publishing, LLC, Houston, pp 208–222
Rao GP, Singh M, Rishi N, Bhargava KS (2002) Century status of sugarcane virus diseases research in India. In: Singh SB, Rao GP, Easwaramoorthy S (eds) Sugarcane crop management. SCI Tech Publishing, LLC, Houston, pp 223–225
Rao GP, Srivastava S, Gupta PS, Sharma SR, Singh A, Singh S, Singh M, Marcone C (2008) Detection of sugarcane grassy shoot phytoplasma infecting sugarcane in India and its phylogenetic relationships to closely related phytoplasmas. Sugar Tech 10:74–80
Rao GP, Sharma SK, Singh D, Arya M, Singh P, Baranwal VK (2014) Genetically diverse variants of Sugarcane bacilliform virus Infecting sugarcane in India and evidence of a novel recombinant badnavirus variant. J Phytopathol. doi:10.1111/jph.12270
Ricaud E, Autrey LJE (1989) Gumming disease. In: Ricaud BC, Egan BT, Gillaspie Jr, Hughes CG (eds) The diseases of sugarcane: major diseases. Elsevier Science Publishers BV, Amsterdam, pp 21–38
Ricaud C, Ryan CC (1989) Leaf scald. In: Ricaud BC, Egan BT, Gillaspie Jr, Hughes CG (eds) The diseases of sugarcane: major diseases. Elsevier Science Publishers BV, Amsterdam, pp 39–58
Rishi N (2009) Significant plant virus diseases in India and a glimpse of modern disease management technology. J Gen Plant Pathol 75:1–18
Rishi N, Chen CT (1989) Grassy shoot and white leaf disease. In: Ricaud BC, Egan BT, Gillaspie Jr, Hughes CG (eds) The diseases of sugarcane: major diseases. Elsevier Science Publishers BV, Amsterdam, pp 289–300
Rodriguez-Lema E, Rodriguez D, Fernandez E, Acevedo R, Lopez D (1985) Reporte de un nuevo virus de la cana de azucar. Cienc Agric 23:130
Rott P, Davis MJ (2000a) Leaf scald. In: Rott P, Bailey RA, Comstock JC, Croft BJ, Saumtally AS (eds) A guide to sugarcane diseases. CIRAD-ISSCT, France, pp 38–44. ISBN 2-87614-386-0
Rott P, Davis MJ (2000b) Red stripe (top rot). In: Rott P, Bailey RA, Comstock JC, Croft BJ, Saumtally AS (eds) A guide to sugarcane diseases. CIRAD, ISSCT, France, pp 58–62

Rott P, Mohamed IS, Klett P, Soupa D, de Saint-Albin A, Feldmann P, Letourmy P (1997) Resistance to leaf scald disease is associated with limited colonization of sugarcane and wild relatives by *Xanthomonas albilineans*. Phytopathology 87:1202–1213

Rott P, Bailey RA, Comstock JC, Croft BJ, Saumtally AS (2000) A guide to sugarcane diseases. CIRAD-ISSCT, France, ISBN 2-87614-386-0

Saumtally AS (2000) Mottled stripe. In: Rott P, Bailey RA, Comstock JC, Croft BJ, Saumtally AS (eds) A guide to sugarcane diseases. CIRAD, ISSCT, France, pp 45–48

Scagliusi S, Lockhart B (2000) Transmission, characterization and serology of a luteovirus associated with yellow leaf syndrome of sugarcane. Phytopathology 90:120–124

Schaad NW, Frederick RD (2002) Real-time PCR and its application for rapid plant disease diagnostics. Can J Plant Pathol 24:250–258

Schenck S (1990) Yellow leaf syndrome- a new disease of sugarcane. Annual report, Experimental station, Hawaiian Sugar planters Association, 1990, p 38

Schenck S, Hu J, Lockhart B (1997) Use of a tissue blot immunoassay to determine the distribution of sugarcane yellow leaf virus in Hawaii. Sugar Cane 4:5–8

Sdoodee R (2000) Recent studies on white leaf and grassy shoot phytoplasma of sugarcane. In: Sugarcane pathology, Vol. II: virus, phytoplasma disease. Science Publishers, Enfield, pp 235–244

Singh V, Baitha A, Sinha OK (2002) Transmission of grassy shoot disease of sugarcane by a leaf-hopper (*Deltocephalus vulgaris* Dash and Viraktamath). Ind J Sugar Cane Technol 17:60–63

Smith GR, Van De Velde R (1994) Detection of sugarcane mosaic virus and fiji disease virus in diseased sugarcane using the polymerase chain reaction. Plant Dis 78:557–561

Srivastava S, Sinha OK (2009) Molecular indexing of diseases in seed cane. In: Shrivastava AK, Solomon S, Chandra A, Singh P, Rai RK, Sawnani A, Shukla SP (eds) Quality seed cane in sugarcane. Indian Institute of Sugarcane Research, Lucknow, pp 107–109

Srivastava S, Singh V, Gupta PS, Sinha OK (2003) Detection of phytoplasma of GSD of sugarcane based on PCR assay using ribosomal RNA sequences. In: Singh V, Sinha OK (eds) National seminar on emerging trends in plant disease research and management. Indian Institute of Sugarcane Research, Lucknow, p 37

Srivastava S, Singh V, Gupta PS, Sinha OK, Baitha A (2006) Nested-PCR assay for detection of sugarcane grassy shoot phytoplasma in the leafhopper vector *Deltocephalus vulgaris* Dash & Viraktamath: a first report. Plant Pathol 55:25–28

Suma S, Jones P (2000) Ramu stunt. In: Rott P, Bailey R, Comstock JC, Croft B, Saumtally S (eds) A guide to sugarcane diseases. CIRAD/ISSCT, Montpellier, pp 226–230

Thompson N, Choi Y, Randles JW (1998) Sugarcane striate mosaic disease: development of a diagnostic test. Proceedings of the 7th International Congress of Plant Pathology, 3.3.21. Edinburgh

Vega J, Scagliusi SMM, Ulian E-C (1997) Sugarcane yellow leaf disease in Brazil: evidence of association with a luteovirus. Plant Dis 81:21–26

Viswanathan R (2002) Sugarcane yellow leaf syndrome in India: incidence and effect on yield parameters. Sugar Cane Int 5:17–23

Viswanathan R (2012) Sugarcane diseases and their management. Sugarcane Breeding Institute, Coimbatore, p 140

Viswanathan R, Rao GP (2011) Disease scenario and management of major sugarcane diseases in India. Sugar Tech 13:336–353

Viswanathan R, Alexander KC, Garg ID (1996) Detection of sugarcane bacilliform virus in sugarcane germplasm. Acta Virol 40:5–8

Viswanathan R, Padmanaban P, Mohanraj D, Viswanathan R, Padmanaban P, Mohanraj D, Nallathambi P (1998) Immunological techniques for the detection of leaf scald disease of sugarcane. Ind Phytopathol 51:187–189

Viswanathan R, Balamuralikrishnan M, Karuppaiah R (2007) Association of sugarcane mosaic virus and sugarcane streak mosaic virus with sugarcane mosaic in India. Sugar Cane Int 25:10–18

Viswanathan R, Balamuralikrishnan M, Karuppaiah R (2008) Characterization and genetic diversity of sugarcane streak mosaic virus causing mosaic in sugarcane. Virus Genes 36:553–564

Viswanathan R, Karuppaiah R, Malathi P, Ganesh Kumar V, Chinnaraja C (2009) Diagnosis of sugarcane yellow leaf virus in asymptomatic sugarcane by RT-PCR. Sugar Tech 11:368–372

Viswanathan R, Poongothai M, Malathi P (2011) Pathogenic and molecular confirmation of *Fusarium sacchari* causing wilt in sugarcane. Sugar Tech 13:68–76

Wang ZK, Comstock JC, Hatziloukas E, Schaad NW (1999) Comparison of PCR, BIO-PCR, DIA, ELISA and isolation on semiselective medium for detection of *Xanthomonas albilineans*, the causal agent of leaf scald of sugarcane. Plant Pathol 48:245–252

Ward E, Foster SF, Fraaije AB, Mccartney HA (2004) Plant pathogen diagnostics: immunological and nucleic acid-based approaches. Ann Appl Biol 145:1–16

Wongkaew P, Poosittisak S (2015) Diagnosis of sugarcane white leaf disease using the highly sensitive DNA based voltammetric electrochemical determination. AJPS 5:2256–2268

Wongkaew P, Hanboonsong Y, Sirithorn P (1997) Differentiation of phytoplasmas associated with sugarcane and graminaceous weed white leaf disease and sugarcane grassy shoot by RFLP and sequencing. Theor Appl Genet 95:660–663

Wongwarat T, Sakuanrungsirikul S, Theerakulpisut P (2011) Effective methods of preserving SCWL-diseased sugarcane leaves for genomic DNA extraction and molecular detection of phytoplasma. Afr J Biotechnol 10:10871–10876

Physiological and Molecular Signalling Involved in Disease Management Through *Trichoderma*: An Effective Biocontrol Paradigm

Veena Pandey, Alok Shukla, and J. Kumar

Introduction

Diseases are the most important biotic causes for low crop yields. The disruption in normal physiology of plant, usually with some kind of negative effect on survival or fitness by some causal agent is known as plant disease. These infectious causal agents include bacteria, fungi, oomycetes, virus, nematodes, phytoplasmas or parasitic higher plants. These pathogens cause disease either by secreting enzymes, which catalyze the breakdown of host tissues or often benefit by producing toxins, which kill the tissue in advance of enzymatic degradation. Viruses are able to force the plant to produce pathogen gene products from pathogen genetic material and disrupt their function. The interference of these agents with one or more of a plant's essential physiological or biochemical systems elicits characteristic pathological symptoms, and cause diseases. Though plants carry some inherent disease resistance, but there are numerous examples of devastating plant disease impacts, as well as recurrent severe plant diseases, which severely affects yield. Plant diseases often substantially reduce quality and quantity of agricultural commodities and post-harvest infestation can affect the health of humans and livestock, especially if the contaminating organism produces toxic residues on consumable products (Cheeke 1995).

V. Pandey • A. Shukla
Department of Plant Physiology, College of Basic Sciences and Humanities,
G.B. Pant University of Agriculture and Technology, Pantnagar,
Uttarakhand 263145, India
e-mail: aloks99@yahoo.com

J. Kumar (✉)
Department of Plant Pathology, College of Agriculture,
G.B. Pant University of Agriculture and Technology,
Pantnagar, Uttarakhand 263145, India
e-mail: jkumar56@gmail.com

P. Kumar et al. (eds.), *Current Trends in Plant Disease Diagnostics and Management Practices*, Fungal Biology, DOI 10.1007/978-3-319-27312-9_14

It is well accepted that agricultural production must be increased considerably in the future to meet the food and feed demands of a rising population. The control of these plant pathogens or diseases is thus crucial to the reliable production of food. Loss of crops from plant diseases may result in hunger and starvation, especially in less developed countries where access to disease-control methods is limited (Sewariya et al. 2012). Prevention and early diagnosis are critical to limiting damage by plant pathogens. Thus, plant diseases need to be controlled to maintain the quality and abundance of food, feed, and fiber produced by growers around the world. Crop losses due to these harmful organisms/diseases can be prevented, or reduced, by crop protection measures.

Crop protection plays a key role in safeguarding crop productivity against competition from weeds, animal pests, pathogens and viruses (Oerke and Dehne 2004). Though there are many chemical products that are available for the crop protection, but they are considered as a major pollutant in the environment, and responsible for disturbing the ecological harmony of the planet. Soil fumigation is one of the effective measures in eradicating the resident inoculum but is expensive and poses environmental and safety concerns. Another effective way of crop protection is the use of biological control or biocontrol agents (BCA). They are perceived to have specific advantages over synthetic fungicides/pesticides due to their increased level of safety, reduced probability of resistance development and minimal environmental impacts. This is a method in which natural enemies of pests or pathogens are used to eradicate or control target population. The induction of plant resistance using non-pathogenic or incompatible micro-organisms is also a form of biological control.

Biological Control

The terms "biological control" has been used in different fields of biology, most notably entomology and plant pathology. In entomology, it has been used to describe the use of live predatory insects, entomopathogenic nematodes, or microbial pathogens to suppress populations of different pest insects. In plant pathology, the term applies to the use of microbial antagonists to suppress diseases as well as the use of host specific pathogens to control weed populations. In both fields, the organism that suppresses the pest or pathogen is referred to as the biological control agent (BCA) (Pal and McSpadden Gardener 2006). More broadly, biocontrol is the use of specific microorganisms that interfere with plant pathogens and pests, and sustain organisms useful to human. The success of biocontrol depends on the nature of antagonistic properties and the mechanism of action acquired by the organisms used. There are a number of bacterial and fungal isolates that have been reported as biocontrol agents although the fungus *Trichoderma* clearly dominates.

Trichoderma as Biological Control Agent

Trichoderma species, belonging to division Ascomycota and class Sordariomycetes, are beneficial plant symbionts that act as natural biocontrol agents and antagonist of several important phytopathogenic organisms. These beneficial fungi are called antagonists, because they infect, attack or compete against (antagonize) the pathogens that cause plant diseases. Faster metabolic rates, anti-microbial metabolites, and physiological conformation are key factors which chiefly contribute to antagonism of these fungi (Verma et al. 2007). They are free-living that are worldwide in occurrence and are characterized by green conidia and possess repetitively branched conidiophore structure. These fungi colonize woody and herbaceous plant materials, in which the sexual teleomorph (genus Hypocrea) has most often been found. The beneficial activities of *Trichoderma* as a biocontrol agent (BCA) have been known since 1930 and since then there have been extensive efforts to use them for plant disease control. Weindling in 1932, for the first time implicated the role of *Trichoderma lignorum* in the biological control of citrus seedling disease caused by *Rhizoctonia solani*. Since this pioneering work, several reports on successful biocontrol by *Trichoderma* spp. have accumulated. Among several species, *T. harzianum, T. viride and T. virens* are the most widely used for biological control.

Trichoderma species are soil borne fungi and show significant activity against a wide range of plant pathogenic fungi (Elad et al. 1982). Although they have been considered soil inhabitants, based on in situ diversity studies using a taxon specific metagenomic approach, Friedl and Druzhinina (2012) suggested that only a relatively small number are adapted to soil as a habitat. Their presence in the root zone creates a symbiotic relationship with the host plant, and can cause the plant to generate an immune response. Properly selected strains interact with the plant by colonizing roots, establishing chemical communication and systemically altering the expression of numerous plant genes. They establish long-lasting colonizations of root surfaces of plant and penetrate into the epidermis and a few cells below. They are reported effective in controlling root rots/wilt complexes and foliar diseases in several crops (Singh et al. 2004, 2006; Zaidi and Singh 2004a, b). They have been most commonly used for biocontrol agent against plant pathogen that contain chitin and glucans in their cell wall. In addition, the non-biocontrol agent *Trichoderma reesei* is a biotechnological factory for the production of secreted cellulases, and a model for basic studies on protein secretion (Peterson and Nevalainen 2012; Saloheimo and Pakula 2012).

Mechanisms Involved in Biocontrol Activities

It is well known that one particular strain of *Trichoderma* may be differentially effective against different plant pathogens. Since a variety of mechanisms of action may be brought into play within the interaction of *Trichoderma* with different fungi and with the plant (Harman 2000; Howell 2003), it is conceivable that differences in the efficacy of one particular strain against different plant pathogens might be due to its potential to express high levels of one or another mechanism of action. The mechanism of biocontrol by *Trichoderma* includes direct competition with the target organism, antibiosis, parasitism of the target organism and induced resistance of the host plant.

Mycoparasitism

It is a complex process that finally results in dissolution of the target cell wall by lytic enzymes released by *Trichoderma*. A typical mycoparasitic interaction involves different steps, such as sensing of the host/prey fungus, attraction, attachment, coiling around and lysis brought about by hydrolytic enzymes, in many cases, in conjunction with secondary metabolites. Therefore, the main component of antagonism of *T. harzianum* is the extracellular secretion of chitinases, β-1,3-glucanases and proteases (Geremia et al. 1993). Chitin and β -1, 3 glucan are the main structural components of the fungal cell wall and thus, chitinases, and β -1, 3 glucanases have been proposed as the key enzymes in the degradation of cell wall during mycoparasitism against phytopathogenic fungi. The proteolytic activity of *T. harzianum* is a prerequisite for the lysis of the protein matrix of the pathogen cell wall, and for inactivation of the hydrolytic enzymes secreted by the pathogen, which decreases its pathogenicity. Trypsin and chymotrypsin, β -1,4-N-acetyl-glucosaminidase (NAGase) are also secreted by *Trichoderma*.

Trichoderma species with a high potential for the secretion of hydrolytic enzymes can be obtained through the transformation by insertion of genes which encode lytic enzymes. However, this can affect the production of antibiotics and other enzymes involved in the mycoparasitism, as well as the growth rate and colonization properties of the BCA (Flores et al. 1997). But best alternative to transformation is the use of *Trichoderma* isolates obtained naturally from different sources.

Competition for Nutrient and Space

This process is considered to be an indirect interaction whereby pathogens are excluded by depletion of a food base or by physical occupation of site (Lorito et al. 1994). Competition for infection site has been believed to have an important role in disease suppression. The competition for nutrient also plays a role in biological control of plant diseases. Generally, *Trichoderma* species are considered to be aggressive competitors, grow very fast and rapidly colonize substrates to exclude pathogens. It may grow faster or use its food source more efficiently than the pathogen, thereby crowding out the pathogen and taking over. *Trichoderma* also has a superior capacity to mobilize and take up soil nutrients compared to other organisms. The efficient use of available nutrients is based on the ability of

Trichoderma to obtain ATP from the metabolism of different sugars, such as those derived from polymers wide-spread in fungal environments: cellulose, glucan and chitin and others, all of them rendering glucose (Chet et al. 1997). The efficiency of glucose transport system may be crucial in competition, as supported by the isolation of a high affinity glucose transporter, Gtt1, in *Trichoderma harzianum* CECT 2413. This strain is present in environments very poor in nutrients, and it relies on extracellular hydrolases for survival. Gtt1 is only expressed at very low glucose concentrations, i.e. when sugar transport is expected to be limiting in nutrient competition. This suggests the role of Gtt1, allowing the fungus to transport sugar rapidly into the cells (Delgado-Jarana et al. 2003).

Antibiosis

During antibiosis, Trichoderma releases products that slow down or kill the pathogens in the vicinity. Studies revealed that antimicrobial metabolites produced by *Trichoderma* are effective against a wide range of fungal phytopathogens e.g., *Fusarium oxysporum*, *Rhizoctonia solani*, *Curvularia lunata*, *Bipolaris sorokiniana*, *Colletotrichum lagenarium*, *Colletotrichum acutatum* and *Colletotrichum gloeosporioides* (Svetlana et al. 2010).

Trichoderma species are known to produce a number of such antibiotics, as trichodermin, viridian and harzianolide (Simon and Sivasithamparam 1988; Schirmbock et al. 1994). Claydon et al. (1987) reported antifungal properties of volatile metabolites (alkyl pyrones) produced by *T. harzianum*. Pandey and Upadhyay (1997) have also reported the effectiveness of diffusible volatile metabolites of *T. harzianum* and *T. viride* in vitro. Dal Bello et al. (1997) studied the volatile compounds produced by *Trichoderma hamatum* against various phytopathogenic fungi and suggested the inhibitory volatiles of *Trichoderma hamatum* as one of the possible mechanism of biological control. It is also found that there is large variety of volatile secondary metabolites produced by *Trichoderma* such as ethylene, hydrogen cyanide, aldehydes and ketones which play an important role in controlling the plant pathogens (Vey et al. 2001). Antibiosis occurs during interactions with other microorganisms involving low molecular weight diffusible volatile and nonvolatile toxic metabolite compounds or antibiotics like harzianic acid, alamethicins, tricholin, peptaibols, antibiotics, 6-penthyl-pyrone, massoilactone, viridin, gliovirin, glisoprenins, heptelidic acid and others (Sharma et al. 2012a) that inhibit the growth of other microorganisms.

Induced Systemic Resistance in Plants

The term induced resistance is a generic term for the induced state of resistance in plants triggered by biological or chemical inducers, which protects nonexposed plant parts against future attack by pathogenic microbes and herbivorous insects

(Kuc 1982). Specific strains of *Trichoderma* colonize and penetrate plant root tissues and initiate a series of morphological and biochemical changes in the plant, considered to be part of the plant defense response, which subsequently leads to induced systemic resistance (Bailey and Lumsden 1998). Yedidia et al. (1999) provided evidence that *T. harzianum* may induce systemic resistance mechanisms in cucumber plants. *Trichoderma* releases a variety of compounds that induce localised and systemic resistance in plants and protect host plant. Several studies have shown that root colonization by *Trichoderma harzianum* results in increased level of plant enzymes, including various peroxidases, chitinases, β -1, 3-glucanases, lipoxygenase-pathway hydro peroxide lyase and compounds like phytoalexins and phenols to provide durable resistance against stress (Gachomo and Kotchoni 2008). Induction of immunity in host plants is the consequence of interactions between different elicitors released by microorganism and plant receptors, leading to the activation of signalling pathways, triggering physiological and biochemical changes in plants (Contreras-Cornejo et al. 2011; Mastouri et al. 2010). Hormones also play a major regulatory role in these interconnected signaling pathways (Pieterse et al. 2012).

Management of Soil Born Diseases

Soils contain diverse communities of microscopic organisms that are capable of damaging plants and create major economical losses in many important crops. These soil organisms include fungi, bacteria, viruses, nematodes and protozoa. Some pathogens of the above ground parts of plants (leaves, stems) survive in the soil at various stages in their life cycles. Therefore, a soil phase of a plant pathogen may be important, even if the organism does not infect roots. In particular, soil-borne pathogens cause important losses, fungi being the most aggressive. *Trichoderma spp.* are reported to inhibit a number of soil borne bacteria like *Escherichia coli*, *Klebsiella*, *Staphylococcus aureus*, *Pseudomonas aeruginosa*, *Micrococcus luteus* and fungi like *Rhizoctonia*, *Alternaria*, *Pythium*, *Phytophthora*, *Sclerotinia*, *Sclerotium*, *Fusarium spp.*, *Macrophomina* etc. and root knot nematode like *Meloidogyne* spp., and significantly decreased infection and disease (Claydon et al. 1987). Kumar et al. (2015) evaluated eight species of *Trichoderma* viz. *T. viride*, *T. harzianum*, *T. atroviride*, *T. longibrachiatum*, *T. koningii*, *T. asperllum and T. virens* in vitro against the most widely occurring soil inhabiting plant pathogens viz., *Fusarium oxysporum* f. sp. *ciceri*, *Alternaria solani*, *Phytopathora infestans*, *Pythium aphanidermatum*, *Sclerotium rolfsii*, *Bipolaris sorokiyana* and *Rhizoctonia solani*, and concluded that all the *Trichoderma* spp. restricted the growth of all seven phytopathogen in its own way. But *T. Reesei*, *T. harzianum and T. viride* showed highest antagonistic potential.

Management of Fungal Diseases

Most of the studies on *Trichoderma* species have been conducted with respect to their activity as biocontrol agents against fungal pathogens. They are known to control numerous fungal diseases, such as those caused by *Pythium ultimum* (Naseby et al. 2000), *Sclerotinia sclerotiorum* (Inbar et al. 1996), and *Fusarium oxysporum* (Sivan and Chet 1993) etc. and considered as an important commercial source of several enzymes and as biofungicides. As an antagonist, *Trichoderma* spp. produce antibiotics and antifungal toxic metabolites and also inhibit fungal pathogens by secreting enzymes like glucanase, cellulase, chitinase, protease etc., which disintegrate the cell wall of pathogen and they also act as a competitor for mineral nutrients.

Biocontrol Against *Fusarium*

Diseases caused by *Fusarium* are one of the major yield limiting factors of agricultural crops. *Fusarium* spp. are known to cause seed rot, damping off, wilting and root rot diseases resulting in severe losses to a variety of crop plants (Miller 1994). In several crops *Fusarium* diseases are generally controlled by fumigation with methyl bromide (Awuah and Lorbeer 1991). But due to environmental issue, there emerge need for some environmental friendly and efficient alternative.

Fakhrunnisa et al. (2006) suggested that *Trichoderma hamatum* could be efficiently used in the biological control of the diseases caused by *Fusarium* spp. Many species including *T. viride* has been observed to be a good BCA with an efficient and effective control on diseases caused by *Fusarium* (John et al. 2010; Basak and Basak 2011). Other *Trichoderma spp.* viz. *T. atroviride*, *T. harzianum*, *T. longibrachiatum, and T. virens* were also found to reduce disease incidence and inhibited mycelial growth of various *Fusarium sp.* viz *F. oxysporum* (Muriungi et al. 2013; Sundaramoorthy and Balabaskar 2013), *F. solani* (Morsy et al. 2009), *F. culmorum and F. graminearum* (Matarese et al. 2012). Sharma (2011) categorized the sequence of events during the interaction between *Trichoderma* and *Fusarium* as pre-contact antagonistic interaction, chemo-attractive intermediate phase and, finally, parasitic interaction. *Trichoderma* species either formed hook or bunch like structures around the hyphae of FOL (*Fusarium oxysporum* f. sp. *lentis*) before penetration, or sometimes entered them directly. Microscopically, the hyphal interaction indicated that antagonistic hyphae of *Trichoderma* coiled around the hyphae of pathogen, denatured the mycelia and finally killed them (Dolatabadi et al. 2012).

During an in vitro biocontrol test, *Trichoderma* showed mycoparasitism and destructive control against fungal pathogens *Fusarium oxysporum* f. sp. *adzuki*, infecting soybean and simultaneously increased the growth and yield of *Fusarium* infected plants treated with *Trichoderma* and proved as avirulent opportunistic symbiont in rhizosphere of soybean plant (John et al. 2010). Kataoka et al. 2010 showed that application of *T. viride* to the soil in the greenhouse resulted in consistent and effective control of yellow disease of *Brassica campestris* caused by *Fusarium oxysporum*. Rajeswari and Kannabiran (2011) observed that culture filtrate of *T. viride* and *T. harzianum* inhibited conidial germination and mycelial growth of *Fusarium oxysporum* infecting *Arachis hypogaea* and concluded that inhibition is due to the volatile and non volatile metabolites and cell wall degrading enzymes produced by *Trichoderma* spp. Inhibition of growth of *F. solani* sp. *dalbergiae* by the antagonist fungi, *T. viride* and *T. harzianum* was observed by Basak and Basak (2011). They concluded that the pathogenic fungus was successfully destroyed by *T. viride* and *T. harzianum*. Antagonist *Trichoderma* sp. penetrated and coiled the wilt fungus and these were direct antagonistic actions of T. *viride* and *T. harzianum*. *T. viride* produced viridin which is antifungal and gliotoxin, an antibiotic which is antibacterial and antifungal. Ru and Di (2012) examined 146 isolates of *Trichoderma* spp. by means of dual culture method for antagonism against *Fusarium sambucinum*, which causes potato dry rot. Via screening, ten isolates showed the best antagonism against *F. sambucinum* and they were identified as *T. hazianum*, *T. longibrachiatum*, *T. atroviride, and T. virens*. Among these isolates, *D-3-1* (*T. longibrachiatum*) showed the strongest inhibition of the growth of *Fusarium sambucinum*.

Meraj-ul-Haque and Nandkar (2012) examined seven isolates *of Trichoderma spp*. viz. *T. harzianum Rifai*, *T. hamatum Rifai*, *T. longibrachiatum Rifai*, *T. atroviride Karsten*, *T. viride Pers*, *T. ressei* Simmons and *T. virens* Miller, for antagonism against *Fusarium oxysporum* f.sp. *lycopersici*, which causes tomato damping-off. Among these isolates, *T. virens* showed the strongest inhibition of the growth *of Fusarium* oxysporum f.sp. lycopersici. *T. viride* also exerted a significant effect on the growth of *Fusarium circinatum* causing pitch canker disease, in the in vitro assay, reducing the length of the pathogen colony by half (Martínez-Álvarez et al. 2012).

Antagonistic variability of various isolates of *T. atroviride* revealed significant suppression in the radial growth of three legume pathogens *Fusarium oxysporum* f.sp. *ciceri*, *F. oxysporum* f.sp. *lentis and Fusarium udum* was observed by Singh et al. (2013). They found that the efficacy of these strains against various pathogen species is different and concluded that antagonistic and molecular variability exist among eight isolates of *T. atroviride*. Bhagat et al. (2013) evaluated 12 isolates of three species viz. *T. harzianum*, *T. viride and T. hamatum*, *against Fusarium oxysporum* f. sp. *lycopersici* for their biocontrol potential under in vitro and field conditions. They revealed that all isolates of *Trichoderma* spp. have significantly inhibited mycelial growth of F.o. f. sp. lycopersici as compared to control but the isolate Th-CARI-50, TvCARI-85, Th-CARI-61, Tv-CARI-100, Tv-CARI-73 were most efficient in per cent inhibition of test pathogen. The seed and soil application of *Trichoderma* spp. was found to be most effective in the reduction of disease incidence of fusarial wilt of tomato under both greenhouse and field conditions than that of either seed or soil application alone.

Bernal-Vicente et al. (2009) developed different formulations of *Trichoderma harzianum* T-78 on different carriers: liquids (spore suspension, guar gum and carboxymethylcellulose) and solids (bentonite, vermiculite and wheat bran) against

Fusarium wilt on melon plants and concluded that most effective treatments were the solid treatments bentonite and superficial vermiculite, each gave the lowest percentage of infected plants and the greatest *T. harzianum* population.

Thus, most of the *Trichoderma* isolates showed considerably good antibiosis and parasitism against *Fusarium*. The isolate of *Trichoderma* which can overcome the inhibition posed by the pathogen and parasitize it in a short span of time can be considered as the most efficient biocontrol agent. *Trichoderma* isolate T1 (*T. atroviride*) showed quicker action as compared to the others. After reaching the verge of inhibition zone posed by *Fusarium* in dual culture plates, this agent took only about 2 days to parasitize the pathogen (Sharma 2011). Thus, it is necessary to find most aggressive strains of *Trichoderma* which can parasitize the fungus efficiently and quickly so as to completely control the disease.

Various *Trichoderma* isolates produce different compounds for antibiosis against same or different pathogen. Volatile metabolites produced by *Trichoderma* are considered more effective against *Fusarium* as compared to non-volatile compounds. Role of diffusible volatile compounds produced by *T. viride* and *T. harzianum* in the inhibition of germination and mycelial growth of *Fusarium oxysporum* in vitro was reported by Pandey and Upadhyay (1997). Volatile metabolites produced by *T. harzianum* caused maximum radial growth inhibition of *Fusarium oxysporum f. sp. lentis* (FOL), followed by *T. viride* in dual culture plate. *Trichoderma* species differentially limited the colonical growth of the pathogen, overgrew the pathogen colony and parasitized it (Dolatabadi et al. 2012).

Some members of the *F. oxysporum* species complex are known to synthesize a number of biologically active compounds like enniatin (EN), beauvericin (BEA) and fusaric acid (FA) (Bacon et al. 2006; El-Hasan et al. 2008; Meca et al. 2010; Wang and Xu 2012). An inhibitory effect of EN against *Trichoderma harzianum* Rifai has been reported by Meca et al. (2010). As a result, the efficacy of biocontrol of *F. oxysporum* by *Trichoderma* decreases. Various strategies can be used to improve the efficacy of biological control agents. One of these is the genetic manipulation of wild-type strains. For instance, strains of *Trichoderma* transformed to overexpress hydrolytic enzymes have been shown to be better biocontrol agents than their corresponding parental strains (Herrera-Estrella and Chet 2003). Marzano et al. (2013) showed that FA, phytotoxin involved in pathogenesis of *Fusarium* wilts, has a strong inhibitory effect to *T. harzianum* strain ITEM 908 growth. They improved the tolerance to FA of ITEM 908 and its biocontrol performance *against F. oxysporum* f. sp. *lycopersici* through UV mutation.

Control of *Botrytis*

B. cinerea is a ubiquitous pathogen that causes severe losses in many fruits, vegetables and ornamental crops, and which can damage and even kill plants and affect the quality of the product. It is a major pathogen of grapes and greenhouse crops, on which it causes grey mould. The first biocontrol agent to be commercialized, registered and used for effectively controlling *Botrytis* diseases in greenhouse crops and in vineyards was T39 isolate of *T. harzianum* (TRICHODEX 20P) (Elad 1994; Elad et al. 1994). It was found to be effective in controlling *Botrytis cinerea* in tomato plants (Fiume and Fiume. 2006). An isolate of *T. reesei* studied by El-Naggar et al. (2008) showed a 30 % reduction in the growth of *B. cinerea*, 40.2 % in the growth of *B. fabae* and only 4 % in the growth of *B. allii* after 5 days of incubation. The antagonistic ability of *T. harzianum, T. viride and T. longibrachiatum* against *Botrytis fabae and B. cinerea*, disease-causing agents of 'chocolate spot' in bean, through inhibiting its sporulation and mycelial growth, was confirmed by Bendahmane et al. (2012).

Fiume and Fiume (2006) suggested that the antagonistic effect of *T. harzianum* isolates against *Botrytis* is based on the competition for niche and nutrients and not on a chemical aggressiveness or classic antibiosis. This antagonistic effect can also be explained by the ability of *Trichoderma* to produce volatile substances that are able to limit and even stop the development and sporulation of this pathogen (Bendahmane et al. 2012). Cheng et al. (2012) hypothesized that the mycoparasitic process of *Trichoderma spp.* against *B. cinerea* involves two steps; an initial hyphal coiling stage and a subsequent hyphal coiling stage, with different coiling rates. They identified an l-amino acid oxidase (Th-l-AAO) from *T. harzianum* ETS 323, which effectively inhibited *B. cinerea* hyphal growth, caused cytosolic vacuolization in the hyphae, and led to hyphal lysis during in vitro assay. Th-l-AAO also showed disease control against the development of *B. cinerea* on postharvest apple fruit and tobacco leaves and is capable of inducing apoptosis-like response, including the generation of reactive oxygen species in *B. Cinerea*, suggesting that Th-l-AAO triggers programmed cell death in *B. cinerea*. Bogumił et al. (2013) revealed that the ability of *T. atroviride* to supress *Botrytis* is due to the fact that *T. atroviride* produce siderophores, indole-3-acetic acid and chitinases. These enzymes are connected with mycoparasitism that is initiated against phytopathogenic fungi. Recently, Soliman et al. (2015) suggested that *T. harzianum* and *T. viride* were better in inhibiting disease incidence and improving plant defense against *Botrytis* in cucumber and concluded that application of these antagonistic fungi might be an easily applied, safe and cost effective alternative control method to control grey mold.

Control of *Colletotrichum*

Anthracnose caused by *Colletotrichum acutatum* is one of the major fungal diseases occurring worldwide. Various isolates of *Trichoderma*, including *T. harzianum* isolate T-39 from the commercial biological control product TRICHODEX, T-118, T-165 and T-166 were effective in controlling anthracnose in strawberry (Freeman et al. 2001). *Trichoderma* strains, namely *T. harzianum* (T-39), *T. atroviride* (T-161) and *T. longibrachiatum* (T-166), were evaluated in large-scale experiments using different timing application and dosage rates for reduction of strawberry anthracnose. They were effective in reduction of anthracnose disease, while the concentration of 0.8 % was superior (Freeman et al. 2004). In dual plate culture, the isolate T3 of *T. harzinum* showed the highest 89.44 % inhibition of radial growth of C. *dematium* causing anthracnose of soybean (Shovan et al. 2008).

Trichoderma species have also been applied to control *Colletotrichum* species in citrus (Moretto et al. 2001), with concomitant disease reduction. Deshmukh et al. (2010) evaluated *T. viride*, *T. harzianum*, *T. longibrachyatum*, *Gliocladium virens* (synonym, *T. virens*), by dual culture method to monitor antagonistic effect against *C. gloeosporioides* Penz. and Sacc in vitro and revealed that out of various bioagents used, *T. viride* and *T. harzianum* were able to inhibit the growth of the pathogen mycelia significantly. Ghosh and Chakraborty (2012) screened various isolates of *Trichoderma viride* and observed that T1, T10, T12 isolates have best hyperparasitic activity. Rahman et al. (2013) evaluated different *Trichoderma* strains against *C. capsici* under laboratory conditions and found *T. harzianum* as potential antagonist for inhibition of the mycelial growth, conidial germination, germ tube elongation and disease severity of *C. capsici*.

Kushwaha and Verma (2014) showed that *T. harzianum*, *T. viride*, *T. hamantum* all exibit strong antagonistic activity against *Colletotrichum spp.* causing red rot of sugarcane. In another study, in vitro antagonism test carried out between *T. viride* and *C. alienum* showed a radial growth inhibition of the pathogen by 75 % and greenhouse studies also confirmed that *T. viride* significantly controlled the pathogen at par with fungicide treatment (Liju et al. 2014). Incidence and severity level of bean anthracnose was found to be lower in seeds treated with *T. harzianum* and *T. viride* than from untreated seed, with some positive influence in treated bean seed yield (Amin et al. 2014).

The non volatile antibiotic extract of *T. viride* has the capacity to reduce the biomass and synthesis of DNA, RNA and protein of *C. capsici* (Rajathilagam and Kannabiran 2001). Ajith and Lakshmidevi (2010) investigated volatile and nonvolatile (culture filtrate) compounds from four *Trichoderma* species namely *T. saturnisporum*, *T.harzianum*, *T. viride*, *T. reesei* against *C. capsici* causing anthracnose disease in bell peppers and showed that the volatile compounds produced form all the selected *Trichoderma* species showed 30–67 % inhibition, while non-volatile compounds or culture filtrate from *Trichoderma viride* at 3–4 % concentration showed complete mycelial inhibition of the test fungi. Thus, suggesting the non-volatile secondary metabolites from *Trichoderma* species to be more effective in suppressing the mycelial growth of *C.capsici* as compared to volatile compounds. On the contrary, Tapwal et al. (2015) reported the influence of volatile and non-volatile antibiotics of *T. viride* and *T. harzianum* against *C. gloeosporioides* in dual culture technique and described the major advantage of toxic volatile metabolites as they may diffuse through air filled pores in soil and help in checking the root rot pathogen without establishing actual physical contact with pathogen.

Control of *Pythium*

Pythium damping off is a very common problem in fields and greenhouses, where the organism kills newly emerged seedlings and also cause root rot. Several studies address the application of *Trichoderma* spp. with particular emphasis on biological control of *Pythium*. Treatment with various *Trichoderma* strains increased the emergence, wet and dry shoot and root weights (g), number of lateral roots and significantly reduced the number of lesions of pea plants affected with *Pythium ultimum*. Among all strains, N47 had the greatest beneficial characteristics, as it consistently improved the growth measurements in the absence of plant pathogens and reduced the damage in presence of pathogen (Naseby et al. 2000). Aerts et al. (2002) showed that application of *Trichoderma* spp. *T. asperellum* (Biofungus), *T. harzianum* (Tri 003) and *Trichoderma* sp. (KHK) in the presence of *Pythium* spp. increased the germination percentage of tomato seeds sown in soilless growing media. *T. asperellum* strains are also antagonistic and aggressive mycoparasites of *P. myriotylum* and could reduce infection by 50 % (Mbarga et al. 2012).

John et al. (2010) assessed the efficiency of *Trichoderma viride* as biocontrol agent against *Pythium arrhenomanes* in vivo and in vitro. During in vitro test, *Trichoderma* showed mycoparasitism and destructive control against it. During pot assay, along with biocontrol activity, *Trichoderma* enhanced plant height and yield of *Pythium* infected soyabean plants. In greenhouse experiment, *Trichoderma* isolate T-105 significantly reduced the pre- and post-emergence damping-off disease incidence under artificial infection with *P. aphanidermatum*. The volatile compounds produced by *Trichoderma* were found to be important in suppressing the mycelial growth of this pathogen (Kamala and Indira 2011). Studies revealed that application of diffusible and volatile metabolites of *Trichoderma* in vitro reduces the growth of *Pythium*. Among these metabolites, volatile metabolites showed broad-spectrum inhibition of *Pythium* as compared to diffusible metabolites (Patil et al. 2012). In assay for volatile metabolites, *Trichoderma harzianum* revealed significantly higher inhibition on *P. aphanidermatum*, but in assay for nonvolatile metabolites *Trichoderma viride* showed higher inhibition. Interestingly, growth of both *Trichoderma* spp. was induced by nonvolatile metabolites of *P. aphanidermatum* (Jeyaseelan et al. 2012). Antagonism against *P. aphanidermatum* was also supported by Muthukumar et al. (2011) and Singh et al. (2014) in Chilli. *T. harzianum* (Th Azad) was found to be significantly superior and effective against pathogen and in increasing per cent germination, root length, shoot length, seedling length, dry weight and vigour index (Singh et al. 2014). Recently, Kumar et al. 2014 observed that *T. viride and T. harzianum* completely colonized and restricted the mycellial growth of *Pythium aphanidermatum* in plates. They concluded that this might be due to the secretion of some secondary metabolites which diffused in the culture medium and inhibited the growth of pathogen.

Mycoparasitism is another explanation for how *Trichoderma* controls plant diseases caused by *Pythium*. Mycoparasitism of *Pythium* by *Trichoderma* involves fungus-fungus interaction and hostpathogen cross-talk with participation of G proteins (Rocha-Ramirez et al. 2002; Mukherjee et al. 2004; Reithner et al. 2005; Zeilinger et al. 2005), protein kinases (Reithner et al. 2007) and signaling molecules such as cyclic AMP (Omero et al. 1999). The involvement of signal transduction pathway components such as G proteins in control of Cell Wall Degrading Enzyme (CWDE) expression and coiling processes has been suggested (Reithner et al. 2005; Zeilinger et al. 2005; do Nascimento et al. 2009). Monteiro et al. (2015) suggested that the production of some CWDEs during mycoparasitism by T. *reesei* against

P. ultimum can be mediated by GNA1 (a G-alpha protein that belongs to αi group in CWDEs) activity or cAMP levels. cAMP can stimulate coiling/recognition in *Trichoderma*, so the cAMP pathway seems to have antagonistic roles in mycoparasitism-relevant coiling response.

Control of *Penicillium*

The genus *Penicillium* includes many ubiquitous species which are considered antagonist of plant pathogen. However, some species cause major postharvest losses of fruits. For example, *P. italicum* and *P. digitatum*, causing blue mold or green mold, are the most common postharvest pathogens of citrus fruits. Postharvest fungal diseases of apple are mainly caused by *P. expansum*. *Trichoderma* was suggested as potential biological agent to prevent the losses caused by this postharvest pathogen (Díaz and Vila 1990; Zamani et al. 2006; Sharma et al. 2012b).

Navelina oranges protected with aqueous suspension of *T. viride*, showed an increase in resistance toward *P. digitatum*. These oranges, inoculated with *P. digitatum*, did not produce lesions after 5 days when *T. viride* was applied 48 h or 72 h before inoculation (Díaz and Vila. 1990). Batta (2004) used the invert emulsion formulation (water-in-oil type) of *T. harzianum* and showed that it significantly reduced the decay-lesion diameter caused by *P. expansum*. In vitro, *T. viride* inhibited the radial growth *Penicillium* sp. (54 %) in dual culture (Rajendiran et al. 2010). Recently, *T. harziaum* and *T. viride* were found to inhibit the growth of *P. notatum* (Patale and Mukadam 2011; Agarwal et al. 2011) and *P. chrysogenum* (Jat and Agalave. 2013).

Control of *Sclerotium*

Sclerotium is a destructive soil-borne and postharvest plant pathogen. Use of the antagonistic fungus *Trichoderma* to control this pathogen has been reported by many studies. Weindling (1934) reported the parasitism of *Trichoderma lignorum* (Tode) Harz on *Sclerotium rolfsii* Sacc. Coley-Smith et al. (1974) by means of microtome sections have shown that medulla of infected sclerotia of *S. delphinii* were completely replaced by hyphae and chlamydospores of *T. hamatum* on agar plates. Backman and Rodriguez- Kabana (1975) controlled *S. rolfsii* in peanuts by using molasses enriched clay granules as a food base for *T. harzianum*. Under greenhouse conditions, incorporation of the wheat-bran inoculum preparation of *T. harzianum* in pathogen-infested soil significantly reduced bean diseases caused by *S. rolfsii* (Elad et al. 1980). Henis et al. (1982) reported mycoparasitism of *Trichoderma* spp. against *S. rolfsii*, where chlamydospores were abundantly produced in contrast to conidia within the infected fungal sclerotia.

Mukherjee and Raghu (1997) demonstrated that *Trichoderma* is not very effective in suppressing *S. rolfsii* at temperatures above 30 °C. In dual culture, *Trichoderma* overgrew *S. rolfsii* at 25 °C and 30 °C, but at 35 °C and 37 °C, *S. rolfsii* overgrew the colony of *Trichoderma*. Poosapati et al. (2014) identified two potential thermotolerant and saline tolerant isolates of *Trichoderma* viz., *T. asperellum*, TaDOR673 and *T. asperellum*, TaDOR7316, which controlled the collar rot disease in groundnut by 79.7 % when screened in vitro and in vivo. Kumar et al. (2014) evaluated the in vitro antagonistic potential and effective strains of *Trichoderma* against *S. rolfsii*, by dual culture plate techniques and found that *T. viride* (01PP) and *T. harzianum* (Th. azad) completely colonized the pathogen and completely restricted its mycellial growth.

Control of *Rhizoctonia*

Rhizoctonia is one of the phytopathogens causing damping-off, root and crown rot of tomato. *Trichoderma* was found to be an effective biological control agent for preventing damage induced by *R. solani* under both greenhouse and field conditions (Elad et al. 1980), and can be used as an alternative to chemical fungicides. *T. harzianum* and *T. koningii* were found to be capable of parasitizing and destroying *R. solani* mycelium (Melo and Faull 2000). Barakat et al. (2007) showed that *T. harzianum* (Jn14) and *T. hamatum* (T36) were the most effective isolates at 25 °C and inhibited *R. solani* mycelial growth by 42 % and 78 %, respectively, due to fungitoxic metabolites production and increased height, fresh and dry weights of bean seedling. The variation between these isolates was due to genetic variation, mycelium-coiling rate, sporulation rate, fungitoxic metabolites, induced growth response and temperature effect Application of *T. harzianum* to inoculated tomato seedlings resulted in disease suppression and significantly higher yield, associated with the accumulation of high phosphorus levels in tissues of tomato plants (Amer MA and Abou-El-Seoud 2008). Biocontrol of *R. solani* in tomatoes cultivated under greenhouse and field conditions was observed using the *T. harzianum* mutants Th650-NG7, Th11A80.1, Th12A40.1, Th12C40.1, Th12A10.1 and ThF2-1, which prevented the mortality and increased development of fresh and dry weights of tomato plants (Montealegre et al. 2010). In another study, *Trichoderma* treatment increased plant growth in rice and decreased the percentage of rate of infection (Anitha and Das 2011).

Demirci et al. (2011) tested *T. harzianum* against *R. solani* on potato and found that it was able to produce an inhibition zone in front of the *R. solani* and overgrow the mycelium. In the dual culture assay, the percentage inhibition of growth by *T. viride*, *T. harzianum*, on *R. solani* were 70 %, 67 %, respectively. This inhibitory effect was caused by the hyphal interaction between the biocontrol agent and the pathogen causing lysis of pathogen hyphae and reduction of the mycelial growth of the *R. Solani* (Seema and Devaki 2012). Reduction of disease incidence (31.2 %) and disease severity (18.8 %) of bean root rot under green house conditions by *T. harzianum* was observed by Matloob and Juber (2013). The biocontrol abilities of water-soluble and volatile metabolites of three different isolates of *Trichoderma* (*T. asperellum, T. harzianum and Trichoderma* spp.) against *R. solani* were investigated both in vitro and in vivo by Asad et al. (2014). They showed for the first time that mycelial growth inhibition of the pathogen was 74.4–67.8 % with water-soluble metabolites as compared to 15.3–10.6 % with volatile metabolites in vitro. They concluded that *T. asperellum* was more effective and consistent, lowering disease incidence up to 19.3 % in laboratory and 30.5 % in green house conditions.

Management of Bacterial Diseases

In contrast to fungi, *Trichoderma* spp. have been reported to have limited applications in biocontrol of pathogenic bacteria. An immediate explanation would be that bacteria generally have a faster metabolic rate than fungi. Thus, antagonism via physical interaction such as, mycoparasitism would be too slow to be effective from BCA point of view, where faster action is a must. However, if the formulated metabolites from *Trichoderma* spp. were considered, the BCA potential of antagonist fungi would be considerably higher (Verma et al. 2007). Altogether, antibacterial action of *Trichoderma* spp. is based only on the action of antibiotic compounds produced and there is no physical interaction between antagonist and pathogen.

T. asperellum SKT-1, a microbial pesticide was found to be highly effective against various bacterial diseases of rice: bacterial seedling blight caused by *Burkholderia plantarii*, bacterial grain rot caused by *Burkholderia glumae*, and bacterial brown stripe caused by *Acidovorax* spp. (Kumakura et al. 2003). The antibacterial activity of *Trichoderma* was observed against various gram positive and gram negative species: *Pseudomonas aeruginosa* ATCC 27853, *Staphylococcus aureus* ATCC 29213, *Staphylococcus epidermidis* NCIMB 8853, *Escherichia coli* ATCC 5218 and *Micrococcus luteus* NCIMB 8166 by using disc diffusion and liquid dilution methods (Khethr et al. 2008). Recently, the antibacterial activity of *T. harzianum* was confirmed against *Staphylococcus aureus*, *E. coli* and *Klebsiella* and the minimum inhibitory concentration of *T. harzianum* on these bacterial isolates range from 50 to 100 μl/ml of media (Leelavathi et al. 2014).

Bacterium-degrading ability is highly variable among *Trichoderma* strains. Trichoviridin was the first antibiotic compound isolated from a strain of *Trichoderma koningii*. It was active against *E. coli* and *Trichophyton usteroides* (Yamano et al. 1970). The secretion of NAGase (EC 3.2.1.52), trypsin-like (EC 3.4.21.4) and chymotrypsin-like (EC 3.4.21.1) protease and muramidase (EC 3.2.1.17) enzymes by *Trichoderma*, play an important role in the degradation of bacteria. Manczinger et al. (2002) fractionated inductive ferment broths of *T. harzianum* T19 strain, having suspensions of *B. subtilis* cells, and it was found to produce trypsin-like proteases (approx. 19, 13 and 5 kDa in size), chymotrypsin-like proteases in the molecular weight range between 12 and 43 kDa, and NAGases, along with Muramidase-like activities. They provided evidence that along with other compounds, muramidases also have great importance in the degradation of bacterial cells.

The butanolic extract of *Trichoderma* has strong antimicrobial activities and is effective against many bacterial species. Chemical composition of the butanolic extract of *Trichoderma* sp. contained essentially a terpenoid compound identified as limonene (92.6 %) along with weak proportions of alcohols, ketones, hydrocarbons and fatty acid esters families (Khethr et al. 2008). Recently, silver nanoparticles have been successfully synthesized from *Trichoderma harzianum*, which were found to significantly reduce the bacterial growth of *Staphylococcus aureus* and *Klebsiella aeruginosa* in a dose dependent manner (Ahluwalia et al. 2014).

Management of Viral Diseases

A very few studies reveal the efficacy of *Trichoderma* against viral pathogens. Potato mop-top disease, which is caused by the virus potato mop-top virus (PMTV) is vectored by the pathogen *Spongospora subterranea* f. sp. *subterranea* that causes powdery scab disease in potatoes. Nielsen and Larsen (2004) indicated a potential use of seed dressing by *T. harzianum* to control spread of *S. subterranea* and potato mop-top virus from seed potatoes. *Trichoderma* has been considered to have the potential to reduce powdery scab, presumably via reducing resting spore viability or zoospore activity and infectivity.

Arabidopsis plants treated with culture filtrate (CF) of *Trichoderma asperellum* SKT-1, a microbial pesticide, showed reduced *Cucumber mosaic virus* (CMV) titre and disease severity as compared with control plants. The barley grain inoculum treatment of SKT-1 induced systemic resistance against CMV through SA signaling cascade in *Arabidopsis* plants. Treatment with CF of SKT-1 mediated the expression of a majority of the various pathogen related genes, which led to the increased defense mechanism against CMV infection (Elsharkawy et al. 2013).

However, the biochemical mechanism of their antiviral activity remains largely unknown. Recently, trichokonins, antimicrobial peptaibols isolated from *T. pseudokoningii* SMF2, was found to induce defence responses and systemic resistance (Song et al. 2006) in tobacco against tobacco mosaic virus (TMV) infection by significantly increasing activities of pathogenesis-related enzymes PAL and POD, and expression of several plant defence genes (Luo et al. 2010).

Management of Nematodes

The root knot nematodes (*Meloidogyne* spp.) are sedentary endoparasites and are among the most damaging agricultural pests attacking a wide range of crops. Several attempts have been made to use *Trichoderma* spp. to control plant parasitic nematodes. Windham et al. (1989) reported reduced egg production in the root-knot nematode *Meloidogyne arenaria* following soil treatments with *T. harzianum* and *T. koningii* preparation. Rao et al. (1998) evaluated aqueous extracts of neem (*Azadirachta indica*), caster (*Ricinus communis*) and pingamia (*Pingamia harzianum*) as substrates for the mass production of *T. harzianum* which was used in the management of *M. incognita* in eggplant under field conditions. They reported that caster cake extracts showed the best biocontrol activity. *T. harzianum* Rifai has been reported to be an effective bioagent for the management of the citrus nematode (Rao et al. 1998; Seifullah and Thomas 1996; Sharon et al. 2001).

Extracellular enzymes such as chitinase and protease which display antifungal activities appear to participate in the *Meloidogyne javanica* × *Trichoderma* spp. interaction (Sharon et al. 2001). Two mechanisms of action are thought to be responsible for the reduction in nematode infection following root treatment with *Trichoderma* spp. (1) direct parasitism of eggs and larva through the increase in chitinase and protease activities, which would be indicators of eggs infection capability (Sharon et al. 2001; Suarez et al. 2004); and (2) inducing plant defence mechanisms leading to systemic resistance (Sahebani and Hadavi 2008).

Most studies on nematodes concurred that the *Trichoderma* spp. had multiple modes of action. For example, *T. virens* invaded, ramified, grooved and vacuolated the root-knot nematode eggs (Eapen et al. 2005). In vitro studies demonstrated that *T. harzianum* and *T. viride* were effective in causing second-stage juvenile mortality and reduced the incidence and pathogenicity of the root-knot nematode *Meloidogyne incognita* on tomato (Dababat and Sikora 2007). The antagonistic action of *Trichoderma* spp. is chiefly attributed to chitinolytic activity of the fungi on cellular structure of nematodes, which is rich in chitin. Additionally, unlike bacteria, nematodes are mainly antagonized by parasitism and antibiosis akin to fungal pathogens (Verma et al. 2007).

Naserinasab et al. (2011) found that inhibition of the hatching of *M. javanica* eggs was positively correlated with increasing concentrations (standard, 1:1, 1:10, and 1:100) of culture filtrates of *T. harzianum* BI. They concluded that in addition to direct antagonism, induction of defense-related enzymes involved in peroxidase pathway contributed to enhance resistance against invasion of *M. javanica* in tomato. *T. harzianum* strain ESALQ-1306 was confirmed for its potential biological control against *M. incognita* under in vitro and greenhouse conditions (Mascarin et al. 2012). *Trichoderma* spp. also controls other plant-parasitic nematode, *C. elegans*. Five *Trichoderma* species (*T. atroviridae*, *T. harzianum*, *T. rossicum*, *T.tomentosum*, *T. virens*) were tested using i*n vitro* monoculture growth rate tests, dual confrontation assays and comparison of strain specific egg-parasitic index and the results revealed that *T. harzianum* strains possess the strongest egg-parasitic ability against *C. elegans* (Szabó et al. 2012).

Molecular Signalling Involved During Biocontrol

Trichoderma biocontrol relevant processes such as the formation of infection structure like appresoria during mycoparasitism, production of hydrolytic enzymes, antimicrobial metabolites, and triggering systemic resistance in plants, rely on various signalling pathways, which are activated by binding of host-derived ligands to receptors. Recently, it was shown that heterotrimeric G-proteins and mitogen-activated protein (MAP) kinases, cAMP are important factors involved in these processes (Zeilinger and Omann 2007).

Role of Heterotrimeric G Proteins

During mycoparasitism, an interaction is normally mediated through cross-talk between biomolecules secreted from one fungal species that acts as messenger and perceived by receptors present on the other interacting fungal species. A signal is generated at the perception site and further downstream actions follow through a signal transduction mechanism using various receptors (Sarma et al. 2014). Heterotrimeric G proteins act as signal transducers that couple cell-surface receptors to cytoplasmic effector proteins. The signaling via heterotrimeric G protein needs basically three components such as a G protein-coupled receptor (GPCR), a heterotrimeric G protein (α,β,ϒ, subunits), and an effector (Neer 1995). GCPR proteins have seven common transmembrane domains where the N-terminus reside outside and the C-terminus inside the cytoplasm. When ligands bind to these receptors a conformational change occurs and they release the G- α subunit from the G protein for exchanging GDP with GTP. Subsequently, the GTP bound G- α subunit dissociates from their G- β and ϒ- subunits and these two signaling units then regulate the downstream activities of the effectors (Sarma et al. 2014).

Fungal G subunits can be divided into three major subgroups according to a phylogenetic tree generated by multiple alignments of fungal G-protein sequences: subgroup I, subgroup II¸ and subgroup III (Omann and Zeilinger 2010). Biochemical and genetic studies with *T. troviride* provided evidence for the importance of a G-protein α subunit, *tga1*, in the parasitism of *R. solani* hyphae (Omero et al. 1999; Rocha-Ramírez et al. 2002). When accumulation of *T. atroviride* Gαi homolog, *tga1*, was blocked by antisense expression, hyphal extension growth was inhibited and the mutant colonies underwent conidiation profusely. Whereas, over expression had the opposite effect, and promoted vegetative proliferation and increased mycoparasitism associated coiling (Rocha-Ramírez et al. 2002). A more profound functional characterization of Tga1 was performed by Reithner et al. (2005) who extended the involvement of this G-protein subunit to the production of antifungal metabolites and the formation of extracellular chitinases. tga1 mutant was unable to overgrow and lyse host fungi and reduced extracellular chitinase activities and a decreased transcription of the chitinase-encoding genes nag1 (N-acetyl-glucosaminidase-encoding) and ech42 (endochitinase 42-encoding) was observed. However, formation of infection structure was not affected (Reithner et al. 2005). Contrary to *T. atroviride* Tga1, its homologue TgaA in *T. virens* does not influence growth or conidiation. Mukherjee et al. (2004) observed that mutant of two G-protein α subunit genes, *tgaA* and *tgaB*, from the biocontrol fungus *T. virens*, parasitized the mycelia of *R. solani*, but *tgaA* mutants had reduced ability to colonize *S. rolfsii* colonies. Thus indicating that *tgaA* is involved in antagonism against *S. rolfsii*, but neither G protein subunit is involved in antagonism against *R. Solani*. Further characterization of mutants of *T. atroviride* bearing a gpr1-silencing construct of a GPCR revealed that Gpr1 is important for growth, conidial production and germination (Brunner et al. 2008).

Role of cAMP

cAMP, involved in growth, germination, mycoparasitism and secondary metabolism in *Trichoderma*, is produced from ATP by adenylate cyclise. The cAMP signaling cascades seem to be conserved among fungi (D'Souza and Heitman 2001). In these organisms, cAMP activates a cAMP-dependent protein kinase (PKA) that phosphorylates enzymes involved in intermediary metabolism (particularly carbohydrate metabolism) and transcription factors which are key regulators of stress-responsive gene expression (Firmino et al. 2002). Antisense-mediated silencing of the Gα protein Tga1 resulted in reduced cAMP level in *Trichoderma atroviride* IMI 206040 (Rocha-Ramírez et al. 2002). While deletion of Tga1 in *T. atroviride* ATCC 78058 resulted in elevated internal cAMP levels (Reithner et al. 2005). The deletion of another Gα protein, Tga3, resulted in reduced intracellular cAMP levels (Zeilinger et al. 2005). Deletion of tac1, an adenylate-cyclase-encoding gene, brought intracellular cAMP levels below the detection limit and the mutants did not overgrow the test plant pathogens *Pythium* sp., *R. solani* and *S. rolfsii* and showed reduced secondary metabolite production. This suggested the role of cAMP signalling in a Trichoderma biocontrol (Mukherjee et al. 2007).

Many of the known fungal G proteins influence the intracellular level of cAMP by either stimulating or inhibiting adenylyl cyclase (Bölker 1998). An elevated internal steady-state cAMP level in the tga1 mutants compared to the parental strain confirmed that Tga1 represents a member of the adenylyl cylcase inhibiting subgroup I of fungal G subunits (Reithner et al. 2005). The stimulatory role of the subgroup III G proteins Tga3 and Gna3 on the activity of adenylyl cyclase was confirmed by its ability to increase intracellular cAMP levels (Schmoll et al. 2009). Comparative genome analyses of three *Trichoderma* species viz. *T. atroviride* and *T. virens* and *T. reesei* revealed a great diversity of putative GPCRs with genus- and species- specific differences (Gruber et al. 2013).

Role of MAP Kinases

MAP kinases are dual phosphorylated protein kinases, which control adaptation to environmental stress in pathogenic fungi. In *T. virens* in addition to heterotrimeric G proteins, a MAP kinase was found to affect mycoparasitism-related processes (Mendoza-Mendoza et al. 2003; Mukherjee et al. 2003) as well as plant systemic resistance. MAP kinase cascade is highly conserved and comprises three kinases that function by sequential phosphorylation. Upon receiving a stimulatory signal, the MAP kinase kinase kinase (MAPKKK) is phosphorylated and it triggers the phosphorylation of the MAP kinase kinase (MAPKK), which in turn phosphorylates the MAP kinase (MAPK). MAPKs are simultaneously phosphorylated and the signal is finally transferred to downstream effectors, normally a transcription factor or other molecules, which either activate or relieve the repression of the corresponding target genes. The expression of these target genes is essential in the adaptive response of the cell to an activating stimulus (Roman et al. 2007).

MAPK signaling plays important role during *Trichoderma*-plant interaction and induces plant systemic resistance. Role of a mitogen-activated protein kinase (MAPK) TmkA in inducing systemic resistance in cucumber against the bacterial pathogen *Pseudomonas syringae* pv.*lacrymans* was confirmed by Viterbo et al. (2005). In silico reconstruction of signal transduction pathways of *T. reesei* suggested the presence of three MAPKs: Tmk1 that is homologues to yeast Fus3, Tmk2 that is homologues to yeast Slt2 and Tmk3 that is homologous to yeast Hog1 (Schmoll 2008). Recently, Wang et al. (2014) suggested that Tmk2 is involved in cell wall integrity and sporulation. Tmk3 participates in high osmolarity resistance and in promoting cellulase production while Tmk2 is involved in repressing cellulase formation.

However, some contrasting reports are also available regarding the role of MAP Kinases in parasitism by *Trichoderma* species (Sarma et al. 2014). A mitogen-activated protein kinase encoding gene, *tvk1*, from *Trichoderma virens* was suggested to acts as a negative modulator during host sensing and sporulation. *tvk1* null mutants showed a clear increase in the level of the expression of mycoparasitism-related genes under simulated mycoparasitism and during direct confrontation with the plant pathogen *Rhizoctonia solani* (Mendoza-Mendoza et al. 2003). MAPKs homologous to *T. atroviride* Tmk1 were previously shown to be involved in the regulation of appressorium formation and invasive growth (Xu and Hamer 1996). Recently, Reithner et al. (2007) suggested the negative regulatory role of Tmk1 on *Trichoderma*-triggered plant resistance. They found that mycoparasitism-related processes like infection structure formation, coiling as well as chitinase and antifungal metabolite production were unaltered or even enhanced upon *tmk1* gene deletion. In greenhouse experiments also, the examined Δ*tmk1*-12 mutant was able to protect bean plants against *R. solani* infection. Thus, indicating that the deletion of a MAPK gene can generate a more aggressive parasite and, consequently, a better biocontrol agent.

Physiological Signalling Involved During Biocontrol

Trichoderma colonization triggers a wide array of plant responses, which may result in an enhanced defensive capacity of the plant (Morán-Diez et al. 2012). This effect of *Trichoderma* on the plant defense system is not only restricted to the root, but they also manifest in aboveground plant tissues (Mathys et al. 2012), suggesting the involvement of systemic resistance, which is likely the result of modulation of the plant defense network that may translate *Trichoderma*-induced early signaling events into a more efficient activation of defense responses (Martínez-Medina et al. 2013). Various plant hormones like jasmonic acid (JA), ethylene (ET), abscisic acid (ABA) and salicylic acid (SA) act as signal transduction molecules during *Trichoderma* mediated induced systemic resistance (TISR).

Generally, pathogen-induced systemic acquired resistance (SAR), is dependent on the SA-regulated signaling pathway (Durrant and Dong 2004) and leads via the actions of NPR1 (Cao et al. 1994) and WRKY TFs (Jaskiewicz et al. 2011) to the activation of pathogenesis-related proteins (PRs), while ISR by beneficial microorganisms like *Trichoderma* usually relies on JA signaling (Van der Ent et al. 2009). Korolev et al. (2008) using multiple *Arabidopsis* mutant lines showed that the induction of resistance by *T. harzianum* Rifai T39 against *Botrytis cinerea* requires JA, ET, and ABA signaling, while SA was not required. Expression studies on marker genes linked to the main defence signaling pathways suggested that TISR might involve the direct activation of both SA- and JA-related pathways (Mathys et al. 2012). During ISR-priming, SA was synthesized from chorismate rather than via the phenylalanine pathway, suggesting chorismate way as an important alternative route for the production of SA required for defense responses. This *T. hamatum* T382-induced ISR-prime in *A. thaliana* was also characterized by upregulation of phenylpropanoid pathway leading to the production of anthocyanins and showed that this pathway is clearly involved in ISR (Mathys et al. 2012). The phenotypic analysis of disease development in the JA and SA impaired mutants demonstrated that *T. harzianum*-induced systemic resistance against *B. cinerea* requires not only the JA but also the SA signaling pathways, as these mutant lines developed similar level of disease than non-induced control plants. It is also found that *T. harzianum*-mediated systemic resistance against *B. cinerea* does not rely on systemin, a plant peptide hormone, signalling (Martínez-Medina et al. 2013).

These plant growth regulators were suggested to act as signal component as well as to affect enzyme production (endo-1,4-β-glucanase, cellulase) by *Trichoderma*, which might promote biocontrol activities of this beneficial fungus (Gemishev et al. 2005). Along with inducing host plant to produce various hormones, *Trichoderma* themselves synthesize various plants hormones or growth-factors like gibberellic acid (GA3), indol -3- acetic acid (IAA) and abscisic acid (ABA). This can be considered as the direct mechanism to limit pathogens and increase plant growth response induced by *Trichoderma* (Hassanein 2012).

Conclusion

Plant pathogens like bacteria, fungi, oomycetes, virus, nematodes etc. affect survival and fitness of plant and cause severe diseases. Though there are many methods that are available for crop protection from these pathogens, but they are either considered as a major pollutant in the environment or they are less efficacious and expensive.

Another effective way is the use of biocontrol agent *Trichoderma* which assures increased level of safety and minimal environmental impacts. *Trichoderma* species show significant biocontrol activity against a wide range of plant pathogenic organisms through mechanism like mycoparasitism, competition for nutrients and space, producing cell wall degrading enzymes (CWDE) and other antimicrobial compounds. They also induce systemic resistance in host plant for better disease control, through alteration in gene expression of plant, responsible for synthesis of defence related proteins. *Trichoderma* triggered signalling cascade include various receptors, components like phytohormone and various secondary messenger. Heterotrimeric G-proteins, mitogen-activated protein (MAP) kinases and cAMP are another important factors involved in biocontrol signalling. Due to the effective control of plant pathogens/diseases and improvement of plant growth, several *Trichoderma* species can be considered as a promising measure in disease management.

References

Aerts R, De Schutter B, Rombouts L (2002) Suppression of *Pythium* spp. by *Trichoderma* spp. during germination of tomato seeds in soilless growing media. Meded Rijksuniv Gent Fak Landbouwkd Toegep Biol Wet 67(2):343–351

Agarwal T, Malhotra A, Biyani M, Trivedi PC (2011) In vitro interaction of *Trichoderma* isolates against *Aspergillus niger*, *Chaetomium* sp. *and Penicilium* sp. Ind J Fundam Appl Life Sci 1(3):125–128

Ahluwalia V, Kumar J, Sisodia R, Shakil NA, Walia S (2014) Green synthesis of silver nanoparticles by *Trichoderma harzianum* and their bio-efficacy evaluation against *Staphylococcus aureus* and *Klebsiella pneumonia*. Ind Crop Prod 55:202–206

Ajith PS, Lakshmidevi N (2010) Effect of volatile and non-volatile compounds from *Trichoderma* spp. against *Colletotrichum capsici* incitant of anthracnose on bell peppers. Nat Sci 8(9):265–269

Amer MA, Abou-El-Seoud (2008) Mycorrhizal fungi and *Trichoderma harzianum* as biocontrol agents for suppression of *Rhizoctonia solani* damping-off disease of tomato. Commun Agric Appl Biol Sci 73(2):217–232

Amin M, Teshele J, Tesfay A (2014) Evaluation of bioagents seed treatment against *Colletotrichum Lindemuthianum*, in haricot bean anthracnose under field condition. Res Plant Sci 2(1):22–26

Anitha A, Das MA (2011) Activation of rice plant growth against *Rhizoctonia solani* using *Pseudomonas fluorescens*, *Trichoderma* and salicylic acid. Res Biotechnol 2(4):7–12

Asad SA, Ali N, Hameed A, Khan SA, Ahmad R, Bilal M, Shahzad M, Tabassum A (2014) Biocontrol efficacy of different isolates of *Trichoderma* against soil borne pathogen *Rhizoctonia solani*. Pol J Microbiol 63(1):95–103

Awuah RT, Lorbeer JW (1991) Methyl bromide and steam treatment of an organic soil for control of *Fusarium* yellows of celery. Plant Dis 75:123–125

Backman PA, Rodriguez-Kabana R (1975) A system for the growth and delivery of biological control agents to the soil. Phytopathology 65:819–821

Bacon CW, Hinton DM, Hinton J (2006) Growth-inhibiting effects of concentrations of fusaric acid on the growth of *Bacillus mojavensis* and other biocontrol *Bacillus species*. J Appl Microbiol 100:185–194

Bailey BA, Lumsden RD (1998) Direct effects of *Trichoderma* and *Gliocladium* on plant growth and resistance to pathogens. In: Kubicek CP, Harman GE, Ondik KL (eds) *Trichoderma* and *Gliocladium*: enzymes, biological control and commercial applications. Taylor and Francis, London, pp 185–204

Barakat RM, Al-Mahareeq F, Ali -Shtayeh MS, AL- Masri M (2007) Biological control of *Rhizoctonia solani* by indigenous *Trichoderma* spp. isolates from Palestine. Hebron Univ Res J 3:1–15

Basak AC, Basak SR (2011) Biological control of *Fusarium solani* sp. dalbergiae, the wilt pathogen of *Dalbergia sissoo*, by *Trichoderma viride* and *T. harzianum*. J Trop For Sci 23(4):460–466

Batta YA (2004) Effect of treatment with *Trichoderma harzianum* Rifai formulated in invert emulsion on postharvest decay of apple blue mold. Int J Food Microbiol 96:281–288

Bendahmane BS, Mahiout D, Benzohra IE, Benkada MY (2012) Antagonism of Three *Trichoderma* species against *Botrytis fabae* and *B. cinerea*, the causal agents of chocolate spot of faba bean (*Vicia faba* L.). In Algeria. World Appl Sci J 17(3):278–283

Bernal-Vicente A, Ros M, Pascual JA (2009) Increased effectiveness of the *Trichoderma harzianum* isolate T-78 against *Fusarium* wilt on melon plants under nursery conditions. J Sci Food Agr 89:827–833

Bhagat S, Bambawale OM, Tripathi AK, Ahmad I, Srivastava RC (2013) Biological management of fusarial wilt of tomato by *Trichoderma* spp. in Andamans Indian. J Hortic 70(3):397–403

Bogumił A, Paszt LS, Lisek A, Trzciński P, Harbuzov A (2013) Identification of new *Trichoderma* strains with antagonistic activity against *Botrytis cinerea*. Folia Hortic 25(2):123–132

Bölker M (1998) Sex and crime: heterotrimeric G proteins in fungal mating and pathogenesis. Fungal Genet Biol 25(3):143–156

Brunner K, Omann M, Pucher ME, Delic M, Lehner SM, Domnanich P, Kratochwill K, Druzhinina I, Denk D, Zeilinger S (2008) *Trichoderma* G protein coupled receptors: functional characterisation of a cAMP receptor like protein from *Trichoderma atroviride*. Curr Genet 54(6): 283–299

Cao H, Bowling SA, Gordon AS, Dong X (1994) Characterization of an *Arabidopsis* mutant that is nonresponsive to inducers of systemic acquired resistance. Plant Cell 6:1583–1592

Cheeke PR (1995) Endogenous toxins and mycotoxins in forage grasses and their effects on livestock. J Anim Sci 73:909–918

Cheng CH, Yang CA, Peng KC (2012) Antagonism of *Trichoderma harzianum* ETS 323 on *Botrytis cinerea* mycelium in culture conditions. Phytopathology 2(11):1054–1063

Chet I, Inbar J, Hadar I (1997) Fungal antagonists and mycoparasites. In: Wicklow DT, Söderström B (eds) The Mycota IV: environmental and microbial relationships. Springer, Berlin, pp 165–184

Claydon N, Allan M, Hanson JR, Avent AG (1987) Antifungal alkyl pyrenes of *Trichoderma harzianum*. Trans Br Mycol Soc 88:503–513

Coley-Smith JR, Ghaffar A, Javed ZUR (1974) The effect of dry conditions on subsequent leakage and rotting of fungal sclerotia. Soil Biol Biochem 6:307–312

Contreras-Cornejo HA, Macías-Rodríguez L, Beltrán-Peña E, Herrera-Estrella A, López-Bucio J (2011) *Trichoderma*-induced plant immunity likely involves both hormonal- and camalexin-dependent mechanisms in *Arabidopsis thaliana* and confers resistance against necrotrophic fungi *Botrytis cinerea*. Plant Signal Behav 6:1554–1563

Dababat AFA, Sikora RA (2007) Use of *Trichoderma harzianum* and *Trichoderma viride* for the biological control of *Meloidogyne incognita* on tomato. Jordan J Agric Sci 3(3):297–309

Dal Bello GM, Mónaco CI, Cháves AR (1997) Study of the effect of volatile metabolites of *Trichoderma hamatum* on the growth of phytopathogenic soil borne fungi. Rev Iberoam Micol 14(3):131–134

Delgado-Jarana J, Moreno-Mateos MA, Benítez T (2003) Glucose uptake in *Trichoderma harzianum*: role of gtt1. Eukaryot Cell 2:708–717

Demirci E, Dane E, Eken C (2011) In vitro antagonistic activity of fungi isolated from sclerotia on potato tubers against *Rhizoctonia solani*. Turk J Biol 35:457–462

Deshmukh AJ, Mehta BP, Patil VA (2010) In-vitro evaluation of some known bioagents to control *C. gloeosporioides* Penz, and Sacc, causing *Anthracnose* of Indian bean. Int J Pharm Biol Sci 1(2):1–6

Díaz BA, Vila AR (1990) Biological control of *Penicillium digitatum* by *Trichoderma viride* on postharvest citrus fruits. Int J Food Microbiol 11(2):179–183

Dolatabadi HK, Goltapeh EM, Mohammadi N, Rabiey M, Rohani N, Varma A (2012) Biocontrol potential of root endophytic fungi and *Trichoderma* species against *Fusarium* wilt of lentil under in vitro and greenhouse conditions. J Agric Sci Technol 14(407–420):407

D'Souza CA, Heitman J (2001) Conserved cAMP signaling cascades regulate fungal development and virulence. FEMS Microbiol Rev 25:349–364

Durrant WE, Dong X (2004) Systemic acquired resistance. Annu Rev Phytopathol 42:185–209

Eapen SJ, Beena B, Ramana KV (2005) Tropical soil microflora of spice-based cropping systems as potential antagonists of root-knot nematodes. J Invertebr Pathol 88:218–225

Elad Y (1994) Biological control of grape grey mould by *Trichoderma harzianum*. Crop Prot 13:35–38

Elad Y, Chet I, Katan J (1980) *Trichoderma harzianum*: a biocontrol agent effective against *Sclerotium rolfsii* and *Rhizocionia solani*. Phytopathology 70:119–121

Elad Y, Chet I, Henis Y (1982) Degradation of plant pathogenic fungi by *Trichoderma harzianum*. Can J Microbiol 28:719–725

Elad Y, Köhl J, Fokkema NJ (1994) Control of infection and sporulation of *Botrytis cinerea* on bean and tomato by saprophitic bacteria and fungi. Eur J Plant Pathol 100:315–336

El-Hasan A, Walker F, Buchenauer H (2008) *Trichoderma harzianum* and its metabolite 6-pentyl-alpha-pyrone suppress fusaric acid produced by *Fusarium moniliforme*. J Phytopathol 156:79–87

El-Naggar M, Kövics GJ, Sándor E, Irinyi L (2008) Mycoparasitism and antagonistic efficiency of *Trichoderma reesei* against *Botrytis* spp. Contrib Bot 43:141–147

Elsharkawy MM, Shimizu M, Takahashi H, Ozaki K, Hyakumachi M (2013) Induction of systemic resistance against *Cucumber mosaic virus* in *Arabidopsis thaliana* by *Trichoderma asperellum* SKT-1. Plant Pathol J 29(2):193–200

Fakhrunnisa, Hashmi MH, Ghaffar A (2006) In vitro interaction of *Fusarium* spp., with their fungi. Pak J Bot 38(4):1317–1322

Firmino AAP, Ulhoa CJ, Sousa MV, Filho EXF, Ricart CAO (2002) Involvement of G proteins and camp in the production of chitinolytic enzymes by T*richoderma harzianum*. Braz J Microbiol 33(2):169–173

Fiume F, Fiume G (2006) Biological control of *Botrytis* gray mould on tomato cultivated in greenhouse. Commun Agric Appl Biol Sci 71:897–908

Flores A, Chet I, Herrera-Estrella A (1997) Improved biocontrol activity of *Trichoderma harzianum* by over-expression of the proteinase-encoding gene prb1. Curr Genet 31:30–37

Freeman S, Barbul O, Rav David D, Nitzani Y, Zveibil A, Elad Y (2001) *Trichoderma* spp. for biocontrol of *Colletotrichum acutatum* and *Botrytis cinerea* in strawberry. Biocontrol of fungal and bacterial plant pathogens. IOBC/WPRS Bull 24:147–150

Freeman S, Minz D, Kolesnik I, Barbul O, Zveibil A, Maymon M, Nitzani Y, Kirshner B, Rav-David D, Bilu A, Dag A, Shafir S, Elad Y (2004) *Trichoderma* biocontrol of *Colletotrichum acutatum* and *Botrytis cinerea* and survival in strawberry. Eur J Plant Pathol 110:361–370

Friedl MA, Druzhinina IS (2012) Taxon-specific metagenomics of *Trichoderma* reveals a narrow community of opportunistic species that regulate each other's development. Microbiology 158:69–83

Gachomo EW, Kotchoni SO (2008) The use of *Trichoderma harzianum* and *T. viride* as potential biocontrol agents against peanut microflora and their effectiveness in reducing aflatoxin contamination of infected kernel. Biotechnol J 7:439–447

Gemishev OT, Vaseva II, Atev AP (2005) Abscisic acid and ethylene influence on endo-1,4-b glucanase activity in *Trichoderma reesei* I-27. Biotechnol Biotechnol Equip 19(3):106–112

Geremia RA, Goldman GH, Jacobs D, Aviles W, Vila SB, Van Montagu M, Herrera-Estrella A (1993) Molecular characterization of the proteinase encoding gene, prb1, related to mycoparasitism by *Trichoderma harzianum*. Mol Microbiol 8:603–613

Ghosh SK, Chakraborty N (2012) In vitro biological control of *Colletotrichum gloeosporioides*, causal organism of anthracnose of sarpagandha (*Roulvolfia serpentina*). Agric Biol J N Am 3(8):306–310

Gruber S, Omann M, Zeilinger S (2013) Comparative analysis of the repertoire of G protein-coupled receptors of three species of the fungal genus *Trichoderma*. BMC Microbiol 13:108

Harman GE (2000) Myths and dogmas of biocontrol: changes in perceptions derived from research on *Trichoderma harzianum* T-22. Plant Dis 84:377–393

Hassanein NM (2012) Biopotential of some *Trichoderma* spp. against cotton root rot pathogens and profiles of some of their metabolites. Afr J Microbiol Res 6(23):4878–4890

Henis Y, Adams PB, Papavizas GC, Lewis JA (1982) Penetration of sclerotia of *Sclerotium rolfsii* by *Trichoderma* spp. Phytophathology 72:70–74

Herrera-Estrella A, Chet I (2003) In: Arora D (ed) Handbook of fungal biotechnology. Dekker, New York, in press

Howell CR (2003) Mechanisms employed by *Trichoderma* species in the biological control of plant diseases: the history and evolution of current concepts. Plant Dis 87:4–10

Inbar J, Menendez A, Chet I (1996) Hyphal interaction between *Trichoderma harzianum* and *Sclerotinia sclerotiorum* and its role in biological control. Soil Biol Biochem 28:757–763

Jaskiewicz M, Conrath U, Peterhänsel C (2011) Chromatin modification acts as a memory for systemic acquired resistance in the plant stress response. EMBO Rep 12:50–55

Jat JG, Agalave HR (2013) Antagonistic properties of *Trichoderma* species against oilseed-borne fungi. Sci Res Rep 3(2):171–174

Jeyaseelan EC, Tharmila S, Niranjan K (2012) Antagonistic activity of *Trichoderma* spp. and *Bacillus* spp. against *Pythium aphanidermatum* isolated from tomato damping off. Arch Appl Sci Res 4(4):1623–1627

John RP, Tyagi RD, Prevost D, Brar SK, Pouleur S, Surampalli RY (2010) Mycoparasitic *Trichoderma viride* as a biocontrol agent against *Fusarium oxysporum* f. sp. adzuki and *Pythium arrhenomanes* and as a growth promoter of soybean. Crop Prot 29(12):1452–1459

Kamala T, Indira S (2011) Evaluation of indigenous *Trichoderma* isolates from Manipur as biocontrol agent against *Pythium aphanidermatum* on common beans. Biotechnology 1: 217–225

Kataoka R, Yokota K, Goto I (2010) Biocontrol of yellow disease of *Brassica compestris* caused by *Fusarium oxysporum* with *Trichoderma viridae* under field conditions. Arch Phytopathol Plant Protect 43:900–909

Khethr FBH, Ammar S, Saïdana D, Daami M, Chriaa J, Liouane K, Mahjoub MA, Helal AN, Mighri Z (2008) Chemical composition, antibacterial and antifungal activities of *Trichoderma* sp. growing in Tunisia. Ann Microbiol 58(2):303–308

Korolev N, Rav David D, Elad Y (2008) The role of phytohormones in basal resistance and *Trichoderma*-induced systemic resistance to *Botrytis cinerea* in *Arabidopsis thaliana*. Biocontrol 53:667–683

Kuc J (1982) Induced immunity to plant disease. Bioscience 32:854–860

Kumakura K, Watanabe S, Toyoshima J, Makino T, Iyozumi H, Ichikawa T, Nagayama K (2003) Effect of *Trichoderma* sp. SKT-1on suppression of six different seed borne diseases of rice. Jpn J Phytopathol 69:384–392

Kumar V, Shahid M, Srivastava M, Singh A, Pandey S, Sharma A, Srivastava YK (2014) Antagonistic effect of rhizospheric *Trichoderma* species against soil borne pathogens. J Program Res 9:408–410

Kumar V, Shahid M, Srivastava M, Singh A, Pandey S, Maurya MK (2015) Screening of *Trichoderma* species for virulence efficacy on seven most predominant phytopathogens. Afr J Microbiol Res 9(11):793–799

Kushwaha M, Verma AK (2014) Antagonistic activity of *Trichoderma* spp., (a biocontrol agent) against isolated and identified plant pathogens. Int J Chem Biol Sci 1:1–6

Leelavathi MS, Vani L, Reena P (2014) Antimicrobial activity of *Trichoderma harzianum* against bacteria and fungi. Int J Curr Microbiol Appl Sci 3(1):96–103

Liju AC, Karthikeyan A, Siva Priya NB, Siddiqui S (2014) Control of anthracnose disease in *Swietenia macrophylla* using*Trichoderma viride*as biocontrol agent. Plant Pathol 13:173–176

Lorito M, Peterbauer C, Hayes CK, Harman GE (1994) Synergistic interaction between fungal cell wall degrading enzymes and different antifungal compounds enhances inhibition of spore germination. Microbiology 140:623–629

Luo Y, Zhang DD, Dong XW, Zhao PB, Chen LL, Song XY, Wang XJ, Chen XL, Shi M, Zhang YZ (2010) Antimicrobial peptaibols induce defense responses and systemic resistance in tobacco against tobacco mosaic virus. FEMS Microbiol Lett 313(2):120–126

Manczinger L, Molnar A, Kredics L, Antal Z (2002) Production of bacteriolytic enzymes by mycoparasitic *Trichoderma* strains. World J Microbiol Biotechnol 18:147–150

Martínez-Álvarez P, Alves-Santos FM, Diez JJ (2012) In vitro and in vivo interactions between *Trichoderma viride* and *Fusarium circinatum*. Silva Fenn 46(3):303–316

Martínez-Medina A, Fernández I, Sánchez-Guzmán MJ, Jung SC, Pascual JA, Pozo MJ (2013) Deciphering the hormonal signaling network behind the systemic resistance induced by *Trichoderma harzianum* in tomato. Front Plant Sci 4:206

Marzano M, Gallo A, Altomare C (2013) Improvement of biocontrol efficacy of *Trichoderma harzianum* vs. *Fusarium oxysporum* f. sp. lycopersici through UV-induced tolerance to fusaric acid. Biol Control 67:397–408

Mascarin GM, Bonfim Junior MF, de Araújo Filho JV (2012) *Trichoderma harzianum* reduces population of *Meloidogyne incognita* in cucumber plants under greenhouse conditions. J Entomol Nematol 4(6):54–57

Mastouri F, Björkman T, Harman GE (2010) Seed treatment with *Trichoderma harzianum* alleviates biotic, abiotic, and physiological stresses in germinating seeds and seedlings. Phytopathology 100:1213–1221

Matarese F, Sarrocco S, Gruber S, Seidl-Seiboth V, Vannacci G (2012) Biocontrol of *Fusarium* head blight: interactions between *Trichoderma* and mycotoxigenic *Fusarium*. Microbiology 158:98–106

Mathys J, De Cremer K, Timmermans P, Van Kerckhove S, Lievens B, Vanhaecke M, Cammue BPA, De Coninck B (2012) Genome-wide characterization of ISR induced in *Arabidopsis thaliana* by *Trichoderma hamatum* T382 against *Botrytis cinerea* infection. Front Plant Sci 3:108

Matloob AH, Juber K (2013) Biological control of bean root rot disease caused by *Rhizoctonia solani* under green house and field conditions. Agric Biol J N Am 4(5):512–519

Mbarga JB, Hoopen GMT, Kuaté J, Adiobo A, Ngonkeu MEL, Ambang Z, Akoa A, Tondje PR, Begoude BAD (2012) *Trichoderma asperellum*: a potential biocontrol agent for *Pythium myriotylum*, causal agent of cocoyam (*Xanthosoma sagittifolium*) root rot disease in Cameroon. Crop Prot 36:18–22

Meca G, Soriano JM, Gaspari A, Ritieni A, Moretti A, Manes J (2010) Antifungal effects of the bioactive compounds enniatins A, A1, B, B1. Toxicon 56:480–485

Melo ID, Faull JL (2000) Parasitism of *Rhizoctonia solani* by strains of *Trichoderma* spp. Sci Agric 57:55–59

Mendoza-Mendoza A, Pozo MJ, Grzegorski D, Martínez P, García JM, Olmedo-Monfil V, Cortés C, Kenerley C, Herrera-Estrella A (2003) Enhanced biocontrol activity of *Trichoderma* through inactivation of a mitogen-activated protein kinase. Proc Natl Acad Sci U S A 100(26):15965–15970

Meraj-ul-Haque, Nandkar PB (2012) Antagonistic effect of rhizospheric *Trichoderma* isolates against tomato damping-off pathogen, *Fusarium oxysporum* f.sp. lycopersici. Int J Res Biosci 1(2):27–31

Miller JD (1994) Epidemiology of *Fusarium graminearum* diseases of wheat and corn. In: Miller JD, Trenholm HL (eds) Mycotoxins in grain: compounds other than aflatoxin. Eagan Press, St. Paul, pp 19–36

Montealegre J, Valderrama L, Sánchez S, Herrera R, Besoain X, María Pérez L (2010) Biological control of *Rhizoctonia solani* in tomatoes with *Trichoderma harzianum* mutants. Electron J Biotechnol 13(2):1–11

Monteiro VN, Steindorff AS, Almeida FBR, Lopes FAC, Ulhoa CJ, Félix CR, Silva RN (2015) *Trichoderma reesei* mycoparasitism against *Pythium ultimum* is coordinated by G-alpha protein GNA1 signaling. J Microb Biochem Technol 7:1–7

Morán-Diez E, Rubio B, Domínguez S, Hermosa R, Monte E, Nicolás C (2012) Transcriptomic response of *Arabidopsis thaliana* after 24h incubation with the biocontrol fungus *Trichoderma harzianum*. J Plant Physiol 169:614–620

Moretto KCK, Gimenes-Fernandes N, Dos Santos JM (2001) Influence of *Trichoderma* spp. on *Colletotrichum acutatum* mycelial growth and morphology on infection of 'Tahiti' lime detached flowers. Summa Phytopathol 27(4):357–364

Morsy EM, Abdel-Kawi KA, Khalil MNA (2009) Efficiency of *Trichoderma viride* and *Bacillus subtilis* as biocontrol agents against *Fusarium solani* on tomato plants. Egypt J Phytopathol 37(1):47–57

Mukherjee PK, Raghu K (1997) Effect of temperature on antagonistic and biocontrol potential of *Trichoderma* sp. on *Sclerotium-rolfsii*. Mycopathologia 139(3):151–155

Mukherjee PK, Latha J, Hadar R, Horwitz BA (2003) TmkA, a mitogen-activated protein kinase of *Trichoderma virens*, is involved in biocontrol properties and repression of conidiation in the dark. Eukaryot Cell 2(3):446–455

Mukherjee PK, Latha J, Hadar R, Horwitz BA (2004) Role of two G-protein alpha subunits, TgaA and TgaB, in the antagonism of plant pathogens by *Trichoderma virens*. Appl Environ Microbiol 70:542–549

Mukherjee M, Mukherjee PK, Kale SP (2007) cAMP signalling is involved in growth, germination, mycoparasitism and secondary metabolism in *Trichoderma virens*. Microbiology 153: 1734–1742

Muriungi JS, Mutitu EW, Siboe MG (2013) Biocontrol of *Fusarium* root rot in beans by antagonistic *Trichoderma* fungi. Int J Agric Sci 3(7):550–557

Muthukumar A, Eswaran A, Sanjeevkumas K (2011) Exploitation of *Trichoderma* species on the growth of *Pythium Aphanidermatum* in Chilli. Braz J Microbiol 42(4):1598–1607

Nascimento SR, Steindorff AS, Ulhoa CJ, Félix CR (2009) Involvement of G-alpha protein GNA3 in production of cell wall-degrading enzymes by *Trichoderma reesei* (Hypocrea jecorina) during mycoparasitism against *Pythium ultimum*. Biotechnol Lett 31:531–536

Naseby DC, Pascual JA, Lynch JM (2000) Effect of biocontrol strains of *Trichoderma* on plant growth, *Pythium ultimum* populations, soil microbial communities and soil enzyme activities. J Appl Microbiol 88(1):161–169

Naserinasab F, Sahebani N, Etebarian HR (2011) Biological control of *Meloidogyne javanica* by *Trichoderma harzianum* BI and salicylic acid on tomato. Afr J Food Sci 5(3):276–280

Neer EJ (1995) Heterotrimeric G proteins: organizers of transmembrane signals. Cell 80(2): 249–257

Nielsen SL, Larsen J (2004) Two *Trichoderma harzianum*-based bio-control agents reduce tomato root infection with *Spongospora subterranea* (Wallr.) Lagerh., f. sp. subterranea, the vector of Potato mop-top virus. J Plant Dis Protect 111(2):145–150

Oerke EC, Dehne HW (2004) Safeguarding production losses in major crops and the role of crop protection. Crop Prot 23:275–285

Omann M, Zeilinger S (2010) How a mycoparasite employs g-protein signaling: using the example of *Trichoderma*. J Signal Transduct 123126. doi:10.1155/2010/123126

Omero C, Inbar J, Rocha-Ramirez V, Herrera-Estrella A, Chet I, Horwitz BA (1999) G protein activators and cAMP promote mycoparasitic behaviour in *Trichoderma harzianum*. Mycol Res 103:1637–1642

Pal KK, McSpadden Gardener B (2006) Biological control of plant pathogens. The Plant Health Instructor. 2:1117–1142

Pandey KK, Uapadhyay JP (1997) Selection of potential biocontrol agents based on production of volatile and non volatile antibiotics. Veg Sci 24(2):140–143

Patale SS, Mukadam DS (2011) Management of plant pathogenic fungi by using *Trichoderma* species. Biosci Discov 2(1):36–37

Patil A, Laddha A, Lunge A, Paikrao H, Mahure S (2012) In vitro antagonistic properties of selected *Trichoderma* species against tomato root rot causing *Pythium* species. Int J Environ Sci Techniol 1(4):302–315

Persoon CH (1794) Disposita methodical fungorum. Römers Neues Mag Bot 1:81–128.

Peterson R, Nevalainen H (2012) *Trichoderma reesei* RUT-C30 – thirty years of strain improvement. Microbiology 158:58–68

Pieterse CMJ, Van der Does D, Zamioudis C, Leon-Reyes A, Van Wees SCM (2012) Hormonal modulation of plant immunity. Annu Rev Cell Dev Biol 28:489–521

Poosapati S, Ravulapalli PD, Tippirishetty N, Vishwanathaswamy DK, Chunduri S (2014) Selection of high temperature and salinity tolerant *Trichoderma* isolates with antagonistic activity against *Sclerotium rolfsii*. SpringerPlus 3:641

Rahman MA, Razvy MA, Alam MF (2013) Antagonistic activities of *Trichoderma* strains against chili anthracnose pathogen. Int J Microbiol Mycol 1(1):7–22

Rajathilagam R, Kannabiran B (2001) Antagonistic effects of *Trichoderma viride* against anthracnose fungus *Colletotrichum capsici*. Indian Phytopathol 54(1):135–136

Rajendiran R, Jegadeeshkumar D, Sureshkumar BT, Nisha T (2010) In vitro assessment of antagonistic activity of *Trichoderma viride* against post harvest pathogens. J Agric Technol 6(1):31–35

Rajeswari P, Kannabiran B (2011) In vitro effects of antagonistic microorganisms on *Fusarium oxysporum* (Schlecht. Emend. Synd & Hans) infecting *Arachis hypogaea* L. J Phytology 3(3):83–85

Rao MS, Reddy PP, Nagesh M (1998) Evaluation of plant based formulations of *Trichoderma harzianum* for the management of *Meloidogyne incognita* on egg plant. Nematol Mediterr 26:59–62

Reithner B, Brunner K, Schuhmacher R, Peissl I, Seidl V, Krska R, Zeilinger S (2005) The G protein alpha subunit Tga1 of *Trichoderma atroviride* is involved in chitinase formation and differential production of antifungal metabolites. Fungal Genet Biol 42: 749–760

Reithner B, Schuhmacher R, Stoppacher N, Pucher M, Brunner K, Zeilinger S (2007) Signaling via the *Trichoderma atroviride* mitogen-activated protein kinase Tmk 1 differentially affects mycoparasitism and plant protection. Fungal Genet Biol 44:1123–1133

Rocha-Ramirez V, Omero C, Chet I, Horwitz BA, Herrera-Estrella A (2002) *Trichoderma atroviride* G-protein alpha-subunit gene tga1 is involved in mycoparasitic coiling and conidiation. Eukaryot Cell 1:594–605

Rocha-Ramírez V, Omero C, Chet I, Horwitz BA, Herrera-Estrella A (2002) *Trichoderma atroviride* G-protein α-subunit gene *tga1* is involved in mycoparasitic coiling and conidiation. Eukaryot Cell 1:594–605

Román E, Arana DM, Nombela C, Alonso-Monge R, Pla J (2007) MAP kinase pathways as regulators of fungal virulence. Trends Microbiol 15:181–190

Ru Z, Di W (2012) *Trichoderma* spp. from rhizosphere soil and their antagonism against *Fusarium sambucinum*. Afr J Biotechnol 11(18):4180–4186

Sahebani N, Hadavi N (2008) Biological control of the root-knot nematode *Meloidogyne javanica* by *Trichoderma harzianum*. Soil Biol Biochem 40:2016–2020

Saloheimo M, Pakula TM (2012) The cargo and the transport system: secreted proteins and protein secretion in *Trichoderma reesei* (*Hypocrea jecorina*). Microbiology 158:46–57

Sarma BK, Yadav SK, Patel JS, Singh HB (2014) Molecular mechanisms of interactions of *Trichoderma* with other fungal species. Open Mycol J 8:140–147

Schirmbock M, Lorito M, Wang YL, Hayes C, Arisan-Atac I, Scala F, Harman G, Kubicek C (1994) Parallel formation and synergism of hydrolytic enzymes and peptaibol antibiotics, molecular mechanisms involved in the antagonistic action of *Trichoderma harzianum* against phytopathogenic fungi. Appl Environ Microbiol 60:4344–4370

Schmoll M (2008) The information highways of a biotechnological workhorse-signal transduction in *Hypocrea jecorina*. BMC Genomics 9:430

Schmoll M, Schuster A, Silva RDN, Kubicek CP (2009) The G-alpha protein GNA3 of *Hypocrea jecorina* (anamorph *Trichoderma reesei*) regulates cellulase gene expression in the presence of light. Eukaryot Cell 8(3):410–420

Seema M, Devaki NS (2012) In vitro evaluation of biological control agents against *Rhizoctonia solani*. J Agric Technol 8(1):233–240

Seifullah P, Thomas BJ (1996) Studies on the parasitism of *Globodera rostochiensis* by *Trichoderma harzianum* using low temperature scanning electron microscopy. Afro-Asian J Nematol 6:117–122

Sewariya SVK, Shrivastava R, Prasad G, Arora K (2012) In-vitro evaluation of novel synthetic compounds against *Fusarium*. Int J Pharm Biol Sci 3(3):910–920

Sharma P (2011) Complexity of Trichoderma-Fusarium interaction and manifestation of biological control. Aust J Crop Sci 5(8):1027–1038

Sharma R, Joshi A, Dhaker RC (2012a) A brief review on mechanism of *Trichoderma* fungus use as biological control agents. Int J Innov Bio-Sci 2(4):200–210

Sharma RN, Maharshi RP, Gaur RB (2012b) Biocontrol of post-harvest green mould rot (*Penicillium digitatum*) of kinnow fruits using microbial antagonists. Indian Phytopathol 65(3):276–281

Sharon E, Bar-Eyal M, Chet I, Herrera-Estrella A, Kleifeld O, Spiegel Y (2001) Biological control of the root-knot nematode *Meloidogyne javanica* by *Trichoderma harzianum*. Phytopathology 91:687–693

Shovan LR, Bhuiyan MKA, Begum JA, Pervez Z (2008) In vitro control of *Colletotrichum dematium* causing anthracnose of soybean by fungicides, plant extracts and *Trichoderma harzianum*. Int J Sustain Crop Prod 3(3):10–17

Simon C, Sivasithamparam M (1988) Interactions among *G. graminis* var tritici, *Trichoderma koningii* and Soil Bacteria. Can J Microbiol 34:871–876

Singh US, Zaidi NW, Joshi D, Khan T, John D, Bajpai A, Singh N, Rohilla R (2004) Trichoderma: a microbe with multifaceted activity. Annu Rev Plant Physiol 3:33–75

Singh US, Zaidi NW, Joshi D, Varshney S Khan T (2006) Current status of *Trichoderma* as a biocontrol agent. In: Ramanujam B, Rabindra R J (eds) Current status of biological control of plant diseases using antagonistic organisms. Technical Document No.57. Project Directorate of Biological Control, Banglore, India, pp 13–48

Singh A, Shahid M, Srivastava M, Kumar V, Bansal A (2013) Antagonistic activity of *Trichoderma viride* isolate against different pathogen of *Fusarium oxysporum* isolated from legume crop of U.P. Progress Res 8(1):47–50

Singh A, Srivastava M, Kumar V, Sharma A, Pandey S, Shahid M (2014) Exploration and interaction of *Trichoderma* species and their metabolites by confrontation assay against *Pythium aphanidermatum*. Int J Sci Res 3(7):44–48

Sivan A, Chet I (1993) Integrated control of Fusarium crown and root of tomato with *Trichoderma harzianum* in combination with methyl bromide or soil solarization. Crop Prot 12:380–386

Soliman HM, El-Metwally MA, Elkahky MT, Badawi WE (2015) Alternatives to chemical control of grey mold disease on cucumber caused by *Botrytis cinerea* Pers. Asian J Plant Pathol 9:1–15

Song XY, Shen QT, Xie ST, Chen XL, Sun CY, Zhang YZ (2006) Broad spectrum antimicrobial activity and high stability of Trichokonins from *Trichoderma koningii* SMF2 against plant pathogens. FEMS Microbiol Lett 260:119–125

Suarez B, Rey M, Castillo P (2004) Isolation and characterization of PRA1, a trypsin-like protease from the biocontrol agent*Trichoderma harzianum* CECT 2413 displaying nematicidal activity. Appl Microbiol Biotechnol 65:46–55

Sundaramoorthy S, Balabaskar P (2013) Biocontrol efficacy of *Trichoderma* spp. against wilt of tomato caused by *Fusarium oxysporum* f. sp. Lycopersici. J Appl Biol Biotechnol 1(3): 036–040

Svetlana Z, Stojanovic S, Ivanovic Z, Tatjana popovic VG, balaz J (2010) Screening of antagonistic activity of microorganisms against *Colletotrichum acutatum* and *Colletotrichum gloeosporioides*. Arch Biol Sci Belgrade 62(3):611–623

Szabó M, Csepregi K, Gálber M, Virányi F, Fekete C (2012) Control plant-parasitic nematodes with *Trichoderma* species and nematode-trapping fungi: the role of *chi18-5* and *chi18-12* genes in nematode egg-parasitism. Biol Control 63(2):121–128

Tapwal A, Thakur G, Chandra S, Tyagi A (2015) In-vitro evaluation of *Trichoderma* species against seed borne pathogens. Int J Biol Chem Sci 1(10):14–19

Van der Ent S, Van Wees SCM, Pieterse CMJ (2009) Jasmonate signaling in plant interactions with resistance-inducing beneficial microbes. Phytochemistry 70:1581–1588

Verma M, Brar SK, Tyagi RD, Surampalli RY, Valéro JR (2007) Antagonistic fungi, *Trichoderma spp.*: panoply of biological control. Biochem Eng J 37:1–20

Vey A, Hoagland RE, Butt TM (2001) Toxic metabolites of fungal biocontrol agents. In: Butt TM, Jackson C, Magan N (eds) Fungi as biocontrol agents: progress, problems and potential. CAB International, Bristol, pp 311–346

Viterbo A, Harel M, Horwitz BA, Chet I, Mukherjee PK (2005) *Trichoderma* MAP-kinase signaling is involved in induction of plant systemic resistance. Appl Environ Microbiol 71:6241–6246

Wang Q, Xu L (2012) Beauvericin, a bioactive compound produced by fungi: a short review. Molecules 17:2367–2377

Wang M, Dong Y, Zhao Q, Wang F, Liu K, Jiang B, Fang X (2014) Identification of the role of a MAP kinase Tmk2 in *Hypocrea jecorina* (*Trichoderma reesei*). Sci Rep 4:6732. doi:10.1038/srep06732

Weindling R (1934) Studies on a lethal principle effective in the parasitic action of *Trichoderma lignorum* on *Rhizoctonia solani* and other soil fungi. Phytopathology 24:1153–1179

Windham GL, Windham MT, Williams WP (1989) Effects of *Trichoderma* spp. on maize growth and *Meloidogyne arenaria* reproduction. Plant Dis 73:493–494

Xu JR, Hamer JE (1996) MAP Kinase and cAMP signaling regulate infection structure formation and pathogenic growth in the rice blast fungus *Magnaporte grisea*. Genes Dev 10:2696–2706

Yamano T, Hemmi S, Yamamoto L, Tsubaki K (1970) Trichoviridin. A new antibiotic. Japanese Kokai 15435. Chem Abs 73, 65093

Yedidia I, Benhamou N, Chet I (1999) Induction of defense responses in cucumber plants (*Cucumis sativus* L.) by the biocontrol agent *Trichoderma harzianum*. Appl Environ Microbiol 65:1061–1070

Zaidi NW, Singh US (2004a) Use of farmyard manure for mass multiplication and delivery of biocontrol agents, *Trichoderma harzianum* and *Pseudomonas fluorescens*. Asian Agric Hist 8(4):297–304

Zaidi NW, Singh US (2004b) Mass multiplication and delivery of *Trichoderma* and *Pseudomonas*. J Mycol Plant Pathol 34(3):732–741

Zamani M, Tehrani AS, Ahmadzadeh M, Abadi AA (2006) Effect of fluorescent P*seudomonades* and *Trichoderma* sp. and their combination with two chemicals on *Penicillium digitatum* caused agent of citrus green mold. Commun Agric Appl Biol Sci 71:1301–1310

Zeilinger S, Omann M (2007) *Trichoderma* biocontrol: signal transduction pathways involved in host sensing and mycoparasitism. Gene Regul Syst Biol 1:227–234

Zeilinger S, Reithner B, Scala V, Peissl I, Lorito M, Mach RL (2005) Signal transduction by Tga3, a novel G protein alpha subunit of *Trichoderma atroviride*. Appl Environ Microbiol 71:1591–1597

Biotechnology in the Diagnosis and Management of Infectious Diseases

Amit Kumar, R.S. Sengar, Raj Singh, Anju Rani, Vineet Girdharwal, and Gyanika Shukla

Introduction

The most significant fungus characteristics used for identification are spores and spore-bearing structure (sporophores) and, to some extent, the characteristics of the fungus body (mycelium). These items are examined under a compound microscope directly after removal from the specimen. The specimen is often kept moist for a few days to promote spore development. Alternatively, the fungus may be isolated and grown on artificial media and identification on the basis of spores produced on the media. For some fungi, special nutrient media have been developed that allow selective growth only of the particular fungus, allowing quick identification of the fungus.

In many fungi, hyphae in a colony or in adjacent colonies fuse (hyphal anastomosis). If the hyphae that fuse carry genetically different nuclei, the colony that is produced is a heterocaryon. Many fungi, however, have genetic systems that prevent mating between genetically identical cells. If the hyphae that come in contact belong to different strains of the same species but are of the same mating type, their encounter may result in vegetative incompatibility. Thus, the resulting vegetative incompatibility between colonies of various strains belonging to the same species is

A. Kumar (✉) • G. Shukla
Department of Biotechnology, Swami Vivekanand Subharti University, Meerut, Uttar Pradesh, India
e-mail: amit.agbiotech1581@gmail.com

R. Singh • A. Rani
Department of Botany, Swami Vivekanand Subharti University, Meerut, Uttar Pradesh, India

V. Girdharwal
Department of Zoology, Swami Vivekanand Subharti University, Meerut, Uttar Pradesh, India

R.S. Sengar
Department of Agriculture Biotechnology, Sardar Vallabhbhai Patel University of Agriculture and Technology, Meerut, Uttar Pradesh, India

P. Kumar et al. (eds.), *Current Trends in Plant Disease Diagnostics and Management Practices*, Fungal Biology, DOI 10.1007/978-3-319-27312-9_15

used to type the atrains as belonging to different incompatibility groups constituting different biological species.

In recent years, immunoassay techniques, often involving monoclonal antibodies against specific protein of a fungus conjugated with a fluorescent compound, which is used for the identification and detection of certain fungi.

The advent of molecular techniques, particularly of PCR (polymerase chain reaction), of quick and inexpensive sequencing of DNA, and the accumulation of a relatively large databank of ribosomal DNA sequences have revolutionized both the lower limits of detection of pathogens and the accuracy and rapidity of their identification. These developments have made possible the detection of pathogens within plant tissues in the early stages of infection while there is still a minimal presence of the pathogen and early intervention may prevent an epidemic. They have also made possible a definitive identification of the pathogen by using DNA probes of known pathogens and, furthermore, they have made possible the quantification of the pathogen within, or in a mixture with, plant tissue, such as seed. Most DNA primers are for internal transcribed sequences of ribosomal DNA. The methodology, however, improves constantly and quickly. Much more sensitive and specific sets of primers have been designed which is based on families of highly repeated DNA that were ten times more sensitive than primers directed at internal transcribed spacer sequences for ribosomal DNA.

Plant-Pathogen Interaction

The pathogens themselves are of two basic types: first is necrotrophs, which kill the host and feed on it, by which toxic substance are produced; and second is biotrophs, which complete their life cycle on the living host. Fungi may be of either type, whereas viruses are obligate biotrophs. In between these two extremes are hemibiotrophs; these are initially biotrohic but then switch to being necrotrophic. It is also clear that some organisms are pathogenic to some plants but not other. Why this is the case is beginning to be understood at the molecular level. There are two extremes of diseases resistance exhibited by plants: non host resistant to a specific parasite or pathogen. The resistance mechanism that brings this about is conserved across the species and is complex- involving many inheritable markers – but is consequently robust in nature. This durable resistance is perhaps the ultimate goal of the biotechnologist, contrasting favorably with host resistance. Host resistance is exhibited by single plant cultivars to a specific pathogen. Other members of the species may be susceptible to the infection. This type of resistance is due to interaction between a single plant protein a single protein from the pathogen (gene-gene interaction). Normally it leads to a hypersensitive response that blocks the spread of the pathogen. It is not as robust as non host resistance but at present it is achievable although possible only for a short term benefit. As with the pathogen, between the two extremes of resistance there is a race non specific resistance or a general response. This is also polygenic response but individually the genes would probably not provide a successful defense.

Fungi as a Pathogen

Fungi are the most important plant pathogens. The large number of cellular pathogens of plants are found among the fungi, the losses of food caused by fungal disease have had profound effects on humanity. This is in a number of ways. They can damage plants in the field and after harvest (rooting etc.) and they can contaminate foodstuffs with extreme toxins such as aflotoxins. There are over 100,000 known species of fungi. Most are examples of saprophytes, but the fungi alos include symbionts, such as the mycorrhiza, and perhaps more than 8,000 species that cause plant diseases. Modes of entry include wound sites and natural opening, like stomata, to take entry in to the plants by degrading the macromolecules of cell wall. There are too many fungal pathogens to go into any detail of the diseases they cause or their modes of action, but some specific examples are given in Table 1. Included is ***phytophthora infestans*** (which is a classic example of an Oomycetes-group pathogen), the organism responsible for the potato blight that caused the great Irish famine of 1846. Until recently, Oomycetes were classified as fungi; however, biochemical and molecular studies indicate that these water moulds should now be classified separately. The cell wall of these organisms is composed of cellulosic compounds and glycan, not chitin, and the nuclei within the hypha are diploid, not haploid, as in fungi.

Natural Fungal Resistant Pathway

Plant defence systems are complex but effective. It would be wrong to give the impression that plants have no resistance against pathogen attack, for it is clear that they do. Plant cannot produce specific cell against microbes attack because plants do not have an immune system, But Plants have adopted general defence system because they have many similarities with innate immunity of mammals. There is also a lot of overlap between the plant's response to pathogens and plant pests. The

Table 1 Name of fungal disease and its causing agent

Class	Disease name	Disease causing agent
Oomycetes	Late blight of potato	*Phytophthora infestans*
Oomycetes	Downy mildew of grape	*Plasmopara viticola*
Chytridiomycetes	Black wart of potato	*Synchytrium endobioticum*
Ascomycetes	Peach leaf curl diseases	*Taphrina deformans*
Ascomycetes	Powdery mildew of pea	*Erysiphe polygony*
Deuteromycetes	Ergot of bajra	*Claviceps purphrea*
Deuteromycetes	Apple scab	*Venturia inagnalis*
Deuteromycetes	Wilt disease of potato	*Fusarium oxysporum sub sp. Solani*
Deuteromycetes	Red rot of sugarcane	*Calletotrichum falcatum*
Basidiomycetes	Loose smut of wheat	*Ustilago tritici*
Basidiomycetes	Rust of wheat	*Puccinia graminis tritici*

general cellular damage caused by both can act as a signal to trigger general defence systems. There are advantages to the plant having such general systems because pests often act as vectors for pathogen. So we can say, plant also have systems that respond to specific signals. Four different levels of defence against fungi present in plants.

Anatomical Defence

Many fungi attack on plants though wounds. This is because plants developed structural system that stop fungi access to living cells (the first line of defence). These can be thick layers of protective material, such as cuticle and bark. Once this defence is breached then cascades of defence systems come into play.

Pre-existing Protein and Chemical Protection

The second line of defence is produced by the plants during development, growth and defense like protein which is made up of antimicrobial proteins. The defense proteins are similar to those found in insects and mammals where they play an important role in defence against infectious agents. Their structure has a conserved three dimensional folding pattern, which suggests that they represent a super family of peptides that pre dates the divergence of plants and animals. Some defense cause increased branching in fungi, while others simply slow growth. They are frequently associated with seeds at the time of germination, when they may be released into the environment and create a microenvironment around the seed suppresses fungal growth. Many of the large number of small chemicals made as secondary products may also have antimicrobial properties. These protein and chemical may simply deter pest or pathogen growth, or they may actually be toxic to them.

Inducible Systems

Elicitor Response

The third level of defence is a synthesis of protein de novo by which plants protects from pathogen. It would be costly for the plant to have its defence system permanently switched on, so there are mechanisms by which detect the infection and then turn on the defence system. These mechanisms will be discussed in Fig. 1

When pathogen arrives and gain entry to a plant cell, it may affect resistance to infection. The elicitors are of several types, and the response is complex and

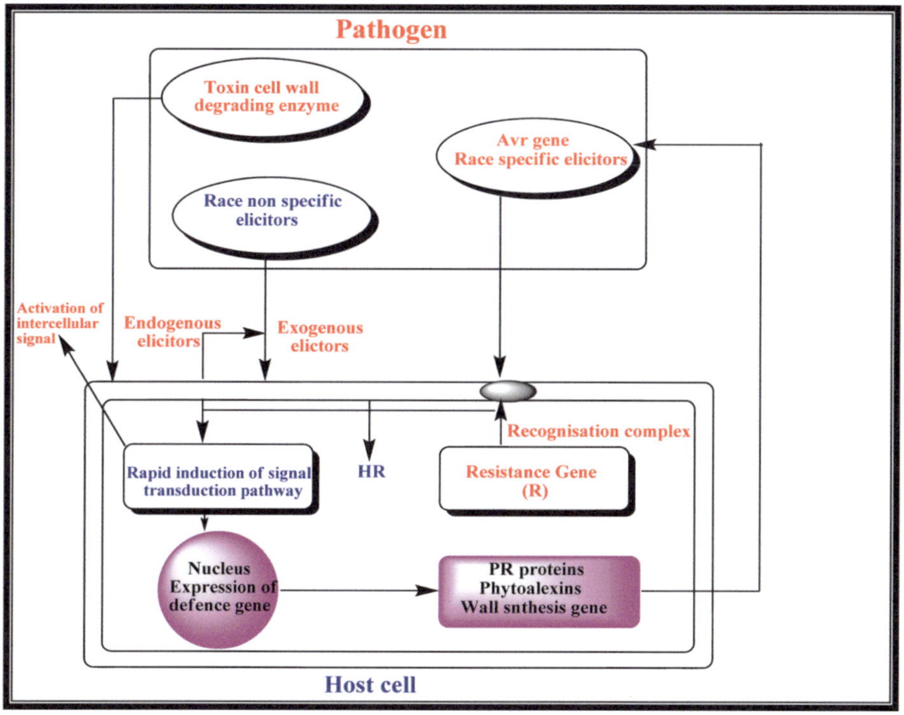

Fig. 1 Representation of host defence systems are induced by endogenous and exogenous elicitors and by the interaction between plant resistance genes (*R*) and the pathogen avirulence gene (Avr)

overlapping in nature. First, there may be a local response that involves interactions with substance which is released by the pathogen or the host itself, in response to pathogen attack (endogenous elicitors), with receptor molecules on the host. These interactions may stimulate general defence systems in both host and non host plants, or they may be race specific interactions, inducing defence responses only in specific cultivars. Both these interactions may lead to a cascade of reactions that may include the hypersensitive response (HR). These interactions are diagrammatically show in Fig. 1 and Table 2 show a range of elicitors and their effects.

General Response

In the first stages of an infection, there is normally damage to the plant cell wall. Fungi release proteolytic enzyme that cause the release of wall fragments such as pectic oligomers. These often act as a general signals endogenous elicitors which bind to specific receptors, setting off a cascade of reaction that lead to the induction of specific defence gene. Many structural components of the cell wall are thick by these defence genes which code for enzymes, enzymes of secondary metabolism,

Table 2 Examples of elicitors

S. No.	Elicitor	Type	Biological response
1.	β- Glucans	Pathogen associated molecular pattern	Induction of defence response in rice, tobacco
2.	Chitin	Pathogen associated molecular pattern	Induction of defence response in tomato, Arabidopsis, rice, wheat and barley
3.	Avirulence protein	Race specific	Hypersensitive response in a range of plants
4.	Peptide toxins	Race specific	Programmed cell death in oat
5.	Endoxylanase	Race specific	Hyper sensitive response and defence genes in tobacco
6.	Pectolytic enzymes	General	Protein inhibitors and defence genes in Arabidopsis and tomato

lectins and many so called pathogenesis related protein (PR). PR proteins include chitinases and β-1,3-glucanase, protease inhibitors, non specific lipid-transfer proteins, ribosomal inhibitor proteins and various antimicrobial proteins. These antimicrobial proteins include defensins such as SN1, which is active against fungal pathogen in potato. One important point of relevance to the use of the gene in transgenic experiments is that a number of the Pathogen related gene has been identified as latex allergens. These are the proteins, which are presents in most of the fruit species and latex products which show allergenic response in many people.

The effect of synthesizing the defence proteins depends, in part, on the pathogen. For instance, if fungal hyphae breach the cells defence, then the chitinases and glucanases may cause some degradation of the pathogen's cell wall. This will lead to the production of chitin and β-1,3-glucan oligomers. These compounds may act as signal molecules (exogenous elicitors), which bid to membrane receptors and re enforce the induction of the defence systems. The cells may also produce phytoalexins (phenolic protein) that kill any pathogens al well as the cell in the vincinity of the infection, therefore limiting the spread.

Race Specific Response

The pathogen carries the avirulence gene (Avr), which codes for a protein that is identified by a specific receptor protein in the plant cell, encoded by the resistance gene (R). The interaction induce the HR, which is manifested as a local necrosis that develops through a NADPH (reduced nicotinamide adenine dinucleotide phosphate) dependent oxidative burst and the release of phenolics and nitric oxide. Most of the pathogenesis related protein active by the activation of signaling pathway. The important feature of this system is that if protein is absent, then the pathogen will cause disease Fig. 2.

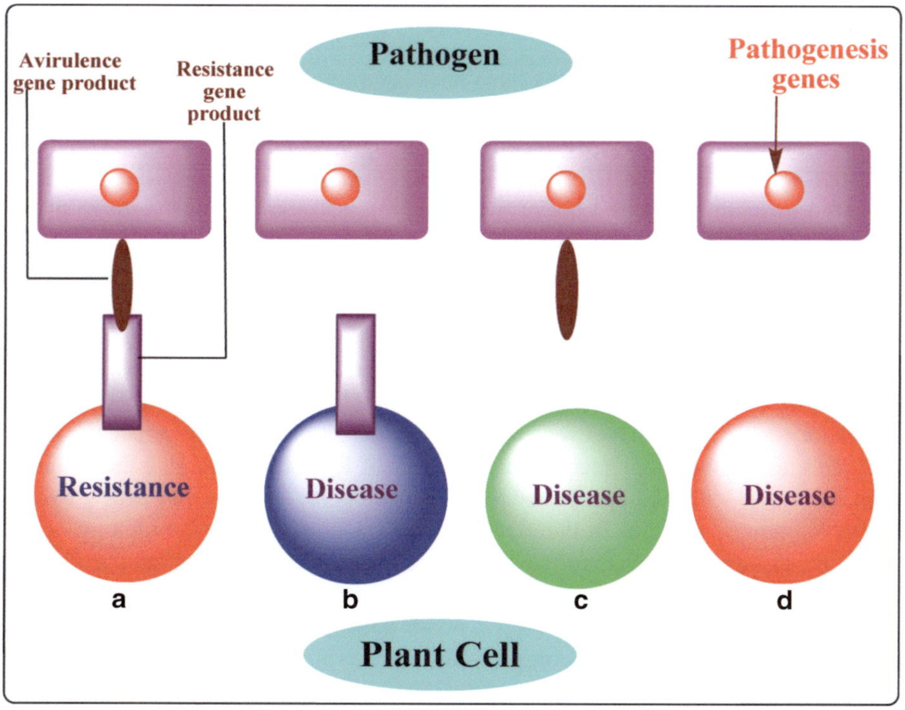

Fig. 2 Host resistant interaction. In (*a*) the plant cell contains the resistance gene (R) and the pathogen produces the avirulence gene product (Avr), defence system are activated and the plant shows resistant to infection. (*b*) Pathogen is not producing Avr protein, HR reaction is not induced and the interaction result in disease. (*c*) Host cell is not producing R protein so disease results. (*d*) Neither the R protein nor the Avr protein is produced so disease caused

The gene system is one of the best ways by which plants have switch on resistance system, and interaction between plants and fungi have been shown. It is becoming clear that the many of the avirulence proteins fall into a class protein transported into the plant cell by the type III protein secretion system (TTSS). It is postulated that they might functions as virulence factors, subverting cellular function through interaction with plant encoded pathogenicity target.

When fungal hyphae break through the cell wall into a cell, they produce a complex cellular structure, the haustoria, that drains the required nutrients from the cell. Interestingly, the haustoria does not actually penetrate the host cell's membrane, yet the fungus is able to highjack the cell's content. Recent work with fungi and oomycete pathogens indicates that these organisms inject small proteins into the host plant cells that are designed to alter the host's metabolism, to the benefit of the pathogen, and to disrupt any defence pathways. It is these proteins that can stimulate the HR response.

Elicitor Receptors

Many examples of the R proteins have now been identified: the annotation of the Arabidopsis genome sequence has indicated there are about 100 R loci distributed throughout the genome. PAMP receptors (Pathogen associated molecular Pattern) and TTSS proteins (Type III protein secretion system) show many similarities in structure to mammalian and insect receptor proteins required for the onset of the innate immune response and involved in the sensing of pathogen derived factors.

Xa21 is an R protein from rice that is an LRR receptor kinase. It confers cultivars-specific resistance to *xanthomonas oryzae* pv. Oryzae strains that express ArvXa21 avirulence protein. It resembles PAMP receptor in structure: both human Toll like receptor and the flagellin receptor FLS2 from *Arabidopsis*. Comparisons like these indicate that plants use receptors of similar structure for the sensing off different types of elicitor and that these have evolved from a limited number of ancient receptors. It is intriguing to consider that such a small number of R proteins can give the diversity required to deal with the large number of plant pathogens, and how they evolve to deal with evolutionary changes in the elicitor protein. One factors in the biology of the pathogen, and as such they do not evolve at high rates. There is also evidence indicating that to meet with the changes that do occurs, some variability in the plant R proteins comes from the fact that the LRR doain does nutate, within a highly variable regions, and that mutations encoding new amino acids are selected for. The R gene loci are very polymorphic in wild populations, so with each R gene allele being present at a low frequency there is limited selection for virulence in the pathogen population.

Systemic Response

The indication of local defence pathways may lead to the induction of intercellular signals that produce a systemic response, termed systemic acquired resistance (SAR). Both avirulent and virulent pathogen may result in the induction of SAR, although it is usually a slower process in the case of virulent pathogen. SAR has two phases: first is the initiation phase and second is the maintenance phase. In the initiation phase, cells at the foci of the infection release signal molecules, typically salicylic acid, into the phloem. These are transported to target cells in other parts of the plant where SAR gene are expressed, thus giving the plant some level of resistance against infection. In the longer term, a maintenance phase is reached that may last for weeks or even the full life of the plant, in which there is quasi steady state resistance against virulent pathogens.

Biotechnological Approach to Fungal Resistance

Biotechnology has rapid emerged as an area of activity having a marked realized as well as potential impact on virtually all domains of human welfare and protecting the environment. In biotechnology various molecular diagnostic tools have been used for the improvement of accuracy, efficacy and identification of disease causing agents, which are also used for the characterization of the diversity of fungal pathogen. In the Biotechnology various technique are involve for the disease diagnosis and management but in the biotechnology various molecular technique involve especially molecular marker for the identification, detection, quantification and characterization of fungal pathogens that incites diseases in plants.

The molecular marker are no longer looked upon as simple DNA fingerprinting markers in variability studies or as mere forensic tools, but, they are constantly being modified to enhance their utility and to bring about automation in the process of genome analysis (Joshi et al. 2011). From the last decade Molecular marker play important role for the detection of phyto-pathogen and also has been used for the identification of unknown species from their nucleotide sequence. Non radioactive probes have been used for the detection of fungal pathogen (Singh 2009). Molecular data of fungi combined with classical characterization provide new thing about the fungal interactions in terrestrial communities (Bahnweg et al. 1998). The application of molecular marker in the plant pathology has improved the ability to detect plant pathogens.

Most of the molecular markers like RFLPs (Botstein et al. 1980), AFLPs (Vos et al. 1995), STS (Powell et al. 1996), RAPD (Williams et al. 1991) are used for the molecular mapping and characterization of disease resistant gene of plant pathogen. Molecular markers are also used for mapping and cloning of various R-proteins. These cloned R- protein have been transferred for the improvement of susceptible cultivars for high yield. The present status of molecular markers is that these are used for disease diagnosis, molecular characterization and molecular mapping of disease resistance gene.

Molecular Approach for Detection of Pathogenic Fungi

Polymerase Chain Reaction (PCR)

There are two important reasons of amplifying DNA. First it provides a limitless supply of the material for analysis. Second the specific amplification of a region of a genome provides a relatively easy way to purify that segment from the bulk. Either these purposes can be solved by performing in vitro amplification of DNA or Polymerase chain reaction. PCR allows the amplification of specific DNA sequences by repeated cycles of denaturation, polymerization and elongation at different temperature using specific primers, DNA polymerases which were isolated from thermostable organisms *Thermus aquaticus* enzyme and deoxyribonucleotide triphosphates (dNTPs) (Mullis and Faloona 1987). The amplified DNA fragment can be visualized by electrophoresis or fluorometric assays (Fraaije et al. 1999). The specific size of the DNA band indicates the presence of target pathogen in the sample. Main advantages of PCR technique are to provide high specificity, sensitivity and reliability.

DNA Extraction from the Sample

Most of the protocols are available for the isolation of DNA from fungi or infected plants (Niu et al. 2008; Chi et al. 2009; Zhang et al. 2009; Feng et al. 2010; González Mendoza et al. 2010; Zelaya Molina et al. 2011) but there are no universally validated protocols for the fungi and infected plant. Isolation of nucleic acid is a time consuming and costly and time consuming method. Many step have been used to isolate the nucleic acid from the fungus. In the Isolation of nucleic acid few microliters of extract loaded and immobilized on FTA cards. Many scientists used FTA cards for the detection of nucleic acid. Nucleic acids of *Aspergillus oryzae* could be detected through FTA card reported (Suzuki et al. 2006). Grund et al. (2010) used FTA card for the identification of plant pathogen including *oomycetes* such as *phytopthora* and some filamentous fungi such as *fusarium*.

Design of Primers and Probes

A primer is a synthetic oligonucleotide which is used in PCR and DNA sequencing. These primers are specific sequence which is the reverse complement of a target DNA. In some real time PCR methods additional specific oligonucleotides are used, named probes, that hybridize with the target DNA between to the two primers. The design of primers and probes is crucial for PCR to be specific and efficient. Primer specificity relies on some points like; primer length of PCR should be 18–22 bp. Primer length affects the PCR efficiency. The primer length also influences the rate at which primer molecules pair with template DNA; this rate decreases as the primer length increases. Therefore, if the primer length are too long, complete pairing of the primer molecules may not occur during the time allowed for annealing. As, a result, PCR efficiency declines with primer length. Primer longer than 30 bases are rarely used for PCR. The annealing temperature is very important since the success and specificity of PCR depend on it. DNA amplified fragments size must be shorts enough to ensure efficiency of the reaction and high sensitivity (Singh and Singh 1997).

The first step for design probes and primers consist in the alignment of the sequence by the BLASTn program (http://blast.ncbi.nlm.nih.gov.Blast.cgi) (Altschul et al. 1997) using sequences from the GeneBank, EMBL and DDBJ databases. Partial or complete nucleotide sequences of many fungal genes are present at the NCBI (National Center of Biotechnology Information) http://www.ncbi.nlm.nih.gov/Genebank/). On the other hand, variable sequences are useful for the differentiation of pathogens at lower taxonomic levels and for the analysis of the molecular variability of fungal population in phylogenetic studies.

PCR Based Method for Detection of Plant Fungi

Conventional PCR

Conventional PCR is the basic method for the detection of fungal pathogen at different taxonomic levels. In this method primers are highly specific for genus and species (Jeeva et al. 2010). Identified *Sclerotium rolfsii* by conventional PCR method and (Torres-Calzada et al. 2011) developed *Colletotrichum capsici*, based in specific sequence of internal transcribed spacer (ITS) region.

Nested PCR

Nested PCR is a variation of the polymerase chain reaction, in that two pairs (instead of one pair) of PCR primers are used to amplify a fragment. The first pair of PCR primer amplify a fragment similar to a standard PCR. However, a second pair of primers called nested primers bind inside the first PCR product fragment to allow amplification of a second PCR product which is shorter that the first one. This method has been used for the identification and characterization of numerous fungi (Langrell et al. 2008; Hong et al. 2010; Meng and Wang 2010; Qin et al. 2011; Wu et al. 2011).

Multiplex PCR

Multiplex PCR approach is used when more than two primer involve in the same reaction. This PCR is highly sensitive for the detection of different DNA targets. This approach is highly useful for the identification of plant pathology since plants are usually infected by more than one pathogen. This PCR has been used for identification and differentiation of *Golovinomyces cichoracearum* and *Podosphaera xanthii* in sunflower (Chen et al. 2008), for detecting *Phytophthora lateralis* in cedar trees and water samples, including detection of an internal control in the same reaction (Dyer et al. 2001).

Reverse Transcription-PCR

This approach can be sued to amplify RNA sequence in DNA duplex. In this technique cDNA copy of the RNA is produced using the enzyme reverse transcriptase, this cDNA is then used for amplification. m-RNA is highly sensitive in the cell so it is degraded rapidly, the identification of m-RNA by RT-PCR is an accurate method for check cell viability (Sheridan et al. 1998). RT-PCR has been used to detect viable population of *Mycosphaerella graminicola* in wheat (Guo et al. 2005). So this approach highly valuable in phytopathology for the analysis of fungal gene expression during disease development (Yang et al. 2010).

Real Time PCR

This approach is highly useful method for the detection of plant pathogens. The real time PCR was developed to quantitate differences is m-RNA production, particularly in such cases where only small amounts of cells/tissue are available. The various methods of mRNA quantitation include northern hybridization, ribonuclease protection assays, situ hybridization, and RT-PCR. RT-PCR is the highly specific method and can discriminate between closely related m-RNAs. But conventional PCR does not yield truly and accurate quantitative result because ethidium bromide does not give a bright fluorescence, and when a band is detectable the logarithmic phase of amplification is already over. Real time PCR has been used to detect more than two pathogens in the same reaction (Moradi et al. 2010; Brandfass and Karlovsky 2006; Schroeder et al. 2006; Okubara et al. 2008).

Role of Molecular Marker in Phytopathology

Molecular marker may be defined as a DNA sequence used for chromosome mapping as it can be located at a specific site in a chromosome. This technique have been also useful for identifying specific sequences fused for the detection of fungi at very low taxonomic level, and even for differentiate strains of the same species with different host range. Some important marker are used for the identification and characterization of plant fungi.

Restriction Fragment Length Polymorphism (RFLP)

RFLP is generated by the presence and absence of a recognition site for the same restriction endonuclease in the same region of a chromosome from the different individuals of a species. As a result, the concerned restriction enzyme produces fragments of different lengths representing the same combined with hybridization with a labeled probe specific for that chromosomal region. Thus RFLP markers result from a combination of a specific restriction endonuclease and a specific DNA sequence used as probe. Largely, RFLPs markers have been used for the genetic diversity of microrrhizal and soil fungal communities (Thies 2007; Kim et al. 2010; Martínez García et al. 2011). This approach is also used for the differentiation of pathogenic fungi (Hyakumachi et al. 2005).

Random Amplified Polymorphic DNAs (RAPD)

RAPD is to use a single ten nucleoptide long randomly designed primer that binds to homologous regions in the genome. When two primers are bound in two strands of a double helix template in opposite orientation within an amplifiable distance, the region is amplified an appears as a band in electrophoretic gel. For allowing high primers template binding the usual annealing temperature in PCR reaction cycle is lowered from 55 °C making the reaction conditions less stringent. Mutation in primer binding site and insertions or deletions in the target fragment leads to polymorphism. However, due to possibility of generation of same fragment from different loci resulting in variation in band intensity and low reproducibility due to effect of PCR conditions under low stringency are the major bottleneck of RAPD analysis. This RAPD primer have been used for the specifically identify *Fusarium subglutinans* (Zaccaro et al. 2007) and *Guignardia citricarpa* (Stringari et al. 2009).

Amplified Fragment Length Polymorphism (AFLP)

AFLP is currently one of the most desired techniques for DNA fingerprinting from any origin and complexity. Being a combination of restriction digestion as well as PCR amplification, it bears the advantages of both RFLP and RAPD, providing higher reproducibility, entire genome coverage, high reproducibility and very high polymorphism. As RFLP, first step of AFLP is digestion of genomic DNA by restriction enzymes, but unlike RFLP, a combination of to restriction enzyme, ore rare cutter and one frequent cutter are used, so that three types of fragments are generated. Some of these will have restriction sites of frequent cutter are generated. Some of these will have restriction sites of frequent cutter at both end, some with rate cutting sites at both end and rest will be fragments having restriction site of frequent cutter at one end and that of rare cutter at the other end. This third group is the target DNA of PCR amplification. AFLP is used to screen a large number of anonymous markers which is distributed throughout the genome. This primer is highly reproducible and sensitive. As a result, AFLP has been used to differentiate fungal isolates at several taxonomic levels e.g. to distinguish *C.gossypii* var. *cephalosporioides* (Schmidt et al. 2004; Stewart et al. 2006) studied that AFLP markers have been used to separate non pathogenic strains of *Fusarium oxysporum* from those of *F. Commune*. AFLP markers have also been used to construct genetic linkage maps e.g. of Phytophthora infestans (VanderLee et al. 1997).

Microsatellites

Microsatellite based marker systems are principally based on either amplification of microsatellite regions by designing primers for the conserved flanking regions of microsatellites and observe the variation in repeat length and number or by amplifying internal regions of two microsatellites by using microsatellite specific primers. Microsatellite markers have high applicability in almost all the fields of molecular marker analysis because they are single locus, codominant and universal in nature as well as provide wide genome coverage. In addition to these, they are non-coding DNA sequence, so more useful in population genetic analysis where neutrality of marker system is an important requirement. Designing of microsatellite primers are time consuming, but provides more accuracy than any other marker system. In addition, once a set of microsatellite markers are developed for a species, it can be used for the study of genetic diversity of plant pathogenic fungi within species e.g. *Macrophomina phaseolina* (Jana et al. 2005), *Puccinia grainis* and *Puccinia triticina* (Barnes and Szabo 2007), *Sclrotinia subarctica* and *S. sclerotiorum* (Winton et al. 2007).

Molecular Detection of Plant Pathogens

Population of India is increasing continuously so high yield crops are need. People continued to suffer from hunger and malnutrition due to partially at least to disease destroyed their crops. Identification and diagnosis of Plant pathogen is a requisite for disease management and high yield potential of crops. In the nineteenth century visual identification was very important and rapid method for the plant disease diagnosis, which is present on the aerial part of the plant. Although from the last few year molecular detection and identification of pathogen using nucleic acids based methods have been used. These molecular methods were superior from microscopical detection of plant pathogen. Most probably DNA markers are more reliable than morphological markers because phenotypic selection is based quantification of metric characters which is not stable, reliable and are subjected to environmental variation and detection of developmental stage of plant pathogen. Most precise detection of plant pathogen can be performed by the specific DNA probes in infected tissues. There are different methods are used for the designing of DNA probes (Sharma et al. 2002). Various non radioactive probes are used for the detection of plant pathogens by dot blot hybridization, which could be developed by polymerase chain reaction (Sharma et al. 2002). Such DNA probes have been used for the detection of plant pathogen including pythium ultimum (Levesue et al. 1994).

In Molecular biology PCR is important tool for the identification and characterization of plant pathogens, which cannot be identify morphologically. Plant pathogen can be detected by PCR, which is mainly dependent on primer. Specific, non-specific and arbitrary primer could be used for the Polymerase chain reaction. DNA sequence of plants is amplified with the primers, which show the specific result. Pathogen specific primers are use for the detection of target sequence in infected tissues. It can be used to develop PCR based diagnostic for *Xanthomonas campestris pv. Phaseoli* and *Xanthomonas compestris* pv. *Phaseoli* verity *fuscans* (Audy et al. 1994), *campestris* pv. *Citris* (Hartung et al. 1996). Gene is a segment of DNA which control the particular character of pathogen, which have also been used

Table 3 List of plant diseases with pathogen and resistance gene tagged with marker

Disease	Pathogen	R-gene tagged	Marker	Reference
Loose smut of wheat	*Ustllago segatium tritici*	T10	SCAR	Mullis and Faloona (1987); Procunier et al. (1997)
Powdery mildew of tomato	*Leveillula taurica*	Lv	RAPD	Chunwongse et al. (1997)
Leaf rust of barley	*Puccinia hordei*	Rph7	RFLP	Graner et al. (2000)
Powdery mildew of wheat	*Erysiphe graminis tritici*	Pm24	RFLP	Huang et al. (2000)
Leaf rust of wheat	*Puccinia graminis tritici*	Lr 47	RAPD	Helguera and Dubcovsky (2000)

RAPD Random amplified polymorphic DNA, *RFLP* Restriction Fragment Length polymorphism, *SCAR* Sequence characterized amplified regions

as target sequences for the particular detection of plant pathogen. PCR amplification of ethylene forming enzyme (*efe*) have been used for the detection of *P. syringae pvs. Cannabina and sesame* (Sato et al. 1997).

Molecular Mapping of Disease Resistance Gene

Molecular mapping play important role for the disease diagnosis. Different types of DNA sequence provide a large number of valuable markers that can be used to map their locations in the chromosome of the various species. Several types of DNA sequences provide a large number of valuable markers that can be used to map heir locations in the chromosomes of various species. A linear map of different chromosomes of a species depicting the locations of various molecular markers is known as molecular map. Molecular mapping is the process of locating gene through linkage is called gene tagging. Molecular marker involved the tagging of disease resistance gene for the evaluation of disease resistance phenotype. The basic need of tagging is the molecular marker which is linked to a resistance gene for the development of mapping population. Many study have been done on molecular mapping of R gene in several host pathogen interactions (Mohan et al. 1997; Sharma et al. 1999). A list of R-gene tagged with DNA marker is given in Table 3.

Conclusion

Biotechnology is truly multidisciplinary in nature and it encompasses several disciplines of basic sciences and engineering. In Biotechnology polymerase chain reaction is important tool for the identification, characterization, and diagnosis of fungal pathogen. Real time PCR is differs from Basic PCR because of the measurement of the amplified PCR product at each cycle throughout the reaction. It is highly sensitivity and specificity method in Biotechnology. It will proliferate in the forthcoming year. This chapter has introduced the different types of plant pathogens and their effects. It is clear that they cause major financial losses to the agriculture productivity, despite the plant having defence mechanisms against them. Initial attempts at enhancing resistance involved the introduction of antifungal genes that were either PR Protein or protein identified as having anti fungal properties.

On the other hand Molecular markers play important role in the Biotechnology because these markers are used to study the phylogenetic structure of fungal populations. It is very commonly used to characterize genetic diversity within or between populations or groups of individuals because they typically detect high levels of polymorphism. This approach has been shown to give some benefit to the plant, but it may be at a cost. It has been also useful for identifying specific sequences which is used for the detection of fungi at very low taxonomic level.

References

Altschul SF, Madden TL, Schaffer AA, Zhang JH, Zhang Z, Miller W, Lipman DJ (1997) Gapped BLAST and PSI-BLAST: a new generation of protein database search programs. Nucleic Acids Res 25:3389–3402

Audy P, Laroche A, Saindon G, Huang HC, Gilbertson RL (1994) Detection of bean common blight bacteria Xanthomonas compestris pv. Phaseoli and X.C. Phaseoli var. fuscans using the polymerase chain reaction. Phytopathology 84:1185–1192

Bahnweg G, Schulze S, Moller EM, Rosenbrock H, Langebartels C, Sandermann H (1998) DNA isolation from recalsistant materials such as tree roots, bark and forest soil for the detection of fungal pathogens by polymerase chain reaction. Anal Biochem 262:79–82

Barnes CW, Szabo LJ (2007) Detection and identification of four common rust pathogens of cereals and grasses using real-time polymerase chain reaction. Phytopathology 97(6): 717–727

Botstein B, White RL, Skolnick M, Davis RW (1980) Construction of a genetic linkage map in man using restriction fragment length polymorphism. Am J Hum Genet 32:314–331

Brandfass C, Karlovsky P (2006) Simultaneous detection of *Fusarium culmorum* and *Fgraminearum* in plant material by duplex PCR with melting curve analysis. BMC Microbiol 6:4–6

Chen RS, Chu C, Cheng CW, Chen WY, Tsay JG (2008) Differentiation of two powdery mildews of sunflower (Helianthus annuus) by a PCR-mediated method based on ITS sequences. Eur J Plant Pathol 121(1):1–8

Chi MH, Park S-Y, Lee YH (2009) A quick and safe method for fungal DNA extraction. Plant Pathol J 25(1):108–111

Chunwongse J, Doganlar S, Crossman C, Jiang J, Tanksley SD (1997) High resolution genetic map of Lv resistance locus in tomato. Theor Appl Genet 95:220–223

Dyer PS, Furneaux PA, Douhan G, Murray TD (2001) A multiplex PCR test for determination of mating type applied to the plant pathogens *Tapesia yallundae* and *Tapesia acuformis*. Fungal Genet Biol 33(3):173–180

Feng J, Hwang R, Chang KF, Hwang SF, Strelkov SE, Gossen BD, Zhou Q (2010) An inexpensive method for extraction of genomic DNA from fungal mycelia. Can J Plant Pathol 32(3): 396–401

Fraaije BA, Lovell DJ, Rohel EA, Hollomon DW (1999) Rapid detection and diagnosis of *Septoria tritici* epidemics in wheat using a polymerase chain reaction PicoGreen assay. J Appl Microbiol 86(4):701–708

González Mendoza D, Argumedo-Delira R, Morales Trejo A, Pulido Herrera A, Cervantes Díaz L, Grimaldo-Juarez O, Alarcon A (2010) A rapid method for isolation of total DNA from pathogenic filamentous plant fungi. Genet Mol Res 9(1):162–166

Graner A, Strengn S, Drescher A, Jin Y, Borovkova I, Steffenson BJ (2000) Molecular mapping of the leaf rust resistance gene Rph7 in barley. Plant Breed 119:389–392

Grund E, Darissa O, Adam G (2010) Application of FTA (R) cards to sample microbial plant pathogens for PCR and RT-PCR. J Phytopathol 158(11–12):750–757

Guo JR, Schnieder F, Beyer M, Verreet JA (2005) Rapid detection of *Mycosphaerella graminicola* in wheat using reverse transcription-PCR assay. J Phytopathol 153(11–12):674–679

Hartung JS, Pruvost UP, Villemont I, Alvarez A (1996) Rapid and sensitive colorimetric detection of Xanthomonas axonopodis pv. Citri by immunocapture and nested polymerase chain reaction assay. Phytopathology 86:95–101

Helguera IA, Dubcovsky KJ (2000) Development of PCR markers for the wheat leaf rust resistance gene Lr 47. Theor Appl Genet 100:1137–1143

Hong SY, Kang MR, Cho EJ, Kim HK, Yun SH (2010) Specific PCR detection of four quarantine *Fusarium Species* in Korea. Plant Pathol J 26(4):409–416

Huang Q, Hsam SLK, Zeller FJ, Wenzel G, Mohler V (2000) Molecular mapping of the wheat powdery mildew resistance gene Pm24 and marker validation for molecular breeding. Theor Appl Genet 101:407–414

Hyakumachi M, Priyatmojo A, Kubota M, Fukui H (2005) New anastomosis groups, AG-T and AG-U, of binucleate *Rhizoctonia* spp. causing root and stem rot of cutflower and miniature roses. Phytopathology 95(7):784–792

Jana T, Sharma TR, Singh NK (2005) SSR-based detection of genetic variability in the charcoal root rot pathogen *Macrophomina phaseolina*. Mycol Res 109:81–86

Jeeva ML, Mishra AK, Vidyadharan P, Misra RS, Hegde V (2010) A species-specific polymerase chain reaction assay for rapid and sensitive detection of *Sclerotium rolfsii*. Australas Plant Pathol 39(6):517–523

Joshi SP, Prabhakar K, Ranjekar PK, Gupta VS (2011) Molecular markers in plant genome analysis. 15:1–19. http://www.ias.ac.in/currsci/jul25/articles

Kim YT, Cho M, Jeong JY, Lee HB, Kim SB (2010) Application of terminal restriction fragment length polymorphism (T-RFLP) analysis to monitor effect of biocontrol agents on rhizosphere microbial community of hot pepper (*Capsicum annuum* L.). J Microbiol 48(5):566–572

Langrell SRH, Glen M, Alfenas AC (2008) Molecular diagnosis of *Puccinia psidii* (guava rust) – a quarantine threat to Australian eucalypt and *Myrtaceae* biodiversity. Plant Pathol 57(4):687–701

Levesue CA, Vrain TC, De Boer SH (1994) Development of species specific probe for phythium ultimum using amplified ribosomal DNA. Phytopathology 84:474–478

Martínez García LB, Armas C, Miranda JD, Padilla FM, Pugnaire FI (2011) Shrubs influence arbuscular mycorrhizal fungi communities in a semi-arid environment. Soil Biol Biochem 43(3):682–689

Meng J, Wang Y (2010) Rapid detection of *Phytophthora nicotianae* in infected tobacco tissue and soil samples based on its *Ypt1* gene. J Phytopathol 158(1):1–7

Mohan M, Nair S, Bhagwat A, Krishna TG, Yano M, Bhatia CR, Sasaki T (1997) Genome mapping, molecular markers and marker assisted selection in crop plants. Mol Breed 3:87–103

Moradi M, Oerke EC, Steiner U, Tesfaye D, Schellander K, Dehne HW (2010) Microbiological and Sybr Green real-time PCR detection of major *Fusarium* head blight pathogens on wheat ears. Microbiology 79(5):646–654

Mullis KB, Faloona FA (1987) Specific synthesis of DNA in vitro via a polymerasecatalyzed chain-reaction. Methods Enzymol 155:335–350

Niu C, Kebede H, Auld DL, Woodward JE, Burow G, Wright RJ (2008) A safe inexpensive method to isolate high quality plant and fungal DNA in an open laboratory environment. Afr J Biotechnol 7(16):2818–2822

Okubara PA, Schroeder KL, Paulitz TC (2008) Identification and quantification of *Rhizoctonia solani* and *R. oryzae* using real-time polymerase chain reaction. Phytopathology 98(7):837–847
Powell W, Machary GC, Provan J (1996) Polymorphism revealed by simple sequence repeats. Trends Plant Sci 1:215–222
Procunier JD, Knox RE, Bernier AN, Grag MA, Howes NK (1997) DNA markers linked to a T10 loose smut resistance gene in wheat (*Triticum aestivum L.*). Genome 40:176–179
Qin L, Fu Y, Xie J, Cheng J, Jiang D, Li G, Huang J (2011) A nested-PCR method for rapid detection of *Sclerotinia sclerotiorum* on petals of oilseed rape (*Brassica napus*). Plant Pathol 60(2):271–277
Sato M, Watanabe K, Yazawa M, Takikawa Y, Nishiyama K (1997) Detection of new ethylene producing bacteria, Pseudomonas syringae pvs. Cannabina and sami, by PCR amplification of gene for the ethylene forming enzyme. Phytopathology 87:1192–1196
Schmidt H, Taniwaki MH, Vogel RF, Niessen L (2004) Utilization of AFLP markers for PCR-based identification of *Aspergillus carbonarius* and indication of its presence in green coffee samples. J Appl Microbiol 97(5):899–909
Schroeder KL, Okubara PA, Tambong JT, Levesque CA, Paulitz TC (2006) Identification and quantification of pathogenic *Pythium* spp. from soils in eastern Washington using real-time polymerase chain reaction. Phytopathology 96(6):637–647
Sharma TR, Prachi, Singh BM (1999) Application of polymerase chain reaction in phytopathogenic microbes. Indian J Microbiol 39:79–91
Sharma TR, Tewari JP, Singh BM (2002) Use of polymerase chain reaction of the synthesis of non radioactive DNA probes in phytopathogenic ungi. In: Sekhawat GS et al (eds) Molecular approaches for plant disease management. Malhotra Publishing House, New Delhi, p 350
Sheridan GEC, Masters CI, Shallcross JA, Mackey BM (1998) Detection of mRNA by reverse transcription-PCR as an indicator of viability in *Escherichia coli* cells. Appl Environ Microbiol 64(4):1313–1318
Singh Y (2009) Non radioactive gene probes. AGROBIOS News Lett 08(02):07–10
Singh M, Singh RP (1997) Potato virus Y detection: sensitivity of RT-PCR depends on the size of fragment amplified. Can J Plant Pathol Rev Can Phytopathol 19:149–155
Stewart JE, Kim MS, James RL, Dumroese RK, Klopfenstein NB (2006) Molecular characterization of *Fusarium oxysporum* and *Fusarium commune* isolates from a conifer nursery. Phytophathology 96:1124–1133
Stringari D, Glienke C, de Christo D, Maccheroni W, de Azevedo JL (2009) High molecular diversity of the fungus *Guignardia citricarpa* and *Guignardia mangiferae* and new primers for the diagnosis of the Citrus Black Spot. Braz Arch Biol Technol 52(5):1063–1073
Suzuki S, Taketani H, Kusumoto KI, Kashiwagi Y (2006) High-throughput genotyping of filamentous fungus *Aspergillus oryzae* based on colony direct polymerase chain reaction. J Biosci Bioeng 102(6):572–574
Thies JE (2007) Soil microbial community analysis using terminal restriction fragment length polymorphisms. Soil Sci Soc Am J 71(2):579–591
Torres-Calzada C, Tapia-Tussell R, Quijano-Ramayo A, Martin-Mex R, Rojas-Herrera R, Higuera-Ciapara I, Perez-Brito D (2011) A species-specific polymerase chain reaction assay for rapid and sensitive detection of *Colletotrichum capsici*. Mol Biotechnol 49(1):48–55
VanderLee T, DeWitte I, Drenth A, Alfonso C, Govers F (1997) AFLP linkage map of the oomycete *Phytophthora infestans*. Fungal Genet Biol 21(3):278–29
Vos P, Hogers R, Bleeker M, Reijans M, Van de Lee T, Hornes M, Frijters A, Pot J, Peleman J, Kuiper M, Zabean M (1995) AFLP- A new technique for DNA fingerprinting. Nucleic Acids Res 23:4407–4414
Williams JGK, Kubelik AR, Livak KJ, Rafalski JA, Tingey SV (1991) DNA polymorphisms amplified by arbitrary primers are useful as genetic markers. Nucleic Acids Res 18:6531–6535
Winton LM, Krohn AL, Leiner RH (2007) Microsatellite markers for *Sclerotinia subarctica* nom. prov., a new vegetable pathogen of the High North. Mol Ecol Notes 7(6):1077–1079, ISSN 1471-8278

Wu CP, Chen GY, Li B, Su H, An YL, Zhen SZ, Ye JR (2011) Rapid and accurate detection of *Ceratocystis fagacearum* from stained wood and soil by nested and realtime PCR. For Pathol 41(1):15–21

Yang F, Jensen JD, Svensson B, Jorgensen HJL, Collinge DB, Finnie C (2010) Analysis of early events in the interaction between *Fusarium graminearum* and the susceptible barley (*Hordeum vulgare*) cultivar Scarlett. Proteomics 10(21):3748–3755

Zaccaro RP, Carareto-Alves LM, Travensolo RF, Wickert E, Lemos EGM (2007) Use of molecular marker SCAR in the identification of *Fusarium subglutinans*, causal agent of mango malformation. Rev Bras Frutic 29(3):563–570

Zelaya Molina LX, Ortega MA, Dorrance AE (2011) Easy and efficient protocol for oomycete DNA extraction suitable for population genetic analysis. Biotechnol Lett 33(4):715–720

Zhang Z, Zhu Z, Ma Z, Li H (2009) A molecular mechanism of azoxystrobin resistance in *Penicillium digitatum* UV mutants and a PCR-based assay for detection of azoxystrobin-resistant strains in packing- or store-house isolates. Int J Food Microbiol 131:157–161

In Vitro Antimycotic and Biosynthetic Potential of Fungal Endophytes Associated with *Schima Wallichii*

Vineet Kumar Mishra, Ajit Kumar Passari, and Bhim Pratap Singh

Introduction

Diversity of Endophytic Fungi

Increasing emergence of drug resistant organisms such as bacteria, virus and fungi has alarmed the necessity of investigating new molecules with better bioactivity (Bhagat et al. 2012). Exploring secondary metabolites from microorganisms isolated from specialized ecological niches or unusual habitats enhances the chances of finding bioactive compounds (Wagenaar et al. 2000; Rodriguez et al. 2008; Wang et al. 2008). Endosphere of plants harbors microbial population in a symbiotic relationship without causing any apparent disease symptoms are reported as endophytic microbial population (Stone et al. 2000). There are estimated about 300,000 reported plant species on earth and almost all have endophytic microbial population, which resides in internal tissues and helps plants in their survival. Interestingly, some rare bioactive compounds produced by plants can also produced by their endophytes (Aly et al. 2010). They reside within tissues of almost all terrestrial plants (Hyde and Soytong 2008; Lin et al. 2010) and are the important components of plant-microbe ecosystem. Endophytes resides in all tissues of plants and showed multifarious relationships with the host, ranging from symbiosis to balanced antagonism (Bacon and White 2000; Rodriguez et al. 2008).

Endophytic fungi are believed to have evolved from weak pathogenic fungi which on due course of time lost virulence and started living as symbiont inside the tissues or

V.K. Mishra • A.K. Passari • B.P. Singh (✉)
Molecular Microbiology and Systematics Laboratory, Department of Biotechnology, Mizoram University, Aizawl, Mizoram 796004, India
e-mail: bhimpratap@gmail.com

P. Kumar et al. (eds.), *Current Trends in Plant Disease Diagnostics and Management Practices*, Fungal Biology, DOI 10.1007/978-3-319-27312-9_16

developed as latent pathogen (Saikkonen et al. 1998). Endophytic fungi can colonize any tissue whereas; mycorrhizal fungi can only colonize only the roots of plants (Carroll 1988; Stone et al. 2004; Verma et al. 2007). Fungal endophytes are mutualistic partners of their host and thus protect the host plant from both biotic and abiotic stresses (Rodriguez et al. 2008; Porras-Alfaro and Bayman 2011; Nagabhyru et al. 2013; Qadri et al. 2014). Additionally, plant-endophyte symbiosis influence host nutrition and growth and survival (Reinhold-Hurek and Hurek 2011; Singh et al. 2011; Iqbal et al. 2013). The host range and geographical distribution of these endophytic symbiont are not yet clear, as some are endophytes of specific host whereas some colonize multiple host species (Doty et al. 2009; Porras-Alfaro and Bayman 2011; Ek-Ramos et al. 2013).

Antimicrobial Secondary Metabolites Produced by Endophytic Fungi

Phytopathogens and pests cause great damage to crop by reducing yield by 30–50 % globally by producing certain mycotoxins (Pimentel 2009). *Fusarium moniliforme* which is pathogenic to maize and rice is known to produce fumonisin B1 which is related with esophageal cancer (Gelderblom et al. 1991). *Aspergillus flavus* causes kernel rot in maize and produce aflatoxin on pre-harvest and in storage (Payne and Widstrom 1992). Similarly, *Fusarium graminearum* is a producer of toxic trichothecenes including deoxynivalenol (Sutton 1982). Pesticides including fungicides used in pest and fungal phytopathogen management are having harmful effect on environment (Mousa and Raizada 2013). An efficient alternative shown by Wang et al. (2008) had isolated six compounds from endophytic *Penicillium* sp. associated with *Hopea hainanensis* having strong antifungal and cytotoxic activity. Fungal biocontrol agents such as *Trichoderma viridi*, has been used to control *Rhizoctonia* stem canker and black scurf of potato (Beagle-Ristaino and Papavizas 1985). Several bioactive compounds have been isolated from endophytic fungi which can be used directly or indirectly as therapeutic agents against various diseases (Strobel et al. 2004; Kusari and Spiteller 2012; Kusari et al. 2012a). Endophytes have produced host specific secondary metabolites with promise for therapeutic potential on a number of occasions such as paclitaxel or taxol (Stierle et al. 1993), podophyllotoxin (Eyberger et al. 2006; Puri et al. 2006), camptothecin and its structural analogs (Puri et al. 2005; Shweta et al. 2010; Kusari et al. 2009, 2011) and azadirachtin (Kusari et al. 2012b). Many antimicrobial compounds such as Trichodermin, Phomenone, Dihydroxycadalene, Cryptocin, Altenusin, Heptelidic acid, Paclitaxel, Periconicins A and B, Peramine, ergot alkaloids, loline alkaloid, Colletotric acids, Phomopsichalasin, Pestalachloride A, 6-O-methylalaternin and nodulosporins from endophytic fungi have been isolated and reported to protect against fungal pathogens (Mousa and Raizada 2013). Paclitaxel isolated from *Taxomyces andreanae* an endophyte of *Taxus brevifolia,* Trichodermin isolated from *Trichoderma harzianum* endophyte of *Ilex cornuta*, phomenone isolated from *xylaria* sp. an endophyte of *Piper aduncum* and nodulisporin isolated from *Nodulisporium sp.* an endophyte of *Juniperus cedrus* have shown strong antifungal activity against fungal plant pathogens (Mousa and Raizada 2013).

Schima wallichii

Schima wallichii D.C. Korth. (Theaceae) is an evergreen, medium-sized to large tree and has been used in traditional practices for the treatment of various ailments including snake bite and insect bite. Its bark is rubefacient, antigonorrhoeic as well as antihelminthic activity whereas the leaf is known to be carminative (Rai and Lalramnghinglova 2010). It has been reported to have antimicrobial (Dewanjee et al. 2008), anticancer (Diantini et al. 2012), anti-plasmodial activity (Barliana et al. 2014) and anti-proliferative activity (Dewanjee et al. 2011). Diantini et al. (2012) has reported cytotoxicity of Kaempferol-3-O-rhamnoside isolated from the leaves of *S. wallichii* Korth. against breast cancer cell line MCF-7 which also promotes apoptosis through activation of caspase cascade pathway. By looking into these studies we targeted *S. wallichii* Korth. for fungal endophytic exploration.

Polyketide Synthase

Polyketides are structurally diverse group of compounds produced by microorganisms and plants and have profound use in health and agriculture sector (Amnuaykanjanasin et al. 2005; Rojas et al. 2012) but reports of polyketides from endophytic fungi are in scanty till date. They have varied range of bioactivity and therapeutic applications including antibacterial activity (erythromycin, tetracycline, griseofulvin), immunosuppresent activity (rapamycin), antiparasitic activity (avermictin), anti-tumor activity (enediyne, daunorubicin, doxorubicin) and anti-cholesteremic activity (lovastatin) (Amnuaykanjanasin et al. 2005; Rojas et al. 2012). Fungal polyketide synthases are iterative and consists of three essential catalytic domains: β-ketosynthase (KS), Acyltransferase (AT) and Acyl-carrier protein (ACP) which can also be supplemented by C-methyltransferase (CMT) domain (Rojas et al. 2012). Fungal iterative PKS I, catalyzes repeatedly the condensation of subunits into polyketide backbone, and thus can be divided into three categories: Non- reducing (NR) or WA-type, with no reduction in structure, partial reducing (PR) and highly reducing (HR) (Bingle et al. 1999; Nicholson et al. 2001; Schumann and Hertweck 2006; Rojas et al. 2012). PCR primers have been designed to detect fungal PKS I gene based on conserved regions and extent of reduction of polyketide structure. LC1/2c primers have been designed to detect NR type PKS, LC3/5c has been designed to amplify PR type whereas, primers KS3/4c has been designed to detect HR – type PKSs (Bingle et al. 1999; Nicholson et al. 2001; Lin et al. 2010; Rojas et al. 2012).

In this study, an attempt was made to investigate the biodiversity of the fungal endophytes associated within endosphere of *S. wallichii* and also to explore their antimycotic potential against fungal plant pathogens with the biosynthetic potential of endophytic fungi was evaluated according to detection of ketosynthase domain of polyketide synthase (PKS) gene.

Materials

1. Host Plant: *Schima wallichii* was collected from Dampa Tiger Reserve
2. Global Positioning System: Garmin GPS 60
3. Microscope: Olympus BX51
4. Laboratory plastic wares, i.e., micropipette tips (Tarsons, India), microcentrifuge tubes (Tarsons, India), 90 mm petri dishes (Hi Media, India) and flat cap PCR tubes (Tarsons, India).
5. Chemicals: Unless mentioned otherwise, all chemicals and reagents were purchased from Hi Media (India).
6. All fungal plant pathogens were purchased from NBAIMCC (National Bureau of Agriculturally Important Microorganisms Culture Collection) and MTCC (Microbial Type Culture Collection, Chandigarh, India).
7. Rotary Evaporator (BUCHI, Switzerland).
8. Megafuge 16R (Thermo Scientific), cooling centrifuge
9. Taq PCR Buffer, Taq polymerase, dNTP mix 100 mM and $Mgcl_2$ were purchased from Bangalore Genei Pvt. Ltd.
10. Primers of ITS and PKS gene were synthesized by Imperial Life Science.
11. PCR Thermal Cycler: Veriti (Applied Biosystem)
12. Bio-Rad Gel Doc XR+ gel documentation system, California, USA

Methods

Collection of Samples

Leaves, small cutting of the stem (twigs) and bark of three symptomless plants of *Schima wallichii* were collected from Dampa Tiger Reserve forest (23°.44′ N 92°.39′ E) in western Mizoram which occupies an area of 500 km^2 along the border between India and Bangladesh. The samples were kept in sterile polythene bags and were transported to the laboratory in ice box. All the samples were stored at 4 °C and processed for surface sterilization within 36 h.

Isolation of Endophytic Fungi

The plant materials were washed thoroughly with running tap water for 10 min to remove dust particles and were cut into small pieces (2–3 cm). Fungal endophytes were isolated by surface sterilization methods as described by Cannon and Simmons (2002) with little modification. Briefly, the plant segments were surface sterilized with 75 % ethanol for 1 min. followed by immersion in 3 % sodium hypochlorite for 3 min and 75 % ethanol for 30 s. before rinsing with sterile distilled water to remove traces of sodium hypochlorite. The outer surface was removed with a sterile scalpel and leaves were cut into 5×5 mm pieces whereas stem and bark were cut into 1×1 cm fragments. To ensure the effectiveness of surface sterilization method the water obtained from the last wash was spreaded on PDA medium and fingerprints of the tissues were also taken on the media plates as a control to check for any epiphytic fungal growth. One hundred and ninety five (195) tissues were evenly plated on 90 mm petri dishes containing PDA (Potato Dextrose Agar), MYA (Malt Yeast Extract Agar) and Czapex Dox Agar (CDA) medium supplemented with streptomycin (50 mg/L) to suppress bacterial growth. The plates were incubated on 27 °C for 3–5 days under 12 h white light: 12 h dark cycles (Bills and Polishook 1991). Petri plates were observed for 2–3 weeks for any hyphae emerging from the tissues. The hyphal tips coming out from the sterile tissues were transferred to fresh plates free from antibiotics. Fifteen days after incubation fungal cultures were preserved in 30 % glycerol at −80 °C. Each fungal endophyte was identified morphologically by preparing slides stained with lactophenol cotton blue and observing under light microscope (Olympus BX 51). Sporulating fungi were identified by morphology based taxonomy (Barnett and Hunter 1998).

Fermentation and Preparation of Fungal Extract

The isolated endophytic fungi were cultured in Potato Dextrose Broth (PDB) medium by inoculating 3 mm mycelial plug of pure culture and incubated for 21 days at 27 °C. The culture was filtered through sterile cheese cloth to remove mycelium and the fermented broth was extracted thrice by equal volume of ethyl acetate in a separating funnel by vigorous shaking for 15 min. Ethyl acetate evaporated to dryness in a rotary evaporator (BUCHI, Switzerland) leaving solid crude extract. The crude ethyl acetate extract was dissolved in DMSO for antimycotic assay.

Evaluation of Antimycotic Activity

Antimycotic activity of the ethyl acetate extract from endophytic fungi was evaluated against several plant pathogenic fungi by agar cup diffusion method with some modifications (Tayung and Jha 2010). The test organisms include *Macrophomina phaseolina* (NAIMCC-F-01261), *Aspergillus flavus* (MTCC 9064) and seven phytopathogens of the genus *Fusarium*: *Fusarium oxysporum* (NAIMCC-F-00809), *Fusarium graminearum* (MTCC 1893) *Fusarium culmorum* (MTCC 2090), *Fusarium tumidum* (MTCC 2462), *Fusarium oxysporum f. sp. pisi* (MTCC 2480), *Fusarium prolifratum* (MTCC 286) and *Fusarium udum* (MTCC 2755). PDA plates were spreaded by inoculating 1.0×10^9 spores of each fungal pathogens respectively. Cork borer (7 mm in diameter) was used to prepare agar cups and each cup was loaded with 100 μl of the crude fungal ethyl acetate extracts dissolved in DMSO. The control cup was filled with DMSO only. The plates were incubated at 27 °C for 4–5 days for evaluating antifungal activity of the endophytes. The experiment was performed in triplicate. The diameter of zones of inhibition were measured and compared with control.

Genomic DNA Extraction and Amplification of ITS (ITS1-5.8S-ITS2) Region

The fungal isolates having significant antimycotic activity were subjected to DNA extraction, amplification and sequencing of the ITS region. Isolation of genomic DNA was performed by procedure previously reported by Cenis (1992). The ITS region was amplified using universal primers ITS1 (5′-TCCGTAGGTGAACCTGCGG-3′) and ITS4 (5′-TCCTCCGCTTATTGATATGC-3′) (White et al. 1990). The 25 μl reaction contained 1X PCR assay buffer, 1.5 mM $Mgcl_2$, 200 μm of each dNTPs, 10 pmols of each primer, 50 ng of template DNA and 1 unit of Taq DNA polymerase. The PCR conditions used were as follows: an initial denaturation step at 95 °C for 5 min, followed by 35 cycles of 1 min denaturation at 95 °C, annealing at 55 °C for 1 min, extension at 72 °C for 1 min 20 s, followed by final extension at 72 °C for 10 min. The amplification was performed in Veriti thermal cycler (Applied Biosystems, Singapore). A negative control without template DNA was included in each batch of PCR. The PCR products were separated on 1 % (w/v) agarose in 1X TBE buffer stained with ethidium bromide1 % (10 mg/ml) and analyzed under a gel documentation system (Bio-Rad Gel Doc XR+ gel documentation system, California, USA).

Sequencing of ITS Region and Phylogenetic Analysis

ITS region (ITS1-5.8S-ITS2) was sequenced and was analyzed using Finch TV v1.40v (http://www.geospiza.com/finchtv). The identifications of all the sequences were made by sequence similarity searches by aligning with the sequences available in NCBI GenBank using BLASTN program so as to find the homology with the closest related organisms. Sequences from closely related organisms were aligned with sequences from this study using Clustal W packaged in MEGA 5.05 (Tamura et al. 2011). A model test was performed using MEGA 5.05 to choose the most appropriate model based on Bayesian Information Criterion and Akaike Information Criterion. The evolutionary history was inferred by using the Maximum Likelihood method based on the Kimura 2-parameter model (Kimura 1980). A discrete Gamma distribution was used to model evolutionary rate differences among sites (five categories (+G, parameter = 0.5322)). The rate variation model permitted for some sites to be evolutionarily invariable ([+I], 12.9649 % sites). Bootstrap analysis was used to test the robustness of the phylogenetic tree. The tree was drawn to scale, with branch lengths measured in the number of substitutions per site. Sequences were deposited in GenBank and accession numbers were obtained.

Data Analysis

Colonization frequency (%CF) of fungal isolates was calculated as % CF = N_{col}/N where, N_t = Total number of segments and N_{col} = number of segments colonized by a specific fungus (Hata and Futai 1995; Verma et al. 2014). Simpson's diversity indices, Shanon-Weiner diversity indices and evenness were calculated by PAST software. Simpson's diversity index was calculated using the formula: $1-\sum(pi)^2$, where pi is the proportion of the frequency of the ith species in a sample. Species evenness (E) was calculated as following: H/log(S), where H = Shanon Weiner diversity and S = Species richness.

Amplification of Ketosynthase Domain Sequence from PKS Gene

Genomic DNA of all the identified strains showing significant bioactivity were amplified by three sets of degenerate primers, LC1 and LC2c, LC3 and LC5c (Bingle et al. 1999), KS3 and KS4c (Nicholson et al. 2001) which are specific primers for the ketosynthase domain of the PKS gene. 25 ul of PCR reaction contained 1X PCR buffer, 8 mM $Mgcl_2$, 0.2 mM dNTPs, 10 pmole of each primer, 2 U of Taq polymerase and 50–100 ng of template DNA. PCR reaction was performed in verity thermal cycler (Applied Biosystems). Conditions for thermal cycler was: Initial denaturation at 94 °C for 5 min and 35 cycles of denaturation at 94 °C for 1 min., annealing of primers from 1 min 20 s at 55 °C for LC1/2 and LC3/5 primers, 1 min 20 s at 50.5 °C for KS3/4 primers and extension for 3 min at 72 °C followed by final extension for 10 min at 72 °C.

Results and Discussion

Diversity and Distribution of Endophytic Fungi

Endophytic fungi were isolated from healthy leaves, stems and bark of traditional medicinal plant *Schima wallichii.* 126 endophytic fungal isolates belonging to 15 genera were recovered from 195 tissue segments using three different mycological media. Maximum endophytes were recovered by using PDA media (n=52, 41.26 %), followed by MYA (n=41, 32.53 %) and CDA (n=33, 26.19 %)

Table 1 Colonization frequency (%CF) of three tissues of *Schima wallichii*

	Stem	Leaf	Bark	Total
No. of samples	65	65	65	195
No. of isolates	49	40	37	126
Colonization frequency (CF %)	75.38	61.53	56.92	**64.61**

Table 2 Colonization frequency and % dominance of endophytic fungal isolates of *Schima wallichii*

Endophytic fungi	No. of endophytes	Colonization frequency (CF)%	Dominance (%)	Class
Nodulisporium sp.	06	3.07	4.76	**Sordariomycetes**
Phomopsis sp.	23	11.79	18.25	
Xylaria sp.	01	0.51	0.79	
Hypoxylon sp.	02	1.02	1.58	
Diaporthe sp.	06	3.07	4.76	
Chaetomium sp.	09	4.61	7.14	
Fusarium sp.	05	2.56	3.96	
Colletotrichum sp.	12	6.15	9.52	
Penicillium sp.	07	3.58	5.55	**Eurotiomycetes**
Talaromyces sp.	01	0.51	0.79	
Aspergillus sp.	01	0.51	0.79	
Alternaria sp.	21	10.76	16.66	**Dothidiomycetes**
Phoma sp.	19	9.74	15.07	
Corynespora sp.	06	3.07	4.76	
Leptosphaeria sp.	07	3.58	5.55	
Total no. of isolates	**126**			

Total no. of plant segments plated = 195

respectively. The colonization frequency was higher in stem (75.38 %) followed by leaf (61.53 %) and bark (56.92 %) (Table 1). All fungal isolates belonged to different classes of phylum Ascomycota. Sordariomycetes (50.79 %) was the most dominant class represented by genera *Phomopsis*, *Colletotrichum*, *Chaetomium*, *Nodulisporium*, *Diaporthe*, *Fusarium*, *Hypoxylon* and *Xylaria* followed by Dothideomycetes (42.06 %). Similar results are reported by Arnold et al. (2007) from foliar endophytic fungi of *Pinus taeda*. The abundant endophytes observed of class Dothideomycetes were *Alternaria*, *Phoma*, *Corynespora* and *Leptosphaeria*. Eurotiomycetes (7.14 %) were the least dominant class represented by genera *Penicillium*, *Talaromyces and Aspergillus*. Colonizing frequency of *Phomopsis* was found to be highest (11.79 %), followed by *Alternaria* (10.76 %), *Phoma* (9.74 %), *Colletotrichum* (6.15 %), *Chaetomium* (4.61 %), *Penicillium* (3.58 %) and *Leptosphaeria* (3.58 %) (Table 2). The endophytes isolated from all three tissue

Table 3 Diversity indices of endophytic fungal isolates recovered from different tissues

	Stem	Leaf	Bark
Taxa S	14	13	10
Individuals	49	40	37
Dominance_D	0.1437	0.1138	0.1322
Shannon-Wiener index_H	2.226	2.339	2.149
Simpson index (1-D)	0.8563	0.8862	0.8678
Evenness ^ H/S	0.6617	0.7981	0.8576

types were analyzed using diversity indices (Shanon-Weiner index, Simpson's diversity index, evenness and dominance). Shanon-Weiner index was greater in Leaf (2.339) followed by stem (2.226) and bark (2.149) whereas, Simpson's diversity index was also higher in leaf (0.8862), but it was found to be more in case of bark (0.8678) than stem (0.8563) (Table 3). Similar results were obtained by Gond et al. (2012), where Simpson's dominance was higher in stem tissues than leaf tissues of *Nyctanthes arbor-tristis*, but Shanon-Weiner index and Simpson's diversity index were higher in leaf tissues than stem tissues. However there was little difference with respect to evenness in tissues. Kayini and Pandey (2010) have also reported endophytic fungi from *Schima wallichii* but the bioactive potential of the fungal isolates were not evaluated.

Antimycotic Potential, Phylogenetic Analysis and Polyketide Synthase (PKS) Gene

Endophytic fungi are known as source of various bioactive compounds which may be used as potential therapeutic agents (Kusari et al. 2012a). Fungal ethyl acetate extracts were tested against 9 fungal phytopathogens of which 12 endophytes displayed bioactivity against at least two pathogens (Fig. 1, Table 4). Isolates EF18 and EF49 displayed significant antimycotic activity against seven out of nine tested plant pathogens and may be used as potential biocontrol. Similar activity was reported by Qadri et al. (2014) using dual plate method in which endophytic fungal isolates associated with *Pinus wallichiana* inhibited seven fungal pathogens of different genus. Highest inhibition zone diameter (18 mm) was observed against *Fusarium proliferatum.* The most susceptible pathogens were *Fusarium proliferatum* and *Fusarium oxysporum* f. sp. *pisi*, each being inhibited by nine isolates whereas, *Aspergillus flavus* was inhibited only by EF 18.

All the 12 isolates having antimycotic potential were sequenced by amplifying ITS1-5.8S-ITS2 region and identified by finding best match using BLASTN program. Sequences of some closely related organisms were retrieved from NCBI GenBank for phylogenetic analysis. The maximum likelihood tree with the highest log likelihood (−2051.4907) is shown in Fig. 2 which depicts phylogenetic

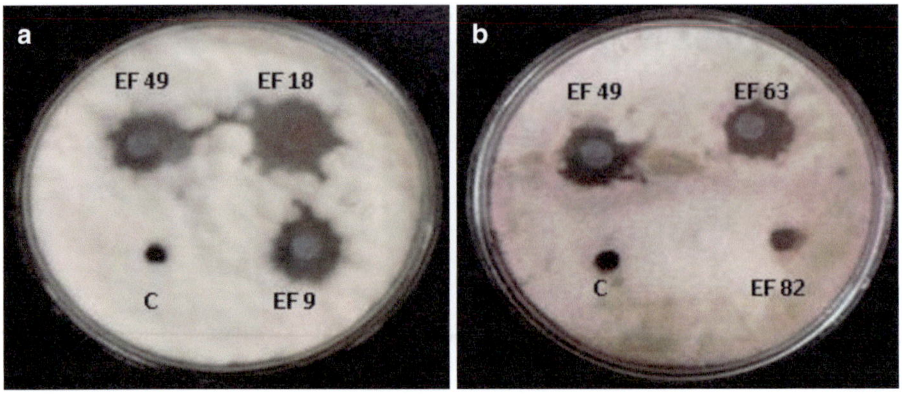

Fig. 1 Antimycotic activity of endophytic fungi extracts against (**a**) *Fusarium proliferatum* and (**b**) *Fusarium tumidum*. *C* control (DMSO only)

placement of endophytic fungal isolates of *Schima wallichii* having bioactive potential. The percentage of trees in which the associated taxa clustered together is shown next to the branches. The sequences were deposited in NCBI GenBank with accession numbers KJ826506-KJ826515 and KJ826516. Endophytic fungal isolates EF 18 and EF 49 which displayed antimycotic activity against seven plant pathogens were identified as *Penicillium simplicissimum* (KJ826510) and *Talaromyces verruculosus* (KJ826513). There are some reports which are in accordance to our result. Komai et al. (2006) had isolated six penicillide derivatives from *Penicillium simplicissimum* having significant antifungal activity against, *Aspergillus fumigatus*, *Aspergillus niger*, *Candida albicans* and *Cryptococcus neoformans*, whereas, Miao et al. (2012) had isolated two compounds from *Talaromyces verruculosus* which showed antifungal activity against *Alternaria solani*, *Valsa mali*, *Curvularia lunata* and *Botryosphaeria berengeriana.*

The potential isolates were screened for detecting iterative type I PKS using LC1–LC2c, LC3–LC5c and KS3–KS4c primers for non-reduced, partially reduced and highly reduced KS domain. Polyketides have found widespread use in both pharmaceuticals and agriculture industries (Rojas et al. 2012). Nine isolates have shown to contain at least two KS domain amplified by LC1–LC2c and KS3–KS4c primers and most number of isolates were amplified by LC1–LC2c primers which are specific for non-reducing KS domains. Similar results were found by Lin et al. (2010). As per our knowledge, there is no report describing detection of polyketide synthase gene from either from *Schima wallichii* or by its endophytes. This study concludes that Endophytic fungi of *Schima wallichii* have significant antimycotic activity and genetic repertoire for producing polyketides.

Table 4 Antimycotic activity of ethyl acetate extract of endophytic fungi represented in zone of inhibition (mm) against fungal plant pathogens

Endophytic fungi	Accession no	Zone of inhibition (mm)								
		Macrophomina phaseolina	*A. flavus*	*F.oxysporum*	*F.graminearum*	*F.culmorum*	*F. tumidum*	*F.oxysporum f. sp. pisi*	*F. proliferatum*	*F.udum*
Nodulisporium sp. (EF2)	KJ826506	–	–	–	–	12.33 ± 0.57	–	12 ± 0.5	11 ± 0.5	–
Aspergillus niger (EF9)	KJ826507	–	–	–	14.66 ± 0.57	–	–	16 ± 0.57	17.33 ± 0.57	14 ± 0.5
Alternaria sp. (EF12)	KJ826508		–	14.83 ± 0.28	–	–		14.16 ± 0.28	15 ± 0.5	–
Penicillium chermesinum (EF16)	KJ826509	–	–	15 ± 0.57		12 ± 0.5	–	15 ± 0.00	13.16 ± 0.28	–
***Penicillium simplicissimum* (EF18)**	**KJ826510**	**–**	**10 ± 0.5**	**–**	**12.66 ± 0.57**	**14.66 ± 0.28**	**17.66 ± 0.57**	**13 ± 0.5**	**18 ± 0.57**	**13 ± 0.5**
Hypoxylon fragiforme (EF21)	KJ826511	11.33 ± 0.57	–	09.66 ± 0.57	–	13.83 ± 0.28	–	–	11 ± 0.57	–
Phomopsis sp. (EF32)	KJ826512	10.83 ± 0.28	–	–	–	–	11 ± 0.5	13 ± 0.5	–	10 ± 0.5
***Talaromyces verruculosus* (EF49)**	**KJ826513**	**09.66 ± 0.57**	**–**	**12.66 ± 0.28**	**–**	**15 ± 0.5**	**16.33 ± 0.57**	**14 ± 0.5**	**17 ± 0.5**	**14 ± 0.5**
Talaromyces sp. (EF63)	KJ826514	–	–	12 ± 0.28		–	17 ± 0.5	13 ± 0.00	15 ± 0.5	–
Nodulisporium sp. (EF82)	KJ826516	–	–	–	–	15.33 ± 0.57	–	11.33 ± 0.57	–	–
Penicillium shearii (EF97)	KJ826517	–	–	12 ± 0.5	–	16 ± 0.57	13 ± 0.5	–	–	13 ± 0.5
Penicillium cf. *janthinellum* (EF102)	KJ826518	–	–	15.83 ± 0.28	–	17.83 ± 0.28	–	–	16 ± 0.5	–

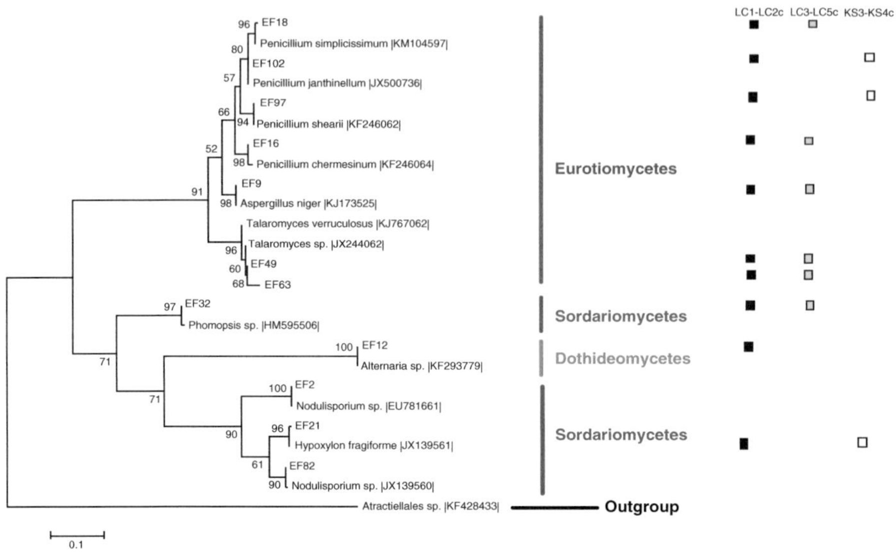

Fig. 2 Phylogenetic relationship of endophytic fungi isolated from *Schima wallichii* having antimycotic potential and diversity of KS domain of PKS I gene among them

Acknowledgements BPS thankful to University Grant Commission (UGC) for financial support as major research project. VKM is thankful to Department of Science and Technology (DST), New Delhi for providing fellowship under DST INSPIRE fellowship (IF 130374). Authors are also thankful to the Department of Biotechnology, New Delhi for establishment of DBT-BIF centre and DBT-State Biotech hub in the Department of Biotechnology, Mizoram University which has been used for the present study.

References

Aly AH, Debbab A, Kjer J, Proksch P (2010) Fungal endophytes from higher plants: a prolific source of phytochemicals and other bioactive natural products. Fungal Divers 41:1–16

Amnuaykanjanasin A, Punya J, Paungmoung P, Rungrod A, Tachaleat A, Pongpattanakitshote S et al (2005) Diversity of type I polyketide synthase genes in the wood-decay fungus *Xylaria* sp. BCC 1067. FEMS Microbiol Lett 251:125–136

Arnold AE, Henk DA, Eells RL, Lutzoni F, Vilgalys R (2007) Diversity and phylogenetic affinities of foliar fungal endophytes in loblolly pine inferred by culturing and environmental PCR. Mycologia 99(2):185–206

Bacon CW, White JF (2000) Microbial endophytes. Dekker, New York, pp 341–388

Barliana MI, Suradji EW, Abdulah R, Diantini A, Hatabu T, Nakajima-Shimada J et al (2014) Antiplasmodial properties of kaempferol-3-O-rhamnoside isolated from the leaves of *Schima wallichii* against chloroquine-resistant *Plasmodium falciparum*. Biomed Rep 2:579–583

Barnett HL, Hunter BB (1998) Illustrated genera of imperfect fungi, 4th edn. American Pathological Society, St. Paul

Beagle-Ristaino JE, Papavizas GC (1985) Biological control of Rhizoctonia stem canker and black scurf of potato. Phytopathology 75:560–564

Bhagat J, Kaur A, Sharma M, Saxena AK, Chadha BS (2012) Molecular and functional characterization of endophytic fungi from traditional medicinal plants. World J Microbiol Biotechnol 28:963–971

Bills GF, Polishook ID (1991) Microfungi from *Carpinus caroliniana*. Can J Bot 69:1477–1482

Bingle LE, Simpson TJ, Lazarus CM (1999) Ketosynthase domain probes identify two subclasses of polyketide synthase genes. Fungal Genet Biol 26:209–223

Cannon PF, Simmons CM (2002) Diversity and host preference of leaf endophytic fungi in the Iwokrama forest reserve, Guyana. Mycologia 94:210–220

Carroll G (1988) Fungal endophytes in stems and leaves: from latent pathogen to mutualistic symbionts. Ecology 69:2–9

Cenis JL (1992) Rapid extraction of fungal DNA for PCR amplification. Nucleic Acids Res 20:2380

Dewanjee S, Maiti A, Majumdar R, Majumdar A, Mandal SC (2008) Evaluation of antimicrobial activity of hydroalcoholic extract *Schima wallichii* bark. Pharmacol Online 1:523–528

Dewanjee S, Mandal V, Sahu R, Dua TK, Manna A, Mandal SC (2011) Anti-inflammatory activity of a polyphenolic enriched extract of *Schima wallichii* bark. Nat Prod Res 25:696–703. doi:10.1080/14786410802560732

Diantini A, Subarnas A, Lestari K, Halimah E, Susilawati Y et al (2012) Kaempferol-3-*O*-rhamnoside isolated from the leaves of *Schima wallichii* Korth. inhibits MCF-7 breast cancer cell proliferation through activation of the caspase cascade pathway. Oncol Lett 3: 1069–1072

Doty SL, Oakley B, Xin G, Kang JW, Singleton G, Khan Z, Vajzovic A, Staley JT (2009) Diazotrophic endophytes of native black cottonwood and willow. Symbiosis 47:23–33

Ek-Ramos MJ, Zhou W, Valencia CU, Antwl JB, Kalns LL, Morgan GD, Kerns DL, Sword GA (2013) Spatial and temporal variation in fungal endophyte communities isolated from cultivated cotton (Gossypium hirsutum). PLoS ONE 8:e66049

Eyberger AL, Dondapati R, Porter JR (2006) Endophyte fungal isolates from *Podophyllum peltatum* produce podophyllotoxin. J Nat Prod 69:1121–1124

Gelderblom WCA, Kriek NPJ, Marasas WFO, Thiel PG (1991) Toxicity and carcinogenicity of the *Fusarium moniliforme* metabolite, Fumonisin B1 in rats. Carcinogenesis 12:1247–1251

Gond SK, Mishra A, Sharma VK, Verma SK, Kumar J, Kharwar RN et al (2012) Diversity and antimicrobial activity of endophytic fungi isolated from Nyctanthes arbor-tristis, a well-known medicinal plant of India. Mycoscience 53:113–121

Hata K, Futai K (1995) Endophytic fungi associated healthy pine needle infested by pine needle gall midge *Thecodiplosis japonensis*. Can J Bot 73:384–390

Hyde KD, Soytong K (2008) The fungal endophyte dilemma. Fungal Divers 36:69–88

Iqbal J, Nelson JA, McCulley RL (2013) Fungal endophyte presence and genotype affect plant diversity and soil-to-atmosphere trace gas fluxes. Plant Soil 364:15–27

Kayini A, Pandey RR (2010) Phyllosphere fungi of *Alnus nepalensis*, *Castanopsis hystrix* and *Schima wallichii* in a subtropical forest of north east India. J Am Sci 6(3):118–124

Kimura M (1980) A simple method for estimating evolutionary rate of base substitutions through comparative studies of nucleotide sequences. J Mol Evol 16:111–120

Komai S, Hosoe T, Itabashi T, Nozawa K, Yaguchi T, Fukushima K et al (2006) New penicillide derivatives isolated from *Penicillium simplicissimum*. J Nat Med 60:185–190

Kusari S, Spiteller M (2012) Metabolomics of endophytic fungi producing associated plant secondary metabolites: progress, challenges and opportunities. In: Roessner U (ed) Metabolomics. In Tech, Rijeka, pp 241–266

Kusari S, Zuhlke S, Spiteller M (2009) An endophytic fungus from *Camptotheca acuminata* that produces camptothecin and analogues. J Nat Prod 72:2–7

Kusari S, Zuhlke S, Spiteller M (2011) Effect of artificial reconstitution of the interaction between the plant *Camptotheca acuminata* and the fungal endophyte *Fusarium solani* on camptothecin biosynthesis. J Nat Prod 74:764–775

Kusari S, Hertweck C, Spiteller M (2012a) Chemical ecology of endophytic fungi: origins of secondary metabolites. Chem Biol 19:792–798

Kusari S, Verma VC, Lamshoft M, Spiteller M (2012b) An endophytic fungus from *Azadirachta indica* A. Juss. that produces azadirachtin. World J Microbiol Biotechnol 28:1287–1294
Lin X, Huang YJ, Zhong HZ, Su WJ, Qian XM, Shen YM (2010) Endophytes from the pharmaceutical plant, *Annona squamosa*: isolation, bioactivity, identification and diversity of its polyketide synthase gene. Fungal Divers 41:41–51
Miao F, Yang R, Chen DD, Wang Y, Qin BF, Yang XJ et al (2012) Isolation, identification and antimicrobial activities of two secondary metabolites of *Talaromyces verruculosus*. Molecules 17:14091–14098
Mousa WK, Raizada MN (2013) The diversity of anti-microbial secondary metabolites produced by fungal endophytes: an interdisciplinary perspective. Front Microbiol 4:1–18
Nagabhyru P, Dinkins RD, Wood CL, Bacon CW, Schardl CL (2013) Tall fescue endophyte effects on tolerance to water-deficit stress. BMC Plant Biol 13:127
Nicholson TP, Rudd BA, Dawson M, Lazarus CM, Simpson TJ, Cox RJ (2001) Design and utility of oligonucleotide gene probes for fungal polyketide synthases. Chem Biol 8:157–178
Payne GA, Widstrom NW (1992) Aflatoxin in maize. Crit Rev Plant Sci 10:423–440
Pimentel D (2009) Pesticides and pest control. In: Peshinand R, Dhawan AK (eds) Integrated pest management: innovation-development process. Springer, Dordrecht, pp 83–87
Porras-Alfaro A, Bayman P (2011) Hidden fungi, emergent properties: endophytes and microbiomes. Annu Rev Phytopathol 49:291–315
Puri SC, Verma V, Amna T, Qazi GN, Spiteller M (2005) An endophytic fungus from *Nothapodytes foetida* that produces camptothecin. J Nat Prod 68:1717–1719
Puri SC, Nazir A, Chawla R, Arora R, Riyaz-Ul-Hasan S, Amna T et al (2006) The endophytic fungus *Trametes hirsuta* as a novel alternative source of podophyllotoxin and related aryl tetralin lignans. J Biotechnol 122:494–510
Qadri M, Rajput R, Abdin MZ, Vishwakarma RA, Riaz-UL-Hassan S (2014) Diversity, molecular phylogeny and bioactive potential of fungal endophytes associated with the Himalayan blue pine (*Pinus wallichiana*). Microb Ecol 67:877–887
Rai PK, Lalramnghinglova H (2010) Ethnomedicinal plant resources of Mizoram, India: implication of traditional knowledge in health care system. Ethnobot Leafl 14:274–305
Reinhold-Hurek B, Hurek T (2011) Living in side plants: bacterial endophytes. Curr Opin Plant Biol 14:435–443
Rodriguez RJ, Henson J, Van Volkenburgh E, Hoy M, Wright L, Beckwith F et al (2008) Stress tolerance in plants via habitat-adapted symbiosis. ISME J 2:404–416
Rojas JD, Sette LD, De Araujo WL, Lopes MSG, Da Silva LF, Furlan RLA et al (2012) The diversity of polyketide synthase genes from sugarcane-derived fungi. Microb Ecol 63:565–577
Saikkonen K, Faeth SH, Helander M, Sullivan TJ (1998) Fungal endophytes: a continuum of interactions with host plants. Annu Rev Ecol Syst 29:319–343
Schumann J, Hertweck C (2006) Advances in cloning, functional analysis and heterologous expression of fungal polyketide synthase genes. J Biotechnol 124:690–703
Shweta S, Zuehlke S, Ramesha BT, Priti V, Mohana Kumar P, Ravikanth G et al (2010) Endophytic fungal strains of *Fusarium solani*, from *Apodytes dimidiate* E. Mey. ex Arn (Icacinaceae) produce camptothecin, 10-hydroxycamptothecin and 9-methoxycamptothecin. Phytochemistry 71:117–122
Singh LP, Gill SG, Tuteja N (2011) Unraveling the role of fungal symbionts in plant abiotic stress tolerance. Plant Signal Behav 6:175–191
Stierle A, Strobel GA, Stierle D (1993) Taxol and taxane production by *Taxomyces andreanae* an endophytic fungus of Pacific yew. Science 260:214–216
Stone JK, Bacon CW, White JF Jr (2000) An overview of endophytic microbes: endophytism defined. In: Bacon CW, White JF Jr (eds) Microbial endophytes. Marcel Dekker, New York, pp 3–30
Stone JK, Polishook JD, White JRJ (2004) Endophytic fungi. In: Mueller G, Bills GF, Foster MS (eds) Biodiversity of fungi: inventory and monitoring methods. Elsevier, Burlington, pp 241–270
Strobel GA, Daisy B, Castillo U, Harper J (2004) Natural products from endophytic microorganisms. J Nat Prod 67:257–268

Sutton JC (1982) Epidemiologyof wheat head blight and maize ear rot caused by *Fusarium graminearum*. Can J Plant Pathol 4:195–209

Tamura K, Peterson D, Peterson N, Stecher G, Nei M, Kumar S (2011) MEGA5: molecular evolutionary genetics analysis using maximum likelihood, evolutionary distance, and maximum parsimony methods. Mol Biol Evol 28(10):2731

Tayung K, Jha DK (2010) Antimicrobial endophytic fungal assemblages inhabiting bark of *Taxus baccata* L. of Indo-Burma mega biodiversity hotspot. Indian J Microbiol 50:S74–S81

Verma VC, Gond SK, Kumar A, Kharwar RN, Strobel GA (2007) Endophytic mycoflora from leaf, barks, and stems of *Azadirachta indica* A Juss. from Varanasi India. Microbial Ecol 54: 119–125

Verma SK, Gond SK, Mishra A, Sharma VK, Kumar J, Singh DK et al (2014) Impact of environmental variables on the isolation, diversity and antibacterial activity of endophytic fungal communities from *Maduca indica* Gmel. at different locations in India. Ann Microbiol 64:721–734

Wagenaar MM, Corwin J, Strobel G, Clardy J (2000) Three new cytochalasins produced by an endophytic fungus in the genus *Rhinocladiella*. J Nat Prod 63:1692–1695

Wang FW, Hou ZM, Wang CR, Li P, Shi DH (2008) Bioactive metabolites from *Penicillium* sp., an endophytic fungus residing in *Hopea hainanensis*. World J Microbiol Biotechnol 24:2143–2147

White T, Burns T, Lee S, Taylor J (1990) Amplification and direct sequencing of fungal ribosomal RNA genes for phylogenetics. In: Innis MA, Gelfand DH, Sninsky JJ, White TT (eds) PCR protocols: a guide to methods and applications. Academic, San Diego, pp 315–322

The Biological Control Possibilities of Seed-Borne Fungi

Nuray Özer and Arzu Coşkuntuna

Introduction

Fungi constitute the largest group among seed-borne pathogens. Many fungal pathogens infect the developing and maturing seeds and reduce yield of seeds both quantitatively and qualitatively. They cause different types of disease and disorders such as seed abortion, shrunken seed, reduced seed size, seed rot, seed necrosis, seed discoloration, reduction or elimination of germination capacity, and physiological alterations in seed as a result of metabolic products produced by pathogens. Furthermore, seed-borne fungi can transmit from seed to plants and infect seedlings or plants even in the absence of symptoms on seeds if they are present in the seed. Some of them are also soil-borne and have the ability to survive in soil for many years, and be transmitted from infested seeds to the soil (Neergaard 1977).

Treatments for managing seed-borne fungal diseases include the use of cultural practices, development of resistant cultivars, application of biological agents, hot water and chemical fungicides. Among them, chemical control is still common and farmers have been finding themselves more confident in the use of seeds treated with fungicides. However, because of demand from society for foods with fewer chemical residues and consciousness for preservation of the environment, extensive research studies to develop non-chemical treatments for seeds, such as biological control, have been ongoing for nearly 30 years, and were even proposed half a century ago. Unfortunately few commercial biopesticides recommended for seed-borne diseases are available in the global market (Table 1), although several biopesticides have been registered for plant diseases and pests (Mishra et al. 2015).

N. Özer (✉) • A. Coşkuntuna
Department of Plant Protection, Faculty of Agriculture, Namık Kemal University, Tekirdağ 59030, Turkey
e-mail: nurayozer62@gmail.com

P. Kumar et al. (eds.), *Current Trends in Plant Disease Diagnostics and Management Practices*, Fungal Biology, DOI 10.1007/978-3-319-27312-9_17

Table 1 Commercially available biopesticides for seed-borne fungal diseases in the global market (Anonym 2013; Mishra et al. 2015)

Biopesticides as fungicides	Common name or trade name	Target
Bacillus subtilis MBI 600	Integral	*Fusarium* sp., *Aspergillus* sp.
Gliocladium catenulatum J1446	Prestop	Seed rot, Seed-borne diseases
	Prestop mix	
Pseudomonas chlororaphis 63-28	AtEze™	Seed-borne fungi
P. chlororaphis MA342	Cedomon	*Dreschslera* spp., *Bipolaris* sp., *Fusarium* spp., *Ustilago* spp.
	Cerall	*Tilletia caries*, *Stagonospora* (*Septoria*) *nodorum*, *Fusarium* spp./*Microdochium nivale*
Streptomyces griseoviridis K61	Mycostop Mix	Seed-rot, seed damping off
Trichoderma harzianum Rifai T-22	PlantShield	Seed rot
T. harzianum KRL-AG2	RootShield	
	T-22 Planter box	

This chapter reviews the disease symptoms caused by seed-borne fungi and provides an overview of 30 years of research activities related to treatment with biocontrol agents and formulations for seeds infested with seed-borne fungi. The chapter includes the results of biological control pot and field experiments using seeds naturally or artificially infested with seed-borne pathogens; it does not cover *in vitro* antagonist screening programs and experiments based on soil inoculations of pathogens.

Seed Borne Fungi in Field Crops Subjected to Biological Control Studies

Antagonist bacteria and fungi were screened to determine the possibility of biological control for seed-borne fungi such as *Bipolaris sorokiniana*, *Drechslera* spp, *Fusarium* spp, *Stagonospora nodorum*, *Ustilago* spp, *Tilletia* spp., *Bipolaris oryzae*, *Fusarium fujikuroi*, *Gerlachia oryzae*, *Colletotrichum graminicola*, *C. truncatum*, *C. lindemuthianum*, *Ascochyta pisi*, *Botrytis cinerea* on cereals and legumes.

B. sorokiniana (Sacc.) Shoemaker (=*Helminthosporium sativum* Pammel, C.M. King & Bakke, teleomorph *Cochliobolus sativus* (S. Ito & Kurib.) Drechsler ex Dastur) causes diseases such as common root rot, spot blotch, seedling blight and crown rot on wheat and barley (Sivanesan and Holliday 1981). Short brown stripes with purple edges appear on the emerging leaves of oat as a result of infection by seed-borne *Drechslera avenae* (Eidam) Scharif (=*H. avenae* Eidam, telemorph *Pyrenophora chaetomioides* Speg.). *D. teres* (Sacc.) Shomaker (=*H. teres* Sacc, teleomorph *P. teres* Drechsler), is the agent of net blotch disease on barley. It is present as mycelium within the seed coat or as conidia on the seed surface (Arnst 1978). *D. graminea* (Rabenh. ex

Schltdl.) (=*H. gramineum* Rabenh. ex Schltdl., teleomorph *P. graminea* S. Ito & Kurib.) is another seed-borne fungus in barley and causes long stripes with pale green colour at first. As the infection progresses, the colour of the stripes becomes yellow and then dark brown. The disease severity increases when seeds infected with this pathogen are used and the seeds are not treated with fungicides (Zad et al. 2002). Seed-borne *Fusarium culmorum* (W.G. Smith) Sacc., *F. graminearum* Schwabe (teleomorph *Gibberella zeae* (Schwein.) Petch) and *Microdochium nivale* (Fr.) Samuels (formerly *F. nivale* and *M. nivale* var *nivale*) cause seedling blight on wheat and barley. Significant yield losses occur as a result of poor germination if the seeds are significantly infected with these pathogens (Humpreys et al. 1995; Hare et al. 1999; Haigh et al. 2009; Sherm et al. 2013). *Stagonospora nodorum* [=*Septoria nodorum* (Berk.) Castell. and Germano, teleomorph *Phaeosphaeria* (=*Leptosphaeria*) *nodorum* (Müll.) Hedjar.] is an important pathogen of wheat and related cereals, and causes a necrotic leaf blotch as well as discoloration of the head in the symptom known as glume blotch (Solomon et al. 2006). *Ustilago spp.* is the cause of loose smut disease on wheat and barley. The pathogen is present within seed embryos as dormant mycelium until the infected seeds are sown and germinate. *Tilletia leavis* Kühn in Rabenh [=*T. foetida* (Wallr.) Liro] and *T. tritici* (Bjerk.) G. Wint. in Rabenh [= *T. caries* (DC.) Tul.&C.Tul] cause common bunt disease on winter wheat. *T. tritici* is common worldwide where wheat is grown. *T. leavis* is prevalent throughout the European mainland and central and eastern North America. Seeds filled with masses of teliospores are the characteristic symptoms of this disease (Wiese 1977). *B. oryzae* (Breda de Haan) Shoemaker [=*D. oryzae* (Breda de Haan) Subramanian and Jain, =*H. oryzae*, teleomorph *Cochliobolus miyabeanus* (Ito and Kuribayashi) Drechsler ex Dastur] causes brown spot on the leaves of rice. This pathogen develops on seedlings; however, it is most noticeable when leaves and heads of older plants become infected. *F. fujikuroi* Nirenberg (=*F. moniliforme* J. Sheld., teleomorph *Gibberella fujikuroi* (Sawada) Wollenw = *G. fujikuroi* Swada mating population C [MP-C]) is the agent of Bakanae disease in rice. Typical symptoms of the disease include yellowing, stunting, stem elongation, and root and crown rot (Webster and Gunnell 1992). *Gerlachia oryzae* (Hashioka and Yokogi) W. Gans is the agent of leaf scald. This disease causes a decrease in seed germination as well as leaf symptoms (Webster and Gunnell 1992). *Colletotrichum truncatum* Schw. (Andrus Moore) is the agent of brown blotch disease in cowpea. It causes reduction in germination of seeds and purple brown discolouration on pods extending to petioles, leaf veins and peduncles. The maldevelopment and distortion of pods may occur in the case of pod infection (Allen et al. 1998; Adegbite and Amusa 2008). This pathogen also causes anthracnose disease on soybean (Wrather et al. 1997). *C. graminicola* (Ces) Wils (syn. *C. sublineolum* P. Henn. in Kabat and Bubák), *C. lindemuthianum* (Sacc. & Magnus) Lams.-Scrib., and *Ascochyta pisi* Lib. are the agents of anthracnose disease in sorghum, bean and pea, respectively. *C. lindemuthianum* infects seedlings through the seed coat and cotyledons. *A. pisi* penetrates into the inner parts of seeds (Tivoli and Banniza 2007), *B. cinerea* Pers. ex Fr. generally causes foliar infection in chickpea. However the pathogen is also seed-borne in chickpea and is responsible for poor germination and soft rot of the lower stem (Brefag and Mebalds 1987).

Some researchers have examined the ability of biological control agents to control multiple pathogens such as *B. sorokiniana*, *Drechslera* spp., *Fusarium* spp., and *Ustilago* spp., in the same study. Some of them have worked with individual pathogens such as *Tilletia* spp., *B. oryzae*, *F. fujikuroi*, *G. oryzae*, *C. graminicola*, *C. truncatum*, *C. lindemuthianum*, *A. pisi*, *B. cinerea*, and these studies will be presented as separate sections based on pathogen throughout the chapter.

Studies on Biological Control of Multiple Pathogens Causing Seed-Borne Fungal Diseases in Field Crops

Knudsen et al. (1995) applied antagonist fungi with different origins to barley and wheat seeds naturally infested with *B. sorokiniana* and *F. culmorum*, respectively, and cultivated these seeds in pots with sand and field soil, and also under field conditions. The antagonist fungus *Idriella bolleyi* 454 gave the highest control of diseases caused by *B. sorokiniana* (57 %) and *F. culmorum* (76 %) in pots with sand. *Gliocladium roseum* 726 was found to be the most effective antagonist against *B. sorokiniana* in pots containing field soil, showing an effectiveness of 60 %. In field experiments, wheat and barley seeds, naturally and artificially infected, respectively, with *F. culmorum*, and barley seeds naturally infested with *B. sorokiniana* were used. *I. bolleyi*, *G. roseum* and a Finnish isolate (J76) reduced the disease by both pathogens under field conditions. But only reductions in disease caused by *F. culmorum* on wheat and barley were significant. The dry weight of plants, number of tillers per row, yield and 1,000 grain weight were higher in *G. roseum*-treated plots than in the untreated control.

Etheridge (1997) investigated the possibilities of biological control of seedling blight in winter wheat caused by *F. culmorum* and *M. nivale* by artificial inoculation of seeds with these pathogens, then utilizing several antagonist fungi and a commercial product, Mycostop (a.i.: *Streptomyces griseoviridis* K61), which is registered for the diseases of damping-off, stem, and crown rots. A Finnish isolate (J76) and *G. roseum* (IMI 040222) were the most effective in reducing pre-emergence death caused by *F. culmorum*. All biological seed treatments reduced disease by *M. nivale* with an effectiveness of approximately 40 %. However, none of the seed treatments controlled foot rot assessed at harvest, although they increased yield.

Hökeberg et al. (1997) screened bacterial strains isolated from roots of wild and cultivated plants, collected at various places in Sweden, for effectiveness against seed-borne diseases caused by *D. teres* and *M. nivale* under greenhouse conditions. For screening against *D. teres*, naturally and heavily infected barley seeds were used; against *M. nivale*, artificially infected wheat seeds were used. Bacterial strains (scraped from the agar surface of Triptic Soy Agar-TSA and mixed with SNB nutrient broth and 2 % sodium-carboxymethyl cellulose) were used to treat the naturally or artificially infected seeds. Two *Pseudomonas* strains, MA342 and MA250, were found to significantly suppress the diseases in screening tests. The antagonist strain MA342 was tested against two pathogens; MA250 was tested against *M. nivale* under field conditions using a bacterial culture prepared from Tryptic Soy Broth 50 % (TSB50). MA342 controlled disease caused by *D. teres* and *M. nivale* with an effectiveness of 99 % and 67 %, respectively. The authors recommended the use of TSB culture for antagonist bacteria which can be stored for up to 1 month or deep frozen for even longer without decreasing the biocontrol effect.

The following year, the strain MA342 (*Pseduomonas chlororaphis*), grown on TSB 50, was used to treat barley seeds naturally infected with *B. sorokiniana*, *Drechslera* spp. and *Ustilago nuda*, seeds of barley, wheat and rye infected with *M. nivale* and oat seeds infected with *D. avenae*. Seeds were treated by mixing them vigorously 4–5 min with MA 342 at a dosage of 200–300 ml/kg seed (Johnsson et al. 1998). This culture was also applied to oat and barley seeds artificially infected with *Ustilago avenae* and *U. hordei*, respectively. The experiments were conducted under field conditions in different locations of Sweden and different years. This strain had a consistent suppressive effect on disease caused by *D. graminea*, *D. teres*, *D. avenae*, *U. avena, and U. hordei*, not only over time, but also under different climatic conditions. Application of this bacterial strain to wheat seeds naturally infected with *S. nodorum* also had good effect in greenhouse experiments. However, these authors reported that the bacterization of seeds did not control the disease caused by *U. nuda* and gave less than full effects against diseases caused by *B. sorokiniana* and *M. (Fusarium) nivale* under field conditions. The commercial products of MA 342 are presently named Cedemon and Cerall in the European market.

Teperi et al. (1998) described a screening system for identifying fungal antagonists from soils of Finland that have the ability to control seed-borne *F. culmorum* on wheat. This screening system consisted of three separate pot tests with sand, peat and field soil carried out under controlled conditions in a greenhouse, and field experiments run in 1991–1995. Wheat seeds heavily infected with pathogen were used in the experiments. Antagonist isolates were applied by drenching their spore suspensions onto seeds in pots with sand, or by shaking seeds in a conidia suspension for pot tests with peat and field soil and also in field experiments. These authors suggested that some of the isolates belonging to *Gliocladium* spp. were superior to other fungal isolates in both greenhouse and field tests.

Mycoparasitic *Pythium* species were investigated for biological control of *F. culmorum* in barley (Davanlou et al. 1999). The pathogen was inoculated by pipetting its spore suspension over the seeds, and then the spore suspension of potential antagonists was added to seeds in the pots under controlled conditions. An isolate of *P. acanthophoron* IMI 330381 and *P. mycoparasiticum* IMI341972 from JW Deacon, *P. periplocum* 1048 from Indonesian soils, and two isolates (1004 and MM9) of *P. oligandrum* from Denmark significantly suppressed disease severity on barley seedlings.

Jensen et al. (2000) applied the conidia of *Clonostachys rosea* (IK 726) (formerly *G. roseum*), which were freshly harvested, dried and stored, to the seeds of barley and wheat artificially inoculated with *F. culmorum* in pots filled with sand and under field conditions in Denmark. In this study, freshly harvested conidia were obtained from liquid culture containing Potato Dextrose Broth (PDB). A stored formulation of IK 726 was prepared from growth on a mixture of sphagnum peat, wheat bran and water. After incubation of the antagonist on this medium for 14 days, the inoculum was air-dried for 2 days, milled in a blender and stored in glass petri dishes at 4 °C for 8 weeks (8×10^8 cfu/g) and 32 weeks (6×10^8) in field experiments with barley, and for 30 weeks (6×10^8 cfu/g) in field experiments with wheat. Seed treatment with dried and stored conidia of *C. rosea* reduced the disease as effectively as freshly harvested conidia, with reduction in disease index varying from 47 % to 54 %, from 37 % to 68 % and from 70 % to 94 % in barley field, wheat field, and sand experiments, respectively. Sowing dates did not influence the effectiveness of this biocontrol agent against *F. culmorum* in barley. Dosage of the antagonist with a density of $>5 \times 10^3$ conidia/seed for both types of conidia was recommended to control the disease. Mamarabadi et al. (2009) suggested that this antagonist secreted chitin-hydrolysing agent in order to target the cell wall of *F. culmorum*.

In Italy, the treatment of durum wheat seeds with *C. rosea* (CR 47) along with *Trichoderma atroviride*, *T. harzianum*, *T. longibranchiatum* and *Penicillium frequentans* increased emergence and yield, and reduced disease incidence and severity in plants developed from seeds naturally infested with *F. culmorum* under field conditions during 1994–1996 (Roberti et al. 2000). The isolate *C. rosea* CR 47 was also found to be the least sensitive to fungicides with active ingredients such as carboxin, thiram, triticonazole and guazatine used as seed treatment to control foot rot disease, and it was also compatible with herbicides such as flufenacet, chlorotoluron, chlorosulfuron and pendimethalin used at pre-emergence stage in Italian wheat cultivation (Roberti et al. 2006). The same authors reported that several chitinase isoforms were induced by CR47 treatment of wheat seeds both in coleoptiles and roots (Roberti et al. 2008).

The seeds of eight wheat cultivars inoculated with *F. graminearum*, were treated with a solution of bacterial or fungal antagonists prepared with sodium carboxymethylcellulose (CMC), and sown in pots filled with non-sterilized cultivation soil by Dal Bello et al. (2002). Three weeks after sowing, seedling stand, disease percentage on emerging seedlings, plant height and dry weight of seedlings were evaluated. Among the antagonists, *Stenotrophomonas maltophilia* had good performance for criteria tested in all cultivars, although it caused a non-significant decrease in the percentage of diseased plants. Three strains of *Bacillus cereus* and one isolate of *T. harzianum* controlled the disease in some cultivars with an effectiveness of <50 %. The authors recommended these isolates be tested for their effects on seedling blight by *F. graminearum* under field conditions.

Several bacterial strains were applied to wheat seeds naturally or artificially infested with *F. culmorum* and *M. nivale*, and were tested for their effects on seedling blight, number of surviving plants, seedling emergence and yield (Johansson et al. 2003). The authors inoculated wheat seeds with *F. culmorum* and *M. nivale* by soaking them in mung bean liquid medium containing spores and mycelia of the pathogens for 0.5 h. Bacterial strains were suspended in an isotonic water after being washed from the surface of cultures grown on TSA; the bacterial suspensions were separately poured over artificially or naturally infested seeds. Bacterial strains were first screened under greenhouse and field conditions, and four strains (three of them fluorescent pseudomonas MF30, MF416 and MF588; one a species of *Pantoea* sp, MF 626) were selected for other studies in different years and locations in Sweden. This study showed that the selected strains suppressed infection by both pathogens during 5 years of experimentation and that seed treatment with *Pantoea* sp. increased yield at different locations.

Koch et al. (2006) examined the numerous products for controlling *D. graminea* and *D. teres* on barley, and *F. culmorum* and *Fusarium* spp. on wheat under controlled conditions. They used seeds naturally infected with these pathogens. The strains FZB 53 and W 490 (*Streptomyces* spp.) were found to be effective for reducing infection by *D. graminea* at rates of 51.9–75.9 % and 44.1–57.6 %, respectively, in experiments conducted in four trays. Strains or preparations of *B. subtilis*, *Streptomyces* spp. or *Trichoderma* were not effective on net blotch caused by *D. teres*. The strain FZB53 increased the number of healthy plants raised from seeds naturally infested with *F. culmorum* as compared with fungicides.

Sjöberg et al. (2007) multiplied arbuscular mycorrhizal (AM) fungi collected from fourteen Swedish arable soils, in which different crops were cultivated, in trap cultures using a mixture of plant species such as Alexandrian clover (*Trifolium alexandrium*), corn (*Zea mays*), leek (*Allium porrum*), marigold (*Tagetes erecta*), pea (*Pisum sativum*), sunflower (*Helianthus annuus*), tomato (*Lycopersicum esculentum*), wheat (*Triticum aestivum*) and white clover (*Trifolium repens*). These were added to pot soil in which barley seeds naturally infected with *B. sorokiniana* were sown. The commercial mycorrhizal preparations, *Glomus intraradices* (BCCM™/MUCL, Belgium, culture no. 43194) and Vaminoc$^{(R)}$ (Becker Underwood, Littlehampton, UK) were also used in the experiments. Mycorrhiza soil inocula from six trap cultures reduced the transmission of this pathogen from seeds to the stem base. Among them, two soil inocula of AM from semi-natural grassland and barley cultivated soils significantly inhibited leaf lesions, the proportion of infected barley plant nodes and stem bases. Vominoc was only effective on the node infection. Additionally, the roots of barley seedlings raised from seeds infected with *B. sorokiniana* were treated with AM fungal spores and spores of *G. intraradices*. In this experiment, AM spore inocula obtained from soil in which barley was cultivated suppressed disease symptoms on the base and upper half of leaves as well as did *G. intraradices*.

Hasan et al. (2012) inoculated wheat seeds with a spore suspension of *B. sorokiniana*, *F. graminearum*, *Aspergillus* spp. and *Penicillium* spp. separately; after drying, the seeds were treated with *T. harzianum* RUT 103. The treated seeds were sown in a field in Bangladesh. Three foliar sprays of *T. harzianum* spore suspension were used at tillering, heading and grain filling stages in addition to the seed treatment. Treatment with *T. harzianum* completely controlled seedling infection by the pathogens tested, except for *F. graminearum* and *B. sorokiniana*, and was also effective on leaf blight severity. At harvest, number of tillers per plant, plant height, ear length, grain number in an ear, number of healthy grains in an ear, 1,000-seed weight and grain yield increased in plants treated with *T. harzianum*. Hasan (2013) tested the same antagonist isolate against *B. sorokiniana*, *F. graminearum* and *Aspergillus flavus* by using different wheat cultivars under field conditions in Bangladesh. In this test, the author inoculated seeds by soaking them in spore suspensions prepared as combinations of *T. harzianum* with each of the pathogens, separately. Foliar sprays were applied as described in Hasan et al. (2012). Few seedlings infected with the pathogens were recorded in plots using these treatments. Seed treatment along with foliar spray of *T. harzianum* increased the number of tillers per plant, plant height, ear length, grain number in an ear, healthy grain number in an ear, 1,000-seed weight and grain yield compared with control plots for this experiment.

Several researches have investigated the potential of biocontrol agents when *Fusarium graminearum* or *F. culmorum* were inoculated to the ear at anthesis stage and antagonists were applied to the ear at the same stage to evaluate their effectiveness on head blight. Results of using different antagonists for anthesis inoculation are listed in Table 2.

Studies on the Biological Control of **Tilletia** *spp. on Wheat*

In all studies on biological control of *Tilletia* spp. given below, seeds were artificially inoculated with the pathogen before antagonist treatment.

Wheat seeds artificially inoculated with *T. leavis* were coated with a suspension of *Pseudomonas* strain 2–79 (Pf2–79r) obtained from a nutrient broth yeast extract (NBYA) culture containing methyl cellulose (200 ml/kg seeds), air-dried overnight and used for field experiments conducted in 2 years. Seedlings were also sprayed with this stain at 14 days after planting. The strain reduced common bunt incidence caused by *T. leavis* at the rate of 65 % and 50 % during consecutive seasons when the antagonist was applied to wheat seeds and 2-week-old seedlings (McManus et al. 1993).

Hökeberg et al. (1997) used the bacterial strain MA 342 (*P. chlororaphis*) for control of *T. caries*, applying the strain to wheat seeds after inoculation of the pathogen as described above for *D. teres* and *M. nivale*. The bacterial strain completely controlled the disease under field conditions. The same strain (MA342), prepared and applied to seeds as above for *B. sorokiniana*, *Drechslera* spp. and *U. nuda*, effectively controlled seed-borne *T. caries* under different climatic conditions and in different years (Johnsson et al. 1998).

Borgen and Davanlou (2001) examined the effect of biological control on *T. tritici* using different doses of two biocontrol agents (*P. chlororaphis* MA 342; *C. rosea* IK 726) and several commercial products (Mycostop, EM1, Effective microorganism, a.i.: 80 species of beneficial microorganism, the major part being yeast and lactic acid bacteria; Symbioplex, a.i.: *Lactobacillus acidophilus*, *Bifidobacterium bifidus* and *Streptococcus thermophillus*; Supresivit, a.i.: *T. harzianum*) under field conditions in Denmark. The authors applied these biocontrol agents to seeds with and without milk powder. This research showed that treatment of winter wheat seeds (Cv. Pepital) with a combination of MA 342 (40×10^9 bacteria/kg seed) and *C. rosea* (2.5×10^9 cfu/kg seed) with 2 % milk powder controlled bunt disease by *T. tritici* at the rates of 97.2 % and 86.6 %, respectively.

Table 2 Biological control agents developed to reduce disease severity or mycotoxin caused by *Fusarium graminearum* in anthesis stage of wheat

Antagonist	Reduction in disease severity or mycotoxin	Application methods	Reference
Bacillus subtilis (AS.43.3)	77 %	SI, by dropping, greenhouse	Khan et al. (2001)
Bacillus subtilis (AS 43.4)	93 %		
Cryptococcus nodaensis (OH 182.9)	56 %		
Bacillus mycoides (S-07-01)	48.1 %	DI (4 h), seed coating + injection	Fernando et al. (2002)
B. cereus (L-07-01)	48.1 %		
B. subtilis (H-08-02)	49.1 %		
Bacillus subtilis (AS.43.3)	92 %, 78 % (depending on the cultivars, Cv.)	SI, by dropping, greenhouse	Schisler et al. (2002)
Cryptococcus sp. (OH 71.4)	18.6–57.1 % (depending on the Cv., locations)	SI, by dropping, field	
Cryptococcus sp. (OH 181.1)	20.0–59.3 % (depending on the Cv., locations)		
Cryptococcus sp. (182.9)	0–57.1 % (depending on the Cv., locations)		
Cryptococcus sp. (71.4)	50 % (for disease incidence)	SI, by spraying, field	Khan et al. (2004)
Cryptococcus sp. (181.1)	36 %		
Cryptococcus nodaensis (OH 182.9)	36 %		
Fusarium equiseti (62)	92 % (includes *F. culmorum* also)	DI, by spraying, field	Dawson et al. (2004)
	94 % (for DON)		
Streptomyces sp. (3)	~50.0 %	DI (6 h), by spraying, greenhouse	Nourozian et al. (2006)
Lysobacter enzymogenes (C3)	70 %	DI (1 day), by spraying, greenhouse	Jochum et al. (2006)
	0, 14, 40.7 % (depending on the Cv., locations)		
Brevibacillus sp. (263)	70.1 %, 32.0 % (for DON)	SI, by dropping, greenhouse	Palazzini et al. (2007)
Bacillus subtilis (218)	53.6 %, 57.0 % (for DON)		
Bacillus sp. (166)	47.4 %, 53.0 % (for DON)		
Bacillus sp. (127)	36.0 %, 36.0 % (for DON)		

(continued)

Table 2 (continued)

Antagonist	Reduction in disease severity or mycotoxin	Application methods	Reference
Pseudomonas fluorescens (MKB158, MKB249) and *P. frederiksbergensis* (MKB202)	(all for *F. culmorum*)	DI (24 h), by spraying, greenhouse	Khan and Doohan (2009)
	22.0–47.0 % (in wheat), 60.0–65.0 % (in barley),		
	82.0 % (for DON by MKB158 in wheat), 98.0 % (for DON by MKB158 in barley)		
	≥44.0 % (in wheat), ≥55.0 % (in barley)	DI (24 h), by spraying, field	
	74.0–78.0 % (for DON by MKB 158 and 249, in wheat and barley)		
Clonostachys rosea (ACM941)	65 %	DI (2 days), by spraying, greenhouse	Xue et al. (2009)
C. rosea, biofungicide (CLO-1)	68–92 % (depending on the concentrations)	DI (3 h), by spraying, greenhouse	Xue et al. (2014)
	51–95 % (for DON, depending on the concentrations)		

SI Simultaneous inoculation with pathogen, *DI* Inoculation before pathogen, *Cv* Cultivar, *DON* Deoxynivalenol

Koch et al. (2004) selected microbial antagonists effective against *T. caries* by evaluating chlorotic flecking on leaf sheaths and blades at an early stage of wheat growth in pots under controlled conditions. The authors re-tested the selected strains under the same conditions and obtained consistent control of the pathogen with *Trichoderma* isolates and some actinomycetes. The following year, five actinomycete and four *Trichoderma* isolates were applied to seed and were tested under controlled and field conditions (Koch et al. 2006). Efficacy of the treatments was calculated based on plants with leaf symptoms under controlled conditions and on plants with infected heads in the field experiments. *Streptomyces* strain FZB 53 treatment had the greatest effects on bunt disease under both controlled and field conditions. The commercial product Polyversum (a.i.: *Pythium oligandrum*), which was previously found to have 60 % effect on common bunt in field trials by Beneda and Pospisil (1999), showed some control in the field experiments. FZB 53 and Polyversum were tested in second field trials with five different wheat varieties. FZB 53 was more effective than Polyversum in this experiment although its effectiveness varied between 11 % on a sensitive variety and 77 % on a resistant variety. Another experiment by these authors included a comparison of pot experiments with field experiments to test for efficacy of *Streptomyces* spp., *Trichoderma* sp., *B. subtilis* and Polyversum on the disease by applying antagonists to the seeds. Among these treatments, *Streptomycess* strains FZB53 and W490 provided control of the disease with low efficacy (40 %) under field conditions, but were highly effective (90 %) under pot conditions.

Culture filtrates of *Cylindrocarpon olidum* var. *olidum*, isolated from the rhizosphere of *Liqidambar orientalis* Mill. (Oriental sweet gum, Syn. levant storax) grown in Turkey, was supplemented with methylcellulase to enhance stickiness and was used for control of common bunt disease (*T. leavis*) on wheat by Yolageldi and Turhan (2005). Wheat seeds were wetted with this filtrate and sown in the field. These authors recorded that this treatment controlled the disease with an effectiveness of 51.3 % and 48.4 % in two growing seasons.

Muscador albus isolate C2620, which originated as an endophyte from a cinnamon tree (*Cinnamomum zeylanicum*), was applied in two ways: by mixing the seeds with powder obtained from rye grain culture of *M. albus* (125 mg/g seed) and by placing particles from the rye grain formulation of the antagonist in furrows at the rate of 4 g/m of row, along with wheat seeds infested with *T. tritici*. These applications were made during planting for two growing seasons and two planting dates in Aberdeen, Idaho (Goates and Mercier 2011). In this study, seed and in-furrow treatments had 11.8 % and 8.5 % bunted spikes, respectively, for the early planting date (April 6) when the percentage of bunted spike was at its highest rate (43.8 %) in the untreated control. Both treatments were also effective for later planting dates and years although the disease was in low incidence in controls. The authors suggested that *M. albus* may have potential for common bunt control in organic wheat production where options for managing the disease are very limited.

Studies on Biological Control of B. oryzae and G. oryzae on Rice

Khalili et al. (2012) applied *Trichoderma* isolates to rice seeds artificially infested with *B. oryzae* by soaking them in spore suspensions of *Trichoderma* isolates for 2 h. The treated seeds were sown in pot soil and evaluated under greenhouse conditions. Among the *Trichoderma* spp., *T. harzianum* 1 and *T. harzianum* 20 gave high control of disease on seedlings with an efficacy of 61.2 % and 58.1 %, respectively.

Moura et al. (2014) applied the antagonist strains *Pseudomonas synxantha* (DFs 185), *P. fluorescens* (DFs 223), *Bacillus* sp. (DFs 418) and an unidentified strain (DFs 306) to seeds naturally infested with *B. oryzae* and *G. oryzae*, by immersing the seeds in a suspension of each bacterium for 24 h. After a stirring period of 30 min. at 10 °C, the treated seeds were sown in pots containing sterilized vermiculite. These authors evaluated transmission of the pathogens to seedlings under controlled conditions, and measured intensity of symptoms. They reported that DFs185 and DFs 306 reduced transmission of both pathogens to the seedlings and that DFs306 significantly increased growth of the plants.

Studies on the Biological Control of F. fujikuroi on Rice

Kumakura et al. (2003) applied *Trichoderma* sp. (SKT1) to rice seeds naturally infested with *F. fujikuroi*. This treatment gave high control of Bakanae disease in pot experiments. The antagonist isolate SKT1 was identified as *Trichoderma asperellum* by Watanabe et al. (2005).

B. subtilis, *T. harzianum* and *T. virens*, isolated from paddy soil samples at different locations in Guilan province, controlled the disease when rice seeds were treated with antagonists prior to inoculation of the pathogen; however, the effect of antagonists was lower than that of fungicide treatment (Dehkaei et al. 2004).

An isolate of *Talaromyces* sp. (KNB422), discovered from stem tissue of an apparently healthy rice plant in a nursery box with Bakanae diseased plants, was effective as a seed treatment against Bakanae disease caused by *F. fujikuroi* (Tateishi et al. 2006). This isolate was registered as a biopesticide in 2010 as Momi-Keeper (Central Glass, Co., Ltd.). Kato et al. (2012) reported that the mode of action of this antagonist was mycoparasitic, causing collapsed cell walls in hyphal cells and deterioration of the cytoplasm of *F. fujikuroi* after contact with *Talaromyces* sp.

Rice seeds infested with *F. fujikuroi*, collected from plants in the field after artificial inoculation with this pathogen at flowering stage, were sown in soil to which biopesticide KNB422 was added at doses of $1 \times 10^{3-6}$ cfu/g, 260 g to examine the effect of this biopesticide on Bakanae disease at seedling stage (Miyake et al. 2012). Hot water (10 min. at 60 °C) treated seeds were also tested in the same application of biopesticide. Treatment of soil with the biopesticide at doses of

1×10^6 cfu/g, 1×10^5 cfu/g and 1×10^4 cfu/g controlled the disease at percentages of 96.6 %, 96.9 % and 88.4 %, respectively. However a combination of KNB422-soil treatment and hot-water submersion of infested rice seeds with the pathogen completely suppressed the disease.

Matic et al. (2014) investigated the effect of seed treatments with epiphytic yeasts isolated from rice seeds, and the combination of antagonists with thermotherapy (dipping seeds in hot water) on Bakanae disease, measuring disease index at 28 days after sowing under controlled conditions. Seeds naturally infected with *F. fujikuroi* were used in this study. Among the yeast isolates, *Pichia guilliermondii* (R9) and *Metschnikowia pulcherrima* (R23) significantly controlled the disease. The disease index was lowest (5 %), when a combination of these antagonist yeasts and thermotherapy at 60 °C for 10 min was applied to the seeds. Biofungicides tested in this study, such as Serenade, Mycostop mix and microorganism mixture, had low effect for controlling the disease.

Studies on the Biological Control of Colletotrichum spp.

Bankole and Adebanjo (1996) tested the efficacy of a *T. viride* isolate (TH31) obtained from cowpea phylloplane against *C. truncatum* in cowpea using seeds naturally infested with the pathogen and different treatment methods with an antagonist conidia suspension (10^8 conidia/ml) under greenhouse conditions. Percentages of seed germination and disease incidence were recorded in this experiment. Highest seed germination rates were obtained by dipping seed in the antagonist spore suspension. Percentages of infected seedling (disease incidence) were reduced to 27.6 % and 45.2 % by seed dipping in spore suspension and soil drenching with spore suspension of this antagonist, respectively, as compared with 85.5 % in infected control.

Tinivella et al. (2009) applied several commercially formulated microorganisms and non-formulated selected strains of different microorganisms (fungi, bacteria and yeasts) to the seeds of bean and pea naturally infested with *C. lindemuthianum* and *Ascochyta* spp., respectively. Percentages for emergence, diseased and healthy plants were recorded under greenhouse conditions. Among the treatments, application of *C. rosea* (IK726-F) to seeds by dipping them in a suspension (40 ml/10 g seed) of the antagonist in a clay preparation (10^8 cfu/g) with sterile distilled water gave good control only of *Ascochyta* spp. on pea for the criteria evaluated. *Bacillus subtilis*-based formulations, FZB24 and Serenade, some strains of *P. putida* (E183; G12; G53), applied by dipping seeds in a bacterial cell suspension in nutrient yeast dextrose broth medium for 15 min., and a disease-suppressive saprophytic isolate of *F. oxysporum* (MSA35), applied by dipping seeds in a conidia suspension prepared from Potato Dextrose Agar (PDA) for 15 min., were successful for controlling anthracnose disease in bean, although plant-based products were more effective than antagonists.

Soybean seeds were treated with spore suspensions of *Pseudomonas aeruginosa*, *T. harzianum*, *T. virens* and a mixture of *T. harzianum* and *T. virens* after inoculation with *C. truncatum* (Begum et al. 2010). Percentage of pre- and post-emergence damping-off, seed germination, and final seedling stand were recorded in seedlings raised from treated seeds that were sown in two fields at different months (February and August) in Malaysia. Treatment of seeds with *P. aeruginosa* was the most effective treatment. It reduced the pre-emergence damping-off rates by 48.6 % and 51.9 % for February and August experiments, respectively, post-emergence damping-off 65.0 % and 97.2 %, respectively. However, increases in seed germination and healthy seedling stands were higher in the February experiment than in the August experiment. *T. virens* was the least effective of the bio-primed treatments.

T. viride, *T. harzianum*, *T. hamatum*, *G. virens*, isolated from the rhizosphere of anthracnose-infected bean seedlings, were multiplied on wheat bran and applied to bean seeds naturally infected with *C. lindemuthianum* by smearing the seeds with antagonists for 15 min. and adding antagonists to the pot soil where the same seeds were used (Padder and Sharma 2011). These experiments, conducted under controlled conditions for 30 days, showed that applications of *T. viride* and *T. harzianum* to seeds had maximum potentiality to suppress seed-borne infection of *C. lindemuthianum*, with efficacies of 86.4 % and 83.6 %, respectively.

Three fungal isolates, *Chaetomium globosum* 57, *T. harzianum* 184 and *F. oxysporum* NSF9 from the rhizosphere and rhizoplane of *Cynodon dactylon*, *Heteropogon contortus*, and *Alloteraropsis cimicina* were tested for control of *Colletotrichum graminicola* in sorghum under pot and field conditions in India (Vasanthakumari and Shivanna 2014). Inocula of antagonists grown on sorghum grain for 6 months were added to the soil in both experiments at different rates (1 %, 1.5 % and 2.0 %, w/w). *C. graminicola* spore coated seeds were sown in the soil treated with antagonists. Various parameters (pre- and post-emergence seedling mortalities, disease incidence, disease severity, growth promotion, number of seedlings with intact roots for pot and field experiments, additional seed infection percentage, and yield for field experiments) were evaluated. All antagonists, tested at the application rate of 2 %, reduced post emergence seedling mortality, disease incidence and severity in pots. Post-emergence mortalities in the field were 53.84 %, 61.53 % and 69.23 % as a result of 2 % applications with *F. oxysporum*, *C. globosum* and *T. harzianum*, respectively. Disease incidence and severity were also significantly reduced at 120 days after sowing with the same application rate of antagonists.

Studies on the Biological Control of B. cinerea on Chickpea

Khan et al. (2011) obtained beneficial microorganisms from commercial preparations such as Biowilt-X, PBAT1 Trichodex (a.i.: *T. harzianum*), AU Derma, Sanjeevni (a.i.: *T. virens*), Biocomp-X and PBAP-2 (a.i.: *P. fluorescens*) as mycelial mats and bacterial pellets by dilution methods from Potato Dextrose Broth (PDB) and nutrient broth (NB) for fungi and bacteria, respectively. Chickpea seeds were coated with mycelium (4 g mycelium/kg seed) of *B. cinerea* grown on PDB. The mycelial mats/bacterial pellets of beneficial organisms were applied to seeds (4 g/kg seed) and soil (4 g for each pot containing 2.5 kg sterilized soil mixture). Experiments were conducted under controlled conditions. All antagonists significantly reduced disease severity by *B. cinerea* when chickpea seeds were simultaneously treated with mycelia/pellets of biocontrol agents, commercial *Rhizobium* (4 g/kg seed) and *B. cinerea*. Additionally, seed treatment with PBAT-1 significantly increased both dry weight of plant and yield; however, Biowilt-X treatment enhanced yield only. Other biocontrol agents did not result in any significant effect on growth.

Seed Borne Fungi in Vegetable Crops Subjected to Biological Control Studies

Studies on biological control of diseases affecting cabbage, carrot, onion and muskmelon caused by seed borne fungi *Alternaria brassicicola*, *Alternaria* spp., *Aspergillus niger* and *Phoma cucurbitacearum*, respectively, are summarized below.

Alternaria dauci (Kühn) Graves&Skolko and *A. radicina* Meier, Drechsler&Eddy are the agents of leaf blight and black rot, respectively, in carrot. *A. radicina* also causes foliar blight on parsley and stalk/root rot on celery (Chen and Wu 1995). It is the primary agent of root and crown rot disease of carrot (Farrar et al. 2004). *Alternaria brassicicola* (Schwein.) Wiltshire causes dark leaf spot in crucifers. Symptoms of the disease are small dark brown spots surrounded by a halo of chlorotic tissues on the leaves. A mat of spores the colour of dark olive brown appears on the older lesions. The pathogen is present as spores and mycelium on the surface of seed coats, and as internal mycelium within the testa and occasionally in the embryo tissue (Humpherson-Jones 1988). *Aspergillus niger* Van Tieghem is the causal agent of black mould disease in onion. It is present in all seed parts, but mostly on the seed coat. No visual symptoms are observed on set bulbs developing from the contaminated seeds. Symptoms can be seen on mature bulbs as small black spore masses under the outer dry scales of the bulbs, then spreading as strips from the base to the neck (Özer and Köycü 2004). *Phoma cucurbitacearum* (Fr. Fr) Sacc) (teleomorph *Didymella bryoniae* Auersw) on cucurbits is located on and in the seed coat and transmits from seeds to seedling in cucumber and pumpkin (Lee et al. 1984; Sitterly and Kenath 1996). Gummy stem blight by the pathogen causes defoliation in late production stages (Wehner and Amand 1993).

Studies on the Biological Control of Alternaria spp. on Carrot and Cabbage

B. subtilis T99 increased carrot seed germination and survival of seedlings when it was applied to seeds naturally contaminated with *A. radicina* (Hentschel 1991).

Pseudomonas strain W24 had no positive effect on emergence of seedlings from seeds naturally infested with *Alternaria* spp. (Jahn and Puls 1998). Applications of Mycostop and T22 (a.i.: *T. harzianum* KRL-AG2) to seeds naturally infested with *A. dauci* were also unsuccessful in promoting field emergence (Hermansen et al. 1999).

Carrot seeds artificially infested with *A. radicina* were treated with suspensions of several bacterial antagonists grown on King's B medium and tested in vermiculite and greenhouse for effects on seedling emergence in a study by Chen and Wu (1999). Among the antagonist strains, *Psedomonas cepaica* 229 promoted emergence and length of seedling in both experiments.

Jensen et al. (2004) used fungal antagonists isolated from different carrot and cereal habitats and also several biocontrol products to control seed-borne *Alternaria* spp. in carrot. Seed lots with *A. radicina* of 4 and 29 %, and *A. dauci* of 7 and 11 % were treated with conidia suspensions of antagonist fungi and biocontrol products, and evaluated in pots with sand under controlled conditions. Biopriming of seeds with isolates of *C. rosea* controlled pre- and post-emergence infection and provided healthy seedling stands. The authors reported that biopriming of *Alternaria* spp.-infected seeds with *C. rosea* IK 726 had no adverse effect on seedling establishment.

Szopinśka et al. (2010) applied isolates of *C. rosea* IK1871 and *C. solani* IK1889, isolated from carrot seeds, and *C. rosea* IK726 from barley roots to carrot seeds after inoculation of *A. radicina* at 10^4 or 10^5 conidia/ml. Seedling emergence in sand and soil was evaluated under controlled conditions. *Clonostachys* spp. increased percentage of emergence and healthy plants in the sand test at low inoculum concentrations of pathogen. However, seed treatment with antagonists did not affect emergence, although they increased percentage of healthy plants in the soil test.

Using carrot seeds naturally infested with *A. dauci* and *A. radicina* at different degrees (high and medium), several formulated or non-formulated microorganisms, resistant inducers, plant derived products, chemicals and physical methods alone and in several combinations were tested to evaluate the effects of seed treatments on emergence and incidence of healthy plants under controlled and field conditions (Koch et al. 2010). Seed treatment with *Pseudomonas* sp. M416, *P. fluorescens* L 18 and *C. rosea* IK726 (seed dipped in microbial culture or spore suspension for 15 min) resulted in a high number of healthy carrot seedlings under controlled conditions. These antagonists and the biopesticide Mycostop Mix were tested again under controlled and field conditions, and the percentages of healthy plants and emergence were measured. Among them, MF416 had the best performance for these criteria. The authors reorganized another experiment under controlled and field conditions using two seed lots treated with combinations such as hot water (seed dressing at 53 °C for 10 min) + IK726 or hot water + MF416 in addition to single applications of antagonists. Seed treatment with a combination of hot water plus antagonist was more effective than hot water treatment or antagonist treatment alone under controlled conditions, but the hot water + MF416 combination was the least effective in the field. In subsequent field experiments, conducted in different countries, the combination of hot water with IK726 (clay formulation, 100 mg/10 g seed) was as effective as hot water and electron seed treatment for an average plant stand, and provided a denser plant stand than application of physical methods alone.

Cabbage seeds naturally infested with *A. brassicicola* were treated with dry powder formulations of commercial microbial products such as FZB24 (a.i.: *Bacillus subtilis*), MBI600 (a.i.: *B. subtilis* MBI600), Serenade (a.i.: *B. subtilis* Q ST 713), Mycostop and F251/2 (a.i.: non-pathogenic *F. oxysporum* 251/2), with a liquid formulation of BA2552 (a.i.: *Pseudomonas chlororapsis* strain MA342) and also with microbial cultures or spore suspensions of various microorganisms (Amein et al. 2011). The disease suppression effect of each treatment was evaluated under controlled and field conditions. Among the commercial products, BA2552 and Mycostop Mix increased the percentage of healthy plants. Two strains of *P. fluorescens* L18 and *T. viride* TV6903 from 15 non-commercialized microorganisms tested were effective, increasing the number of healthy plants at the rate of 29 %.

Studies on the Biological Control of A. niger and D. bryoniae

Antagonist fungi isolated from onion growing soils fungistatic toward *A. niger* were tested under pot conditions for their effect on disease incidence by this pathogen under controlled conditions (Özer 2011). Onion seeds were treated with spore suspensions of the antagonists combined with pathogen (simultaneous inoculation). Four months after sowing, sets raised from seeds inoculated with the pathogen and treated with antagonists were examined for presence of pathogen as well as defence reactions. Effects of the antagonists on shoot length and bulb diameter were also determined. Seed treatment with three antagonists, non-aflotoxigenic *Aspergillus flavus* AS3, *T. harzianum* TRIC7 and TRIC8, protected onion sets against the pathogen and also had the ability to stimulate accumulation of antifungal compounds in sets. These antagonists had no negative effects on shoot length although they did not enhance bulb diameter.

Muskmelon seeds naturally infected with various levels of four cultivars of *D. bryoniae* were treated with a slurry form of *P. fluorescens* and *T. harzianum* at rates of 8 and 10 g/kg seed, respectively, and with pure cultures (1×10^8cfu/ml) by mixing 400 g of seeds with 5 ml cell/conidial suspension of antagonist; efficacy of the treatments was evaluated under field conditions in India (Sudisha et al. 2006). Both types of application significantly reduced disease incidence, but pure culture application to seeds was more effective than mixed cultures in all cultivars. However, both treatments of the antagonists significantly increased fruit weight as compared with untreated control.

References

Adegbite AA, Amusa NA (2008) The major economic field diseases of cowpea in the humid agro-ecologies of south-western Nigeria. Afr J Biotechnol 7:4706–4712

Allen DJ, Thoattappilly G, Emechebe AM, Singh BB (1998) Disease of cowpea. In: Allen DJ, Lenne JM (eds) Pathology food and pasture legumes of cowpea. CAB International, Wallingford, pp 267–324

Amein T, Al WS, Wikström M, Koch E, Schmitt A, Stephan D, Jahn M, Tinivella F, Gullino ML, Forsberg G, Werner S, van der Wolf J, Groot SRC (2011) Evaluation of non-chemical seed treatment methods for control of *Alternaria brassicicola* on cabbage seeds. J Plant Dis Protect 118:214–221

Anonym (2013) MBI 600 (*B. subtilis*) bioactive technical backgrounder. https://agro.basf.ca

Arnst BJ (1978) Two important fungal seed-borne diseases of barley in New Zealand: neet blotch caused by *Drechslera teres* (Sacc.) Shoemaker and leaf stripe caused by *Drechslera graminea* (Rabenh ex Schlecht) Shomaker. N Z J Agric Res 21:697–701

Bankole SA, Adebanjo A (1996) Biocontrol of brown blotch of cowpea caused by *Colletotrichum truncatum* with *Trichoderma viride*. Crop Prot 15:633–636

Begum MM, Sariah M, Puteh AB, Zaina Abidin MA, Rahman MA, Siddiqui Y (2010) Field performance of bio-primed seeds to supress *Colletotrichum truncatum* causing damping –off and seedling stand of soybean. Biol Control 53:18–23

Benada J, Pospišil A (1999) Antagonistic microorganisms and medium moisture as possible sources of variation in common bunt (*Tilletia tritici*) incidence. Plant Prot Sci 35:121–123

Borgen A, Davanlou M (2001) Biological control of common bunt (*Tilletia tritici*). J Crop Prod 3:157–171

Brefag TW, Mebalds MI (1987) Pathogenicity of fungi isolated from *Cicer arietinum* (chickpea) grown in northwestern Victoria. Aust J Exp Agric 27:141–148

Chen TW, Wu WS (1995) Detection of seed-borne fungi of vegetables in Taiwan. Plant Pathol Bull (Taiwan) 4:17–24

Chen TW, Wu WS (1999) Biological control of carrot black rot. J Phytopathol 147:99–104

Dal Bello GM, Mónaco CI, Simón MR (2002) Biological control of seedling blight of wheat caused by *Fusarium graminearum* with beneficial rhizosphere microorganisms. World J Microbiol Biotechnol 18:627–636

Davanlou M, Madsen AM, Madsen CH, Hockenhull J (1999) Parasitism of macroconidia, chlamydospores and hyphae of *Fusarium culmorum* by mycoparasitic *Pythium* species. Plant Pathol 48:352–359

Dawson WAJM, Jestoi M, Rizzo A, Nicholson P, Bateman GL (2004) Field evaluation of fungal competitors of *Fusarium culmorum* and *F. graminearum*, causal agents of ear blight of winter wheat for the control of mycotoxin production in grain. Biocontrol Sci Tech 14:783–799

Dehkaei FP, Mansouri SH, Rouhani H (2004) Effects of paddy soil antagonistic microorganisms of Guilan on the causal agent of rice Bakanae disease. J Agric Sci Technol 8:213–222

Etheridge VJ (1997) Biological control of seedling blight of winter wheat caused by *Fusarium culmorum* and *Microdochium nivale*. Open University, Harper Adams Agricultural College, Ph.D. Thesis, 330 pp

Farrar JJ, Pryor BM, Davis RM (2004) *Alternaria* diseases of carrot. Plant Dis 88:776–784

Fernando WGD, Chen Y, Parks P (2002) Effect of three *Bacillus* sp. from wheat on FHB reduction. In Proceedings of National Fusarium Head Blight Forum, Erlanger, Kentucky 73–75

Goates BJ, Mercier J (2011) Control of common bunt of wheat under field conditions with the biofumigant fungus *Muscodor albus*. Eur J Plant Pathol 131:403–407

Haigh IM, Jenkinson P, Hare MC (2009) The effect of mixture of seed-borne *Microdochium nivale* var majus and *Microdochium nivale* var. *nivale* on *Fusarium* seedling blight severity and subsequent stem colonisation and growth of winter wheat in pot experiment. Eur J Plant Pathol 124:65–73

Hare MC, Parry DW, Baker MD (1999) The relationship between wheat seed weight, infection by *Fusarium culmorum* or *Michrodochium nivale*, germination and seedling disease. Eur J Plant Pathol 105:859–866

Hasan MM (2013) Biological control of wheat diseases caused by *Bipolaris sorakiana*, Fusarium graminearum and Aspergillus flavus with antagonist's of Trichoderma sp. Pers Gulf Crop Prot 3:1–9

Hasan MM, Rahman SME, Kim GH, Abdallah E, Oh DH (2012) Antagonistic potentiality of *Trichoderma harzianum* towards seed-borne fungal pathogens of winter wheat cv. Protiva *in vitro* and *in vivo*. J Microbiol Biotechnol 22:585–591

Hentschel KD (1991) Biocontrol of seedborne *Alternaria radicina* on carrots by antagonistic *Bacillus subtilis*. Bull SROP 14:73–76

Hermansen A, Brodal G, Balvoll G (1999) Hot water treatments of carrot seeds: effects on seed-borne fungi, germination, emergence and yield. Seed Sci Technol 27:599–613

Hökeberg M, Gerhardson B, Johnsson L (1997) Biological control of cereal seed-borne diseases by seed bacterization with greenhouse-selected bacteria. Eur J Plant Pathol 103:25–33

Humpherson-Jones FM (1988) *Alternaria brassicae* (Berk.) Sacc. & *A. brassicola* (Schwein.) Wilthshire. In: Smith IA, Dunez J, Lelliott RA, Philips DH, Archer SA (eds) Europea handbook of plant disease. Blackwell Scientific Publications, Oxford, pp 366–368

Humpreys J, Cooke BM, Storey T (1995) Effects of seed borne *Microdochium nivale* on establishment and grain yield of winter sown wheat. Plant Var Seeds 8:107–117

Jahn M, Puls A (1998) Investigation for development of a combined biological-physical method to control soilborne and seed-borne pathogens in carrot seed. J Plant Dis Protect 105:359–375

Jensen B, Knudsen IMB, Jensen DF (2000) Biological seed treatment of cereals with fresh and long term stored formulating of *Clonostachys rosea*: biocontrol efficacy against *Fusarium culmorum*. Eur J Plant Pathol 106:233–242

Jensen B, Knudsen IMB, Madsen M, Jensen DF (2004) Biopriming of infected carrot seed with antagonist, *Clonostachys rosea*, selected for control of seedborne *Alternaria* spp. Phytopathology 94:551–560

Jochum CC, Osborne LE, Yuen GY (2006) Fusarium head blight biological control with Lysobacter enzymogenes strain C3. Biol Control 39:339–344

Johansson PM, Johnsson L, Gerhardson B (2003) Supression of wheat-seedling diseases caused by *Fusarium culmorum* and *Microdochium nivale* using bacterial seed treatment. Plant Pathol 52:219–227

Johnsson L, Hökeberg M, Gerhardson B (1998) Performance of the *Pseudomonas chlororapsis* biocontrol agent MA 342 against cereal seed-borne diseases in field experiments. Eur J Plant Pathol 104:701–711

Kato A, Miyake T, Nishigata K, Tateishi H, Teraoka T, Arie T (2012) Use of fluorescent proteins to visualize interactions between the Bakanae disease pathogen *Gibberella fujikuroi* and the biocontrol agent *Talaromyces* sp. KNB-422. J Gen Plant Pathol 78:54–61

Khalili E, Sadravi M, Naemi S, Khosravi V (2012) Biological control of rice brown spot with native isolates of three *Trichoderma* species. Braz J Microbiol 43:297–305

Khan MR, Doohan FM (2009) Bacterium-mediated control of Fusarium head blight disease of wheat and barley and associated mycotoxin contamination of grain. Biol Control 48:42–47

Khan NI, Schisler DA, Boehm MJ, Slininger PJ, Bothast RJ (2001) Selection and evaluation of microorganisms for biocontrol of Fusarium head blight of wheat incited by *Gibberella zeae*. Plant Dis 85:1253–1258

Khan NI, Schisler DA, Boehm MJ, Lipps PE, Slininger PJ (2004) Field testing of antagonists of Fusarium head blight incited by *Gibberella zeae*. Biol Control 29:245–255

Khan MR, Anwer MA, Shahid S (2011) Management of gray mold of chickpea, Botrytis cinerea with bacterial and fungal biopesticides using differents modes of inoculation and application. Biol Control 57:13–23

Knudsen IMB, Hockenhull J, Jensen DF (1995) Biocontrol of seedling diseases of barley and wheat caused by *Fusarium culmorum* and *Bipolaris sorakiniana*: effects of selected fungal antagonists on growth and yield components. Plant Pathol 44:467–477

Koch E, Weil B, Eibel P (2004) Development of leaf symptoms based screening method for seed treatment with activity against *Tilletia caries* and application of the method using microbial antagonists. J Plant Dis Protect 111:470–483

Koch E, Weil B, Wächter R, Wohlleben S, Spiess H, Krauthausen HJ (2006) Evaluation of selected microbial strains and commercial alternative products as seed treatments for the control of *T. tritici*, *Fusarium culmorum*, *Drechslera graminea* and *D. teres*. J Plant Dis Protect 113:150–158

Koch E, Schmitt A, Stephan D, Kromphardt C, Jahn M, Krauthausen HJ, Forsberg G, Werner S, Amein T, Wright SAI, Tinivella F, Gullino ML, Roberts SJ, van der Wolf J, Groot SPC (2010) Evaluation of non-chemical seed treatment methods for the control of *Alternaria dauci* and *A. radicina* on carrot seeds. Eur J Plant Pathol 127:99–112

Kumakura K, Watanabe S, Toyoshima J, Makino T, Ichikawa T, Lyozumi H, Nagayama K (2003) Effect *Trichoderma* sp. SKT-1 on suppression of six different seedborne disease of rice (*Oryzae sativa*). Jpn J Phytopathol 69:384–392

Lee DH, Mathur SB, Neergaard P (1984) Detection and location of seed-borne inoculum of *Didymella bryoniae* and its transmission in seedling of cucumber and pumpkin. Phytopathol Z 109:301–308

Mamarabadi M, Jensen DF, Lübeck M (2009) An N-acetyl-βα-glucosaminidase gene, cr-nag1, from the biocontrol agent *Clonostachys rosea* is up-regulated in antagonistic interaction with *Fusarium culmorum*. Mycol Res 113:33–43

Matić S, Spadora D, Garibaldi A, Gullino ML (2014) Antagonistic yeast and thermotherapy as seed treatments to control *Fusarium fujikuori* on rice. Biol Control 73:59–67

McManus PS, Ravenscroft AV, Fulbright DW (1993) Inhibition of *Tilletia leavis* teliospore germination and suppression of common bunt of wheat by *Pseudomonas fluorescens* 2–79. Plant Dis 77:1012–1015

Mishra J, Tewari S, Singh S, Arora NK (2015) Biopesticides: where we stand? In: Arora NK (ed) Plant microbes symbiosis: applied facets. Springer, India pp 37–75

Miyake T, Tateishi H, Sakuma Y, Saishoji T (2012) A novel soil-type biopesticide KNB422-soil against rice seddling diseases. J Pestic Sci 37:129–134

Moura AB, Ludwig J, Santos AG, Schafer JT, Soares UN, Corréa BO (2014) Biocontrol and seed transmission of *Bipolaris oryzae* and *Gerlachia oryzae* to rice seedlings. J Seed Sci 36:407–412

Neergard P (1977) Seed pathology, vol 1. MacMillan Press, London, 4. 11., 1187pp

Nourozian J, Etebarian HR, Khodakaramian G (2006) Biological control of *Fusarium graminearum* on wheat by antagonistic bacteria. J Sci Technol 28:29–38

Özer N (2011) Screening for fungal antagonists to control black mold disease and to induce the accumulation of antifungal compounds in onion after seed treatment. BioControl 56:237–247

Özer N, Köycü ND (2004) Seed-borne fungal diseases of onion and their control. In: Mukerji KG (ed) Disease management of fruits and vegetables: fruit and vegetable diseases. Kluwer Academic Publishers, Dordrecht, pp 281–306

Padder BA, Sharma PN (2011) *In vitro* and *in vivo* antagonism of biocontrol agents against *Colletotrichum lindemuthianum* causing bean anthracnose. Arch Phytopathol Plant Protect 44:961–969

Palazzini JM, Ramirez ML, Torres AM, Chulze SN (2007) Potential biocontrol agents for Fusarium head blight and deoxynivalenol production in wheat. Crop Prot 26:1702–1710

Roberti R, Flori P, Pisi A, Brunelli A, Cesari A (2000) Evaluation of biological seed treatment of wheat for the control of seed-borne *Fusarium culmorum*. J Plant Dis Protect 107:484–493

Roberti R, Badiali F, Pisi A, Veronesi A, Pancaldi D, Cesari A (2006) Sensitivity of *Clonostachys rosea* and *Trichoderma* spp. as potential biocontrol agents to pesticides. J Phytopathol 154:100–109

Roberti R, Veronesi A, Cesari A, Cascone A, di Berardino I, Bertini L, Caruso C (2008) Induction of PR proteins and resistance by the biocontrol agent *Clonostachys rosea* in wheat plants infected with *Fusarium culmorum*. Plant Sci 175:339–347

Schisler DA, Khan NI, Boehm MJ, Slininger PJ (2002) Greenhouse and field evaluation of biological control of Fusarium head blight on Durum wheat. Plant Dis 86:1350–1356

Sherm B, Balmas V, Spanu F, Pani G, Delogu G, Pasquali M, Migheli Q (2013) Pathogen profile, Fusarium culmorum: causal agent of foot and root rot and head blight on wheat. Mol Plant Pathol 14:323–341

Sitterly WR, Keinath AP (1996) Gummy stem blight. In: Ziter TA, Hopkins DL, Thomas CE (eds) Compendium of cucurbit disease. American Phytopathology Society, St Paul, pp 27–28

Sivanesan A, Holliday D (1981) *Cochliobolus sativus*. CMI Descriptions of pathogenic fungi and bacteria Set 71, Nos. 701–710:20 pp

Sjöberg J, Märtensson A, Persson P (2007) Are field population of arbuscular mycorrhizal fungi able to suppress the transmission of seed-borne *Bipolaris sorakiniana* to aerial plant parts? Eur J Plant Pathol 117:45–55

Solomon PS, Lowe RGT, Tan KC, Waters ODC, Oliver RP (2006) *Stagonospora nodorum*: cause of *Stagonospora nodorum* blotch of wheat. Mol Plant Pathol 7:147–156

Sudisha J, Niranjana SR, Umesha S, Prakash HJ, Shetty HS (2006) Transmission of seed-borne infection of muskmelon by *Didymella bryoniae* and effect of seed treatments on disease incidence and fruit yield. Biol Control 37:196–205

Szopinska D, Jensen B, Knudsen IMB, Krystyna T, Dorna H (2010) Non-chemical methods for controlling seedborne fungi in carrot with special reference to *Alternaria radicina*. J Plant Prot Res 50:184–192

Tateishi H, Sakuma Y, Miyake T (2006) Isolation and screening of antagonistic fungi against rice seedling diseases ant their efficacy. Jpn J Phytopathol 72:265–266

Teperi E, Keskinen M, Ketofa E, Tahvonen R (1998) Screening for fungal antagonists of seed-borne *Fusarium culmorum* on wheat using *in vitro* tests. Eur J Plant Pathol 104:243–251

Tinivella F, Hirata LM, Celan MA, Wright SAI, Amein T, Schmitt A, Koch E, Vander Wolf JM, Groot SPC, Stephan O, Garibaldi A, Gullino ML (2009) Control of seed-borne pathogens on legumes by microbial and other alternative seed treatments. Eur J Plant Pathol 123:139–151

Tivoli B, Banniza S (2007) Comparison of the epidemiology of Ascochyta blights on grain legumes. Eur J Plant Pathol 119:59–76

Vasanthakumari MM, Shivanna MB (2014) Biological control of sorghum anthracnose with rhizosphere and rhizoplane fungal isolates from perennial grasses of Western Ghats of India. Eur J Plant Pathol 139:721–733

Watanabe S, Kamakura K, Kato H, Iyozumi H, Togawa M, Nagayama K (2005) Identification of *Trichoderma* SKT-1, a biological agent against seedborne pathogens of rice. J Gen Plant Pathol 71:351–356

Webster RK, Gunnell PS (1992) Compendium of rice disease. The American Phytopathological Society APS Press, St Paul

Wehner TC, St Amand PC (1993) Field tests for cucumber resistance to gummy stem blight in North-Carolina. Hortic Sci 28:327–329

Wiese MV (1977) Compendium of wheat disease. The American Phytopathological Society APS Press, St Paul

Wrather JA, Anderson TR, Arsyad DM, Gai J, Ploper LD, Porta-Puglia A, Ram HH, Yorinori JT (1997) Soybean diseases loss estimates for the top 10 soybean producing countries in 1994. Plant Dis 81:107–110

Xue AG, Voldeng HD, Savard ME, Fedak G, Tian X, Hsiang T (2009) Biological control of Fusarium head blight of wheat with *Clonostachys rosea* strain ACM941. Can J Plant Pathol 31:169–179

Xue AG, Chen Y, Voldeng HD, Fedak G, Savard M, Längle T, Zhang J, Harman G (2014) Concentration and cultivar effects on efficacy of CLO-1 biofungicide in controlling Fusarium head blight of wheat. Biol Control 73:2–7

Yolageldi L, Turhan G (2005) Effect of biological seed treatment with *Cylindrocarpon olidum* var *olidum* on control of common bunt (*Tilletia leavis*) of wheat. Phytoparasitica 33:327–333

Zad J, Aghakhani M, Etebarian R, Okhovat M (2002) Barley leaf stripe disease. Meded Rijksuniv Gent Fak Landbouwkd Toegep Biol Wet 67:279–281

Mycotoxin Menace in Stored Agricultural Commodities and Their Management by Plant Volatiles: An Overview

Abhay K. Pandey, Satish K. Sain, Pooja Singh, Uma T. Palni, and N.N. Tripathi

Introduction

Worldwide estimated post-harvest losses of agricultural food commodities is about 30–50 % of whole produce which not only threats the global food security but this translates to waste 1.47–1.96 Gha of arable land, 0.75–1.25 trillion m^3 of water and 1–1.5 % of global energy (Fox and Fimeche 2013). The postharvest losses in food commodities may occur due to the fungi, bacteria, insects, physical injuries due to insects, mechanical force, chemical force, heat or freezing, non-disease disorders resulting from storage conditions that upset normal metabolism. In tropical hot and humid climates fungal bio-deterioration of stored food commodities is a chronic problem in storage system. Harvested food commodities are invaded by various species of *Alternaria*, *Aspergillus*, *Cladosporium*, *Fusarium*, *Mucor*, *Penicillium*, *Rhizopus*, etc. under such conditions leading to deterioration and can produce poisonous substances called Mycotoxins. The major toxin-producing fungi are *Aspergillus* spp. (*A. carbonarius*, *A. flavus*, *A. ochraceus*, *A. oryzae*, *A. parasiticus*,

A.K. Pandey
National Institute of Plant Health Management, Ministry of Agriculture and Co-operation, Rajendranagar, Hyderabad, Telangana 500030, India

S.K. Sain (✉)
ICAR - Central Institute for Cotton Research, Regional Station Sirsa, Haryana 125055, India
e-mail: sain.skumar@gmail.com

P. Singh • N.N. Tripathi
Bacteriology and Natural Pesticide Laboratory, Department of Botany, DDU Gorakhpur University, Gorakhpur, Uttar Pradesh 273009, India

U.T. Palni
Department of Botany, DSB Campus, Kumaun University, Nainital, Uttarakhand 263002, India

P. Kumar et al. (eds.), *Current Trends in Plant Disease Diagnostics and Management Practices*, Fungal Biology, DOI 10.1007/978-3-319-27312-9_18

A. versicolor), *Penicillium* spp. (*P. nordicum*, *P. expansum*, *P. viridicatum*, *P. verrucosum*), *Fusarium* spp. (*F. culmorum*, *F. graminearum*, *F. oxysporum*, *F. proliferatum*, *F. verticillioides*), *Alternaria* spp. (*A. alternata*, *A. brassicae*, *A. solani*, *A. tenuissima*) and mycotoxins produced by them are aflatoxins, ochratoxin A, sterigmatocystin, patulin, fumonisins, zearalenone, deoxynivalenol, alternariol, alternariol monomethyl ether, tenuazonic acid. Mycotoxins are known to be hazardous influence in human and livestock health, affect the marketability of food commodities, and hence are of great commercial status. More than five billion people in developing countries are constantly bared to mycotoxins by unknowingly means and are consuming contaminated foods. Alimentary intake of mycotoxins by livestock and human being causes of intoxication – mycotoxicosis (Tanaka et al. 2007). Mycotoxicosis exhibited acute and chronic toxicity these include cytotoxicity, hepatotoxicity, neurotoxicity, teratogenicity, mutagenicity, and cancerogenity. Mycotoxins, at cellular level interact with nucleic acids and inhibit the DNA and RNA synthesis (Bhatt et al. 1978). Some important mycotoxins and their toxic effects and chemical structure are summarised in Table 1 and Fig. 1.

Major Classes of Mycotoxins in Agricultural Commodities and Their Biosynthesis

Aflatoxin

Aflatoxin is extremely toxic and worldwide produced by *Aspeigillus flavus* and *A. parasiticus* (Abbas et al. 2008) and is of four major types such as AfB1, AfB2, AfG1, and AfG2. *A. flavus* is found to produce B toxin among which B1 is the most common in food having genotoxic and carcinogenic activity (Payne and Brown 1998). Aflatoxin produced by the *A. flavus* is more prevalent in corn and cotton seeds (Hell et al. 2000); however, aflatoxin produced by *A. parasiticus* is more prevalent in groundnut than other crops (Kaaya et al. 2006). The chemical biosynthesis pathway of aflatoxin in *Aspergillus* species consist of 23 steps of enzymatic reactions and 15 intermediate reactions and determined by 25 recognised genes assembled within a 70-kb DNA region on chromosome III (Bhatnagar et al. 2006; Smith et al. 2007). The primary substrate acetate is used to create polyketides with the first stable pathway intermediate being the anthraquinone norsolorinic acid (NOR) (Bennett et al. 1997). This is followed by anthraquinones, xanthones, and finally aflatoxins synthesis (Yu et al. 2004).

Zearalenone

Zearalenone (ZEA) is also called as F-2 mycotoxin or RAL. Several *Fusarium* species produce deoxynivalenol, T-2 toxin, HT-2 toxin, diacetoxyscirpenol (DAS) and zearalenone which are toxic substances of considerable concern to livestock and poultry producers (Kuiper-Goodman et al. 1987). This toxin is heat stable and

Table 1 List of important mycotoxins, their principal toxic effects and prone food commodities

Mycotoxin and its producing fungi	Food products	Symptoms/toxicology
Aflatoxins: *Aspergillus* spp., *A. parasiticus*, *A. flavus*, *A. nomius*	Grain, strawberries, raspberries, cherries, groundnut, corn, peanuts, cotton, maize, pearl millet, chillies, sorghum, pistachios, cassava, spices, oil seeds, dried fruits etc.	Liver necrosis, liver tumours, reduced growth, depressed immune response, carcinogen
Cyclopiazonic acid: *Aspergillus* and *Penicillium* spp., *A. flavus*, *A. versicolor*, *A. oryzae*, *A. tamarii*, *P. verrucosum*, *P. patulum*, *P. cyclopium*, *P. camembertii*, *P. puberulum*, *Penicillium griseofulvum*	Peanuts, corn, cheese etc.	Neurotoxin
Deoxynivalenol (DON) (Vomitoxin), Zearalenone: *Fusarium* spp., *F. graminearum*, *F. subglutinans*	Wheat, corn, maize, oats, rice, barley, sorghum etc.	Feed refusal, reduced weight gain, diarrhoea, vomiting, infertility abilities
Fumonisin B1 & Fumonisin B2: *Fusarium* spp., *Fusarium verticillioides*, *F. moniliforme*	Corn, wheat etc.	Equine leuko-encephalomalacia, porcine pulmonary edema
Trichothecenes: *F. graminearum*, *F. culmorum*, *F. poae*, *Trichoderma*, *Trichothecium*, *Cephalosporium*	Wheat, maize and oats etc.	Alimentary toxic aleukia, necrosis, hemorrhages, oral lesion in broiler chickens
Ochratoxin: *Aspergillus* spp., *Penicillium* spp., *P. verrucosum*, *A. ochraceus*, *A. carbonarius*	Wheat, grapes, spices, coffee etc.	Porcine nephropathy; various symptoms in poultry
Patulin, Citrinin: *Penicillium* spp., *P. expansum*	Apple and apple products etc.	Kidney damage
Sterigmatocystin: *Aspergillus* and *Penicillium* spp., *A. versicolor*, *A. parasiticus*, *A. flavus*, *A. rugulosus*, *A. nidulans*, *A. chevalieri*, *A. rubber*, *A. amsyelodami*, *P. camembertii*, *P. communer*, *P. griseofulvum*	Corn, rice, wheat, hay etc.	Carcinogen, mutagen

Source: Calvo (2005)

several cereal crops (Table 1) are found to exhibit this toxin. It is also cause of several diseases like infertility, abortion or other breeding problems, especially in swine. Zearalenone is synthesised by the combined action of two PKSs and an isoamyl alcohol oxidase (Lysoe et al. 2006). The gene involved in the biosynthesis is encoded by 39 kb of large gene cluster consisting of four genes FgPKS13, FgPKS4, ZEB1 and ZEB2. The ZEB1 protein contains a basic-region leucine zipper (bZIP) domain and functions as a cluster specific transcription factor that controls the expression of the cluster (Kim et al. 2005).

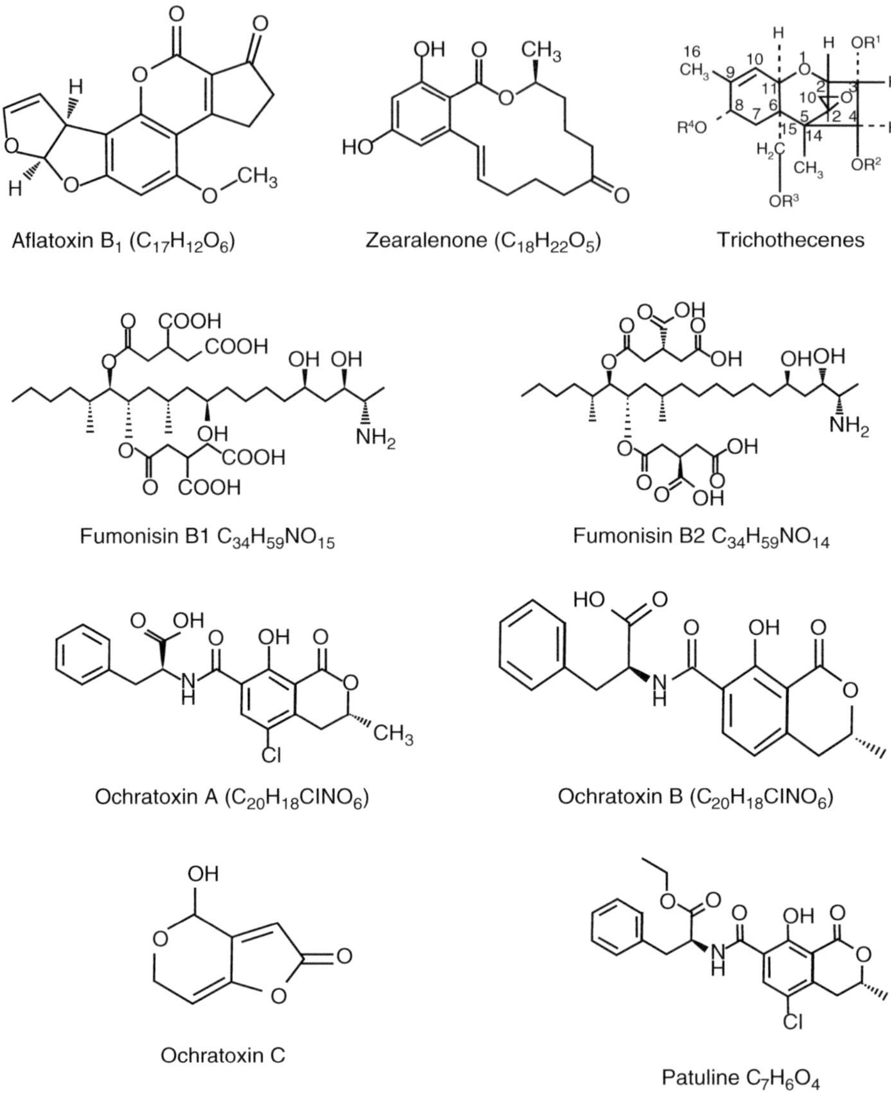

Fig. 1 Chemical formula of major mycotoxins

Trichothecenes

This toxin belongs to sesquiterpene compounds and is a very large family of chemically related mycotoxin formed by several mycofloral organisms as described in the Table 1. The biological activity of trichothecene is mainly due to the presence of 12, 13-epoxyring, hydroxyl or acetyl groups at appropriate positions (Etzel 2002). This mycotoxin is potent inhibitors of protein synthesis (Dohnal et al.

2008). The biochemical and genetic analysis of trichothecenes, (T-2 toxin (T-2), nivalenol (NIV), and deoxynivalenol (DON begins with the cyclization of farnesyl pyrophosphate (FPP) to form trichodiene, which then undergoes a series of oxygenation, isomerization, cyclization, and esterification reactions to form T-2 toxin, nivalenol, or deoxynivalenol. It is found that most of the gene from *Fusarium* species directly take part in the toxins synthesis is positioned at a locus designated the core trichothecene biosynthetic gene (TRI) assembly (Alexander et al. 2009).

Fumonisin

Fumonisin is a mycotoxin derived from *Fusarium* and consists of fumonisin B1 and fumonisin B2. Fumonisin B1 is the most prevalent member of a family of toxins, known as fumonisins, produced by several species of *Fusarium* moulds, such as *F. verticillioides* (Marasas and Paul 1987), which occur mainly in maize (corn), wheat and other cereal crops. *F. verticillioides* and *F. moniliforme* are also found to produce fumonisin B2 which is the structural analog of fumonisin B1. It is found that B2 fumonisin is more cytotoxic than fumonisin B1 and inhibits sphingosine acyltransferase. It contaminates the maize and other commodities. At least 15 different fumonisins have so far been reported, although most of them have not been shown to occur naturally (Jackson et al. 1996). Fumonisin biosynthetic enzymes recognised to date are all determined at one locus i.e. the 17-gene FUM cluster. The FUM cluster is found to also encode a protein that controls the expression of the group of genes and taking part in transport of fumonisins across the cell membranes (Brown et al. 2004). Biochemical and genetic evidence regarding biosynthesis pathway of fumonisin is revealed that during this process inactivation of FUM gene occur and initially a linear, 20-carbon polyketide is formed (Alexander et al. 2009). Subsequently, an amino group, up to four hydroxyl functions, and two tricarboxylic acid moieties are added to various positions along the polyketide backbone.

Ochratoxin

Ochratoxin is of three types i.e. Ochratoxin A, B and C. Ochratoxin A (OTA) is one of the most-abundant mycotoxins in food commodities produced by *A. ochraceus*, *A. carbonarius* and *Penicillium verrucosum* (Al-Anati and Petzinger 2006). Human acquaintance can occur through intake of contaminated food commodities, particularly grains and pork products, coffee as well as grapes and grape products (Richard et al. 1999). OTA is composed of a dihydro-isocoumarin ring joint to phenylalanine, and its biological synthetic pathway has not been fully studied. However, in *Penicillium* species the genetic and enzymatic aspects of OTA synthesis has been explained and it has been found that only PKS gene is involved in biosynthesis pathway of *Aspergillus* species. Concerning the biosynthesis process inactivation of a gene encoding a non-ribosomal peptide synthetase in

OTA-producing *A. carbonarius* ITEM 5010 species has been reported to eliminate the ability of this fungus to produce OTA (Gallo et al. 2012). The pathway involved in synthesis of the isocoumarin group by the catalytic property of polyketide synthase (PKS) which ligated with amino acid phenylalanine through the carboxyl group and finally chlorination step. In addition to this, several school of thoughts have been proposed from different investigations. Synthesis of polyketide for mellein, which carboxylated to OTβ and then transformed through a chloroperoxidase reaction to OTα (Huff and Hamilton 1979). According to this hypothesis, in the subsequent step, OTα would be phosphorylated and joined to the ethyl ester of phenylalanine to form OTC, followed by de-esterification reaction which leads to the formation of OTA in final product form. In *P. nordicum*, a putative OTA synthetic path has been recognized which exhibiting biosynthetic genes encoding a non-ribosomal peptide synthetase (NRPS) (otanpsPN), a PKS (otapksPN) – which is responsible for the formation of peptide bond among the polyketide, the phenylalanine and gene otachlPN are supposed to be involved in the chlorination reaction, and gene otatraPN is involved in OTA export (Geisen et al. 2006).

Patulin

It is a toxin produced by *Aspergillus*, *Byssochlamys* and *Penicillium* commonly found in rotting apples. Patulin has shown antimicrobial properties against some microorganisms. Studies of the health risks due to consumption of patulin by humans have led many countries to regulate its quantity in food. Several countries have instituted patulin restrictions in apple products (Puel et al. 2010). (E)-ascladiol include a mycotoxin, is found to be a direct precursor of patulin in cell-free formation of *P. urticae* patulin-minus mutants i.e. J1 and S11, but not S15. The isomerization of (Z)-ascladiol to a side product, is non-enzymatically catalysed by sulfhydryl compounds (Sekiguchi et al. 1983). Although, chemical structure of patulin has studied, however, molecular biosynthetic pathway of patulin is incomplete, unlike other regulated mycotoxins such as aflatoxins, trichothecenes and fumonisins. The biosynthetic pathway is assumed to be approximately ten steps. Recent studies showed that it includes assembly of 15 genes taking part in patulin biosynthesis and exhibiting several characteristics enzymes, transporter gene and regulation factor (Artigot et al. 2009).

Incidence of Mycotoxins in Commercially Important Agricultural Food Commodities

Agricultural commodities constitute important food and feed sources which are contaminated by various mycotoxigenic fungi. Food products like cereals, pulses, fruits and vegetables are directly contaminated by mycotoxins which are provoked to some extent due to rapidly changing agricultural technology. It seems to create significant problem in the tropics than in the temperate regions, however, no zone of the world can be regarded as mycotoxin-free. This may be due to the transport of various food commodities from one part to the other part of the country. Some of the important food commodities which have been found to be naturally contaminated by the mycotoxins are summarise below:

Rice (Oryza sativa L.)

Rice is one of the important food crops worldwide along with wheat and corn, and has been major food in several countries. Climate and storage conditions play a significant role in the occurrence of mycotoxins. *Aspergillus*, *Fusarium* and *Penicillium* species are reported to be the major mycotoxigenic fungi in rice. The harmful effects of these fungal infections are glume/grain discoloration, loss in viability, quality and toxin contamination. Different mycotoxins such as aflatoxins, ochratoxin A, cyclopiazonic acid, fumonisins, trichothecenes, zearalenone, deoxynivalenol (DON), citrinin, gliotoxin and sterigmatocystin production in rice have been recorded from time to time (Tanaka et al. 2007; Reddy et al. 2008; Gummert et al. 2009). Rice bran and parboiled rice bran samples exhibited the presence of AFB1 up to 35 % (Jayaraman and Kalyansundaram 1990). In a study, aflatoxin quantity is higher in rice samples as compared to wheat and maize (Pande et al. 1990). The levels of aflatoxins in rice can range from 184 to 2,830 μg/kg. From China, Liu et al. (2006) have reported the occurrence of AFB1, AFB2, AFG1, and AFG2 in 36 de-husked brown rice samples which ranged from 0.99 to 3.87 μg/kg. In India rice and paddy samples have been found to be aflatoxin AFB1-positive in 67.8 % of the samples in which the amount ranges of from 0.5 to 38.5 μg/kg (Reddy et al. 2008). Moreover, the majority of stored rice varieties like PAU 201 samples collected from six districts of Punjab were also found to be aflatoxin B1 positive@ <15 μg/kg (Siruguri et al. 2012). Similarly, out of 196 samples collected from Nigeria, the occurrence of aflatoxin (24–1,164 μg/kg) was reported in 97 samples, ochratoxin (20–1,642 μg/kg) in 56 samples and zearalenone (24 and 1,169 μg/kg) in 93 samples (Hussaini et al. 2009a).

Maize (Zea mays L.)

Maize is one of the most important food grains and commonly colonized by several spoilage fungi in pre- and post-harvest conditions, where the relative abundance of those species depends on several abiotic and biotic factors leads to the mycotoxins production. The most common mycotoxins that contaminate maize in Mediterranean countries are *Fusarium* toxins trichothecenes, zearalenone (ZEA) and fumonisins (FB) (Jestoi 2008) while in tropical and subtropical countries occurrence of aflatoxin is reported to be major once (Muriuki and Siboe 1995). The occurrence of different types of mycotoxins has been reported in maize samples collected from Mediterranean basin (Marin et al. 2012). During five consecutive years (1995–1999) observation in Italy, two samples of maize exhibited incidence of aflatoxin with up to 109 or 158 μg AFB1/kg with the mean values in the range of 0.3–4.10 μg AFB1/kg seeds (Pietri et al. 2004). High level (120.3–133 μg/kg) of aflatoxins (AFs) has been also observed in Turkey's samples (Nizamlýoðlu and Oguzz 2003; Giray et al. 2009). The toxin (AFs) level reported from Syria between 2003 and 2005 are higher than 20 μg/kg (Majid 2007). Similarly, the sample from Egypt exhibited mycotoxin level 30 μg/kg (Abdelhamid 1990), from Southern Guinea 77 μg/g (Hell et al. 2003), and from Croatia 224 and 614 μg/kg level during 1996 and 1997, respectively (Jurjevic et al. 2002). Nevertheless, the percentage of positive samples has been very high (25 %) in Croatia with the highest OTA amount 31.7 μg/kg (Segvic et al. 2009). Despite of these two mycotoxins, *Fusarium* mycotoxins (FBs) are the most frequent contaminants, with contamination incidence very often near 100 %. One hundred percent samples of maize grown in Turkey showed contamination with FBs, with mean level of toxin production was 88,240 μg/kg (Oruc et al. 2006). Fifty four samples of french maize imported in between 2004 and 2007 into the UK, are found to have FB1 + FB2 contamination above 10 μg/kg (Scudamore and Patel 2009). During 1996, the percentage of samples exhibiting ZEA above 200 μg/kg increased to 53.8 % (91 % of samples with detectable amounts of ZEA), with a mean contamination of 453 μg/kg and one sample containing 2,531 μg/kg ZEA (Pietri et al. 2004). High aflatoxin contamination levels are also measured in maize sold to the public of West Africa, and range are from 0.4 to 490 μg/g in Ghana, 0.7–110 μg/g in Togo, and 0.2–120 μg/g in Benin (James et al. 2007). In the same study, 40 % of the samples from the Southern Guinea Savanna exceeded the level 20 μg/g internationally recommended safety limit. In India out of 190 analysed samples of maize for mycotoxin contamination, 69 (34.8 %) samples exhibited the contamination by mycotoxin (Janardhana et al. 1999). Maize samples of the *Kharif* crop reported to have a greater incidence of aflatoxins (47 %) than *rabi* crop samples (17 %) in Bihar, India (Sinha 1990). Stored maize grains also had a high incidence of aflatoxins (43 %) and most of the contaminated samples contained aflatoxins at levels above 20 μg/kg. *Fusarium* has been found to be dominant mycoflora in 12 maize samples collected from affected households and found to produces the fumonisin B1 at the level of 0.25–64.7 mg/kg (Bhatt et al. 1997). Again 25 rain-affected maize showed the contamination with fumonisin B1 (00.04–65 mg/kg) and 89 % of normal maize samples also contained aflatoxin B1 (0.38–109 μg/kg).

Wheat (Triticum aesticum L.)

Wheat is also a worldwide important crop and is provoked by variety of fungi during transit and storage. A new *Fusarium* mycotoxin i.e. glucoside, fusarenon X-glucoside (FUXGlc), is reported for the first time in wheat grain that is artificially infected with *Fusarium* fungi (Nakagawa et al. 2011). OTA and DON (deoxinivalenol) are detected in wheat at the levels of 12 and 53 μg/kg respectively. In wheat flours, presence of OTA leads to the suspicion of contamination by the T-2 and AFB1 toxins also (Badiale-Furlong et al. 2003). In other investigation FB1 levels varied in between 0.5 and 3.9 μg/kg for wheat and in between 0.6 and 2.3 μg/kg for the flours (Birck et al. 2003). However, the contamination by DON in wheat (297 samples) from the southern region has been of 74 (24.91 %) with a mean and maximum levels of 603.2–8,504 μg/kg (Mallmann et al. 2003). Analysis of 39 mycotoxins in wheat samples showed the limits of detection range from 0.03 to 220 μg/kg (Sulyok et al. 2006). The occurrence of mycotoxin producing *Fusarium* species and other fungi on wheat kernels in five growing districts of Kenya has been reported and it is found that most isolates of *F. graminearum* are produce zearalenone and deoxynivalenol (Muthomi and Mutitu 2003). The incidence of *Fusarium* mycotoxin in wheat and its flour samples procured from local markets in Egypt has been recorded (Aziz et al. 1997). The deoxynivalenol (DON) is detected in five wheat samples at levels ranging from 103 to 287 μg/kg and one sample each of flour and bread at 188 and 170 μg/kg. Zearalenone (ZEN) is detected in ten wheat samples at levels from 28 to 42 μg/kg and four samples each of flour and bread at 95 and 34 μg/kg respectively. T-2 toxin is detected only in one wheat sample, flour and bread @ 2.9, 2.2 and 2.3 μg/kg respectively. Some *Fusarium* species such as *F. graminearum*, *F. avenaceum* and *F. culmorum* are predominantly found to be associated with *Fusarium* head blight (FHB) in wheat and responsible for the mycotoxin production (Bottalico and Perrone 2002). The level of aflatoxins like AFB1, AFB2, AFG1, AFG2 produced by two aspergilla such *A. flavus* and *A. parasiticus*, is found to be ranging from 10.4 to 643.5 μg/kg in 41 wheat samples used for cultivation and consumed in few regions of Turkey. Fifty nine percent of the samples are found to be positive for total AFs i.e. AFB1, AFB2, AFG1, and AFG2 with a percent of 42, 12, 37, and 12 % respectively (Giray et al. 2007).

Groundnut (*Arachis hypogaea* L.)

Groundnut, also called peanut, is considered as second most vital legume after beans grown throughout the country widely. However, lack of storage technologies leads to the contamination of mycotoxins especially aflatoxin (Kaaya et al. 2006). Production of aflatoxin types B and G and cyclopiazonic acid (CPA) from a new growing peanut region in Argentina (Formosa province) by *A. flavus* has been recorded (Pildain et al. 2004). The level of aflatoxin in peanut cake samples is found to be 10–346 μg/kg and ochratoxin A is @ <LOQ–2 μg/kg (Ediage et al. 2011). Twenty one percent of groundnut samples exhibited aflatoxin in the range of 4–100 μg/kg body wt/day (Vasanthi and Bhat 1998). Occurrence of ochratoxin A and ochratoxin A- containing black species of *Aspergillus* are reported in stored peanut seeds from Córdoba, Argentina. OTA is found in 32 % of the seeds ranging from 0.5 to 170 μg/g. Out of 47 samples studied, 43 isolates (27 %) of *Aspergillus* section Nigri, are OTA producing strains. *A. carbonarius* exhibited

highest percentage of ochratoxigenic strains (57 %) (Magnoli et al. 2006). Presence of ochratoxin A and *Aspergillus* section *Nigri* in groundnut seeds at different months of storage in Córdoba, Argentina has also been reported. One hundred four (32 %) of 322 isolates of *Aspergillus* section Nigri, are OTA producers and the levels of toxin produced is 2–24 μg/ml (Magnoli et al. 2007). From India several peanut samples showed the occurrence of aflatoxin content >30 mg/kg (Kishore et al. 2002). A disquieting aflatoxin amount of 851.9 mg/kg and zearalenone of 98.1–847.3 mg/g are found in samples from Anantapur district of Andhra Pradesh, India. Brazilian groundnut seeds from sowing to harvest also exhibited the presence of mycotoxins. The screening of mycotoxins indicated that aflatoxins and cyclopiazonic acid both are present the highest incidence, being detected in 32 % of the samples from 4.20 to 198.84 μg/kg and from 260 to 600 μg/kg respectively (Gonçalez et al. 2008). Metabolites produced by *Aspergillus* species are the most prevalent toxins. As considers mycotoxins addressed by regulations, aflatoxins exceeded the USDA maximum limit of 20 μg/kg in about 90 % samples (Ezekiel et al. 2012), and from Ethiopia it varied in between 15 and 11,900 μg/kg (Chala et al. 2013).

Sorghum (Sorghum vulgare L.), Barley (Hordeum vulgare L.), Millets (Pennisetum glaucum L.), Oats (Avena sativa L.) etc.

Like wheat, these small grains are also contaminated by several mycotoxins producing fungi during transit and storage as well as in field conditions. These grains are found to be suitable substrates for the occurrence of aflatoxin. Eighty four (72 %) of 116 oat and barley samples in eastern Canada were found to be contaminated with deoxynivalenol (DON) up to 8–9 mg/kg during the year 1991–1998 (Campbell et al. 2000). Of 73 oat samples, 34 % oat samples (47 %) contained DON and 34 % of the barley samples (18/53) and 15 % of the oat samples (4/26) contained nivalenol. Previously, Stratton et al. (1993) reported that 58 % (37/64) and 53 % (49/92) of the barley samples in Nova Scotia and Prince Edward Island, respectively, are found to be contaminated with DON; but none exceeded the concentration of 1 mg/kg. In this study, of 12 barley samples, 6 from Atlantic Canada are contaminated with DON and 1 sample exceeded the concentration of 1 mg/kg (1.73 mg/kg). Barley samples contaminated with DON, exhibited the maximum level up to 15.79 mg/kg (Abramson et al. 1998). Sorghum contained fumonisin B1 in 0.14–7.8 mg/kg range (Bhatt et al. 1997). Indian sorghum exhibited natural occurrence of fumonisin B1 (0.07–8 mg/kg) and its co-occurrence with aflatoxin B1 (5–125 μg/kg) (Shetty and Bhat 1997). Zearalenone is also found in the grains which are having moisture content 20–22 % (Jurjevic et al. 2007). From Nigeria occurrence of aflatoxin B – AFB, ochratoxin A-OTA and one zearalenone-ZEN are spotted in several collected samples of sorghum (Hussaini et al. 2009b). 92.1 μg/kg occurrence of moniliformin level and 414.6 μg/kg beauvericin levels in pearl millet from Africa and Asia are recorded (Wilson et al. 2006).

Pulses and Oil Seeds

Pulses are important source of dietary protein and other essential nutrients. During the survey under FAO (1981) sponsored Food Contamination Monitoring Project in Western Uttar Pradesh it has been found that few samples of pulses such as green gram, black gram and lentil and few samples of cotton seed are contaminated with aflatoxin B1. The ranges of toxin in the oilseeds are 35–200 ppb and in pulses are 4–80 ppb. Aflatoxin B1 (333–10,416 μg/kg) is produced by *Aspergillus* spp. in rice, pulses and oilseeds (Begum and Samajpati 2000). Tseng and Tu (1997) studied mycoflora and mycotoxins in adzuki and mung beans produced in Ontario, Canada. FB1 is detected by TLC in discoloured adzuki bean (*Vigna angularis*) and green gram (*Vigna mungo*) samples but not in the healthy samples. The quantification of FB1 by HPLC revealed that discoloured adzuki and mung bean samples contained 261 and 230 mg/g of FB1, respectively. This investigation highlights the need for more research on mycotoxin contamination in various food commodities including legumes. Out of 66 isolates of *A. flavus*, isolated from mustard seeds during storage, 24 are produced aflatoxins (0.5–22 IJ g/ml), 8 isolates are high toxin producers whereas the remaining 16 isolates are low toxin producers. Thirteen (out of 34) isolates of *F. moniliforme* and 4 (out of 12) isolates *P. citrinum* produced zearalenone (1.2–4.0 IJ g/ml) and citrinin (1.0–3.0 IJ g/ml) (Ahmad and Sinha 2002). Cowpea cultivars from South Africa showed the presence of fumonisin B1 (0.12 and 0.61 μg/g), whereas cultivars from Benin showed no fumonisins. Other scholars investigated 0.8 and 25.30 μg/g total fumonisin in pulses, and the highest level of FB1 detected is 16.86 μg/g (Kritzinger et al. 2003). Embaby et al. (2013) demonstrated that two fungi such as *A. parasiticus* (No .59) isolated from beans seeds and *F. moniliforme* (No. 8) isolated from soybean had the ability to produce mycotoxins in significant concentrations i.e. 196.58 μg/kg aflatoxin and 198 mg/kg fumonisin.

Fruits

Mycotoxins may also occur in raw agricultural and horticulture products during pre- and post-harvest. Though, cereals are amongst the most studied crops for toxin contamination, but fruits and their processed products may also represent a potential source of risk. Worldwide, the species causing toxins production in fruits include *Aspergillus* and *Penicillium* and are recognised as the main concern. Dried vine fruits (e.g., sultanas, raisins) exhibited very high levels of OTA worldwide (Palumbo et al. 2011). The potential production of aflatoxins (0.3 %), OTA (6.0 %), patulin (0.5 %) and trichothecenes (1.2 %) in grapes has been reported (Serra et al. 2005). Sixty samples of retail dried vine fruits from the United Kingdom exhibited the occurrence of Ochratoxin A (53.67 μg/kg) and aflatoxins (MacDonald et al. 1999). Most of the reports showed the highest percentage of contaminated samples with an average OTA level over 2 μg/kg, with maximum values up to 100 μg/kg (Magnoli et al. 2004; Aksoy et al. 2007). OTA levels in grape juice which is consumed by children, are found to be higher than allowed (Chulze et al. 2006). Avery high incidence (100 %) of PAT (patulin), even though at low levels (5–75 μg/ml), is found in 44 samples of Turkish apple juice (Karadenizm and Eksi 1997). Formation of PAT in pear inoculated with *P. expansum* is found and its diffusion in the apparently sound flesh, in concentration surpassing the accepted maximum European limits (50 mg/kg) (Laidou et al. 2001). Three hundred fifty one samples of seven different varieties of apples with small rotten areas, collected throughout Portugal, the occurrence of PAT (up to 80.5 mg/kg) is recorded in 89 % samples, and OTA and trichothecene production in grapes by *A. carbonarius* and *T. roseum*, respectively, prior to harvest time (Martins et al. 2002). Mogensen et al. (2010) studied on 10 selected fumonisin producing *A. niger* strains and reported that they are able to produce fumonisin B2 and fumonisin B4 on grapes in the range 171–7,841 μg fumonisin B2/kg and 14–1,157 μg fumonisin B4/kg.

PAT is detected in 79, 86 and 43 % in the different tested samples of apples (Tangni et al. 2003). However, no contaminated sample exceeded the safe level of 50 μg PAT/l. Data of Beretta et al. (2000) reported that patulin intake with apple derivatives is usually below the tolerable level of 0.4 μg/kg bw/day. Occurrence of trichothecenes (nivalenol, deoxynivalenol, 3- and 15-acetyldeoxynivalenol, neosolaniol, diacetoxyscirpenol, T-2 tetraol, T-2 and HT-2 toxins), zearalenone and zearalenols, and fumonisin B1 from bananas have been reported (Jimenez and Mateo 1997). Ochratoxin A is found to present in all the samples of apricot (50–110 μg/kg), fig (60–120 μg/kg) and plum (210–280 μg/kg) collected from Egypt (Zohri and Abdel-Gawad 1993).

Essential Oils in the Management of Naturally Occurring Mycotoxins in Stored Food Commodities

Several synthetic fungicides have been used for the management of mycoflora incursion of commercially importance agricultural food commodities. However, reduction and detoxification of mycotoxins in food by physical and chemical methods have not yet proven to be an effective or desirable practice. This is due to their residual toxicities and adverse effects on the food chain (Gurney et al. 2014). Therefore, essential oil based botanical detoxification offers promising alternative for eliminating mycotoxins and safe guarding the quality of the food and feed. The existing dispute on the negative effects of synthetic preservatives has also transformed the interest of users towards natural food protectant for improving the quality and shelf life of food commodities and protecting them from biodeterioration by toxic microbes (Pandey et al. 2013a, b, 2014). Since these compounds are eco-friendly and harmless to humans, there is increasing attention, both in industry and academic research, to herbal, medicinal and aromatic plants for their antifungal activities against food spoilage and mycotoxigenic fungi. Essential oils isolated from higher plants are made from a very complex mixture of volatile molecules and

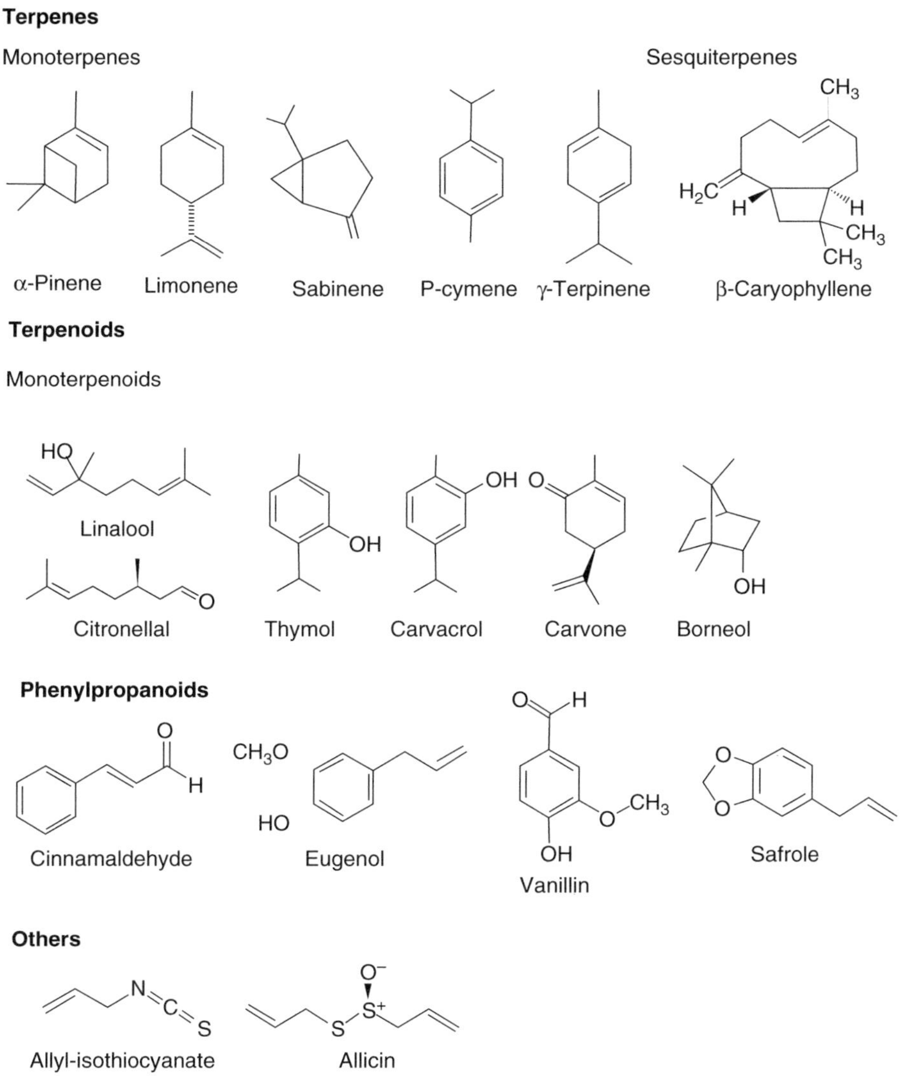

Fig. 2 Chemical configuration of some constituents of essential oil

are produced by the secondary metabolism of aromatic and medicinal plants. These essential oils can be obtained by different methods including hydrodistillation and low or high pressure distillation methods. In essential oils terpene and terpenoid constituted the main group followed by aromatic and aliphatic components. The major group of plant natural products is terpenes which are characterized by an extensive variety of structural types and are divided into monoterpenes (C_{10}), sesquiterpenes (C_{15}), diterpenes (C_{20}), hemiterpenes (C_5), triterpenes (C_{30}) and tetraterpenes (C_{40}). The structural formulas of some important toxic compounds are given in Figure 2.

EOs Against Aflatoxin

The application of essential oils (EOs) extracted from herbal, medicinal and aromatic plants against the fungal growth and aflatoxin production of *A. flavus* and *A. parasiticus* have been suggested by many researchers (Maraqa et al. 2007; El-Nagerabi et al. 2012). EOs and flavonoids have inhibitory effect on aflatoxin activity of several food borne fungi (Alpsoy 2010). Cinnamon and clove oils are inhibitory for aflatoxin production at 200–250 ppm, cinnamic aldehyde at 150 ppm and eugenol at 125 ppm (Bullerman et al. 1977). In maize, EOs from *Azadirachta indica* seeds extract are found to be completely inhibited the aflatoxin production at 500 and 1,000 ppm concentration while *Morinda lucida* showed complete inhibition at 1,000 ppm (Bankole 1997). Essential oils of Iranian medicinal plants are also proved as novel aflatoxin (AF) inhibitors in food system. *Satureja hortensis* and its active components are reported as an ideal inhibitor of aflatoxins B 1 (AFB 1) and G1 (AFG 1) production by *A. parasiticus* NRRL 2999. The IC_{50} values for the inhibition of microbial growth are found as 0.79 and 0.86 mM for the essential oil components carvacrol and thymol respectively. For AFB1 and AFG1, IC_{50} values are 0.50 and 0.06 mM for carvacrol and 0.69 and 0.55 mM for thymol respectively. After their observation it is found that carvacrol and thymol are the effective constituents of *S. hortensis* and they may be useful to manage the aflatoxin contamination in crops (Razzaghi-Abyaneh et al. 2008). Different concentrations of baobab (*Adansonia* sp.) seeds' EOs (0.5, 1, 3 and 5 % v/v) had led the highest inhibition levels of total aflatoxin and aflatoxin B1 secretion by *A. flavus* (47.2–95.7 %; 28.1–89.7 %) and *A. parasiticus* (42.7–93.3 %; 25.9–80.2 %) (El-Nagerabi et al. 2013). At concentrations of 1–3 %, *Nigella sativa* oil caused 47.9–58.3 % reduction in aflatoxin B1 for *A. flavus* and 32–48 % for *A. parasiticus* strains (El-Nagerabi et al. 2012). The results of antiaflatoxinogenic assay of Adjou et al. (2012) showed that EO of *Ageratum conyzoides* has important aflatoxin inhibition potential on toxigenic strains *A. parasiticus* (Ab2242) at 2.0 μl/ml and *A. flavus* (La3228) at 1.5 μl/ml. Aflatoxin B1 production by NKD-208 isolates of *A. flavus* is strongly inhibited by EO of *Callistemon lanceolatus* (Shukla et al. 2012) and the EO of *Zataria multiflora* at 150 ppm inhibited the aflatoxin production up to 99.4 % (Gandomi et al. 2009). The growth of *A. parasiticus* is significantly decreased ($P<0{,}001$) by marjoram and clary sage EOs (Gömöri et al. 2013). Similarly, inhibition of aflatoxin production has been reported by *Thymus eriocalyx* and *T. X-porlock* EOs (250 ppm) (Rasooli and Owlia 2005), *Rosmarinus officinalis* (450 ppm) and *Trachyspermum copticum* (450 ppm) EOs (Rasooli et al. 2008), *Ocimum gratissimum* EO (Prakash et al. 2011) and for Turmeric leaf oil (95.3 % and 100 % inhibition of toxin production) (Sindhu et al. 2011). Fungal growth and aflatoxin B1 production are inhibited by EOs at 50, 30, 15, and 10 μl dosage, but the *Ageratum conyzoides* oil is more effective in soybeans than that of *Origanum vulgare* (Esper et al. 2014). Recently, EOs of *Cymbopogon martinii*, *Foeniculum vulgare*, and *Trachyspermum ammi* are found to be totally inhibiting the mycotoxin production by *A. niger* and *A. flavus* at 0.5 and 0.75 μ L/mL, respectively (Gemeda et al. 2014). Correspondingly, EOs from *Cymbopogon citratus* (Sonker et al. 2014) and *Artemisia nilagirica* (Sonker et al. 2015) are found to be inhibited the aflatoxin production in grapes at 0.8 and 1.6 μl/ml respectively.

EOs Against Zearalenone

Perusal of the literature showed that little work has been carried our regarding the activity of EOs against zearalenone production. Essential oils of cinnamon, palmarosa, oregano, clove and lemongrass are inhibited the production of two mycotoxins i.e. deoxynivalenol (DON) and zearalenone (ZEA) in non-sterilized naturally contaminated maize grain produced by *Fusarium graminearum* at 0.995 and 0.950 aw and at 20 °C and 30 °C is evaluated at a 500 mg/kg level (Marin et al. 2004). Efficacy of EOs is found to be less effective and among all clove EO is proved as better protectant for maize grains. Palmarose and clove EOs are significantly inhibited zearalenone and deoxynivalenol production as well as the growth rate, by *F. graminearum* under different environmental conditions in maize grain. At 0.995 aw all the EOs tested had inhibitory effect on the growth rate of *Fusarium* (Velluti et al. 2004).

EOs Against Fumonisin

The essential oils from plants deal a faith in the prevention and detoxification of several mycotoxins. In maize grains fumonisin B1 production by *F. proliferatum* is significantly inhibited by the lemongrass, cinnamon, clove, palmarose and oregano EOs (Velluti et al. 2003) at 0.995 aw at both temperatures, while at 0.950 aw only cinnamon, clove and oregano oils are effective in inhibiting growth of *F. proliferatum* at 20 °C and none of them at 30 °C. Out of four EOs of aromatic plants *Origanum vulgare*, *Aloysia triphylla*, *A. polystachya* and *Mentha piperita*, EO of *O. vulgare* reduced the production level of FB1 ($P<0.01$) by *F. verticillioides* while *A. triphylla* EO increased it ($P<0.001$). Comparatively, *A. triphylla* and *O. vulgare* EOs at 250 and 500 epsilonl/l have the better inhibitory effects on the *F. verticillioides* mycelia development (López et al. 2004). The microbial growth of *F. culmorum* and *F. graminearum* are significantly inhibited by 500 μg/g cinnamon oil at 0.955 aw/25 °C yet toxin production is enhanced (Hope et al. 2005). Dambolena et al. (2008) found that out of four monoterpenes (limonene, menthol, menthone and thymol) at 75 ppm, thymol is the most active inhibitor on FB1 biosynthesis by *F. verticillioides*. The EO of *Zingiber officinale* has the inhibitory effect on fumonisin B1 (FB1) and fumonisin B2 (FB2) production at 4,000 and 2,000 μg/ml, respectively (Yamamoto-Ribeiro et al. 2013). In recent investigation *Litsea cubeba*, clove, cinnamon, citral, spearmint, eucalyptus, anise and camphor EOs showed their degrading nature on the production level of FB1 by *F. proliferatum* (Xing et al. 2014). The order of efficacy of EOs were cinnamon > citral> eugenol > eucalyptus > anise > camphor oils. Cinnamon oil reduced FB1 from 15.03 to 0.89 μg/ml (94.06 %) at 120 h time and the 280 μg/ml concentration. Thus, these oils could be ideal agents for the detoxification and management of FB1 in the crops.

EOs Against Ochratoxin A (OTA)

From time to time studies are conducted in order to control the ochratoxin in food commodities by plant essential oils. A 1,000 ppm dose of oregano (*Origanum vulgare*) and mint (*Mentha arvensis*) EOs completely inhibited the mycotic growth and OTA production by *A. ochraceus* NRRL 3174 up to 21 days, while basil is only effective up to 7 days (Basílico and Basílico 1999). Other scholars have also reported the reduction of ochratoxin A production by the microrganisms through the use of EOs of spices (Soliman and Badeaa 2002). The reduction efficiency of both the essential oil and the aqueous extracts are found to fallen in between 66.7 % and 95.7 %. Clove leaf, cinnamon and bay leaf EOs at 50 ppm cause stimulation of OTA production regardless of aw or temperature. However, at 500 ppm significant control of OTA production by *A. ochraceus* is observed with the best essential oil treatments. Indian borage oil (*Plectranthus amboinicus*) is reported to completely inhibit the ochratoxin (OTA) production by the toxigenic strain *A. ochraceus* at 500 ppm (Cairns and Magan 2003). Also, the application of oil at 100 mg/g in food samples inhibits the growth of *A. ochraceus* in food systems such as groundnut, maize and poultry feed even at a high moisture level of 30 %, after 7 days (Murthy et al. 2009). Similarly, ochratoxigenic activity of *Aframomum danielli* EO at from 500 to 2,000 ppm has been reported to decrease OTA contents of cocoa bean (Aroyeun et al. 2009). Additionally, 0.10 % of basil EO reduces the production level of ochratoxin A (OTA) from 135 to 98 μg/ml (Mohamed et al. 2012). Recently, EOs from *Cymbopogon citratus* (Sonker et al. 2014) and *Artemisia nilagirica* (Sonker et al. 2015) completely inhibited the OTA production in table grapes at 0.8 and 1.6 μl/ml respectively.

Conclusion

Over population has necessitated the need to store large amount of food commodities for use in near future. But storage of such food commodities accompanies problems of storage pests and pest based toxins like mycotoxins. The knowledge of the past decade indicates that the agricultural food commodities are contaminated by different types of mycotoxins. Combating these fungal pathogens/contaminants by xenobiotics and synthetic chemicals, pose a severe and complex symptom at anthropogenic and ecological level. The studied publications have dealt with the inhibition of mycotoxigenic species by varied natural plant products especially essential oils.

Most of them showed a high efficacy in the management of mycotoxins as an anti-aflatoxic, antiochratoxic, anti-fumonotoxic agent. Ecofriendly mycotoxin management using essential oil based botanicals are safer to the user and the environment. Botanicals are cost effective, sustainable sources which have a plethora of activities against storage pests including mycotoxin producing organisms. They are also very close chemically to those plants from which they are isolated, so they are easily biodegradable and are renewable in nature. Because of greater consumer awareness and harmful side effects towards synthetic fungicides, protection of commercially importance agricultural food commodities using botanical fungicides is becoming more popular. There is an extensive scope for plant-based pesticides usage in the integrated management of different agricultural pests. Use of such technologies can be one of the aims of sustainable agriculture and a cost effective management tool in poor economies but which is rich in plant biodiversity. However, finally, before application safety issues should be fully addressed prior to the widespread application of such plant products.

References

Abbas HK, Zablotowicz RM, Bruns HA (2008) Modelling the colonization of maize by toxigenic and non-toxigenic *Aspergillus flavus* strains: implications for biological control. World Mycotoxin J 1:333–340

Abdelhamid AA (1990) Occurrence of some mycotoxins (aflatoxins, ochratoxin A, citrinin, zearalenone and vomitoxin) in various Egyptian feeds. Arch Anim Nutr 40:647–664

Abramson D, Clear RM, Usleber E, Gessler R, Nowicki TW, Märtlbauer E (1998) Fusarium species and 8 keto-trichothecene mycotoxins in Manitoba barley. Cereal Chem 75:137–141

Adjou ES, Dahouenon-Ahoussi E, Degnon R, Soumanou MM, Sohounhloue DCK (2012) Investigations on bioactivity of essential oil of *Ageratum conyzoides* L., from benin against the growth of fungi and aflatoxin production. Int J Pharm Sci Rev Res 13:143–148

Ahmad MS, Sinha KK (2002) Mycotoxin contamination of mustard seeds during storage. Indian Phytopathol 55:299–302

Aksoy UR, Eltem KB, Meyvaci AA, Karabat S (2007) Five-year survey of Ochratoxin A in processed sultanas from Turkey. Food Addit Contam 24:292–296

Al-Anati L, Petzinger E (2006) Immunotoxic activity of ochratoxin A. J Vet Pharmacol Ther 29:79–90

Alexander NJ, Proctor RH, McCormick SP (2009) Genes, gene clusters, and biosynthesis of trichothecenes and fumonisins in *Fusarium*. Toxin Rev 28:198–215

Alpsoy L (2010) Inhibitory effect of essential oil on aflatoxin activities. Afr J Biotechnol 17:2474–2481

Aroyeun SO, Adegoke GO, Varga J, Teren J (2009) Reduction of aflatoxin B1 and ochratoxin a in cocoa beans infected with *Aspergillus* via ergosterol value. World Rev Sci Technol Sustain Dev 6:75–89

Artigot MP, Loiseau N, Laffitte J, Mas-Reguieg L, Tadrist S, Oswald IP, Puel O (2009) Molecular cloning and functional characterization of two CYP619 cytochrome P450s involved in biosynthesis of patulin in *Aspergillus clavatus*. Microbiology 155:1738–1747

Aziz NH, Attia ESA, Farag SA (1997) Effect of gamma-irradiation on the natural occurrence of *Fusarium* mycotoxins in wheat, flour and bread. Food Nahrung 41:34–37

Badiale-Furlong E, Dors GC, Oliveira M dos S, de Souza MM, Kuhn RC (2003) Avaliação da qualidade de farinha de trigoe produtos de panificação comercializadas no Rio Grande do Sul. In: Simpósio de Ciências de Alimentos e Saúde. Florianópolis-SC/UFSC, Anais, pp 1–4

Bankole SA (1997) Effect of essential oils from two Nigerian medicinal plants (*Azadirachta indica* and *Morinda lucida*) on growth and aflatoxin B_1 production in maize grain by a toxigenic *Aspergillus flavus*. Lett Appl Microbiol 24:190–192

Basílico MZ, Basílico JC (1999) Inhibitory effects of some spice essential oils on *Aspergillus ochraceus* NRRL 3174 growth and ochratoxin A production. Lett Appl Microbiol 29:238–241

Begum H, Samajpati N (2000) Mycotoxin production on rice, pulses and oilseeds. Naturwissenschaften 87:275–277

Bennett JW, Chang PK, Bhatnagar D (1997) One gene to whole pathway: the role of norsolorinic acid in aflatoxin research. Adv Appl Microbiol 45:1–15

Beretta B, Gaiaschib A, Galli CL, Restani P (2000) Patulin in apple-based foods: occurrence and safety evaluation. Food Addit Contam 17:399–406

Bhatnagar D, Cary JW, Ehrlich K, Yu J, Cleveland TE (2006) Understanding the genetics of regulation of aflatoxin production and *Aspergillus flavus* development. Mycopathologia 162:155–166

Bhatt RU, Nagarajan U, Tulpute PG (1978) Health hazards of mycotoxins in India. lGMR, New Delhi, p 58

Bhatt RV, Shetty PH, Amruth RP, Sudershan RV (1997) A foodborne disease outbreak due to the consumption of moldy sorghum and maize containing fumonisin mycotoxins. Clin Toxicol 35:249–255

Birck NMM, Lorini I, Scussel VM (2003) Sanitary conditions and mycotoxins in wheat grains (*Triticum aestivum*) and flour (common and special) through milling processing. In: IV Congreso Latino americano De Mycotoxicologia. LaHabana, Cuba. Anais

Bottalico A, Perrone G (2002) Toxigenic *Fusarium* species and mycotoxins associated with head blight in small-grain cereals in Europe. Eur J Plant Pathol 108:611–624

Brown DW, Dyer RB, McCormick SP, Kendra DF, Plattner RD (2004) Functional demarcation of the *Fusarium* core trichothecene gene cluster. Fungal Genet Biol 41:454–462

Bullerman LB, Lieu FY, Seier SA (1977) Inhibition of growth and aflatoxin production by cinnamon and clove oils, cinnamic aldehyde and eugenol. J Food Sci 42:1107–1109

Cairns V, Magan N (2003) Impact of essential oils on growth and ochratoxin A production by *Penicillium verrucosum* and *Aspergillus ochraceus* on a wheat-based substrate. In: Credl and PF, Armitage DM, Bell CH, Cogan PM, Highley E (eds) Advances in stored product protection. Proceedings of the 8th international working conference on stored product protection, York, 22–26 Jul 2002. pp 479–485

Calvo AM (2005) Mycotoxins. In: D'browski WM, Sikorski ZE (eds) Toxins in food. CRC Press, Boca Raton

Campbell H, Choo TM, Vigier B, Underhill L (2000) Mycotoxins in barley and oat samples from Eastern Canada. Can J Plant Sci 80:977–980

Chala A, Mohammed A, Ayalew A, Skinnes H (2013) Natural occurrence of aflatoxins in groundnut (*Arachis hypogaea* L.) from eastern Ethiopia. Food Control 30:2602–2605

Chulze SN, Magnoli CE, Dalcero AM (2006) Occurrence of Ochratoxin A in wine and ochratoxigenic mycoflora in grape and dried vine fruits in South America. Int J Food Microbiol 111(Suppl 1):S5–S9

Dambolena JS, López AG, Cánepa MC, Theumer MG, Zygadlo JA, Rubinstein HR (2008) Inhibitory effect of cyclic terpenes (limonene, menthol, menthone and thymol) on *Fusarium verticillioides* MRC 826 growth and fumonisin B1 biosynthesis. Toxicon 51:37–44

Dohnal V, Jezkova A, Jun D, Kuca K (2008) Metabolic pathways of T-2 toxin. Curr Drug Metab 9:77–82

Ediage EN, Di Mavungu JD, Monbaliu S, Peteghem CV, De Saeger S (2011) A validated multianalyte LC–MS/MS method for quantification of 25 mycotoxins in cassava flour, peanut cake and maize samples. J Agric Food Chem 59:5173–5180

El-Nagerabi SAF, Al-Bahry SN, Elshafie AE, AlHilali S (2012) Effect of *Hibiscus sabdariffa* extract and *Nigella sativa* oil on the growth and aflatoxin B1 production of *Aspergillus flavus* and *Aspergillus parasiticus* strains. Food Control 25:59–63

El-Nagerabi SAE, Elshafie AE, AlKhanjari SS, Al-Bahry SN, Elamin MR (2013) The potential of Baobab (*Adansonia digitata* L.) extracts as biocontrol on the growth and aflatoxin production by *Aspergillus flavus* and *A. parasiticus*. J Food Res 2:93–103
Embaby EM, Reda M, Abdel-Wahhab MA, Omara H, Mokabel AM (2013) Occurrence of toxigenic fungi and mycotoxins in some legume seeds. J Agric Technol 9:151–164
Esper RH, Gonçalez E, Marques MOM, Felicio RC, Felicio JD (2014) Potential of essential oils for protection of grains contaminated by aflatoxin produced by *Aspergillus flavus*. Front Microbiol 5:269
Etzel RA (2002) Mycotoxins. J Am Med Assoc 287:425–427
Ezekiel CN, Sulyok M, Warth B, Odebode AC, Krska R (2012) Natural occurrence of mycotoxins in peanut cake from Nigeria. Food Control 27:338–342
FAO (1981) Annual report, Indian Grain Storage Institute, Hapur. 1981–82
Fox T, Fimeche C (2013) Global food waste not, want not. IMechE, England
Gallo A, Bruno KS, Solfrizzo M, Perrone G, Mulè G, Visconti A, Baker SE (2012) New insight into the ochratoxin A biosynthetic pathway through deletion of a non-ribosomal peptide synthetase gene in *Aspergillus carbonarius*. Appl Environ Microbiol 78:8208–8218
Gandomi H, Misaghi A, Basti AA, Bokaei S, Khosravi A, Abbasifar A, Javan AJ (2009) Effect of *Zataria multiflora* Boiss. essential oil on growth and aflatoxin formation by *Aspergillus flavus* in culture media and cheese. Food Chem Toxicol 47:2397–2400
Geisen R, Schmidt-Heydt M, Karolewiez A (2006) A gene cluster of the Ochratoxin A biosynthetic genes in *Penicillium*. Mycotoxin Res 22:134–141
Gemeda N, Woldeamanuel Y, Asrat D, Debella A (2014) Effect of *Cymbopogon martinii*, *Foeniculum vulgare*, and *Trachyspermum ammi* essential oils on the growth and mycotoxins production by *Aspergillus* species. Int J Food Sci Article ID 874135, 9 pages
Giray B, Girgin G, Engin AB, Aydın S, Sahin G (2007) Aflatoxin levels in wheat samples consumed in some regions of Turkey. Food Control 18:23–29
Giray B, Atasayar S, Sahin G (2009) Determination of ochratoxin A and total aflatoxin levels in corn samples from Turkey by enzyme-linked immunosorbent assay. Mycotoxin Res 25:113–116
Gömöri C, Nacsa-Farkas E, Kerekes EB, Kocsubé S, Vágvölgyi C, Krisch J (2013) Evaluation of five essential oils for the control of food spoilage and mycotoxin producing fungi. Acta Biol Szeged 57:113–116
Gonçalez E, Nogueira JHC, Fonseca H, Felicio JD, Pino FA, Corrêa B (2008) Mycobiota and mycotoxins in Brazilian peanut kernels from sowing to harvest. Int J Food Microbiol 123:184–190
Gummert M, Balingbing CB, Barry G, Estevez LA (2009) Management options, technologies and strategies for minimised mycotoxin contamination of rice. World Mycotoxin J 2:151–159
Gurney SMR, Scott KS, Kacinko SL, Presley BC, Logan BK (2014) Pharmacology, toxicology, and adverse effects of synthetic cannabinoid drugs. Forensic Sci Rev 26:53
Hell K, Setamou M, Cardwell KF, Poehling HM (2000) The influence of storage practices on aflatoxin contamination in maize in four agroecological zones in Benin, West Africa. J Stored Prod Res 36:365–382
Hell K, Cardwell KF, Poehling HM (2003) Distribution of fungal species and aflatoxin contamination in stored maize in four agroecological zones in Benin, West-Africa. J Phytopathol 151:690–698
Hope R, Cairns-Fuller V, Aldred D, Magan N (2005) Use of antioxidants and essential oils for controlling mycotoxins in grain. BCPC Crop Sci Technol 5B:429–436
Huff WE, Hamilton PB (1979) Mycotoxins-their biosynthesis in fungi: ochratoxins-metabolites of combined pathways. J Food Protect 42:815–820
Hussaini AM, Timothy AG, Olufunmilayo HA, Akanya H, Ezekiel AS, Ogbadu GH (2009a) Fungi and some mycotoxins contaminating rice (*Oryza sativa*) in Niger State, Nigeria. Afr J Biotechnol 6:099–108
Hussaini AM, Timothy AG, Olufunmilayo HA, Ezekiel AS, Godwin HO (2009b) Fungi and some mycotoxins found in mouldy sorghum in Niger state, Nigeria. World J Agric Sci 5:05–17
Jackson Lauren S, Jonathan W, Vries D, Lloyd BB (1996) Fumonisins in food. Plenum Press, New York, p 289

James B, Adda C, Cardwell K, Annang D, Hell K, Korie S, Edorh M, Gbeassor F, Nagatey K, Houenou G (2007) Public information campaign on aflatoxin contamination of maize grains in market stores in Benin, Ghana and Togo. Food Addit Contam 24:1283–1291

Janardhana GR, Raveesha KA, Shetty HS (1999) Mycotoxin contamination of maize grains grown in Karnataka (India). Food Chem Toxicol 37:863–868

Jayaraman P, Kalyanasundaram I (1990) Natural occurrence of toxigenic fungi and mycotoxins in rice bran. Mycopathologia 110:81–85

Jestoi M (2008) Emerging fusarium-mycotoxins fusaproliferin, beauvericin, enniatins, and moniliformin-a review. Crit Rev Food Sci Nutr 48:21–49

Jimenez M, Mateo R (1997) Determination of mycotoxins produced by *Fusarium* isolates from banana fruits by capillary gas chromatography and high-performance liquid chromatography. J Chromatogr A 778:363–372

Jurjevic J, Solfrizzo M, Cvjetkovic B, De Girolamo A, Visconti A (2002) Occurrence of beauvericin in corn from Croatia. Food Technol Biotechnol 40:91–94

Jurjevic Z, Wilson JP, Wilson DM, Casper HH (2007) Changes in fungi and mycotoxins in pearl millet under controlled storage conditions. Mycopathologia 164:229–239

Kaaya AN, Eigel W, Harris C (2006) Peanut aflatoxin levels on farms and in markets of Uganda. Peanut Sci 33:68–75

Karadenizm F, Eksi A (1997) The correlation of fumaric acid with patulin, L-lactic acid and HMF in apple juice concentrate. Fruit Process 12:475–478

Kim Y, Lee Y, Jin J, Han K, Kim H, Kim J, Lee T, Yun S, Lee Y (2005) Two different polyketide synthase genes are required for synthesis of zearalenone in *Gibberella zeae*. Mol Microbiol 58:1102–1113

Kishore GK, Pande S, Manjula K, Rao JN, Thomas D (2002) Occurrence of mycotoxins and toxigenic fungi in groundnut (*Arachis hypogaea* L.) seeds in Andhra Pradesh, India. Plant Pathol J 18:204–209

Kritzinger Q, Aveling TA, Marasas WF, Rheeder JP, Van Der Westhuizen L, Shephard GS (2003) Mycoflora and fumonisin mycotoxins associated with cowpea (*Vigna unguiculata* (L.) Walp) seeds. J Agric Food Chem 51:2188–2192

Kuiper-Goodman T, Scott PM, Watanabe H (1987) Risk assessment of the mycotoxin zearalenone. Regul Toxicol Pharmacol 7:253–306

Laidou IA, Thanassoulopoulos CC, Liakopoulou-Kyriakides M (2001) Diffusion of patulin in the flesh of pears inoculated with four post-harvest pathogens. J Phytopathol 149:457–461

Liu Z, Gao J, Yu J (2006) Aflatoxins in stored maize and rice grains in liaoning province, china. J Stored Prod Res 42:468–479

López AG, Theumer MG, Zygadlo JA, Rubinstein HR (2004) Aromatic plants essential oils activity on *Fusarium verticillioides* Fumonisin B_1 production in corn grain. Mycopathologia 158:343–349

Lysoe E, Klemsdal SS, Bone KR, Frandsen RJN, Johansen T, Thrane U, Giese H (2006) The PKS4 gene of *Fusarium graminearum* is essential for zearalenone production. Appl Environ Microbiol 72:3924–3932

MacDonald S, Wilson P, Barnes K, Damant A, Massey R, Mortby E, Shepherd MJ (1999) Ochratoxin A in dried vine fruit: method development and survey. Food Addit Contam 16:253–260

Magnoli C, Astoreca A, Ponsone L, Combina M, Palacio G, Rosa CAR, Dalcero AM (2004) Survey of mycoflora and ochratoxin A in dried vine fruits from Argentina markets. Lett Appl Microbiol 39:326–331

Magnoli C, Astoreca A, Ponsone L, Fernández-Juri MG, Chiacchiera S, Dalcero A (2006) Ochratoxin A and the occurrence of ochratoxin A-producing black aspergilli in stored peanut seeds from Córdoba, Argentina. J Sci Food Agric 86:2369–2373

Magnoli C, Astoreca A, Ponsone ML, Fernández-Juri MG, Barberis G, Dalcero AM (2007) Ochratoxin A and *Aspergillus* section *Nigri* in peanut seeds at different months of storage in Córdoba, Argentina. Int J Food Microbiol 119:213–218

Majid I (2007) Studying the mycotoxins found in wheat and maize under conditions of storage. Ph.D. Dissertation. Damascus University, Syria

Mallmann CA, Dilkin M, Mürmann L, Dilkin P, Almeida CAA (2003) Avaliação da contaminação por deoxivalenol em trigo utilizado na alimentação humana. In: Congresso Brasileiro de Farmácia. Anais

Maraqa A, Alsharoa NF, Farah H, Albjeirami WM, Shakya AK, Sallal AJ (2007) Effect of *Nigella sativa* extract and oil on aflatoxin production by *Aspergillus flavus*. Turk J Biol 31:155–159

Marasas WFO, Paul EN (1987) Mycotoxicology: introduction to the mycology, plant pathology, chemistry, toxicology, and pathology of naturally occurring mycotoxicoses in animals and man. The Pennsylvania State University Press, University Park, p 47

Marin S, Velluti A, Ramos AJ, Sanchis V (2004) Effect of essential oils on zearalenone and deoxynivalenol production by *Fusarium graminearum* in non-sterilized maize grain. Food Microbiol 21:313–318

Marin S, Ramos AJ, Ho GC, Sanchis V (2012) Reduction of mycotoxins and toxigenic fungi in the Mediterranean basin maize chain. Phytopathol Mediterr 51:93–118

Martins ML, Gimeno A, Martins HM, Bernardo F (2002) Co-occurrence of patulin and citrinin in Portuguese apples with rotten spots. Food Addit Contam 19:568–574

Mogensen JM, Thrane JC, Nielsen KF (2010) Production of fumonisin B2 and B4 by *Aspergillus niger* on grapes and raisins. J Agric Food Chem 58:954–958

Mohamed SR, Deabes MM, Abdel-Motaleb NM, Mohamed SS, Badr AN, Shaban EM, Khalil AF (2012) Antifungal activity of basil and mustard essential oils against spoilage toxigenic fungi in Egyptian pan bread and its economic evaluation. J Appl Sci Res 8:5536–5542

Muriuki GK, Siboe GM (1995) Maize flour contaminated with toxigenic fungi and mycotoxins in Kenya. Afr J Health Sci 2:236–241

Murthy PS, Ramalakshmi K, Srinivas P (2009) Fungitoxic activity of Indian borage (*Plectranthus amboinicus*) volatiles. Food Chem 114:1014–1018

Muthomi JW, Mutitu EW (2003) Occurrence of mycotoxin producing *Fusarium* species and other fungi on wheat kernels harvested in selected districts of Kenya. African crop science conference proceedings, vol. 6. pp 335–339

Nakagawa H, Ohmichi K, Sakamoto S, Sago Y, Kushiro M, Nagashima H, Yoshida M, Nakajima T (2011) Detection of a new *Fusarium* masked mycotoxin in wheat grain by high-resolution LC–Orbitrap™ MS. Food Addit Contam Part A 28:1447–1456

Nizamlýoðlu F, Oguzz H (2003) Occurrence of aflatoxins in layer feed and corn samples in Konya province, Turkey. Food Addit Contam 20:654–658

Oruc HH, Cengiz M, Kalkanli O (2006) Comparison of aflatoxin and fumonisin levels in maize grown in Turkey and imported from the USA. Anim Feed Sci Technol 128:337–341

Palumbo JD, O'Keeffe TL, Vasquez SJ, Mahoney NE (2011) Isolation and identification of ochratoxin A-producing *Aspergillus* section *Nigri* strains from California raisins. Lett Appl Microbiol 52:330–336

Pande N, Saxena J, Pandey H (1990) Natural occurrence of mycotoxins in some cereals. Mycoses 33:126–128

Pandey AK, Palni UT, Tripathi NN (2013a) Evaluation of *Clausena pentaphylla* (Roxb.) DC oil as fungitoxicant against storage mycoflora of pigeon pea seeds. J Sci Food Agric 93:1680–1686

Pandey AK, Singh P, Palni UT, Tripathi NN (2013b) Application of *Chenopodium ambrosioides* Linn. essential oil as botanical fungicide for the management of fungal deterioration in pulse. Biol Agric Hort 29:197–208

Pandey AK, Singh P, Palni UT, Tripathi NN (2014) *In vivo* evaluation of two essential oil based botanical formulations (EOBBF) for the use against stored product pests, *Aspergillus* and *Callosobruchus* (Coleoptera: Bruchidae) species. J Stored Prod Res 59:285–291

Payne GA, Brown MP (1998) Genetics and physiology of aflatoxin biosynthesis. Annu Rev Phytopathol 36:329–362

Pietri A, Bertuzzi T, Pallaroni L, Piva G (2004) Occurance of mycotoxins and ergosterol in maize harvested over 5 year in the Northern Italy. Food Addit Contam 21:479–487

Pildain MB, Vaamonde G, Cabral D (2004) Analysis of population structure of *Aspergillus flavus* from peanut based on vegetative compatibility, geographic origin, mycotoxin and sclerotia production. Int J Food Microbiol 93:31–40

Prakash B, Shukla S, Singh P, Mishra PK, Dubey NK, Kharwar RN (2011) Efficacy of chemically characterized *Ocimum gratissimum* L. essential oil as an antioxidant and a safe plant based antimicrobial against fungal and aflatoxin B_1 contamination of spices. Food Res Int 44:385–390

Puel O, Galtier P, Oswald IP (2010) Biosynthesis and toxicological effects of patulin. Toxins (Basel) 2:613–631

Rasooli I, Owlia P (2005) Chemoprevention by thyme oils of *Aspergillus parasiticus* growth and aflatoxin production. Phytochemistry 66:2851–2856

Rasooli I, Fakoor MH, Yadegarinia D, Gachkar L, Allameh A, Rezaei MB (2008) Antimycotoxigenic characteristics of *Rosmarinus officinalis* and *Trachyspermum copticum* L. essential oils. Int J Food Microbiol 122:135–139

Razzaghi-Abyaneh M, Shams-Ghahfarokhi S, Yoshinari T, Rezaee M, Jaimand K, Nagasawa H, Sakuda S (2008) Inhibitory effects of *Satureja hortensis* L. essential oil on growth and aflatoxin production by *Aspergillus parasiticus*. Int J Food Microbiol 123:228–233

Reddy KRN, Reddy CS, Muralidharan K (2008) Potential of botanicals and biocontrol agents on growth and aflatoxin production by *Aspergillus flavus* infecting rice grains. Food Control. doi:10.1016/j.foodcont.2008.03.009

Richard JL, Plattner RD, Mary J, Liska SL (1999) The occurrence of ochratoxin A in dust collected from a problem homehold. Mycopathologia 146:99–103

Scudamore KA, Patel S (2009) Occurrence of *Fusarium* mycotoxins in maize imported into the UK, 2004–2007. Food Addit Contam Part A 26:363–371

Šegvić M, Cvetnić Z, Pepeljnjak S, Kosalec I (2009) Co-occurrence of aflatoxins, ochratoxin A, fumonisins, and zearalenone in cereals and feed, determined by competitive direct enzyme-linked immunosorbent assay and thin- layer chromatography. Arch Ind Hyg Toxicol 60:427–434

Sekiguchi J, Shimamoto T, Yamada Y, Gaucher GM (1983) Patulin biosynthesis: enzymatic and non-enzymatic transformations of the mycotoxin (E)-ascladiol. Appl Environ Microbiol 45:1939–1942

Serra R, Bragab A, Venâncio A (2005) Mycotoxin-producing and other fungi isolated from grapes for wine production, with particular emphasis on ochratoxin A. Res Microbiol 156:515–521

Shetty PH, Bhat RV (1997) Natural occurrence of fumonisin B_1 and its co-occurrence with aflatoxin B_1 in Indian sorghum, maize, and poultry feeds. J Agric Food Chem 45:2170–2173

Shukla R, Singh P, Prakash B, Dubey NK (2012) Antifungal, aflatoxin inhibition and antioxidant activity of *Callistemon lanceolatus* (Sm.) sweet essential oil and its major component 1,8-cineole against fungal isolates from chickpea. Food Control 25:27–33

Sindhu S, Chempakam B, Leela NK, Suseela Bhai R (2011) Chemoprevention by essential oil of turmeric leaves (*Curcuma longa* L.) on the growth of *Aspergillus flavus* and aflatoxin production. Food Chem Toxicol 49:1188–1192

Sinha KK (1990) Incidence of mycotoxins in maize grains in Bihar State, India. Food Addit Contam 7:55–61

Siruguri V, Kumar UP, Raghu P, Rao VVM, Sesikeran B, Toteja GS, Gupta P, Rao S, Satyanarayana K, Katoch VM, Bharaj VH, Mangat GS, Sharma N, Sandhu JS, Bhargav VK, Rani S (2012) Aflatoxin contamination in stored rice variety PAU 201 collected from Punjab, India. Indian J Med Res 136:89–97

Smith CA, Woloshuk CP, Robertson D, Payne GA (2007) Silencing of the aflatoxin gene cluster in a diploid strain of *Aspergillus flavus* is suppressed by ectopic aflR expression. Genetics 176:2077–2086

Soliman KM, Badeaa RI (2002) Effect of oil extracted from some medical plants on different mycotoxigenic fungi. Food Chem Toxicol 4:1669–1675

Sonker N, Pandey AK, Singh P, Tripathi NN (2014) Assessment of *Cymbopogon citratus* (DC.) Stapf essential oil as herbal preservatives based on antifungal, antiaflatoxin and antiochratoxin activities and in vivo efficacy during storage. J Food Sci 79:628–634

Sonker N, Pandey AK, Singh P (2015) Efficiency of *Artemisia nilagirica* (Clarke) Pamp essential oil as a mycotoxicant against postharvest mycobiota of table grapes. J Sci Food Agric 95: 1932–1939

Stratton GW, Robinson AR, Smith HC, Kittilsen L, Barbour M (1993) Levels of five mycotoxins in grains in Atlantic Canada as measured by high performance liquid chromatography. Arch Environ Contam Toxicol 24:399–409

Sulyok M, Berthiller F, Krska R, Schuhmacher R (2006) Development and validation of a liquid chromatography/tandem mass spectrometric method for the determination of 39 mycotoxins in wheat and maize. Rapid Commun Mass Spectrom 20:2649–2659

Tanaka K, Sago Y, Zheng Y, Nakagawa H, Kushiro M (2007) Mycotoxins in rice. Int J Food Microbiol 119:59–66

Tangni EK, Theys R, Mignolet E, Maudoux M, Michelet JY, Larondelle Y (2003) Patulin in domestic and imported apple-based drinks in Belgium: occurrence and exposure assessment. Food Addit Contam 20:482–489

Tseng TC, Tu JC (1997) Mycoflora and mycotoxins in adzuki and mung beans produced in Ontario, Canada. Microbios 90:87–95

Vasanthi S, Bhat RV (1998) Mycotoxins in foods – occurrence, health & economic significance & food control measures. Indian J Med Res 108:212–224

Velluti A, Sanchis V, Ramos AJ, Egido J, Mari S (2003) Inhibitory effect of cinnamon, clove, lemongrass, oregano and palmarose essential oils on growth and fumonisin B_1 production by *Fusarium proliferatum* in maize grain. Int J Food Microbiol 89:145–154

Velluti A, Sanchis V, Ramos AJ, Turon C, Marín S (2004) Impact of essential oils on growth rate, zearalenone and deoxynivalenol production by *Fusarium graminearum* under different temperature and water activity conditions in maize grain. J Appl Microbiol 96:716–724

Wilson JP, Jurjevic Z, Hanna WW, Wilson DM, Potter TL, Coy AE (2006) Host-specific variation in infection by toxigenic fungi and contamination by mycotoxins in pearl millet and corn. Mycopathologia 161:101–107

Xing F, Hua H, Selvaraj JN, Yuan Y, Zhao Y, Zhou L, Liu Y (2014) Degradation of fumonisin B1 by cinnamon essential oil. Food Control 38:37–40

Yamamoto-Ribeiro MM, Grespan R, Kohiyama CY, Ferreira FD, Mossini SA, Silva EL, Filho BA, Mikcha JM, Machinski M Jr (2013) Effect of *Zingiber officinale* essential oil on *Fusarium verticillioides* and fumonisin production. Food Chem 141:3147–3152

Yu J, Chang PK, Ehrlich KC, Cary JW, Bhatnagar D, Cleveland TE, Payne GA, Linz JE, Woloshuk CP, Bennett JW (2004) Clustered pathway genes in aflatoxin biosynthesis. Appl Environ Microbiol 70:1253–1262

Zohri AA, Abdel-Gawad KM (1993) Survey of mycoflora and mycotoxins of some dried fruits in Egypt. J Basic Microbiol 33:279–288

Diagnostics of Seed-Borne Plant Pathogens for Safe Introduction and Healthy Conservation of Plant Genetic Resources

Jameel Akhtar, A. Kandan, Baleshwar Singh, Pardeep Kumar, Z. Khan, B.H. Gawade, Sundeep Kumar, and S.C. Dubey

Introduction

Any country in the world is not self-sufficient in plant genetic resources (PGR) for developing new varieties of crops to overcome various types of threats viz., insurgence of new/more virulent pests, weather calamities, extreme temperatures, as well as to enhance the national productivity, etc. Therefore, international exchange of PGR is the backbone of any crop improvement programme for developing new varieties. However, exchange of the seeds and other planting materials always carries an inherent risk of introducing new exotic pathogens or their more virulent races into new areas. It is evident from several examples that seeds are the most efficient means of long distance dissemination or transboundary movement of pathogens in the history (Khetarpal et al. 2006). There are about 1500 species of seed-borne fungi (~331 not reported from India), ~302 species of bacteria (~270 not reported from India) worldwide that affect 534 crops of 109 plant families (Richardson 1990). Here, the quotation given by (Kandan et al. 2015) (Stakman and Harrar 1957) "the responsibilities of the plant pathologists do not end with the harvest of satisfactory yields of plant products and that harvesting marks the termination of one phase of plant protection and the beginning of another" emphasizes that seeds and other plant propagating material is important in second phase of crop protection. So, supply of pathogen/disease free PGR is the primary means to restrict the introduction of exotic pathogens into the country. Therefore, quarantine testing for associated seed-borne/transmitted organisms is essentially required. In India, ICAR-National Bureau of Plant Genetic Resources (ICAR-NBPGR) has been empowered by Department of Agriculture and Co-operation

J. Akhtar (✉) • A. Kandan • B. Singh • P. Kumar • Z. Khan • B.H. Gawade • S. Kumar • S.C. Dubey
Division of Plant Quarantine, ICAR-National Bureau of Plant Genetic Resources,
New Delhi 110012, India
e-mail: jakhtar@nbpgr.ernet.in; jameelnbpgr@gmail.com

P. Kumar et al. (eds.), *Current Trends in Plant Disease Diagnostics and Management Practices*, Fungal Biology, DOI 10.1007/978-3-319-27312-9_19

(DAC), Government of India through Destructive Insects and Pests Act, 1914 (DIP Act, 1914) which has been revised from time to time and presently known as the "Plant Quarantine Order 2003 (Regulation of Import into India)" to undertake the quarantine processing of all PGR including transgenic planting material under exchange for research purposes, both for public and private sectors. NBPGR has a mandate for acquisition, management of indigenous and exotic PGR for pest-free conservation towards food security and sustainability.

In order to enrich PGR diversity, ICAR-NBPGR, New Delhi, India, imports every year ~70,000 samples of germplasm and trial material for research use both by public and private sectors. About 8000–10,000 accessions are also being added each year from indigenously collected/multiplied PGR to the base collection in the National Gene Bank. Therefore, the Division of Plant Quarantine at ICAR-NBPGR has developed procedures for systematic and stepwise processing for interception/ detection of associated plant pathogens and making exotic as well as indigenous PGR pest-free for quarantine clearance as well as conservation in National Gene Bank (Fig. 1) such as visual and stereo-binocular examination to detect presence of smut and bunt balls, ergot sclerotia, rust pustules, spores on the seed; washing test

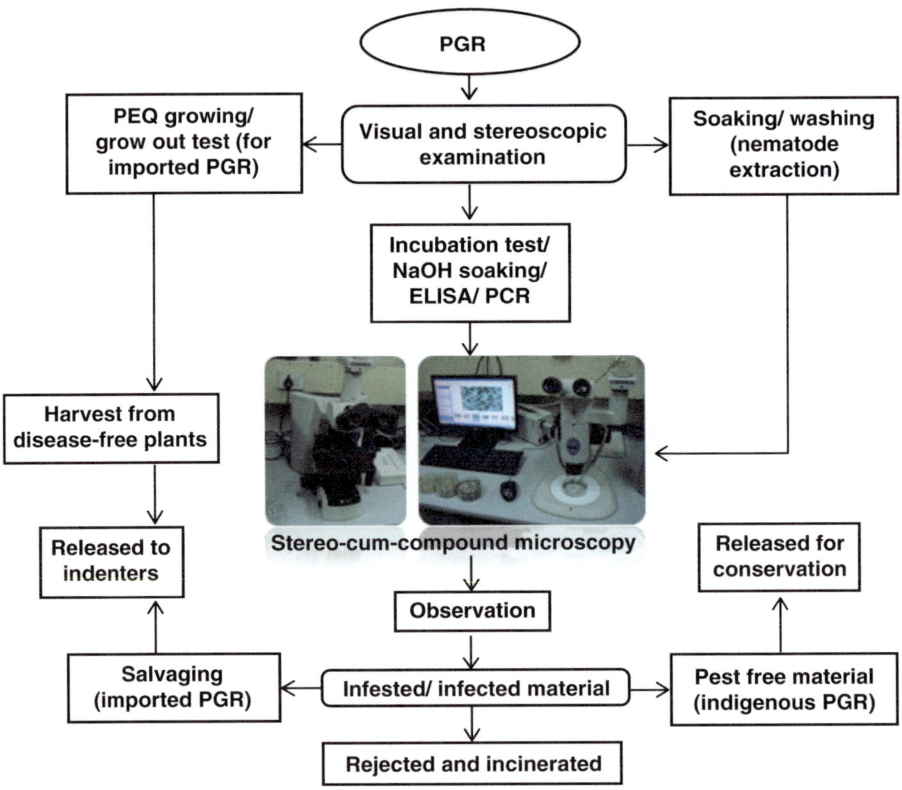

Fig. 1 Flowchart of seed health testing for making PGR disease-free

for presence of rusts and downy mildews; seed soaking/washing for nematode extraction; blotter method, immunological and molecular tests to detect seed-borne pathogens including fungi, bacteria and nematodes. Under soaking/washing test of samples for nematode detection, seeds as well as vegetative propagules are soaked in water for 24 h to extract nematodes, rooted materials washed with tap water to detach nematode adhering to root surface.

Seed health testing of PGR under quarantine processing and pest-free conservation in Gene Bank resulted in interception/detection of a variety of seed-borne/seed transmitted fungi, bacteria and nematodes of quarantine potential from different sources/countries during past three and half decade. Some of which are not yet reported/reported once in the country; those have wide host range and causing great economic losses; have more virulent/large number of physiological races, etc. Seed-borne/seed transmitted pathogen(s) may result in poor quality seed, loss in germination, development of epiphytotics and distribution of new strains or physiological races along with the seeds or other planting materials to new geographical areas.

Examination of Dry Seeds and Washing Test

Preliminary examination with naked eye or with the help of a magnifier to detect abnormalities such as discoloration, deformation shriveling, pigmentation, malformation of seeds with fungal growth like mycelial mats or fructifications like chlamydospores, acervuli, pycnidia, perithecia and other impurities associated with a seed lot such as sclerotia, bunts/smut balls, or spore masses, soil clods, plant debris, etc. and washing test for the presence of rusts and downy mildew spores are being followed. Dry examination of seeds/washing test results in the detection/interception of important pathogens such as *Claviceps purpurea* (ergot of cereals), *Peronospora manshurica* (downy mildew of soybean) and *Uromyces betae* (rust of sugarbeet), are not yet reported from India, and *Puccinia carthami* (rust of safflower) and *P. helianthii* (rust of sunflower) with restricted report in India. Some of those pathogens intercepted from different sources/countries are listed (Table 1).

Seed Soaking Method/NaOH Test

Seed soaking method can be used for detection of bunt (*Tilletia barclayna*) from rice seeds and Karnal bunt (*Tilletia indica*) from wheat seeds. For detection of *T. barclayna* (bunt) in rice, seeds are soaked in 0.2 % sodium hydroxide to soften the tissues. When soaked seeds are examined under stereo-binocular microscope, infected seeds show shiny jet black discolouration. The infected seeds when ruptured in a drop of water, release a stream of bunt spores.

Table 1 Selected pathogenic fungi bacteria and nematode intercepted in introduced germplasm

Fungi/bacteria/nematodes	Major host	Country/Source
Alternaria brassicae (Berk.) Sacc.[2]	*Brassica* spp.	Many countries
A. brassicicola (Schwein.) Wiltshire[2]	*Brassica* spp.	Many countries
A. padwickii (Ganguly) Ellis[2]	*Oryza sativa*	Many countries
Bipolaris oryzae (van Breda de Haan) Subram. & Jain[2]	*O. sativa*	Many countries
B. maydis (Nisikado & Miyabe) Subram. & Jain[2]	*Zea mays*	Many countries
B. sorokiniana (Sacc.) Subram. & Jain[2]	Several hosts	Many countries
Botrytis cinerea Pers.: Fr.[2]	Several hosts	Many countries
Colletotrichum dematium (Pers. ex Fr) Grove[2]	Many hosts	Many countries
C. gloeosporioides Penz. (Sacc.)[2]	Several hosts	Many countries
Dendryphion penicillatum (Corda) Fr.[#,2]	*Papaver* spp.	Germany and UK
Fusarium verticillioides (Sacc.) Nirenberg[2]	Many hosts	Many countries
F. solani (Martius) Sacc.[2]	Many hosts	Many countries
F. nivale Ces. ex Sacc.[Ø,1]	*Triticum aestivum* and *Hordeum vulgare*	GDR, Hungary, Italy, Mexico, Sweden, Turkey, UK and USA
Macrophomina phaseolina (Tassi) Goid[2]	Several hosts	Many countries
Peronospora manshurica (Naumov) Syd. Ex Gaum.[Ø,0,1]	*Glycine max*	Many countries
Puccinia carthami Corda[0,1]	*Carthamus* spp.	Many countries
Puccinia helianthi Schwein.[1]	*Helianthus annuus*	Many countries
Rhizoctonia solani Kuhn[2]	Several hosts	Many countries
Uromyces beticola (Bellynck) Boerema et al.[Ø,0,1]	*Beta vugaris*	Many countries
Ustilago nuda f.sp. *tritici* Schaffnit[3]	*Triticum* spp.	Many countries
U. nuda f.sp. *hordei* Schaffnit[3]	*H. vulgare*	Many countries
U. hordei (Pers.) Lagerh.[3]		
Xanthomonas campestris pv. *campestris* (Pammel) Dowson[2]	*Brassica* spp.	Many countries
X. vesicatoria (Doidge) Dowson[4]	*Solanum lycopersicum*	Thailand
Aphelenchoides besseyi[5]	*O. sativa*	Many countries
Helicotylenchus dihystera[5]	*Annona squamosa*	Taiwan
Pratylenchus penetrans[5]	*Malus domestica*	The Netherlands
Rotylenchus minutus[5,Ø]	*Hypoxis hemerocallidea*	Swaziland

SHT methods used: [0]visual examination; [1]washing test; [2]incubation test; [3]grow-out test; [4]ELISA; [5]seed soaking; [Ø]not yet reported from India; [#]only one report of occurrence in the country

Incubation Test

Incubation is a simple method commonly used for detection of mycoflora accompanied as mycelium, spores, or fruiting structures capable of growing on the seed during incubation of seed on wet blotter. Surface sterilization of the seeds using a 4 % sodium hypochlorite (NaOCl) solution is carried out before incubation to eliminate fast growing saprophytes if the seeds are heavily contaminated. Blotter test, generally referred as the standard blotter test, is the most efficient means of detecting a large number of seed-borne fungal pathogens. The examination under stereo-binocular microscope enables the observation of pathogens as developed on their hosts in situ, undisturbed and in a condition of natural growth. The fungi are identified on the basis of the growth and colour of fungal colonies. The identification is confirmed up to species level by making slides for examining the structure, size and colour of fruiting bodies/conidiophores/conidia under compound microscope. A critical stereoscopic and microscopic examinations of seeds on 8th day after incubation resulted in detection of seed-borne pathogens including fungi and bacteria viz. *Bipolaris maydis*, *Diplodia maydis*, *Fusarium oxysporum*, *Verticillium albo-atrum*; (with limited distribution); *Dendryphion penicillatum* (reported only once in the country); *B. sorokiniana*, *Botrytis cinerea*, *Colletotrichum capsici*, *C. graminicola*, *Macrophomina phaseolina*, *Phoma glomerata*, *P. herbarum*, *Rhizoctonia solani* (having wide host range); *Alternaria brassicae*, *A. brassicicola*, *A. helianthi*, *D. oryzae*, *F. verticillioides* (*F. moniliforme*), *F. solani*, *Sclerotinia sclerotiorum* (causing significant economic losses) and *Xanthomonas campestris* pv. *campestris*, *X. vesicatoria* (having physiological races), etc. Infected samples were either incinerated or salvaged prior to release depending on the category of pathogen(s) detected (Agarwal et al. 1998).

Grow-Out Test

The plants are observed for disease symptoms caused by various seed-borne pathogens including fungi, beyond seedling stage. This is generally undertaken by sowing seeds in post-entry quarantine (PEQ) greenhouses/PEQ nursery in isolation/containment facility. Plants are observed for disease symptoms for certain stipulated period. Most of the systemic infections of fungi can be detected by this method. Growing on procedure is of great importance in quarantine where imported plants are grown in confinement for a specified period of time or till the seed production in a glass house, screen house, poly house, or isolated field or an off-shore island that is established in accordance with guidelines/standards and are duly approved and certified by an inspection authority notified under Plant Quarantine (Regulation of Import into India) Order, 2003. Loose smut of wheat (*U. segetum* var. *tritici*), loose smut of barley (*Ustilago segetum* var. *nuda*) and covered smut of barley (*U. hordei*) can be detected in the pesticide treated exotic germplasm and international trial material grown in post-entry quarantine nursery during quarantine processing (Agarwal et al. 1989; Dev et al. 2003).

Immunological Techniques

For serological diagnosis of plant pathogens including fungi and bacteria, enzyme-linked immuno-sorbent assay (ELISA), a relatively simple, rapid and sensitive technique is used. ELISA has been shown to detect *Penicillium islandicum* in discolored rice grains (Dewey et al. 1990) and *Phomopsis longicolla* in soybean seeds (Gleason et al. 1987). Banks et al. (Wang and Yu 1998) have developed two monoclonal antibodies that could react with the antigens of several field and storage fungi. Indirect ELISA and polyclonal antibodies has been successfully used to detect *P. aurantiogriseum* var. *melanoconidium* in barley seeds. *Aspergillus parasiticus*, *Penicillium citrinum* and *Fusarium oxysporum* in rice and corn have been detected by DAS-ELISA (Kumar et al. 1998). Immuno-sorbent assays have also been demonstrated to be useful in detecting mycotoxin produced by seed-borne fungi such as *Aspergillus* spp., *Claviceps* spp., and *Fusarium* spp. In addition, seed immunoblot assay (SIBA) is able to detect viable *P. longicolla* propagules and is also effective in detecting *Tilletia indica* causing Karnal bunt disease in wheat seeds (Banks et al. 1992).

Bacterial pathogens are more amenable for detection by immunoassay. Both polyclonal and monoclonal antibodies specific to bacterial species/pathovars have been used for the detection of seed-borne bacterial pathogens. In comparison to polyclonal antibodies, monoclonal antibodies provide greater sensitivity and specificity for the immunological techniques that are particularly useful in detecting infection. Using polyclonal antiserum, *X. vesicatoria*, the causal agent of bacterial spot, stem and leaf blight in tomato and pepper, has been detected in *Lycopersicon esculentum* imported from Thailand (Dev et al. 2012). Agglutination tests have been adopted for the detection of *Pseudomonas syringae* pv. *phaseolicola* (Van Vuurde and Van den Bovenkamp 1981), *Xanthomonas campestris* pv. *phaseoli* (Trujillo and Saettler 1979) in bean seeds and *P. syringae* pv. *phaseoli* in peas (Ball and Reeves 1992). *P. syringae* pv. *phaseolicola* and *P. syringae* pv. *pisi* could be detected in bean and peas respectively (Lyons and Taylor 1990). The presence of *Erwinia stewartii* in maize seeds could be detected by applying ELISA (Lamka et al. 1991). Detection and identification of seed-borne *Ralstonia solanacearum* in tomato by using specific monoclonal antibodies (Rajeshwari et al. 1998) and *X. oryzae* pv. *oryzae* in rice seed (Gnanamanickam et al. 1994) could be done reliably.

The orthodox seed is the most prevalent form of exchanged or traded agricultural commodity which also play key role in spreading the seed-transmitted pests from one place to another. Exotic consignments of seeds, soil as contaminant of seeds and packing material may be contaminated with nematodes. Nematological techniques like visual examination, soaking and teasing of seeds and foliage, examination of a small quantity of accompanying soil clods and packing materials are being used in quarantine laboratories for detection of nematodes associated with the germplasm of agricultural crops (Lal and Lal 2006). During last two decades, the examination of samples revealed the presence of plant nematodes of quarantine importance, viz., *Anguina tritici*, *Aphelenchoides besseyi*, *A. arachidis*, *Ditylenchus angustus*, *D. destructor*, *D. dipsaci*, *Heterodera schachtii*, *Pratylenchus crenatus*, *P. penetrans*, *Rhadinaphelenchus cocophilus* and *Rotylenchus minutes* (Lal and Lal 2005, 2006; Khan et al. 2012). All of these species except *A. tritici*, *A. besseyi* and *P. penetrans*, are not reported to occur in India. Some of the nematodes intercepted are new host records and some are new geographical records. The results emphasize the importance of plant quarantine and necessitate processing all agricultural commodities (seeds, plants, planting material etc.) under exchange for detection of pests/pathogens including nematodes.

Molecular Technique

Detection and identification of the plant pathogen is of paramount importance. Development of molecular diagnostics were started after the introduction of polymerase chain reaction (PCR) in the mid 1980s and the first PCR-based detection of a pathogen in diseased plants was published in the beginning of 1990s (Rasmussen and Wulff 1991). PCR and Real-time PCR now facilitate high-throughput identification for many plant pathogens including fungi, bacteria and nematodes.

Many gene sequences of fungi, viz., Internal Transcribed Spacer (ITS) regions, IGS, tef-1α, β-tubulin and calmodulin can be used for the detection of fungi at species level and have been successfully applied in the characterization of several fungal species (Pileggi et al. 2009; Wang et al. 2010; Dubey et al. 2010, 2014; Durai et al. 2012; Ganeshamoorthi and Dubey 2013; Priyanka et al. 2014; Upadhyay et al. 2015). In contrast to conventional methods, samples can be tested directly for the presence of any pathogens and pathogens do not require isolation and culturing. These techniques are rapid, highly specific and can be used to detect minute quantities of DNA from PGR samples. PCR technology can also provide very accurate quantitative data with the necessary additional information required on PGR samples for quarantine processing or seed health testing for conservation.

Dot-blot hybridization has used for the detection of *Peronosclerospora sorghi* causing downy mildew of sorghum (Yao et al. 1990). *Plasmopara halstedii*, causing sunflower downy mildew in sunflower seeds (Says-Lesage et al. 2001), could be detected by PCR assay. *Tilletia indica*, causing Karnal bunt disease in wheat seeds (Frederick et al. 2000), *Rhynchosporium secalis*, causing scald disease in barley seeds, *Alternaria alternata*, *A. radicina*, and *A. dauci* in carrot seeds (Konstantinova et al. 2002), *A. brassicae*, *A. brassicola*, and *A. japonica*, causal agents of black spot in crucifers (Iacomi-Vasilescu et al. 2002) could be effectively detected by PCR assays. Furthermore, a number of *Fusarium* species, such as *F. culmorum*,

F. graminearum, *F. poae*, *F. crookwellense*, *F. sporotrichoides*, *F. sambucinum*, *F. avenaceum*, *F. trinctum* and *F. nivale*, causing fusarium head blight/scab disease in cereals have been differentiated by PCR assay using trichodiene synthase gene (Tri5) (Edwards et al. 2001).

Real-time PCR assay developed by Bluhm et al. (2004) could be used for the group specific detection of trichothecene and fumonisin-producing *Fusarium* spp. and for the identification of *F. graminearum* and *F. verticillioides* in barley and corn seeds. *A. brassicola* and *A. japonica* in radish, *A. alternata* in radish and cabbage, *Stemphylium botryosum*, *Penicillium* sp. and *Aspergillus* sp. in cabbage, and *Verticillium* sp. in tomato seeds have been detected by the quantitative real-time PCR. The nested-PCR assay is proved to be sensitive and specific for the detection of *Ustilaginoidea virens* using primers designed from rDNA ITS and the 5.8S rRNA gene (Zhou et al. 2003). DNA Detection Test StripsTM has been developed for detection of *F. graminearum* (Knoll et al. 2002). Similarly, rep-PCR (repetitive-sequence based polymerase characterization) assay was demonstrated to differentiation of *Tilletia* spp. and has shown the potential for application as a diagnostic tool (McDonald et al. 2000).

Restriction fragment length polymorphism (RFLP) analysis of PCR amplification products could be used to differentiate *Phomopsis longicolla* and *Diaporthe phaseolorum*, both causing seed decay of soybean (Zhang et al. 1997). Amplified fragment length polymorphism (AFLP) could also be employed to detect *Claviceps africana*, causing sorghum ergot (Tooley and Englander 2002). Genetic diversity analysis of different seed-borne pathogens namely *Alternaria alternata* (Kandan et al. 2014), *Bipolaris oryzae* (Kandan et al. 2015) using different molecular markers viz. universal rice primers (URPs), inter-simple sequence repeats (ISSR) and RAPD-PCR was able to group the pathogenic isolates based on their geographical origin.

Specific DNA probes made it possible to detect, identity, differentiate, and quantify the seed-borne bacterial pathogens very rapidly and reliably. DNA hybridization technique has been adopted for the detection of *Pseudomonas syringae* pv. *phaseolicola*, *Xanthomonas campestris* pv. *phaseoli* in beans, *X. oryzae* pv. *oryzae*, *X. oryzae* pv. *oryzicola* and *P. glumae* in rice seed (Schaad et al. 1995; Gilbertson et al. 1990; Cottyn et al. 1994). A rapid and sensitive PCR-based protocol was also developed for detection of *X. campestris* pv. *phaseoli* (Xcp) causing bean common blight disease and *P. syringae* pv. *phaseolicola* causing halo blight disease in bean (Audy et al. 1996). PCR along with primer pairs developed from sequences of cloned random amplified polymorphic DNA (RAPD) fragments could be able to detect *X. campestris* pv. *carotae* (Xcc), causing bacterial leaf blight disease in carrot seeds, leaves, and stem tissues (Meng et al. 2004).

Several Real-time PCR protocols have also been developed to detect and identify several phytopathogenic bacteria from seeds including *Xanthomonas* spp., *Clavibacter michiganensis* subsp. *sepedonicus*, *Ralstonia solanacearum* and *Agrobacterium* spp. (Weller et al. 2007). Using restriction fragments length polymorphism (RFLP) analysis, strains of *X. oryzae* pv. *oryzae* (*Xoo*) from rice were determined (Leach and White 1991; Yashitola et al. 1997). Polymerase chain reaction (PCR) alone or in combination with other molecular diagnostic methods has been very useful for the detection of seed-borne pathogens which may also be adopted for detection of pathogens in other planting materials. The BIO-PCR technique developed by Schaad et al. is capable of detecting DNA sequences of target bacteria such as *P. syringae* pv. *phaseolicola* (Mosqueda-Cano and Herrera-Estrella 1997).

For identification of nematodes, morphological characters supplemented with molecular diagnostics can be the best option, which can be used in decision making regarding quarantine clearance of germplasm. Molecular method becomes useful for identification especially when less number or immature stages of nematodes are intercepted, which otherwise are difficult to identify by morphological characters. The molecular methods based on DNA are gaining importance in recent time. The methods like restriction fragment length polymorphisms (RFLPs), satellite DNA probes, PCR and real-time PCR (RT-PCR) are being used for identification of nematodes (Blok and Powers 2009). The Internal Transcribed Spacer (ITS) region I and II of ribosomal DNA genes is used as a taxonomic marker for identification of nematodes (Powers et al. 1997). The PCR-RFLP profile of the region has been used for comparison of different species of nematode. The molecular identification methods for few species of nematodes are available for their identification. e.g. *Globodera pallida* and *Bursaphelenchus xylophilus* (Skantar et al. 2007; Ye and Giblin-Davis 2013).

Conclusion

Conventional methods for detection and identification of seed-borne plant pathogens often rely on symptoms, morphology, cultural, physiological and biochemical characteristics, etc. Among these, the most useful and widely used procedure are based on symptoms and morphological characters. These methods, although the cornerstone of pathogen diagnostics, may lead to improper identification due to contamination resulting incorrect diagnosis leading to ineffective disease management. Thus, the accurate identification and early detection of plant pathogens are the cornerstones of successful disease management. The morphological identification of plant pathogens is often difficult and time-consuming and requires extensive knowledge of taxonomy and experience in recognizing detailed fungal features. Compared to conventional diagnostic methods, molecular methods offer the possibility of faster, more reliable and efficient techniques. PCR based technique offers several advantages, because organisms do not need to be cultured prior to detection; moreover it is highly sensitive, relatively simple and faster to perform. There has been a shift towards DNA-based protocols developed for diagnostic purpose which will be helpful in decision making with respect to phytosanitary requirements for import/ export of seed and other planting material. Further, comparative studies of morpho-anatomical characters/features supplemented by molecular techniques in a range of closely related species should be made to evaluate their importance in taxonomic studies.

References

Agarwal PC, Nath R, Majumdar A, Khetarpa RK, Usha D, Lambat AK, Kaur P, Varshney JL, Singh S (1989) Quarantine processing of chemically treated wheat, barley and triticale germplasm. Indian J Agric Sci 59:42–43

Agarwal PC, Singh B, Gautam PL (1998) Plant quarantine as an aid to safer exchange, conservation and utilization of plant genetic resources. Indian J Pl Genet Resour 11(2):165–171

Audy P, Braat CE, Saindon G, Huang HC, Laroche A (1996) A rapid and sensitive assay for concurrent detection of bacteria causing common and halo blights in bean seed. Phytopathology 86:361–366

Ball S, Reeves JC (1992) Application of rapid techniques to seed health testing–prospects and potential. In: Duncan JM, Torrance L (eds) Techniques for the rapid detection of plant pathogens. Blackwell Scientific Publications, Oxford, pp 193–207

Banks JN, Cox SJ, Clarke JH, Shamsi RH, Northway BJ (1992) Towards the immunological detection of field and storage fungi. In: Samson RA, Hocking AD, Pitt JI, Kind AD (eds) Modern methods in food mycology. Elsevier, Amsterdam, pp 247–252

Blok VC, Powers TO (2009) Biochemical and molecular identification. In: Perry RN, Moens M, Starr JL (eds) Root-knot nematodes. CABI Publishing, Wallingford, pp 98–118

Bluhm BH, Cousin MA, Woloshuk CP (2004) Multiplex real-time PCR detection of fumonisin producing and trichothecene-producing groups of *Fusarium* species. J Food Protect 67:536–543

Cottyn B, Bautista AT, Nelson RJ, Leach JE, Swings J, Mew TW (1994) Polymerase chain reaction amplification of DNA from bacterial populations of rice using specific oligonucleotide primers. Int Rice Res Note 19:30–32

Dev U, Singh B, Agarwal PC, Singh S, Maurya AK, Chand D, Joshi KD, Khetarpal RK (2003) Quarantine processing of chemically treated wheat (*Triticum* species), barley (*Hordeum vulgare*) and triticale germplasm imported during 1992–2001. Ind J Agric Sci 73:462–465

Dev U, Singh B, Agarwal PC, Chand D, Chalam VC, Maurya AK, Joshi KD (2012) Biosecurity concerns with respect to fungal and bacterial pathogens in germplasm introduced into India during 2007–10. Indian J Agric Sci 82:1083–9

Dewey FM, Macdonald MM, Phillips SI, Priestley RA (1990) Development of monoclonal-antibody-ELISA and monoclonal-antibody-dip-stick immunoassays for *Penicillium islandicum* in rice grains. J Gen Microbiol 136:753–760

Dubey SC, Tripathi A, Singh RS (2010) ITS-RFLP fingerprinting and molecular marker for detection of *Fusarium oxysporum* f. sp. *ciceris*. Folia Microbiol 55:629–634

Dubey SC, Kumari P, Upadhyay BK (2014) Development of molecular markers and probes based on TEF-1α, β-tubulin and ITS gene sequences for quantitative detection of *Fusarium oxysporum* f. sp. *ciceris* by using real-time PCR. Phytoparasitica 42:355–366

Durai M, Dubey SC, Tripathi A (2012) Genetic diversity analysis and development of SCAR marker for detection of Indian populations of *Fusarium oxysporum* f. sp. *ciceris* causing chickpea wilt. Folia Microbiol 57:229–235

Edwards SG, Pirgozliev SR, Hare MC, Jenkinson P (2001) Quantification of trichothecene-producing *Fusarium* species in harvested grain by competitive PCR to determine efficacies of fungicides against Fusarium head blight of winter wheat. Appl Environ Microbiol 67:1575–1580

Frederick RD, Synder KE, Tooley PW, Berthier-Schaad Y, Peterson GL, Bonde MR, Schaad NW, Knorr DA (2000) Identification and differentiation of *Tilletia indica* and *T. walkeri* using the polymerase chain reaction. Phytopathology 90:951–960

Ganeshamoorthi P, Dubey SC (2013) Phylogeny analysis of Indian strains of *Rhizoctonia solani* isolated from chickpea and development of sequence characterized amplified region (SCAR) marker for detection of the pathogen. Afr J Microbiol Res 7:5516–5525

Gilbertson RL, Rand RE, Hagerdorn DJ (1990) Survival of *Xanthomonas campestris* pv. *phaseoli* and pectolytic strains of *Xanthomonas campestris* in bean debris. Plant Dis 74:322–327

Gleason ML, Ghabrial SA, Ferriss RS (1987) Serological detection of *Phomopsis longicolla* in soybean seeds. Phytopathol 77:371–375

Gnanamanickam SS, Shigaki T, Medalla ES, Mew TW (1994) Problem in detection of *Xanthomonas oryzae* pv. *oryzae* in rice seed and potential for improvement using monoclonal antibodies. Plant Dis 78:173–178

Iacomi-Vasilescu B, Blancard D, Guénard M, Molinero-Demilly V, Laurent E, Simoneau P (2002) Development of a PCR- based diagnostic assay for detecting pathogenic *Alternaria* species in cruciferous seeds. Seed Sci Technol 30:87–95

Kandan A, Akhtar J, Singh B, Dev U, Goley R, Chand D, Roy A, Rajkumar S, Agarwal PC (2014) Genetic diversity analysis of *Alternaria alternata* isolates infecting different crops using URP and ISSR markers. Indian J Plant Protect 42:229–236

Kandan A, Akhtar J, Singh B, Dixit D, Chand D, Roy A, Rajkumar S, Agarwal PC (2015) Molecular diversity of *Bipolaris oryzae* infecting *Oryza sativa* in India. Phytoparasitica 43:5–14

Khan Z, Mahamood M, Lal A, Akhtar J, Parakh DB (2012) Interception of an exotic plant parasitic nematode (*Rotylenchus minutus*) on African potato (*Hypoxis hemerocallidea*) from Swaziland. Indian J Agric Sci 82:1090–1092

Khetarpal RK, Lal A, Varaprasad KS, Agarwal PC, Shashi B, Chalam VC, Gupta K (2006) Quarantine for safe exchange of plant genetic resources. In: Singh AK, Srinivasan K, Saxena S, Dhillon BS (eds) Hundred years of plant genetic resources management in India. National Bureau of Plant Genetic Resources, New Delhi, pp 83–108

Knoll S, Vogel RF, Niessen L (2002) Identification of *Fusarium graminearum* in cereal samples by DNA Detection Test Strips™. Lett Appl Microbiol 34:144–148

Konstantinova P, Bonants PJM, van Gent-Pelzer MPE, van der Zouwen P, van den Bulk R (2002) Development of specific primers for detection and identification of *Alternaria* spp. in carrot material by PCR and comparison with blotter and plating assays. Mycol Res 106:23–33

Kumar A, Singh A, Garg GK (1998) Development of seed immunoblot binding assay for detection of Karnal bunt (*Tilletia indica*) of wheat. J Plant Biochem Biotech 7:119–120

Lal A, Lal R (2005) Nematodes intercepted in introduced germplasm of horticultural crops. Indian J Plant Protect 33:282–285

Lal R, Lal A (2006) Plant parasitic nematodes intercepted from seeds, soil clods and packing material under import quarantine. J New Seeds 8:49–60

Lamka GL, Hill JH, McGee DC, Braun EJ (1991) Development of an immunosorbent assay for seed-borne *Erwinia stewartii* in corn seeds. Phytopathol 81(8):839–846

Leach JE, White FW (1991) Molecular probes for disease diagnosis and monitoring. In: Khush GS, Toenniessen GH (eds) Rice biotechnology. CAB International and The International Rice Research Institute, Wallingford, pp 281–307

Lyons NF, Taylor JD (1990) Serological detection and identification of bacteria from plants by conjugated *Staphylococcus aureus* slide agglutination test. Plant Pathol 39:584–590

McDonald JG, Wong E, White GP (2000) Differentiation of *Tilletia* species by rep-PCR genomic fingerprinting. Plant Dis 84:1121–5

Meng XQ, Umesh KC, Davis RM, Gilbertson RL (2004) Development of PCR based assays for detecting the carrot bacterial leaf blight pathogen *Xanthomonas campestris* pv. *carotae* from different substrates. Plant Dis 88:1226–1234

Mosqueda-Cano G, Herrera-Estrella L (1997) A simple and efficient PCR method for the specific detection of *Pseudomonas syringae* pv *phaseolicola* in bean seeds. World J Microbiol Biotechnol 13:463–467

Pileggi SAV, de Oliveira SFV, Andrade CW, Vicente VA, Dalzoto PR, da Cruz GK, Gabardo J, Massola NS, Tozze HJ, Pileggi M, Kava-Cordeiro V, Galli-Terasawa LV, Pimentel IC, Stringari D, Glienke C (2009) Molecular and morphological markers for rapid distinction between 2 *Colletotrichum* species. Can J Microbiol 55:1076–1088

Powers TO, Todd TC, Burnell AM, Murray PC, Fleming CC, Szalanski AL, Adams BA, Harris TS (1997) The rDNA Internal transcribed spacer region as a taxonomic marker for nematodes. J Nematol 29:441–450

Priyanka K, Dubey SC, Singh AK (2014) Intergenic spacer region based marker for identification and quantification of *Fusarium oxysporum* f. sp. *ciceris* in chickpea plant using real time PCR assay. Res J Biotech 9:36–40

Rajeshwari N, Shylaja MD, Krishnappa M, Shetty HS, Mortensen CN, Mathur SB (1998) Development of ELISA for detection of *Ralstonia solanacearum* in tomato: its application in seed health testing. World J Microbiol Biotech 14:697–704

Rasmussen OF, Wulff BS (1991) Detection of *Pseudomonas syringae* pv *pisi* using PCR. In: Proceedings 4th international working group on *Pseudomonas syringae pathovars*. Kluwer Academic Publishers, Dordrecht, pp 369–376

Richardson MJ (1990) An annotated list of seed-borne diseases, 4th edn. International Seed Testing Association, Zurich, p 387

Says-Lesage V, Meliala C, Nicolas P, Roeckel-Drevet P, De Labrouhe DT (2001) Molecular diagnosis of mildew (*Plasmopara halstedii*) infection in sunflower seeds. OCL-OL-CORP 8(3):258–260

Schaad NW, Cheong SS, Tamaki E, Hatziloukas E, Panopulos NJA (1995) Combined biological and enzymatic amplification (BIO-PCR) technique to detect *Pseudomonas syringae* pv. *phaseolicola* in bean seed extracts. Phytopathology 85:243–248

Skantar AM, Handoo ZA, Carta LK, Chitwood DJ (2007) Morphological and molecular identification of *Globodera pallida* associated with potato in Idaho. J Nematol 39(2):133–144

Stakman EC, Harrar JG (1957) Principles of plant pathology. Ronald Press, New York

Tooley PW, Englander L (2002) Infectivity of *Phytophthora ramorum* on selected ericaceous host species. Phytopathology 92:S81

Trujillo GE, Saettler AW (1979) A combined semi-selective medium and serology test for the detection of *Xanthomonas* blight bacteria in bean seed. J Seed Tech 4:35–41

Upadhyay BK, Dubey SC, Singh R, Tripathi A (2015) Diversity analysis and comparison of ITS and SCAR based molecular markers to detect *Rhizoctonia solani* using conventional and real-time PCR. J Pure Appl Microbiol 9:1281–1293

Van Vuurde JWL, Van den Bovenkamp GW (1981) Routine methods for the detection of halo-blight (*Pseudomonas phaseolicola*) in beans. Report on the 17th international workshop on seed pathology, ISTA, Zurich, p 22

Wang YC, Yu RC (1998) Detection of toxigenic *Aspergillus* spp. in rice and corn by ELISA. J Chin Agric Chem Soc 36:512–520

Wang Y, Geng Y, Pei YF, Zhang XG (2010) Molecular and morphological description of two new species of *Stemphylium* from China and France. Mycologia 102:708–717

Weller SA, Bresford-Jones NJ, Hall J, Elphinstone JE, Thwaites R, Parkinson N (2007) Detection of *Xanthomonas fragariae* and presumptive detection of *Xanthomonas arabicola* pv. *fragariae*, from strawberry leaves by real-time PCR. J Microbiol Methods 70:379–83

Yao CL, Frederiksen RA, Magill CW (1990) Seed transmission of sorghum downy mildew: Detection by DNA hybridization. Seed Sci Technol 18:201–207

Yashitola J, Krishnaveni D, Reddy APK, Sonti RV (1997) Genetic diversity within the population of *Xanthomonas oryzae* pv. *oryzae* in India. Phytopathology 87:761–765

Ye W, Giblin-Davis RM (2013) Molecular characterization and development of real-time PCR assay for pine-wood nematode *Bursaphelenchus xylophilus* (Nematoda: Parasitaphelenchidae). PLoS ONE 8(11):e78804. doi:10.1371/journal.pone.0078804

Zhang AW, Hartman GL, Riccioni L, Chen WD, Ma RZ, Pedersen WL (1997) Using PCR to distinguish *Diaporthe phaseolrum* and *Phomopsis longicolla* from other soybean fungal pathogens to detect them in soybean tissue. Plant Dis 81:1143–1149

Zhou YL, Izumitsu K, Sonada R, Nakazaki T, Tanaka E, Tsuda M, Tanaka C (2003) PCR-based specific detection of *Ustilaginoidea virens* and *Ephelis japonica*. J Phytopathol 151:513–518

Aflatoxigenic Fungi in Food Grains: Detection, Its Impact on Handlers and Management Strategies

Sana Ali

Introduction

Fungal deterioration of stored seeds and grains is a common problem especially in tropical hot and humid climate. Harvested grains are infested by various species of *Aspergillus*, leading to their damage and mycotoxin production (Reddy et al. 2008).

There are hundreds of mycotoxins known, but only few have been extensively studied and even fewer have good methods of analysis. They can be produced in the field, during handling, and in storage. Mycotoxicoses are diseases caused by exposure to foods or feeds contaminated with mycotoxins (Nelson et al. 1993).

Aflatoxins are naturally occurring mycotoxins that are produced primarily by some strains of *Aspergillus flavus* and by most strains of *A. parasiticus*, and some other species like *A. nomius*, *A. bombycis*, *A. ochraceoroseus*, *A. pseudotamari* are also aflatoxin producing species but they are less common (Goto et al. 1996; Klich et al. 2000; Peterson et al. 2001). Economically the most important aflatoxin producer is *A. flavus* (Baranyi et al. 2013).

Aflatoxins are difuranocoumarin derivatives produced by a polyketide pathway. B_1, B_2, G_1, and G_2 are major aflatoxins named on the basis of their fluorescence under UV light (blue or green) and relative chromatographic mobility during thin-layer chromatography. However, various other aflatoxins (e.g., M1, M2 P_1. Q_1, B_{2a}, and G_{2a}) are known (Bbosa et al. 2013).

Aflatoxin B_1 is the most potent natural carcinogen known (Squire 1981) and is one of the most toxic mycotoxins (Passone et al. 2010; Sardiñas et al. 2011).

S. Ali
Department of Microbiology, Janakpuri Superspeciality Hospital, New Delhi, India
e-mail: sana.ali2k4@gmail.com

P. Kumar et al. (eds.), *Current Trends in Plant Disease Diagnostics and Management Practices*, Fungal Biology, DOI 10.1007/978-3-319-27312-9_20

Toxigenic Potential/Factors Affecting Toxin Production

There are differences in the toxigenic abilities even between different strains of each aflatoxigenic species. Only about half of *Aspergillus flavus* strains produce aflatoxins (Klich and Pitt 1988), while those that do may produce more than 10^6 μg/kg (Cotty et al. 1994).

Fungal growth as well as mycotoxin production are related to weather extremes (leading to plant stress or excess hydration of stored feed grains), improper storage and poor food grain quality. Adverse environmental conditions leading to plant stress predisposes plants in the field to mycotoxin contamination. Temperature, moisture content, and insect activity are the major post harvest factors affecting mycotoxin contamination of feed grains and foods (Coulumbe 1993). Humans are exposed directly to aflatoxin and other mycotoxins through consumption of contaminated foods. Handling such contaminated feed causes exposure of mycotoxins through the skin and by inhalation (Schiefer 1990). Indirect exposure to aflatoxins occurs through foods mainly milk, liver, and eggs derived from animals that consume contaminated feedstuffs (Hayes 1980).

Aspergillus species usually grow when water content is less and temperature is high as compared to Fusarium species. Hot and humid weather encourages aflatoxin formation (Atanda et al. 2011). The optimal temperature for aflatoxin production by *A. flavus* is 24–30 °C and aflatoxin production decreases as temperatures increase above these levels (Klich 2007). Hence, *A. flavus* and aflatoxin production are usually seen in corn grown in the heat and drought stress. Whereas *Penicillium* species grow at relatively low water activities and low temperatures. Since both *Aspergillus* and *Penicillium* can grow at low water activities, they are included in storage fungi (Christensen et al. 1977). In storage, usually the most important variables are the moisture content of the substrate and the relative humidity of the surroundings (Detroy et al. 1971; Wilson and Payne 1994).

Before the 1970s, aflatoxin contamination of corn was believed to originate after harvest. Improperly stored corn can become contaminated with aflatoxin (Lillehoj and Fennell 1975). However, after aflatoxin was identified in corn before harvest, it is now known that most of the aflatoxin problem in corn originates in the field. Growth of *A. flavus* can occur at 86–87 % equilibrium relative humidity (RH) (Davis and Diener 1983). Field infection of corn with *A. flavus* (Wicklow 1983) occurs when temperatures are high and there is drought stress.

Aflatoxin Contamination

Many substrates support growth and aflatoxin production by aflatoxigenic molds. Aflatoxin contamination of corn, peanuts, tree nuts, cottonseed, and other commodities is a continuing worldwide problem.

Natural contamination of cereals, figs, nuts, tobacco, and various other commodities is a common occurrence (Diener et al. 1987). Crops may be contaminated with aflatoxin in the field even before harvest, especially due to drought stress (Diener et al. 1987); however the fate of crops stored under conditions that favor mold growth is more troublesome.

A. flavus also produces sclerotia, or resting bodies in some circumstances which contain indole alkaloids like aflatrem (Wicklow 1983). Cyclopiazonic acid (CPA), a toxic indole tetramic acid, is also produced by *A. flavus* (CAST Council for Agricultural Science and Technology 1989). However, their role in aflatoxicoses is not well known. *A. flavus* is the predominant fungus in aflatoxin-contaminated corn and cottonseed while *A. parasiticus* is probably more common in peanuts compared to corn (Davis and Deiner 1983).

Corn is susceptible to *A. flavus* infection via the silks (Marsh and Payne 1984) and these stress conditions during anthesis (pollination) results in preharvest contamination of corn with aflatoxin. Early harvest of crop and a decrease in late-season irrigation can decrease contamination.

Detection

Consumption of aflatoxins even at low concentration level creates serious health related problems. Hence, it is imminent to develop new methodologies to detect and quantify the aflatoxins in order to meet the restrictions and legislations assigned for controlling these carcinogenic compounds. Frequent analytical surveillance programs by food controlling agencies necessitate controlling aflatoxin contamination of food grains and ensuring food safety in order to protect health of exposed people.

Several analytical techniques are available for detection and quantification of aflatoxins. These methods can be broadly classified as chromatographic, spectroscopic, electrochemical and immunochemical techniques. All these techniques have their own benefits and limitations.

In fact, these methods need well equipped laboratories, trained personnel, harmful solvents and are time consuming. Therefore, novel methods like biosensors, electrokinetics, electrochemical transduction, amperometric detection, and adsorptive stripping voltammetry have been developed recently (Ali 2014).

Aflatoxins Extraction from Food Samples

Solid phase extraction is one of the significant purification steps. Test extracts are cleaned up before analysis (by thin layer or column liquid chromatography) to remove coextracted materials that are supposed to interfere during the determination of target analytes. The selection of proper solvent for extraction is important as it enhances the specificity of the procedure and isolate the analyte of interest from interfering species.

Aflatoxins are usually soluble in protic polar solvents like methanol, acetone, acetonitrile etc. Hence, such organic solvents are used for aflatoxin extraction (Bertuzzi et al. 2012). The extraction of aflatoxins is usually followed by a cleanup step like immunoaffinity column (IAC) chromatography (Ma et al. 2013). The crude sample extract is applied to the immunoaffinity column containing antibodies specific to aflatoxin immobilized on a solid support (e.g. agarose or silica). When the crude sample moves down the column, the aflatoxin binds to the antibody and is retained onto the column. This is followed by washing to remove impurities and unbound particles. The aflatoxin is ultimately recovered by using an appropriate solvent which can break the bond between the antibody and the aflatoxin. For this purpose acetonitrile has been found most suitable solvent.

Chromatography

The term "chromatography" first introduced by Russian Botanist M. Tswett in 1903 has been one of the most popular methods for selective selection of aflatoxins. The most popular techniques are high performance liquid chromatography (HPLC), gas chromatography (GC), liquid chromatography (LC), and thin layer chromatography (TLC). Fluorescence detection methods coupled with these chromatographic techniques have been found highly sensitive for analysis of aflatoxins (Cavaliere et al. 2006). However, chromatographic techniques often require skilled manpower, extensive sample pretreatment and expensive equipments (Sapsford et al. 2006).

Thin-Layer Chromatography

Thin layer chromatography (TLC), also called as flat bed chromatography or planar chromatography is one of the most widely used separation techniques in the field of biochemical analysis including aflatoxins. This technique has experienced a dramatic surge since its inception in 1938 by Izmailov and Schraiber who separated components present in medicinal plants using thin layer of aluminium oxide (Izmailov et al. 1938).

It has been the method of choice for detection and quantification of aflatoxins at levels of even 1 ng/g (Stroka and Anklarn 2000). TLC is based on the separation of compounds by how far they migrate through selected stationary phase with a specific solvent. The distance that a compound will travel is a unique identifier for specific compounds, and a retention factor (R_F) has been determined for most mycotoxins (Ali 2014).

Advantage of TLC lies in the fact that more than one mycotoxin can be detected for each test sample. However, a positive control containing purified mycotoxins needs to be run in parallel to ensure accuracy, since different chemicals can have a similar R_F values. TLC is the fast, inexpensive and versatile separation technique with many practical considerations that contribute to its effectiveness.

Column Liquid Chromatography

Column liquid chromatography (LC) differs from TLC, as in TLC stationary phase is in planar form. However, both methods are complementary to each other. For an analyst, use of TLC for preliminary work in order to optimize LC separation conditions is not unusual. Liquid chromatography methods for the determination of aflatoxins in foods include normal-phase LC (NPLC), reversed-phase LC (RPLC) with pre- or before-column derivatization (BCD), RPLC followed by postcolumn derivatization (PCD), and RPLC with electrochemical detection.

Immunochemical Methods

Planar and column LC methods for determining aflatoxins in food are laborious and time consuming. Often, these techniques require knowledge and experience of chromatographic techniques to solve separation and interference problems. Through advances in biotechnology, highly specific antibody-based tests are now commercially available. These tests are based on the affinities of the monoclonal or polyclonal antibodies for aflatoxins.

Small molecules, such as mycotoxins, are non-immunogenic and are known as haptens or molecules that will not stimulate antibody production by themselves. However, antibodies can be produced for a specific mycotoxin by conjugating it to a protein carrier, which causes the mycotoxin to become immunogenic. The various known forms of antibodies include polyclonal and monoclonal types. Polyclonal antibodies react with multiple antigens or haptens on a foreign compound. Conversely, monoclonal antibodies react only with specific antigens or haptens. Currently, both polyclonal and monoclonal antibodies have been developed that are available for identifying several types of mycotoxins in test samples by utilizing the ELISA and immuno-affinity chromatography (IAC). The three types of immunochemical methods are:

Radioimmunoassay (RIA)

It is a highly sensitive *in-vitro* assay technique used to measure concentrations of antigens with the aid of antibodies. Though, RIA technique, requiring specialized equipment is extremely sensitive and specific but it requires special precautions and licensing, since radioactive substances are used. The interesting feature of RIA technique is its capability to analyse multiple analytes simultaneously. However, the need of high purity antigen and use of radioactive isotopes has limited its frequent use in the analysis of aflatoxins. Today it has been supplanted by the ELISA method, where the antigen-antibody reaction is measured using colorimetric signals instead of a radioactive signal.

Enzyme-Linked Immunosorbent Assay (ELISA)

ELISA can be performed to evaluate either the presence of antigen or the presence of antibody in a sample. ELISAs are performed in 96-well plates. A general ELISA procedure involves the following five-steps: (1) Coating of the microtiter plate wells with antigen; (2) blocking of all unbound sites to prevent false positive results; (3) adding of primary antibody to the wells; (4) putting of secondary antibody conjugated to an enzyme and (5) the reaction of a substrate with the enzyme to produce a colored product, thus indicating a positive reaction.

ELISA methods for mycotoxin assay have been available for more than a decade. The technology is based on the ability of a specific antibody to distinguish the three-dimensional structure of a specific mycotoxin. The direct competitive ELISA is commonly used in mycotoxin analysis. Though, it requires multiple washing steps but safe to use as there is no inherent health hazards associated with enzyme levels.

Immunoaffinity Column Assay (ICA)

The IAC contains anti-mycotoxin antibody that is immobilized onto a solid support such as agarose gel in phosphate buffer. The sample extract is applied to an IAC. The mycotoxin binds to the antibody and water is passed through the column to remove impurities. By passing a polar solvent such as methanol through the column, the captured mycotoxin is removed from the antibody and eluted from the column. The mycotoxin in eluent is then detected by addition of a chemical substance to either enhance the fluorescence or render the mycotoxin fluorescent before measuring in a fluorometer.

Other Methods

Fluorescence Spectrometry

Fluorescence spectrometry is an important analytical technique for the detection of substances that emit energy at specific wavelengths. All the aflatoxins have a maximum absorption around 360 nm (Akbas and Ozdemir 2006). Letters ‘B’ and ‘G’ of the aflatoxins refer to its blue (425 nm) and green–blue (450 nm) fluorescence colours produced by these compounds under Ultra Violet (UV) light. The fluorescence emission of the G toxin is more than 10 times greater than that for the B toxin (Alcaide-Molin et al. 2009). In general, aflatoxins in the range of 5–5000 ppb can be detected by using this technique.

Ultra-violet Absorption

It has been reported that all the aflatoxins have a maximum absorption around 360 nm with a molar absorptivity of about 20,000 cm^2/mol (Akbas and Ozdemir 2006). Though, aflatoxins could be detected by UV absorption methods but the sensitivity of such systems is not sufficient to detect these compounds at the parts per billion (ppb) levels that are required for food analyses (Alcaide-Molin et al. 2009). The detection limit of UV sensors reaches micromolar ranges (Couderc et al. 1998). This is why fluorescence (FL) techniques have become more popular for AFs detection.

To improve the sensitivity, UV absorption technique is usually combined with HPLC systems. However, HPLC-UV systems are not as sensitive as HPLC-FL systems, especially for detection of AFs at very low levels (Herzallah 2009). HPLC-UV systems have been considered to be accurate, precise, and reliable for rapid determination of aflatoxins in food samples.

Ion-Mobility Spectrometry

The ion-mobility spectrometry is a fascinating technique that is used for the characterization of chemicals on the basis of speed acquired by the gas-phase ions in an electric field. To detect aflatoxins, the sample, is evaporated and mixed with a carrier gas before feeding into the ion mobility spectrometer (IMS) where the mixture is ionized and passed through an electric field gradient, where ions of different substances will travel at different speeds. Using this technique, Sheibani et al. (2008) have quantified aflatoxins at nanogram level (0.25 ng).

Confirmation of Identities of the Aflatoxins

Although analytical methods might consist of different extraction, clean-up, and quantitation steps, the results of the analyses by such methods should be similar when the methods are applied properly. Since the reliability of the quantitative data is not in question, the problem still to be solved is the confirmation of identity of the aflatoxins. The confirmation techniques used involve either chemical derivatization or mass spectrometry (MS).

Monitoring Techniques for Assessing Exposure to Aflatoxins in Humans

New techniques have been devised to monitor individual exposures to aflatoxins accurately especially the analysis of aflatoxin DNA adducts and albumin adducts as surrogates for genotoxicity in humans. The major reactive metabolite is the exo-AFB1 8,9-epoxide, which if not detoxified, can bind to double-stranded DNA to form the promutagenic AFB1-N7-guanine adduct or, following hydrolysis to the AFB1- dihydrodiol, with proteins such as albumin (Baertschi et al. 1988; Johnson et al. 1996). Both urinary AFB1-N7-guanine and serum AF-albumin levels are correlated with dietary intake of aflatoxins (Wild et al. 1990). Since, the half-life of human albumin is about 20 days, Aflatoxin-albumin may theoretically accumulate following chronic exposure to reach levels 30-fold higher than that found after a single dose (Wild et al. 1996). In contrast, urinary AFB1-N7-guanine excretion will be proportionate to intake over the previous few days. Hence, the Aflatoxin-albumin adduct has been widely used to assess exposure in epidemiologic studies, including intervention studies (Turner et al. 2005). The only AFB1 adduct structurally identified in enzymatically digested plasma albumin is AFB1-lysine (Guengerich et al. 2002). This adduct has been measured by ELISA, high-performance liquid chromatography (HPLC) with fluorescence detection and more recently by isotope dilution mass spectrometry (IDMS) (Wang et al. 1996). Among these ELISA has been the method of choice. The sensitivities of the ELISA and HPLC fluorescence assays are comparable, whereas the IDMS method is approximately 10-times more sensitive. HPLC fluorescence, and IDMS methods only detect AF-lysine, the ELISA will probably measure a broader range of AF adducts. AF-albumin adduct levels measured by ELISA and AFlysine adducts measured by HPLC fluorescence in the same human serum samples showed an excellent correlation, but the ELISA indicated ~11-fold higher adduct burdens (Wild et al. 1992).

Wild et al. (1986) used highly sensitive immunoassays for aflatoxins quantification in human body fluids. An enzyme linked immunosorbent assay (ELISA) was used to quantitate aflatoxin B1 over the range of 0.01–10 ng/ml, and was validated in human urine samples. This method showed a positive correlation of aflatoxin-DNA adduct excretion into urine with dietary intake, and the major aflatoxin B1-DNA adduct excreted in urine was found to be an appropriate tool for monitoring aflatoxin dietary exposure.

Effect of Aflatoxin Exposure

Mycotoxicology, the study of mycotoxins, began in early 1960s with the outbreak of Turkey-X disease in the United Kingdom during which approximately 100,000 turkey poults died (Blout 1961). This mysterious turkey X disease was linked to a peanut meal (contaminated with secondary metabolites from *Aspergillus flavus* i.e., aflatoxins) imported from Brazil (Sargeant et al. 1961).

The discovery of aflatoxin and elucidation of its effects paved the way for research on other livestock health and production problems linked with mold contaminated feed and led to the discovery of other mycotoxins produced by other fungi.

Not only aflatoxins affect plant growth, it is also associated with toxicity and carcinogenicity in humans as well as animals. Diseases caused by aflatoxin consumption are called aflatoxicoses which could be acute or chronic. Acute aflatoxicosis can be deadly; chronic aflatoxicosis can cause cancer, immune suppression, and various other pathological conditions. The liver is the primary target organ. Aflatoxin decreases host resistance and interferes with vaccine-induced immunity in animals (Diekman and Green 1992).

The International Agency for Research on Cancer has classified aflatoxin B1 into a group I carcinogen (IARC 1982).

There is also significant evidence of association of aflatoxin with malignancies in extrahepatic tissues, particularly the lungs. Various evidence of air-borne aflatoxin exposure leading to cancer has been reported.

Naturally contaminated feeds are more toxic than those feeds having the same amount of a pure mycotoxin supplemented into the diet. it could be because of presence of more than one mycotoxin present in a naturally moldy feedstuff. In a study by Applebaum et al. (1982), aflatoxin produced from culture was found to be more toxic to dairy cattle than pure aflatoxin added to their diets.

There is substantial problem associated with aflatoxin contamination in maize production because corn is grown in climatic areas that give the mold the greatest opportunity for growth and dispersal, and moreover the population that grow corn consume it as a main part of the diets of both animals as well as humans.

In India, an epidemic due to aflatoxin was reported in 1975 among the Bhils (one of the largest tribal group in India), due to consumption of corn that was heavily contaminated with *A. flavus*. The epidemic was characterized by jaundice, rapidly developing ascites, and portal hypertension. Approximately 400 persons were affected by the epidemic (Krishnamachari et al. 1975). In a study conducted by Hernandez-Vargas et al. (2015) in The Gambia found that even exposure to aflatoxin B1 in utero is associated with methylation in white blood cells DNA of infants.

Because of such contamination of foodstuff food security is threatened especially in resource-poor countries during disease epidemics in staple crops. Also, crop damage contribute directly to malnutrition and indirectly to the spread of infectious diseases among human populations. Moreover, environmental damage occur as a result of shifting of the rural poor population from farming areas that are no more productive to urban areas, forests, or marginal lands (Anderson et al. 2004).

Study to Determine Aflatoxin Contamination of Food Grains

To find the prevalence of *Aspergillus* contamination of stored food grains in and around a region in north India, maize and wheat samples from godowns and home storage were collected. Microbiological culture of samples was done on fungal culture media to study the growth of mold, if any. 37.5 % of the sample grains revealed growth of *Aspergillus* species of which maximum isolates were of *A. flavus* (31.25 %). Of all the samples cultured, 50 % of maize samples and 25 % of wheat samples showed the growth of *Aspergillus* species. *A. niger* was found in 12.5 % of maize and none of the wheat samples.

Study on Effect of Occupational Exposure

To study the effect of aflatoxin contamination of grains on humans we collected bronchoalveolar lavage (BAL) and serum samples from 46 food-grain workers and 44 non-foodgrain workers from the same region. Food grain workers were occupationally exposed to food grains in one form or the other for more than 6 months like farmers, those involved in loading grains in godowns etc. Culture of BAL samples were done to find growth of *Aspergillus* species and aflatoxin was detected in BAL and serum samples using enzyme linked immunosorbent assay (ELISA).

Aflatoxins were detected in 32.6 % of the food-grain workers and 9.1 % of non food grain workers. This association was statistically significant (P<0.01). A significant difference was also observed in BAL culture for *Aspergillus* spp. (P<0.01) between these two groups. Food-grain workers also had higher incidence of chronic respiratory symptoms (47.8 %) when compared to non-food-grain workers (11.4 %) (Malik et al. 2014).

Management Strategies

Prevention

The strategies for preventing aflatoxin contamination can be broadly classified into pre–and post-harvest controls.

Pre-harvest control: Pre-harvest control strategies not only reduce aflatoxin contamination but also enhance agricultural production. These are

Adequate field management using agronomic practices (crop rotation, proper irrigation, soil cultivation, weed control etc.),
Improving host resistance (resistant hybrids)
Biological (using antagonistic fungi or bacteria) and
Chemical control (fungicides, pesticides, insecticides).

Tillage, use of fertilizer, weed control, late season rainfall, irrigation, wind and pest infestation affect the level of fungal inoculum, maintaining a disease cycle in crops like maize (Hell and Mutegi 2011).

Among the various methods available, breeding for mycotoxin-resistant hybrids has been only partially successful and fungicides have shown little efficacy in controlling pre-harvest aflatoxin contamination in corn plants (Duncan et al. 1994).

Post-harvest approaches: These include mycotoxin analysis of foodstuffs and diversion of contaminated feed; ammoniation of corn and cottonseed to damage aflatoxin; dilution with clean feeds; and improved storage technology (Trail et al. 1995). Mycotoxin-contaminated grains may be used in ethanol production (Desjardins et al. 1993). The FDA does not permit dilution of aflatoxin-contaminated feeds, as it is considered adulteration. Hence, the best strategy for postharvest control of mycotoxins is proper storage and handling of feed grains.

Lime application, use of farm yard manure and cereal crop residues for soil improvement have been found to be effective in reducing *A. flavus* contamination and hence aflatoxin levels in feedstuff. Calcium (a part of lime), thickens the cell wall and accelerates pod filling, while manure promotes the growth of microorganisms that suppress soil infections (Diener et al. 1987).

Advanced Techniques

Apart from the aforementioned traditional methods of aflatoxin control, numerous advanced techniques and methods have emerged in recent years that promises better and safer yield in future.

Microbiological Methods

Bacterial and fungal role (*Corynebacterium rubrum*, *Aspergillus niger*, *Trichoderma viride* and *Mucor ambiguous*) in the modification of the aflatoxin structure has been studied. Best known decontamination method is the fermentation process, mainly used in bread production from wheat kernels contaminated with deoxynivalenol.

After fermentation, a reduction in toxins levels is seen, and this results from fermentation and because of the thermal process to which the product is subjected. Decontamination occurs because yeast adsorb toxins (Mallman et al. 2007). Studies have shown promising results in preventing aflatoxin production using microorganisms like Bacillus spp. (98 %), *A. flavus* (90 %), *A. parasiticus* (90 %) and *Trichoderma* spp. (75 %) (Mallman et al. 2007).

Biological Control

Biocompetition uses non toxigenic *Aspergillus* species to competitively exclude toxigenic fungi (Accinelli et al. 2009). In this method the introduced atoxigenic strains competes with and exclude toxigenic strains from colonizing grains thereby reducing aflatoxin production in contaminated grains.

However, the mechanism by which a non-aflatoxigenic strain interferes with aflatoxin accumulation of toxigenic strains has not been completely elucidated (Huang et al. 2011; Chang et al. 2012).

Another method is the use of terrestrial bacteria which are a group of antagonistic microorganisms that can inhibit growth of toxigenic fungus and aflatoxin production. These include bacteria belonging to genera *Bacillus*, *Pseudomonas*, *Agrobacterium* and *Streptomyces* which have worldwide distribution (Ongena and Jacques 2007; Razzaghi-Abyaneh et al. 2011). Their metabolites are potent inhibitors of aflatoxin biosynthesis in laboratory conditions, crop model systems as well as in the fields (Razzaghi-Abyaneh et al. 2011). Research in the field of developing such novel strains of antifungals is being done worldwide (Ranjbarian et al. 2011).

In 2010, a strain *Bacillus megaterium* was evaluated for reducing postharvest decay of peanut kernels caused by *A. flavus* (Blunt et al. 2008). In 2011, study was conducted by Degola et al. (2011) regarding the potential of the different toxin producing *A. flavus* strains, colonizing the corn fields, in reducing aflatoxins accumulation when grown in mixed cultures with non toxigenic strains. Farzaneh et al. (2012), conducted a study in which *Bacillus subtilis* strain (UTBSP1) was isolated from pistachio nuts and evaluated for the degradation of aflatoxin B1 (AFB1). It was found that *B. subtilis* can effectively remediate AFB1 from nutrient broth culture and pistachio nut by 85.66 % and 95 %, respectively. Cell free supernatant fluid caused 78.39 % reduction in the amount of AFB. They also found that destructive AFB1 differed from standard AFB1 chemically, and lost its fluorescence.

Genetic Engineering

Several studies have found some seed varieties with differences related to contamination by *Aspergillus flavus* and its subsequent aflatoxin production. These differences may be because of various factors, and the plant genome can influence the expression of biosynthesis of mycotoxins (Passone et al. 2012).

One of the approaches in the field of AF research with regard to proteomics is to study the aflatoxin resistance proteins in host plants such as corn. The investigation on proteins associated with host resistance can be used for controlling aflatoxin contamination of plants (Razzazi-Fazeli et al. 2011). The use of proteomic tools has made possible to find different categories of resistance associated proteins which can be stress-responsive proteins, storage proteins or antifungal proteins that shows the stress-responsive and storage proteins may play a significant role to enhance stress-tolerance of plant (Razzazi-Fazeli et al. 2011).

Legislation

In developed nations, due to strict regulations for aflatoxin monitoring and ample amount of food help to reduce intake of contaminated food by humans. Unfortunately, in many other countries due to food scarcity or where regulations are either not enforced or does not existent, ingestion of aflatoxin may be a common problem (Cotty et al. 1994).

Early and accurate diagnosis and pathogen surveillance on local, regional, and international levels are required to predict outbreaks and allow time for development and application of strategies to combat them in time (Miller et al. 2009).

In order to minimize the levels of AF and mycotoxins in general, the National Institute of Agricultural Technology of Argentina (INTA), recommends to make early plantings, to grow plant resistant breeds, follow good farming practices, to avoid stress conditions, to minimize insect damage, to avoid damaged kernels and to storage at less of 13 % moisture in a clean, fresh and airy place which is insect free (Iglesias et al. 2011)

In foodstuffs acceptable range of the total aflatoxin is 1–20 ppb worldwide and in the feed the permissible limit is 0–50 ppb (Ashiq et al. 2014). The limits for the aflatoxin M1 in the milk for human consumption are 0.05–0.5 ppb (FAO 2003).

Conclusion

Aflatoxin exposure is a serious health hazard especially in the agricultural setting. Due to its toxicity, aflatoxin is responsible for losses associated with contamination of stored foods and feeds. The aim of the present chapter is to provide an overview of the problem of aflatoxin contamination of feedstuff, its harmful effect on exposed workers and consumers, analytical techniques used in the analysis of aflatoxins in food samples as well as methods to manage this issue. The methods based on chromatographic, spectroscopic and immunochemical characteristics of aflatoxins have been mainly used for aflatoxins determination in food. Though many sensitive methods are available for analysis of aflatoxins, the search for simple and label-free rapid procedures based on immune-biosensors format appears to be felt in near future.

Chronic exposure to aflatoxins as in agricultural sector is a slow and continuous process and usually it is not noticed unless associated health problems emerge. Hence, effective methods to monitor the exposure not only in environment but also in humans is a prerequisite for healthier population.

Advanced agricultural strategies are needed to reduce and prevent aflatoxin exposure especially in developing countries where the problem is grave due to scarcity of food. While numerous studies have been conducted on aflatoxins, much is not known about aflatoxin exposure and the resulting health effects in developing countries.

References

Accinelli C, Saccà ML, Abbas HK, Zablotowicz RM, Wilkinson JR (2009) Use of a granular bioplastic formulation for carrying conidia of a non-aflatoxigenic strain of *Aspergillus flavus*. Bioresour Technol 100:3997–4004

Akbas M, Ozdemir M (2006) Effect of different ozone treatments on aflatoxin degradation and physicochemical properties of pistachios. J Sci Food Agric 86:2099–2104

Alcaide-Molin M, Ruiz-Jimenez J, Mata-Granados J, Luque de Castro M (2009) High through-put aflatoxin determination in plant material by automated solidphase extraction on-line coupled to laser-induced fluorescence screening and determination by liquid chromatography-triple quadrupole mass spectrometry. J Chromatogr A 1216:1115–1125

Ali S (2014) Detection of aflatoxins – a critical review. Glob J Res Anal 3:123–124

Anderson PK, Cunningham AA, Patel NG, Morales FJ, Epstein PR, Daszak P (2004) Emerging infectious diseases of plants: pathogen pollution, climate change and agrotechnological drivers. Trends Ecol Evol 19:535–44

Applebaum RS, Brackett RE, Wiseman DW, Marth EL (1982) Responses of dairy cows to dietary aflatoxin: feed intake and yield, toxin content, and quality of milk of cows treated with pure and impure aflatoxin. J Dairy Sci 65:1503–1508

Ashiq S, Hussain M, Ahmad B (2014) Natural occurrence of mycotoxins in medicinal plants: a review. Fungal Genet Biol 66:1–10

Atanda SA, Pessu PO, Agoda S, Isong IU, Adekalu OA, Echendu MA, Falade TC (2011) Fungi and mycotoxins in stored foods. Afr J Microbiol Res 5:4373–4382

Baertschi SW, Raney KD, Stone MP, Harris TM (1988) Preparation of the 8,9-epoxide of the mycotoxin aflatoxin B1: the ultimate carcinogenic species. J Am Chem Soc 110:7929–7931

Baranyi N, Kocsubé S, Vágvölgyi C, Varga J (2013) Current trends in aflatoxin research. Acta Biol Szeged 57:95–107
Bbosa GS, David Kitya D, Odda J, Ogwal-Okeng J (2013) Aflatoxins metabolism, effects on epigenetic mechanisms and their role in carcinogenesis. Health 5:14–34
Bertuzzi T, Rastelli S, Mulazzi A, Pietri A (2012) Evaluation and improvement of extraction methods for the analysis of aflatoxins B1, B2, G1 and G2 from naturally contaminated maize. Food Anal Methods 5:512–519
Blout WP (1961) Turkey "X" disease. Turkeys 9, 52, 55–58, 61, 77
Blunt JW, Copp BR, Hu WP, Munro MHG, Northcote PT, Prinsep MR (2008) Marine natural products. Nat Prod Rep 25:35–94
CAST (Council for Agricultural Science and Technology) (1989) Mycotoxins: economics and health risks. Task force report no. 116, CAST, Ames
Cavaliere C, Foglia P, Pastorini E, Samperi R, Lagana A (2006) Liquid chromatography/tandem mass spectrometric confirmatory method for determining aflatoxin M1 in cow milk: comparison between electrospray and atmospheric pressure photoionization sources. J Chromatogr A 1101:69–78
Chang PK, Abbas HK, Weave MA, Ehrlich KC, Scharfenstein LL, Cotty PJ (2012) Identification of genetic defects in the atoxigenic biocontrol strain *Aspergillus flavus* K49 reveals the presence of a competitive recombinant group in field populations. Int J Food Microbiol 154:192–196
Christensen CM, Mirocha CJ, Meronuck RA (1977) Molds, mycotoxins, and mycotoxicoses. Cereal Foods World 22:513–520
Cotty PJ, Bayman P, Egel DS, Elias KS (1994) Agriculture, aflatoxins and *Aspergillus*. In: Powell KA, Renwick A, Peberdy JF (eds) The genus *Aspergillus*. Plenum Press, New York, pp 1–27
Couderc F, Caussé E, Bayle C (1998) Drug analysis by capillary electrophoresis and laser-induced fluorescence. Electrophoresis 19:2777–2790
Coulumbe RA (1993) Symposium: biological action of mycotoxins. J Dairy Sci 76:880–891
Davis ND, Deiner UL (1983) Some characteristics of toxigenic and nontoxigenic isolates of *Aspergillus flavus* and *Aspergillus parasiticus*. In: Diener UL, Asquith RL, Dickens JW (eds) Aflatoxin and *Aspergillus flavus* in corn. Craftmaster Southern Coop Series Bull, Opelika, p 279
Degola F, Berni E, Restivo FM (2011) Laboratory tests for assessing efficacy of atoxigenic *Aspergillus flavus* strains as biocontrol agents. Int J Food Microbiol 146:235–243
Desjardins AE, Hohn TM, McCormick SP (1993) Trichothecene biosynthesis in Fusarium species: chemistry, genetics, and significance. Microbiol Rev 57:595–604
Detroy RW, Lillehoj EB, Ciegler A (1971) Aflatoxin and related compounds. In: Ciegler A, Kadis S, Ajl SJ (eds) Microbial toxins, vol VI, Fungal toxins. Academic, New York, pp 3–178
Diekman DA, Green ML (1992) Mycotoxins and reproduction in domestic livestock. J Anim Sci 70:1615–1627
Diener UL, Cole RJ, Sanders TH, Payne GA, Lee LS, Klich MA (1987) Epidemiology of aflatoxin formation by *Aspergillus flavus*. Annu Rev Phytopathol 25:249–270
Duncan HE, Ayers AR, Payne GA, Hagler WM Jr (1994) Lack of fungicidal control of *Aspergillus flavus* in field corn. In: Llewellyn GC, Dashek WV, O'Rear CE (eds) Biodeterioration research, vol 4. Plenum Press, New York, pp 161–175
FAO (Food and Agriculture Organization of United Nations) (2003) FAO corporate repository, Economic and Social Department Worldwide regulations for mycotoxins in food and feed in 2003. Mycotoxin regulations in 2003 and current developments. FAO, Rome, Italy, pp 1–20
Farzaneh M, Shi ZQ, Ghassempour A, Sedaghat S, Ahmadzadeh M, Mirabolfathy M, Javan-Nikkhah M (2012) Aflatoxin B1 degradation by Bacillus subtilis UTBSP1 isolated from pistachio nuts of Iran. Food Control 23:100–106
Goto T, Wicklow DT, Ito Y (1996) Aflatoxin and cyclopiazonic acid production by a sclerotium-producing *Aspergillus* tamarii strain. Appl Environ Microbiol 62:4036–4038
Guengerich FP, Arneson KO, Williams KM, Deng Z, Harris TM (2002) Reaction of aflatoxin B1 oxidation products with lysine. Chem Res Toxicol 15:780–792
Hayes AW (1980) Mycotoxins: a review of biological effects and their role in human diseases. Clin Toxicol 17:45–60

Hell K, Mutegi C (2011) Aflatoxin control and prevention strategies in key crops of Sub Saharan Africa. Afr J Microbiol Res 5:459–466

Hernandez-Vargas H, Castelino J, Silver MJ, Dominguez-Salas P, Cros MP, Durand G, Calvez-Kelm FL, Prentice AM, Wild CP, Moore SE, Hennig BJ, Herceg Z, Gong YY, Routledge MN (2015) Exposure to aflatoxin B1 in utero is associated with DNA methylation in white blood cells of infants in the Gambia. Int J Epidemiol. doi:10.1093/ije/dyv027

Herzallah S (2009) Determination of aflatoxins in eggs, milk, meat and meat products using HPLC fluorescent and UV detectors. Food Chem 114:1141–1146

Huang C, Jha A, Sweany R, Derobertis C, Damann KE Jr (2011) Intraspecific aflatoxin inhibition. *Aspergillus flavus* is thigmoregulated, independent of vegetative compatibility group and is strain dependent. PLoS One 6:e23470

Iglesias J, Presello DA, Fauguel CM, Botta GL (2011) Micotoxinas: Debemos preocuparnos? Presented at the 3 Jornada de Actualización Técnica de Maíz organized by INTA, AIANBA, MAIZAR, InTech, Pergamino

International Agency for Research on Cancer (1982) The evaluation of the carcinogenic risk of chemicals to humans, vol 4, IARC, Monograph Supplement. International Agency for Research on Cancer, Lyon

Izmailov NA, Schraiber MS (1938) Farmatsiya. Application of PLC to inorganics and organometallics. In: Kowalska T, Sherma J (eds) Preparative layer chromatography. CRC Press, Taylor & Francis, Boca Raton, pp 1–7

Johnson WW, Harris TM, Guengerich FP (1996) Kinetics and mechanism of hydrolysis of aflatoxin B1 exo-8,9-oxide and rearrangement of the dihydrodiol. J Am Chem Soc 18:8213–8220

Klich MA (2007) Environmental and developmental factors influencing aflatoxin production by *Aspergillus flavus* and *Aspergillus parasiticus*. Mycoscience 48:71–80

Klich MA, Pitt JI (1988) Differentiation of *Aspergillus flavus* from *Aspergillus parasiticus* and other closely related species. Trans Br Mycol Soc 91:99–108

Klich MA, Mullaney EJ, Daly CB, Cary JW (2000) Molecular and physiological aspects of aflatoxin and sterigmatocystin biosynthesis by *A. tamarii* and *A. ochraceoroseus*. Appl Microbiol Biotechnol 53:605–609

Krishnamachari KAVR, Bhat RV, Nagarajan V, Tilak TBG (1975) Hepatitis due to aflatoxicosis: an outbreak in western India. Lancet 1:1061–1063

Lillehoj EB, Fennell DI (1975) Fungi and aflatoxin in a bin of stored white maize. J Stored Prod Res 11:47–51

Ma F, Chen R, Li P, Zhang Q, Zhang W, Hu X (2013) Preparation of an immunoaffinity column with amino-silica gel microparticles and its application in sample cleanup for aflatoxin detection in agri-products. Molecules 18:2222–2235

Malik A, Ali S, Shahid M, Bhargava R (2014) Occupational exposure to aspergillus and aflatoxins among food-grain workers in india. Int J Occup Environ Health 20:189–193

Mallman CA, Rauber RH, Giacomini L (2007) Factores de formación de las micotoxinas y sus formas de control. [Online]. Engormix. Available from: http://www.engormix.com/MA-micotoxinas/foros/articulo-factores-formacion-micotoxinas-t13355/251-0 htm

Marsh SF, Payne GA (1984) Preharvest infection of corn silks and kernels by *Aspergillus flavus*. Phytopathology 74:1284–1289

Miller SA, Beed FD, Harmon CL (2009) Plant disease diagnostic capabilities and networks. Annu Rev Phytopathol 47:15–38

Nelson PE, Desjardins AE, Plattner RD (1993) Fumonisins, mycotoxins produced by Fusarium species: biology, chemistry and significance. In: Cook RJ (ed) Ann Rev Phytopathol 31:233–249

Ongena M, Jacques P (2007) Bacillus lipopeptides: versatile weapons for plant disease biocontrol. Trends Microbiol 16:115–125

Passone MA, Rosso LC, Ciancio A, Etcheverry M (2010) Detection and quantification of *Aspergillus* section Flavi spp. in stored peanuts by real-time PCR of nor-1 gene, and effects of storage conditions on aflatoxin production. Int J Food Microbiol 138:276–281

Passone MA, Girardi NS, Ferrand CA, Etcheverry MA (2012) Invitro evaluation of five essential oils as botanical fungitoxicants for the protection of stored peanuts from *Aspergillus flavus* and *A. Parasiticus* contamination. Int Biodeterior Biodegrad 70:82–88

Peterson SW, Ito Y, Horn BW, Goto T (2001) *Aspergillus bombycis*, a new aflatoxigenic species and genetic variation in its sibling species, *A Nomius*. Mycologia 93:689–703

Ranjbarian AR, Shams-Ghahfarokhi M, Kalantari S, Razzaghi-Abyaneh M (2011) Molecular identification of antagonistic bacteria from Tehran soils and evaluation of their inhibitory activities toward pathogenic fungi. Iran J Microbiol 3:140–146

Razzaghi-Abyaneh M, Shams-Ghahfarokhi M, Chang PK (2011) Aflatoxins: mechanisms of inhibition by antagonistic plants and microorganisms. In: Guevara-Gonzalez RG (ed) Aflatoxins: biochemistry and molecular biology. INTECH Open Access Publisher, Croatia, pp 285--304

Razzazi-Fazeli E, Rizwan M, Mayrhofer C, Nöbauer K (2011) The use of proteomics as a novel tool in aflatoxin research. In: Guevara-González RG (ed) Aflatoxins- biochemistry and molecular biology. InTech, Rijeka, pp 127–148

Reddy KRN, Reddy CS, Abbas HK, Abel CA, Muralidharan A (2008) Mycotoxigenic fungi, mycotoxins, and management of rice grains. Toxin Rev 27:287–317

Sapsford K, Ngundi M, Moore M, Lassman M, Shriver-Lake L, Taitt C, Ligler F (2006) Rapid detection of food-borne contaminants using an array biosensor. Sensors Actuators B Chem 113:599–607

Sardiñas N, Vázquez C, Gil-Serna J, Gónzalez-Jaén MT, Patiño B (2011) Specific detection and quantification of *Aspergillus flavus* and *Aspergillus parasiticus* in wheat flour by SYBR green quantitative PCR. Int J Food Microbiol 145:121–125

Sargeant KA, Sheridan J, Kelly O, Carnaghan RBA (1961) Toxicity associated with certain samples of groundnuts. Nature 192:1096–1097

Schiefer HB (1990) Mycotoxicosis of domestic animals and their diagnosis. Can J Physiol Pharmacol 68:987–990

Sheibani A, Tabrizchi M, Ghaziaskar H (2008) Determination of aflatoxins B1 and B2 using ion mobility spectrometry. Talanta 75:233–238

Squire RA (1981) Ranking animal carcinogens: a proposed regulatory approach. Science 214:877–880

Stroka J, Anklarn E (2000) Development of a simplified densitometer for the determination of aflatoxins by thin-layer chromatography. J Chromatogr A 904:263–268

Trail F, Mahanti N, Rarick M, Mehigh R, Liang SH, Zhou R, Linz J (1995) Physical and transcriptional map of an aflatoxin gene cluster in *Aspergillus parasiticus* and functional disruption of a gene involved early in the aflatoxin pathway. Appl Environ Microbiol 61:2665–2673

Turner PC, Sylla A, Gong YY, Diallo MS, Sutcliffe AE, Hall AJ, Wild CP (2005) Reduction in exposure to carcinogenic aflatoxins by post-harvest intervention measures in west Africa: a community- based intervention study. Lancet 365:1950–1956

Wang JS, Qian G, Zarba A (1996) Temporal patterns of aflatoxin-albumin adducts in hepatitis B surface antigen positive and antigen-negative residents of Daxin, Qidong Country, People's Republic of China. Cancer Epidemiol Biomarkers Prev 5:252–261

Wicklow DT (1983) Taxonomic features and ecological significance of sclerotia. In: Deiner UL, Asquith RL Dickens JW (eds) Aflatoxin and *Aspergillus flavus* in Corn, vol 279, Southern Coop. Serv. Bull. Craftmaster, Opelika

Wild CP, Garner RC, Montesano R, Tursi F (1986) Aflatoxin B1 binding to plasma albumin and liver DNA upon chronic administration to rats. Carcinogenesis 7:853–858

Wild CP, Jiang Y, Sabbioni G, Chapot B, Montesano R (1990) Evaluation of methods for quantitation of aflatoxin albumin adducts and their application to human exposure assessment. Cancer Res 50:245–251

Wild CP, Hudson GJ, Sabbioni G, Chapot B, Hall AJ, Wogan GN, Whittle H, Montesano R, Groopman JD (1992) Dietary intake of aflatoxins and the level of albumin-boundaflatoxin in peripheral bloodin The Gambia, West Africa. Cancer Epidemiol Biomarkers Prev 1:229–234

Wild CP, Hasegawa R, Barraud L, Chutimataewin S, Chapot B, Ito N, Montesano R (1996) Aflatoxin-albumin adducts: a basis for the comparative carcinogenesis between animals and humans. Cancer Epidemiol Biomarkers Prev 5:179–89

Wilson DM, Payne GA (1994) Factors affecting *Aspergillus flavus* group infection and aflatoxin contamination of crops. In: Eaton DL, Groopman JD (eds) The toxicology of aflatoxins. Human health, veterinary and agricultural significance. Academic, San Diego, pp 309–325

Index

A
Abscissa acid (ABA), 200
Aceria mangiferae, 230
Acervuli, 208, 212
Aflatoxin, 406, 441
  aflatoxin exposure
    effect of, 448–449
    monitoring techniques, 447–448
  chromatography
    column liquid chromatography, 445
    thin-layer chromatography, 444
  confirmation of identities, 447
  contamination, 442–443, 449
  essential oils, 418–419
  extraction of, 443–444
  fluorescence spectrometry, 446
  immunochemical methods
    ELISA, 446
    ICA, 446
    RIA, 445
  ion-mobility spectrometry, 447
  management strategies
    biological control, 451–452
    genetic engineering, 452
    legislation, 452
    microbiological methods, 451
    post-harvest approaches, 450
    pre-harvest control, 450
  occupational exposure, effect of, 449–450
  toxigenic potential/factors, 442
  ultra-violet absorption, 446–447
Altenuene (ALT), 122
Alternaria, 149–150
  A. alternata, 118, 123, 125, 126, 136, 142, 144, 146
    characterization, 135
    genetic variations in, 135
    melanin pigments, 124
    mycotoxin alternariol, 129
  A. arborescens, 134, 136, 139
  A. brassicae, 118, 130, 144
  A. brassicicola, 129, 130, 133, 143
  A. dauci, 141
  A. destruens, 137–138
  A. infectoria, 134
  A. japonica, 116, 118, 122, 129
  A. limicola, 129
  A. mali, 140–141
  A. radicina, 120, 127, 143, 144
  A. raphani, 116
  A. triticina, 145
  biochemical and metabolic profiling, 122
  characteristics features, 115
  conidiogenus mycelia/conidiohores, 115
  disease cycle, 116
  disease management
    biological control, 143–145
    biotechnological approaches, 147
    crop rotation, 142–143
    cultural practices, 140
    forecasting models, 141–142
    fungal disease resistance, 142
    fungicidal management, 145–146
    hot water treatment, 142
    induced host resistance, 148–149
    planting material, source of, 140–141
    tissue culture techniques, 147
    transgenic plants, 147–148
  ELISA, 125–126
  enzyme analysis and protein profiling, 122–123
  epidemiology, 117–118

P. Kumar et al. (eds.), *Current Trends in Plant Disease Diagnostics and Management Practices*, Fungal Biology, DOI 10.1007/978-3-319-27312-9

Alternaria (*Cont.*)
etiology of, 116
in host plant, physiological disorders, 117
immunological assays and serological methods
cell mediated immunity, 124
cryptotypes, 124
epitope, 124
monoclonal antibody, 125
polyclonal antibodies, 123–125
quarantine and phytosanitory certification, 124
molecular markers
ITS regions, 126
LAMP, 127
RAMS, 127
RAPD, 126, 127
RFLP, 126, 127
morphological characterization
A. dauci, 120
A. solani, 120
A. tenuissima, 120–121
DAPI, 119
DRYES, 121
PCA, 120, 121
PDA, 119, 120
PGA, 119
morphological differentiations, 115
nucleic acid based identification
AFLP, 133–134
DNA extraction, 127–130
DNA microarray, 138
FTIR, 137–138
gel electroforetic profiling, 136–137
genomics and proteomics, 138–139
microsatellites markers, 136
molecular techniques, 130
multiplex-PCR assay, 132–133
nested-polymerase chain reaction, 132
RAPD, 134–135
real time PCR, 131
reverse transcription PCR, 131–132
RFLP, 135
soil moisture and relative humidity, 116
symptomatology
Alternariose, 113
apple infection, 112
blight disease, 112
dark brown sunken spots, 114
dark coloured sporulation, 114
environmental conditions, 114
leaf spots, 113, 114
scion image method, 113
Alternariol (AOH), 122, 129
Alternariol monomethyl ether (AME), 122
Altertoxins (ALXs), 122
1-Aminocyclopropane-1-carboxylate (ACC) deaminase, 61
Amplified fragment length polymorphism (AFLP), 175
Alternaria species, 133–134
Fusarium oxysporum, 359
Arachis hypogaea L., 413–414
Arbuscular mycorrhizal (AM) fungi, 55, 389
Arrays, 249–250
DNA arrays, 9, 12–13, 277
electronic nose, 254
micromechanical cantilever array, 252
Atomic force microscopy (AFM), 252
Avena sativa L., 414

B

Bacillus licheniformis, 36
Backward inner primer (BIP), 278
Bacterial diseases
sugarcane disease
gumming disease, 310
leaf scald disease, 309
red stripe, 310
RSD, 309, 310
spindle rot, 309, 310
Trichoderma, 331–332
Biocontrol agents (BCAs), 143
Fusarium graminearum, 391–392
Trichoderma (*see* (*Trichoderma*))
Biopesticides, 230–231, 384
Biosensors
cantilever-based sensors, 252
QCM immune-sensors, 251–252
SPR sensors, 252
Biotechnology, infectious diseases
anatomical defence, 350
disease resistance gene, molecular mapping of, 361
inducible systems
elicitor receptors, 354
elicitor response, 350–352
general response, 351, 352
race specific response, 352–353
molecular detection, plant pathogens, 360, 361
molecular marker, phytopathology
AFLP, 359
microsatellites, 359–360
RAPD, 359
RFLP, 358
pathogenic fungi, molecular approach, 355

DNA extraction, 356
primers and probes, design of, 356
plant fungi
conventional PCR, 357
multiplex PCR, 357
real time PCR, 358
reverse transcription-PCR, 357–358
plant-pathogen interaction, 348
pre-existing protein and chemical protection, 350
systemic response, 354
Bronchoalveolar lavage (BAL), 449, 450

C
Cantilever sensors, 252
Cell wall degrading enzymes (CWDE), 328–329, 338
Cetyltrimethylammonium bromide (CTAB) protocol, 188
Chitinases, 37
Colletotrichum acutatum, 326
Colletotrichum falcatum, 302
Colletotrichum gloeosporioides, 207
asexual stage, 208
biology of infection, 209–211
epidemiology, 211
genomics and transcriptomics, 216
integrated disease management
disease predictive model, 214
fungicide spray, 213
intercropping, 213
irradiation and fruit quality, 214
post-harvest treatments, 213–214
pruning, 213
resistant varieties, 213
life cycle of
disease occurrence, 212
dissemination, 211
pathogen infection and reproduction, 211, 212
pathogen inoculation, 211
MAT genes, 215
molecular identification and characterizations, 214–215
morphology, 208
nutritional strategies, 211
sexual stage, 208
symptomatology
"alligator-skin" effect, 209
Bird's eye spots, 209, 210
conidia production, 209
dead leaves, 208
panicles, 209, 210
ripe fruits, 209
sunken dark brown-black spots, 209, 210
"tear stain" symptom, 209
twigs, 209, 210
Column liquid chromatography, 445
Conditionally dispensable chromosomes (CDCs), 139
Conventional diagnostics, 238
Cooperational polymerase chain reaction (Co-PCR), 245, 274–275
Crop rotation, 34, 142–143
Cryo-scanning electron microscopy (Cryo-SEM), 239
Cumin
Fusarium wilt (*see* (*Fusarium* wilt (FW)))
origin of, 30
production constraints, 31–32
taxonomic classification, 31
uses of, 31
Cyclopiazonic acid (CPA), 443
Cyranose® 320, 254
Cytokinins, 61

D
Deblossoming, 229
Dichloran rose bengal yeast extract sucrose agar (DRYES), 121
1, 8-Dihydroxy naphthalene (DHN), 124
Direct agar plating technique (DAPT), 119, 238
Direct antigen coating ELISA (DAC-ELISA), 125, 241
Direct tissue immuno blot assay (DTIBA), 242
Diterpenes, 417
DNA-DNA hybridization, 174–175
Dot blot hybridization, 245, 435
Dot immuno-binding assay (DIBA), 241–242
Double antibody sandwich ELISA (DAS-ELISA), 125, 241, 434
Double-stranded DNA (dsDNA), 270
Double-stranded RNA (dsRNA), 195

E
Electronic nose (e-nose), 253–254, 287–289
Electron microscopy (EM), 238
Cryo-SEM, 239
ISEM, 240, 306
Electrospray ionization-mass spectrometry (ESI-MS), 289, 290
ELISA. *See* Enzyme-linked immunosorbent assay (ELISA)

Endophytic fungi
antimicrobial secondary metabolites, 368
antimycotic activity, evaluation of, 371–372
data analysis, 373
distribution of, 373–375
diversity of, 367–368, 373–375
ethyl acetate extract of, 377
fungal extract, fermentation and preparation of, 371
isolation of, 370–371
ITS region
genomic DNA extraction and amplification, 372
sequencing and phylogenetic analysis, 372–373
PKS gene
antimycotic potential, phylogenetic analysis, 375–378
ketosynthase domain sequence, amplification of, 373
Enzyme-linked immunosorbent assay (ELISA), 6, 240, 448
Alternaria, 125–126
DAC-ELISA, 241
DAS-ELISA, 241, 434
direct procedure, 241
indirect procedure, 241
mycotoxin assay, 446
PCR-ELISA, 272–273
Essential oils (EOs), 416, 417
aflatoxin, 418–419
antifungal activity, 76–77
Apiaceae, 73
Asteraceae, 71–72
chemical configuration, 417
fumonisin, 419–420
Lamiaceae, 72–73
Lauraceae, 73
Liliaceae, 72
Meliaceae, 74
ochratoxin A, 420
Rutaceae, 72
Theaceae, 74
Verbenaceae, 73
zearalenone, 419
Zingiberaceae, 74

F
Fluorescence imaging, 283–284
Fluorescence resonance energy transfer (FRET), 254
Fluorescence spectroscopy, 282
Fluorescent in situ hybridization (FISH), 249
Food grains, aflatoxigenic fungi. *See* Aflatoxin
Forward inner primer (FIP), 278
Fourier transform infrared spectroscopy (FTIR)
Alternaria, 137–138
phytopathogenic fungi, 280, 282
Fumonisin, 409, 411, 412, 414–416, 419
Fungal diseases, 349
sugarcane disease
PB, 303
red rot disease, 302
sett rot/pineapple disease, 303–304
smut disease, 303
sugarcane rust, 304
wilt disease, 303
Trichoderma
Botrytis, 325–326
Colletotrichum, 326–327
Fusarium, 323–325
Penicillium, 329
Pythium (*see* (*Pythium* species))
Rhizoctonia, 330–331
Sclerotium, 329–330
Fusarium
F. fujikuroi, 394–395
F. graminearum, 368
F. moniliforme, 187, 225, 227, 368
BLASTn analysis, 190–191
cottony growth and pink discoloration of culture, 189
materials and methods, 188
mechanical binding, 188, 189
microscopic examination, 189, 190
molecular detection, 191
PCR analysis, 190
F. oxysporum, 265–266
F. proliferatum, 375, 376
F. tumidum, 376
mango malformation
distribution of, 226
F. mangiferae, 227
F. mexicanum, 226
F. proliferatum, 226
F. pseudocircinatum, 226
F. sporotrichiodes, 228
F. sterilihyphosum, 226
F. subglutinans, 227, 230
F. tupiense, 226
inoculation of, 226
Trichoderma, biological control, 323–325
Fusarium wilt (FW)
cumin
Bacillus, 35–36

biocontrol antagonists, 34–35
biocontrol mechanism, 36
distribution and diversity, 32
F. oxysporum, 31, 32
hyphal tip technique, 31
mycolytic enzyme, 38–40
nutrients and siderophore production, 37
randomly amplified polymorphic DNA (RAPD), 40–41
salicylic acid (SA), in ISR, 37
16s rDNA and 23s ITS region characterization, 41–42
symptomatology, 33
pigeon pea
annual loss, 95
biological agents, 99–103
disease symptoms, 97–98
Fusarium udum, 96–97
in India, 92
pathogen distribution, 97
pathogenic agents, 93
plant health diagnostic, 94
starch store energy, 92

G
Gel electroforetic profiling, 136–137
Gibberellins, 61
β-1,3-Glucanases, 39, 320
G protein-coupled receptor (GPCR), 334, 335
Green Revolution, 165

H
Hemiterpenes, 417
Heterotrimeric G proteins, 334–335
High mobility group (HMG), 215
High performance liquid chromatography (HPLC), 415, 447, 448
Hordeum vulgare L., 414
Host selective toxins (HSTs), 122
Hot water treatment (HWT), 142, 213–214
Hydrogen cyanide (HCN), 101
Hypersensitive response (HR), 351–353
Hyperspectral imaging, 284–285
Hyphopodia, 197

I
ICAR-National Bureau of Plant Genetic Resources (ICAR-NBPGR), 429–430
IDM. *See* Integrated disease management (IDM)
Imaging techniques, 253
fluorescence imaging, 283–284
hyperspectral imaging, 284–285
Immunoaffinity column assay (ICA), 446
Immunosorbent electron microscopy (ISEM), 240, 306
Incubation test, 433
Indoleacetic acid (IAA), 37, 60
Insecticides, 230
Integrated disease management (IDM)
Colletotrichum gloeosporioides
disease predictive model, 214
fungicide spray, 213
intercropping, 213
irradiation and fruit quality, 214
post-harvest treatments, 213–214
pruning, 213
resistant varieties, 213
Macrophomina phaseolina, 198–199
Intercropping, 213
Intergenic spacer (IGS) region, 271
Internal transcribed spacer (ITS) regions, 126, 129–130, 244
C. gloeosporioides, 214–215
endophytic fungi
genomic DNA extraction and amplification of, 372
sequencing and phylogenetic analysis, 372–373
phytopathogenic fungi, 267, 268
R. solani, 173
Ion-mobility spectrometry, 447
Isothermal nucleic acid amplification, 246–247
LAMP, 247–248
NASBA, 247

L
Lab in a box, 255
Lab on a chip, 255
Large subunits (LSU), 173
Lateral flow assay (LFA), 242–243
Lateral flow devices (LFDs), 242
Leaf spots, 113, 114
Loop-mediated isothermal amplification (LAMP), 178, 247–248
Alternaria alternata, 127
Fusarium graminearum, 248
Ganoderma lucidum, 248
phytopathogenic fungi, 278, 279
RSD, 310

M

Macrophomina phaseolina, 193
  binomial nomenclature of, 194
  dsRNA, 195
  epidemiology, 197–198
  infection process, 196–197
  inoculum level of, 194
  management
    ABA, 200
    auxin, 200
    direct seeded treatment, 199
    drought stress, 199
    glyphosate, 200–201
    high microbial diversity agricultural soils, 199
    IDM, 198–199
    in-vitro effect, 201
    microsclerotia, 199
    pathogenesis related (PR) genes, 200
    ROS, 200
    SA pathway, 200
    SAR, 200
    seed rates and soil fertility, 199
    thermotherapy, 200
    Trichoderma koningii, 199
  mature hyaline and pigmented hyphal cells, 195
  phytotoxin, 195
  symptoms, 195–196
Magnetic capture-hybridisation polymerase chain reaction (MCH-PCR), 271, 272
Mango
  flowering responses, 221
  malformation
    biopesticides, 230–231
    cultivars for, 224
    deblossoming, 229
    floral malformation, 222, 223
    fungus, 225–227
    healthy inflorescence, 223
    healthy vegetative tissue, 223
    insecticides and fungicides, 230
    malformed affected inflorescence, 223
    malformed affected vegetative tissue, 223
    mites, role of, 224–225
    molecular diversity and detection, 227–228
    nutrients, 229
    pruning, 229
    severity of, 223
    vegetative malformation, 222, 223
  Mangifera indica L., anthracnose disease (*see* (*Colletotrichum gloeosporioides*))
  mango production, 221
MAP kinase kinase kinase (MAPKKK), 336
Mass absorption spectroscopy (MAS), 136
Microsatellites markers, 136
Mitogen-activated protein (MAP) kinases, 335–336
Molecular inversion probe (MIP) assay, 248–249
Molecular markers
  Alternaria, 126–127
  phytopathology
    AFLP, 359
    microsatellites, 359–360
    RAPD, 359
    RFLP, 358
Monoclonal antibodies (MAbs), 125, 240
Monoterpenes, 417
Mosaic disease, 304–305
Multiplex polymerase chain reaction (Multiplex PCR)
  Alternaria, 132–133
  Golovinomyces cichoracearum, 245
  phytopathogenic fungi, 276–277
  Phytophthora lateralis, 245
  plant fungi, 357
  Podosphaera xanthii, 245
Muscador albus, 393
Mycoparasitism, 320, 323, 326, 328, 329, 331, 333–336, 338
Mycotoxins, 405, 421
  agricultural food commodities, incidence
    barley, 414
    fruits, 415–416
    groundnut, 413–414
    maize, 411–412
    millets, 414
    oats, 414
    pulses and oil seeds, 415
    rice, 411
    sorghum, 414
    wheat, 412–413
  chemical formula of, 408
  classes of
    aflatoxin, 406
    fumonisin, 409
    ochratoxin, 409–410
    patulin, 410
    trichothecenes, 408–409
    zearalenone, 406–408
  essential oils (*see* (Essential oils (EOs)))
  Fusarium graminearum, 391–392
  incidence of, 410–411
  toxic effects, 407

N

Nano-diagnostics
  nanoprobes, 255
  QDs, 254
Nested polymerase chain reaction (Nested PCR)
  Alternaria, 132
  Botryosphaeria dothidea, 244
  Gremmeniella abietina, 244
  Monilinia fructicola, 244
  phytopathogenic fungi, 274
  plant fungi, 357
  Ustilainoidea virens, 244
  Verticillium dahliae, 244
Next-generation sequencing (NGS), 250–251
Nuclear magnetic resonance (NMR) spectroscopy, 283
Nucleic acid sequence-based amplification (NASBA), 247

O

Oat meal agar (OMA), 188
Ochratoxin
  EOs, 420
  OTA, 409, 410, 413, 414, 416, 420
  OTC, 410
Oligonucleotides, 243, 246
Oryza sativa L., 411

P

Pathogen associated molecular pattern (PAMP), 150, 354
Pathogenesis related (PR) proteins, 352
Patulin, 410
Peltier's effect, 243
Pennisetum glaucum L., 414
Percent disease index (PDI), 145
Physical diagnostics, 238–239
Phytopathogenic fungi
  conventional methods, 266
  electronic nose, 287–289
  imaging techniques
    fluorescence imaging, 283–284
    hyperspectral imaging, 284–285
  metabolites as biomarkers, 286–287
  molecular techniques
    Co-PCR, 274–275
    DNA arrays, 277
    fingerprinting based detection, 275–276
    in situ PCR, 273
    ITS primers, 267, 268
    LAMP, 278, 279
    MCH-PCR, 271, 272
    multiplex PCR, 276–277
    nested PCR, 274
    PCR-DGGE, 273–274
    PCR-ELISA, 272–273
    PCR-RFLP, 275
    real-time PCR, 270–272
    RT-PCR, 275
    species-specific PCR primers, 268–270
  PLEX-ID system, 289–290
  spectroscopic techniques, 280, 281
    fluorescence spectroscopy, 282
    NMR spectroscopy, 283
    visible, infrared and near infrared spectroscopy, 280, 282
Phytophthora infestans, 349
  automatic diagnostic, 17
  biosensors, 14
  carboxymethylated dextran-coated sensor surface, 14
  chemical control, 19–20
  cultural control, 18–19
  defoliation, 3
  dipstick assays, 16
  epidemiology
    detection techniques, 6
    DNA array hybridization, 12–13
    ELISA, 6
    field infection, 5
    MALDI-TOF mass spectroscopy (MALDI-TOF MS), 11–12
    molecular identification, 7–8
    monoclonal antibodies, 6
    multiplexed-tandem PCR (MT-PCR), 13
    multiplex PCR, 10
    nuclear-encoded genes, 9–10
    real time PCR assay, 10–11
    seed borne pathogens, 7
    tubers, 5
    zoospores, 5
  foliage infection, 3
  immunofluorescence microscopy assays, 15
  integrated diseases management, 18
  LAMP method, 17–18
  late blight attacks, 3
  late blight forecasting, 20–21
  lateral flow devices, 14–15
  later flow immunoassays, 15–16
  morphology, 2–3
  multiplex immunoassays, 16–17
  resistant varieties, 19
  sexual oospores, 2
  starvation, 2
  symptoms, 4–5

Phytoplasma diseases
SCGS disease, 306–307
SCWL disease, 308
sugarcane leaf yellows phytoplasma, 308
Pigeon pea
annual loss, 95
biological agents
antagonists and efficacy, 103
B. brevis, 102
field application, 102
formulation and methods, 103
genetic manipulation, 103
HCN, 101
soil antagonistic bacteria, 99
T. harzianum, 100
whole-genome analysis, 103
disease symptoms, 97–98
Fusarium udum, 96–97
in India, 92
pathogen distribution, 97
pathogenic agents, 93
plant health diagnostic, 94
starch store energy, 92
Plant genetic resources (PGR), 429–431, 435
Pokkah boeng (PB), 303
Polyclonal antibodies (PAbs), 123–125, 240–241
Polyketide synthase (PKS) gene, 369
antimycotic potential, phylogenetic analysis, 375–378
ketosynthase domain sequence, amplification of, 373
Polymerase chain reaction (PCR), 127, 128, 130, 188, 243–244, 356, 435–437
Co-PCR, 245
multiplex PCR (*see* (Multiplex polymerase chain reaction (Multiplex PCR)))
nested PCR (*see* (Nested polymerase chain reaction (Nested PCR)))
pathogenic fungi, 355
phytopathogenic fungi
in situ PCR, 273
ITS primers, 267, 268
MCH-PCR, 271, 272
PCR-DGGE, 273–274
PCR-ELISA, 272–273
species-specific PCR primers, 268–270
red rot of sugarcane, 302
R. solani, 172–173, 175
RT-PCR (*see* (Real time polymerase chain reaction (RT-PCR); Reverse transcription polymerase chain reaction (RT-PCR)))
SCGS phytoplasma, 307
Polymerase chain reaction-denaturing gradient gel electrophoresis (PCR-DGGE), 273–274
Polyphenol oxidase (PPO), 224
Portable devices and kits
lab in a box, 255
lab on a chip, 255
Phytophthora test kits, 256
Potato carrot agar (PCA), 120, 121
Potato dextrose agar (PDA), 31, 119, 120, 122, 371, 373
Potatoes
in India, 1
Phytophthora infestans (*see* (*Phytophthora infestans*))
production and consumption, 1
Potato glucose agar (PGA), 119
Principal component analysis (PCA), 285
Protein A coating ELISA (PAC-ELISA), 125–126
Pruning, 213, 229
Pseduomonas chlororaphis, 387, 390
Pythium species
damping-off disease, 50
distribution, 51
inoculum, 50
P. aphanidermatum, 50
plant extracts
antifungal activity, 69–70
antimicrobial compounds, 66–67
coumarins, 67
flavones, flavonoids and flavonol, 68
limitations of botanicals, 68–69
phenols, 68
plant extracts, 68
quinone, 68
tannins, 68
pseudomonas fluorescens and bacillus subtilis
antibiotics and lytic enzymes, 63
ethylene, 64
facilitating resource acquisition, 57
induced systemic resistance, 64–65
nitrogen fixation, 58
phosphate solubilization, 58, 63
phytohormone production, 59–61
siderophore, 62–63
sideropore production, 59
Trichoderma species, 51–52
AM fungi, 55
antibiosis, 54
attachment and infection, 53
biocontrol agents (BCAs), 54, 56, 327–329

commercial formulation, 54
host killing, 53
molecular biological techniques, 55
parasitism, 54
plant interactions, 52–53

Q

Quantitative polymerase chain reaction (qPCR), 246
and LAMP, 178
R. solani, 174
Quantum dots (QDs), 254
Quartz crystal microbalance (QCM) immune-sensors, 251–252

R

Radioimmunoassay (RIA), 445
Random amplified microsatellites (RAMS), 127
Random amplified polymorphic DNA (RAPD), 40–41, 127, 175, 270
Alternaria species, 126, 134–135
Fusarium subglutinans, 359
Guignardia citricarpa, 359
mango malformation, 227, 228
Ratoon stunting disease (RSD), 309, 310
Reactive oxygen species (ROS), 200
Real time polymerase chain reaction (RT-PCR), 246, 436
Alternaria species, 131
phytopathogenic fungi, 270–272
plant fungi, 358
R. solani, 174, 246
Red rot disease, 302
Restriction fragment length polymorphism (RFLP), 126, 127, 436
Alternaria, 129, 135
genetic diversity, 358
PCR-RFLP, 275, 437
Reverse transcription polymerase chain reaction (RT-PCR)
Alternaria, 131–132
phytopathogenic fungi, 275
plant fungi, 357–358
Rhizoctonia solani, 166
anastomosis groups, 167–170
characteristic features of, 167
conventional approach, 171–172
DNA-DNA hybridization, 174–175
DNA sequence based approach, 177
host range and diseases, 168–171
PCR based tools, 172–173
PCR fingerprinting techniques, 175
real-time PCR method, 174
rRNA genes, 173
SCAR approach, 176–177
Trichoderma, biological control agent, 330
UP-PCR based cross-hybridization, 176

S

Sanger sequencing technique, 177
Scanning electron microscopy (SEM), 239
Schima wallichii, 369
colonization frequency, 374
materials, 370
phylogenetic relationship, 378
sample collection, 370
Seed-borne fungi, biological control studies, 383
in field crops, 384–386
anthesis stage of wheat, 391–392
Bipolaris oryzae on rice, 394
Botrytis cinerea on chickpea, 396–397
Colletotrichum spp., 395–396
Fusarium fujikuroi on rice, 394–395
Gerlachia oryzae on rice, 394
multiple pathogens, 386–392
Tilletia spp. on wheat, 390, 393
in vegetable crops
Aspergillus niger, 399
carrot and cabbage, *Alternaria* spp. on, 397–399
Didymella bryoniae, 399
Seed-borne plant pathogens
dry seeds and washing test, examination of, 431, 432
grow-out test, 433–434
immunological techniques, 434–435
incubation test, 433
molecular technique, 435–437
seed soaking method/NaOH test, 431
Seed immunoblot assay (SIBA), 434
Seed soaking method, 431
Sequence-characterized amplified region (SCARs), 176–177
Sequence tagged microsatellites (STMs), 136, 276
Sequencing
NGS, 250–251
TGS, 251
Serological diagnostics, 239–240
DIBA, 241–242
ELISA, 240–241
ISEM, 240
LFA, 242–243
TIBA, 242

Sesquiterpenes, 417
Short tandem repeats (STRs), 136
Simpson's diversity index, 373
Single nucleotide polymorphism (SNP), 138, 177, 244, 248–249
Small subunits (SSU), 173
Smut disease of sugarcane, 303
Sodium dodecyl sulphate (SDS), 128
Sorghum mosaic virus (SrMV), 304
Sorghum vulgare L., 414
Spectroscopic techniques, 253, 280, 281
  fluorescence spectroscopy, 282
  NMR spectroscopy, 283
  visible, infrared and near infrared spectroscopy, 280, 282
Sporisorium scitamineum, 303
Stenotrophomonas maltophilia, 388
Sugarcane bacilliform viruses (SCBV), 305, 306
Sugarcane disease
  bacterial diseases
    gumming disease, 310
    leaf scald disease, 309
    red stripe, 310
    RSD, 309, 310
    spindle rot, 309, 310
  diagnosis and detection of pathogen, 300
  fungal diseases
    PB, 303
    red rot disease, 302
    sett rot/pineapple disease, 303–304
    smut disease, 303
    sugarcane rust, 304
    wilt disease, 303
  leaf binding disease in India, *Fusarium moniliforme* (*see* (*Fusarium*))
  molecular diagnostic tools, 300–302
  phytoplasma diseases
    SCGS disease, 306–307
    SCWL disease, 308
    sugarcane leaf yellows phytoplasma, 308
  virus disease
    leaf fleck disease, 305–306
    mosaic disease, 304–305
    yellow leaf disease, 305
Sugarcane grassy shoot (SCGS) disease, 306–307
Sugarcane mosaic virus (SCMV), 304, 305
Sugarcane Research Institute (SRI), 188
Sugarcane streak mosaic virus (SCSMV), 304, 305
Sugarcane white leaf (SCWL) disease, 307, 308
Sugarcane yellow leaf phytoplasmas (SCYLP), 305
Sugarcane yellow leaf virus (SCYLV), 305
Surface plasmon resonance (SPR) sensors, 252
SYBR Green I dye, 278
Systemic acquired resistance (SAR), 200, 337, 354

T

Tentoxin (TEN), 122
Tenuazonic acid (TeA), 122
Tetraterpenes, 417
Thanatephorus praticola, 173
Thin-layer chromatography (TLC), 444, 445
Third generation sequencing (TGS), 251
Tissue immuno blot assay (TIBA), 242
Transgenic plants, 147–148
Trichoderma, 38, 318, 319
  antibiosis, 321
  bacterial diseases, 331–332
  fungal diseases
    Botrytis, 325–326
    Colletotrichum, 326–327
    Fusarium, 323–325
    Penicillium, 329
    Pythium (*see* (*Pythium* species))
    Rhizoctonia, 330–331
    Sclerotium, 329–330
  induced systemic resistance, 321–322
  molecular signalling, 333–334
    cAMP, 335
    heterotrimeric G proteins, 334–335
    MAP kinases, 335–336
  mycoparasitism, 320
  nematodes, 332–333
  nutrient and space, competition for, 320–321
  physiological signalling, 336–337
  T. hamatum, 321
  T. harzianum, 231, 321, 322
  T. koningii, 199
  T. lignorum, 319
  T. reesei, 319
  T. virens, 231
  T. viride, 144, 146, 148, 231, 327
  viral diseases, 332
Trichothecenes, 408–409, 415, 416
Triterpenes, 417
Triticum aestiсum L., 412–413
β-Tubulin, 271
Type III protein secretion system (TTSS), 353, 354

U
Ultra-violet (UV) absorption, 446–447
Universally primed polymerase chain reaction (UP-PCR), 175–177

V
Vegetative compatibility populations (VCPs), 168
Viral disease
  sugarcane disease
    leaf fleck disease, 305–306
    mosaic disease, 304–305
    yellow leaf disease, 305
  Trichoderma, 332
Volatile organic compounds (VOCs), 253–254, 287, 288

W
Washing test, 431, 432
Wilt disease, 303

Y
Yellow leaf syndrome (YLS), 305

Z
Zea mays L., 411–412
Zearalenone (ZEA), 406–408, 411, 413–415, 419

MIX
Papier aus verantwortungsvollen Quellen
Paper from responsible sources
FSC® C105338

If you have any concerns about our products,
you can contact us on
ProductSafety@springernature.com

In case Publisher is established outside the EU,
the EU authorized representative is:
Springer Nature Customer Service Center GmbH
Europaplatz 3, 69115 Heidelberg, Germany

Printed by Libri Plureos GmbH
in Hamburg, Germany